List of Elements With Their Symbols and Atomic Masses

Element	Symbol	Atomic Number	Atomic Mass[a] (amu)
Actinium	Ac	89	(227)
Aluminum	Al	13	26.9815
Americium	Am	95	(243)
Antimony	Sb	51	121.75
Argon	Ar	18	39.948
Arsenic	As	33	74.9216
Astatine	At	85	(210)
Barium	Ba	56	137.33
Berkelium	Bk	97	(247)
Beryllium	Be	4	9.01218
Bismuth	Bi	83	208.9806
Boron	B	5	10.811
Bromine	Br	35	79.904
Cadmium	Cd	48	112.41
Calcium	Ca	20	40.08
Californium	Cf	98	(251)
Carbon	C	6	12.01115
Cerium	Ce	58	140.12
Cesium	Cs	55	132.9055
Chlorine	Cl	17	35.453
Chromium	Cr	24	51.996
Cobalt	Co	27	58.9332
Copper	Cu	29	63.546
Curium	Cm	96	(247)
Dysprosium	Dy	66	162.50
Einsteinium	Es	99	(254)
Element 106	–	106	(263)
Element 107	–	107	(262)
Element 109	–	109	(266)
Erbium	Er	68	167.26
Europium	Eu	63	151.96
Fermium	Fm	100	(253)
Fluorine	F	9	18.998403
Francium	Fr	87	(223)
Gadolinium	Gd	64	157.25
Gallium	Ga	31	69.72
Germanium	Ge	32	72.59
Gold	Au	79	196.9665
Hafnium	Hf	72	178.49
Hahnium[b]	[Ha]	105	(262)
Helium	He	2	4.00260
Holmium	Ho	67	164.9303
Hydrogen	H	1	1.0080
Indium	In	49	114.82
Iodine	I	53	126.9045
Iridium	Ir	77	192.22
Iron	Fe	26	55.847
Krypton	Kr	36	83.80
Lanthanum	La	57	138.9055
Lawrencium	Lr	103	(257)
Lead	Pb	82	207.2
Lithium	Li	3	6.941
Lutetium	Lu	71	174.967
Magnesium	Mg	12	24.305
Manganese	Mn	25	54.9380
Mendelevium	Md	101	(256)
Mercury	Hg	80	200.59
Molybdenum	Mo	42	95.94
Neodymium	Nd	60	144.24
Neon	Ne	10	20.179
Neptunium	Np	93	237.0482
Nickel	Ni	28	58.70
Niobium	Nb	41	92.9064
Nitrogen	N	7	14.0067
Nobelium	No	102	(253)
Osmium	Os	76	190.2
Oxygen	O	8	15.9994
Palladium	Pd	46	106.4
Phosphorus	P	15	30.9738
Platinum	Pt	78	195.09
Plutonium	Pu	94	(244)
Polonium	Po	84	(209)
Potassium	K	19	39.0983
Praseodymium	Pr	59	140.9077
Promethium	Pm	61	(145)
Protactinium	Pa	91	231.0359
Radium	Ra	88	226.0254
Radon	Rn	86	(222)
Rhenium	Re	75	186.207
Rhodium	Rh	45	102.9055
Rubidium	Rb	37	85.4678
Ruthenium	Ru	44	101.07
[Rutherfordium][b]	[Rf]	104	(261)
Samarium	Sm	62	150.4
Scandium	Sc	21	44.9559
Selenium	Se	34	78.96
Silicon	Si	14	28.0855
Silver	Ag	47	107.868
Sodium	Na	11	22.9898
Strontium	Sr	38	87.62
Sulfur	S	16	32.06
Tantalum	Ta	73	180.9479
Technetium	Tc	43	98.9062
Tellurium	Te	52	127.60
Terbium	Tb	65	158.9254
Thallium	Tl	81	204.37
Thorium	Th	90	232.0381
Thulium	Tm	69	168.9342
Tin	Sn	50	118.69
Titanium	Ti	22	47.90
Tungsten	W	74	183.85
Uranium	U	92	238.029
Vanadium	V	23	50.9415
Xenon	Xe	54	131.30
Ytterbium	Yb	70	173.04
Yttrium	Y	39	88.9059
Zinc	Zn	30	65.37
Zirconium	Zr	40	91.22

a Based on the assigned relative atomic mass of ^{12}C = exactly 12; parentheses denote the mass number of the isotope with the longest half-life.

b Name and symbol not officially approved.

BASIC CHEMISTRY

Fourth Edition

William S. Seese
Department of Chemistry
Casper College

Guido H. Daub
Department of Chemistry
University of New Mexico

Prentice-Hall, Inc., Englewood Cliffs, New Jersey   07632

Library of Congress Cataloging in Publication Data

Seese, William S.
  Basic chemistry.

  Includes bibliographies and index.
  1. Chemistry.   I. Daub, Guido H., (date)
II. Title.
QD33.S39   1985   540   84-15066
ISBN 0-13-057811-8

Editorial/production supervision: Maria McKinnon
Interior design: Michael A. Rogondino
Chapter openings and cover design: Judith A. Matz-Coniglio
Cover photo: Jay Freis, The Image Bank
Manufacturing buyer: John Hall

Photo Credits
Chapter 4 (page 71) Mitsuo Ohtsuki, Ph.D., Senior Research Associate of Dr. Albert Crews, Department of Physics, University of Chicago, Chicago, Ill.
Chapter 6 (page 111) Courtesy of Disneyland © Walt Disney Productions
Chapter 12 (page 287) Joseph S. Rychetnik, Photo Researchers, Inc.
Chapter 13 (page 312) NASA
Chapter 17 (page 449) Irvin L. Oakes, National Audubon Society, Photo Reseachers, Inc.
Chapter 19 (page 523) The Monsanto Company, © Walt Disney Productions
Chapter 20 (page 565) Guy Gillete, Photo Researchers, Inc.

Printed in the United States of America

10 9 8 7 6 5 4 3 2

ISBN 0-13-057811-8   01

Prentice-Hall International, Inc., *London*
Prentice-Hall of Australia Pty. Limited, *Sydney*
Editora Prentice-Hall do Brasil, Ltda., *Rio de Janeiro*
Prentice-Hall Canada Inc., *Toronto*
Prentice-Hall of India Private Limited, *New Delhi*
Prentice-Hall of Japan, Inc., *Tokyo*
Prentice-Hall of Southeast Asia Pte. Ltd., *Singapore*
Whitehall Books Limited, *Wellington, New Zealand*

Contents

*Chapter or section that may be omitted without loss of continuity in a brief course.

3 Basic Concepts of Matter   *49*

4 The Structure of the Atom   *71*

5 The Periodic Classification of the Elements   *99*

6 The Structure of Compounds   *111*

10 Calculations Involving Chemical Equations. Stoichiometry   *227*

11 Gases   *254*

12   Liquids and Solids   287

13   Water   312

14 Solutions and Colloids   *338*

15 Acids, Bases, and Ionic Equations   *382*

19   Organic Chemistry II: Derivatives of the Hydrocarbons   *523*

20   Nuclear Chemistry   *565*

Appendices   *593*

Glossary   *631*

Index   *643*

Preface

To the Student As with the previous three editions, this book is written for **you,** not for your instructor. We have assumed that you have had little or no science background, let alone chemistry, and that your mathematics needs to be reviewed. Therefore, we have introduced analogies, cartoons, and summary tables to help you understand some of the principles of chemistry. We have also added a new appendix (Appendix VII) to help you to solve and substitute into linear equations, and thereby to improve your mathematical skills. Like the third edition, this book has four features: (1) problem solving by using the *factor-unit method* (dimensional analysis), (2) drill problems related to sections in the text, (3) the use of tasks and objectives as a method of study, with the objectives keyed to the problems, and (4) chapter summaries.

The *factor-unit* method for solving problems is a general method. It can be applied to word problems you may encounter in any science, not just chemistry.

In this text there are many drill problems. The more problems you do, the more proficient you will become in problem solving and the more your mathematical skills will improve. Problem solving helps you think in a logical manner. General problems are presented at the end of every chapter except chapter 1. These problems usually involve material from previous chapters and are more difficult to solve. You should do these problems last. You may need to refer to previous chapters to solve these problems.

The tasks are things you must *know* to accomplish the objectives. They tell you what facts you must memorize. The objectives give you specific or discrete pieces of information and tell you exactly what we want you to do with that information in order to master the material in the chapter. After each objective a section number or problem example in the text relating to that objective is listed. Study this section or problem example, then work the problem(s) you are directed to do in the objective. All problems, except the general problems, are keyed to the objectives. New terms for each chapter and the section in which they are introduced are listed in the first objective. These terms appear in color in the appropriate chapter section and in the glossary at the end of the text.

Chapter summaries, which briefly give you the highlights of the chapter, are found at the end of each chapter. You should use the summaries as a review to recall the material you read in the text.

We especially welcome your suggestions. Therefore, we would greatly appreciate your writing to us regarding your reaction to this text.

To the Instructor In the fourth edition of *Basic Chemistry*, we have included the following new topics:

1. ideal-gas equation (Section 11-8)

2. viscosity (Section 12-6)

3. colligative properties of solutions (Section 14-12)

4. corrosion (Section 16-8)

5. addition polymers (Section 18-5)

6. condensation polymers (Section 19-11)

7. linear equations (Appendix VII)

In addition new figures, summary tables, and more general problems have been added.

In the contents we have marked with an asterisk (★) those sections that can be omitted without loss of continuity in a brief course. In an average class the first 11 chapters appear to be adequate for one semester. For those taking two semesters of chemistry, that is, those majoring in nursing, agriculture, forestry, home economics, and other applied science students, the remaining chapters are covered.

There are a number of supplementary texts accompanying the fourth edition of *Basic Chemistry*. They are: (1) *Instructor's Manual, Basic Chemistry* (4th ed.), (2) *Solutions Manual, Basic Chemistry* (4th ed.), (3) *Workbook, Basic Chemistry* (4th ed.) by William S. Seese, Guido H. Daub, Mildred Tamminen, and Sylvia J. Gregg, and (4) *Laboratory Experiments, Basic Chemistry* (4th ed.) by Charles H. Corwin. For information on these books you should contact Prentice-Hall, Inc., Englewood Cliffs, New Jersey 07632.

Many of you have made numerous suggestions for improving this text and we have attempted to follow your suggestions in this fourth edition. Some of the teachers who have helped us with this fourth edition are: Michael Blomme, Aquinas College, Grand Rapids; B. Edward Cain, Rochester Institute of Technology; Paul J. Clemmer, Hillsborough Community College, Tampa; Charles H. Corwin, American River College, Sacramento; Keith Edmonson, Kennedy-King College, Chicago; Henry Heikkinen, University of Maryland, College Park; Floyd W. Kelly, Casper College, Casper, Wyoming; Jack Newcomer, Hillsborough Community College, Tampa; Fred Redmore, Highland Community College, Freeport, Illinois; C. A. Vanderwerf, University of Florida, Gainsville; and Mary S. Vennos, Essex Community College, Baltimore.

We would also like to thank our wives, Ann Reeves Seese and Katharine Powell Daub, for their advice and suggestions.

We especially welcome your suggestions and those of your students. Therefore, we would greatly appreciate your writing to us regarding your reaction to this text.

William S. Seese
Casper College
Casper, Wyoming 82601

Guido H. Daub
University of New Mexico
Albuquerque, New Mexico 87131

GUIDO H. DAUB (1920–1984)
A Personal View

Guido Daub died on June 4, 1984. I knew him as a teacher, a researcher, a coauthor, and a friend. In 1950 I met Guido at the University of New Mexico where I was a young pharmacy undergraduate and he was a young chemistry professor. From 1955 to 1958 we rode to the research laboratories at the University together, driving in before 6 A.M. In the afternoon I would have to tear him away from the laboratory so we could go home to eat dinner before we came back in the evening. Guido and I published our first book together in 1960. As coauthors we would throw ideas at each other, using each other as sounding boards, keeping ideas we thought would work, and discarding ideas to which either of us seriously objected. Almost weekly we talked with each other by phone. After we had discussed our business, we always had time for a friendly chat. I would ask about his wife, Kay, and their children, all of whom are chemists, and he would give brotherly advice to Ann and me on raising our two sons.

Guido was a great teacher and researcher, but most of all he was a warm human being interested in all people. Students and colleagues will greatly miss Guido Herman Daub.

William S. Seese

Introduction to Chemistry

Photograph of the painting by Cornelis Bega called The Chemyst. *This painting dates back to about 1660 and is in the collection of Dr. Alfred Bader, Chairperson of Aldrich Chemical Company, Inc.*

OBJECTIVES

1. Given the following terms, define each term and describe the distinguishing characteristic of each:
   (a) science (Section 1-1)
   (b) chemistry (Section 1-1)
   (c) organic chemistry (Section 1-2)
   (d) inorganic chemistry (Section 1-2)
   (e) analytical chemistry (Section 1-2)
   (f) physical chemistry (Section 1-2)
   (g) biochemistry (Section 1-2)
2. Given research interests or research papers of chemists, classify these interests or papers as organic, inorganic, analytical, physical, or biochemical (Section 1-2, Problems 3 and 4).

Why do I have to take chemistry? This question has been asked by thousands of students for nearly 100 years. The standard answer has been that it will be useful in your profession and in the modern world in which you live. But, the real answer for many students is that it is required for a degree in engineering, nursing, home economics, forestry, the agricultural sciences, and other sciences. Now, chemistry can be more to you than a requirement; in fact, some people even find it so fascinating that they devote a lifetime to it as a teacher, a researcher, or a technical sales representative.

We shall attempt throughout this book to point out some of the applications of chemistry. If you happen to be wearing an Orlon or nylon sweater, you are wearing a product developed through chemical research in the last 50 years. Of more recent development is Corfam, a simulated or poromeric (porous) leather, and Qiana, a simulated silk. The research and development of Corfam and Qiana cost the Du Pont Company nearly $175 million and took over 20 years to develop. Among other applications of chemistry are the development of plastics, synthetic fibers, synthetic rubber, fertilizers, and pharmaceuticals. In fact, nearly everything you touch, see, hear, or eat involves chemistry!

1-1

Science, the Scientific Method, and Chemistry

Science can be defined as organized or systematized knowledge. Scientific knowledge is gathered by the scientist through **experimentation**—the basis of the scientific method.

The scientific method consists of

1. the collection of facts and data by observation of natural events under carefully planned and controlled conditions—experimentation

2. the examination and correlation of these facts and the proposal of a hypothesis[1]

3. the planning of further experiments to prove or refute the hypothesis and the possible proposal of a scientific law or theory[1]

Some experiments result in a hypothesis, a few in a theory, and a very few in a scientific law. Many times a hypothesis has to be modified or even discarded after further experimentation. Throughout this book, we consider hypotheses, theories, and a few laws related to chemical principles. If you play bridge you may be loosely applying the scientific method in determining your partner's and opponents' strength and weakness by their bidding.

In its broadest sense as we have just described it, science includes many fields. Science can be divided into the following areas:

1. abstract sciences—mathematics, logic

2. physical sciences[2]—physics, chemistry, astronomy, geology

3. biological sciences[2]—botany, zoology, physiology, microbiology

4. social sciences—history, political science, sociology, economics

FIGURE 1-1

The development of a scientific law.

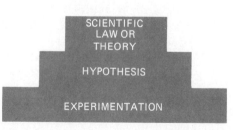

---

[1]A hypothesis is an imagined situation proposed to explain *certain natural phenomena* (events); furthermore, a hypothesis that has undergone *extensive experimentation* and has been found to agree with *all the observed facts* becomes a theory; i.e., a hypothesis is a tentative theory. A scientific law is an accurate statement of the *behavior of nature* derived from extensive experimentation, sometimes expressed in precise terms by mathematics, with no known exceptions to the law.

[2]The physical and biological sciences are usually considered to be the natural sciences.

Basic to all the sciences is mathematics, which attempts to describe natural events in precise terms. Without mathematical descriptions, the usefulness of all the sciences would be very limited. The biological sciences, in their attempt to become more precise, have become more chemically oriented in the explanation of both health and disease.

1-2

Branches of Chemistry

Chemistry can be described as being concerned with the *composition* of matter and the *changes* it undergoes.[3] As an example, chemistry would be concerned with (1) the component parts of water (*composition*) and (2) the reactions of water with other matter and the conditions required to effect such reactions (*transformations*).

Chemistry is divided into the following branches:

1. **Organic chemistry** is mostly concerned with the study of compounds containing the element carbon.[4] The synthesis (preparation) of aspirin ($C_9H_8O_4$—note the symbols C, H, and O for the elements carbon, hydrogen, and oxygen, respectively, in the formula) is an example of a project in organic chemistry.

2. **Inorganic chemistry** is the study of all elements and compounds other than organic compounds. The study of the solubility or reactions of copper salts is an example of a project in inorganic chemistry.

3. **Analytical chemistry** is the study of qualitative (what is present?) and quantitative (how much is present?) *analysis* (examination) of elements and compounds. The determination of what is present in a given antacid preparation such as Tums (which neutralizes excess acid in the stomach) and the determination of how much acid this preparation will neutralize is an example of a project in analytical chemistry.

4. **Physical chemistry** is the study of *reaction rates* (how fast will a reaction go?), *mechanisms* (what path does a reaction take to get to the products?), *bonding* and *structure* of compounds (how are compounds formed, and what are their shapes?), and *thermodynamics* (what makes a reaction go and what is the total energy relationship of a reaction?). The study of the structure of sulfuric acid (H_2SO_4) is an example of a project in physical chemistry. Physical chemistry may overlap with inorganic or organic chemistry, depending on whether the compounds being studied are inorganic or organic. For example, the project above could also be considered physical inorganic chemistry, and if the compound was the organic com-

---

[3]Matter is anything that has mass and occupies space. This term will be further discussed in 2-5

[4]The terms "elements" and "compounds" are defined in 3-3

pound aspirin ($C_9H_8O_4$), then the project could be considered physical organic chemistry.

5.   **Biochemistry** is the study of the chemistry of *biological processes,* such as the *utilization of foods* (carbohydrates, proteins, and fats) that produce energy and the *synthesis* of biologically active compounds *in living organisms.* The study of the breakdown of starch (a carbohydrate) to glucose in the human body is an example of a project in biochemistry.

Work Problems 3 and 4.

For years these divisions were strictly adhered to. Recently, there has been considerable overlap of one branch with another. For example, a physical chemist may be involved in research on the structure of proteins and thus may be a biophysical chemist.

1-3

Brief History of Chemistry

Chemistry as a science did not actually have its beginnings until the seventeenth century. But prior to that time, Chinese, Egyptians, Greeks, and the European alchemists contributed to the development of chemistry.

About 2200 B.C., the Chinese proposed that matter consisted of five elements: metal, wood, earth, fire, and water. The Chinese also proposed *ca.* 1200 B.C. the *Yin-Yang* hypothesis for explaining changes in nature: All changes in nature resulted from a mixing of two opposite elements—a *yang,* which was considered positive, and a *yin,* considered negative. The planets were developed by the combination of the sun (*yang*), symbolizing fire in the five elements, and the moon (*yin*), symbolizing water.

At about the same period of time, the Egyptians' scientific knowledge permitted them to use gold, copper, and lead ores, and pottery and to dye cloth with indigo. The preparation of alcoholic beverages through fermentation was known to both the Chinese and the Egyptians.

The Greek contributions dated from about 600 to 200 B.C. Although Greek contributions were purely speculative, since the Greeks did not carry out any experimentation, at least two are worth considering. Aristotle (384-323 B.C.) developed the idea that earthly matter consisted of four elements: air, earth, fire, and water. He proposed a fifth element—"quintessence" or "ether," which he believed was the composition of heavenly bodies. Democritus (*ca.* 460–370 B.C.) proposed a theory on the atomic structure of matter. His theory preceded by 2200 years John Dalton's atomic theory, which we shall study in 4-2.

One of the greatest contributions (A.D. 825) to our scientific knowledge was the introduction of the *zero* by the Arab mathematician Al-Khowarizmi to our number system. As you may recall, in the Roman numerals there is no zero, and this Arab contribution enhanced our scientific knowledge immensely.

The alchemists (*ca.* A.D. 500–1600) attempted to change various metals into gold. In these attempts they searched for a "philosopher's stone," which they believed was needed for this change. Some alchemists also believed that the possession of the "philosopher's stone" would heal all disease and rejuvenate the owner. The work of the alchemists was more a fantasy than it was a scientific venture, and although they failed in their attempted transformations, they did prepare new substances, develop new techniques, and design many useful types of apparatus. Their attempts to change various metals to gold should not be completely disregarded, since, as we will learn later (19-4) the transformation of the metal platinum to gold has been accomplished on an expensive and limited scale with the advent of nuclear reactions.

One of the great leaders in the development of chemistry as a science was Robert Boyle (1627–1691). With his predecessor, Roger Bacon (1214–1294), Boyle advocated that the only true search for knowledge must be through experimentation. He attempted to destroy the fantasy built up by the alchemists and to establish chemistry as an experimental science. In 11-3, we shall study one of Robert Boyle's contributions concerning the relation of the volume of a gas to changes in pressure.

After Boyle, Georg Ernst Stahl (1660–1734) proposed the *phlogiston theory*, the first great principle in chemistry. Stahl's theory stated that all combustible matter contained phlogiston, which was given off when a substance was burned. The overthrow of this theory, and hence the explanation of combustion as the combination of matter with oxygen from the air, was advanced by the great French physicist-chemist, Antoine Laurent Lavoisier (1743–1794). The work of Lavoisier established chemistry as a modern science. He is often considered the father of modern chemistry. Throughout this book, we shall refer to contributions to chemistry made by famous scientists primarily since Lavoisier's time.

Great contributions to chemistry are not limited to the eighteenth or nineteenth centuries, but are being made now in industrial, government, and private foundation laboratories, and in colleges and universities, by people of *all nationalities* and *ethnic groups*.[5] Prior to 1920, chemistry was dominated primarily by the Europeans, but since 1920 Americans have had an increasing voice in the development of chemistry. Since 1945, the United States has been a world leader in chemistry—research and development and production. Recent Soviet and Japanese efforts have also placed those countries in a position of leadership in many areas of science.

Some Europeans who have had a great influence on the development of chemistry are Erwin Schrödinger (1887–1961), an Austrian physicist who developed a mathematical equation describing energy relations for the behavior

---

[5]Many, but not all, of the people mentioned in this unit have received the Nobel Prize. The Nobel Prize is the greatest single honor that a scientist can receive. It was established in 1901 from the estate of Alfred Bernhard Nobel (1833–1896), inventor of dynamite. Six Nobel Prizes are customarily awarded annually for outstanding contributions to physics, chemistry, physiology or medicine, literature, economic sciences, and the promotion of world peace.

of electrons; Niels Bohr (1885–1962), a Danish physicist who developed a model of the atom (4-4); Otto Hahn (1879–1968), a German chemist credited with discovering nuclear fission (20-5); Sir Robert Robinson (1886–1975), a British chemist who synthesized many useful medicinal agents used to treat various diseases and who postulated the correct structure for the complex strychnine molecule; Albert von Szent-Györgyi (b. 1893), a Hungarian biochemist, now working in the United States, who did fundamental research on the chemistry of muscle contractions; Francis H. C. Crick (b. 1916), British molecular biologist who, with the American James D. Watson (b. 1928), proposed a comprehensive structure for DNA, deoxyribonucleic acid—the substance basic to the genetic code, which controls life; Aaron Klug (b. 1926), a British structural molecular chemist born in South Africa, who identified by crystallographic electron microscopic techniques the structure of the tobacco mosaic virus (TMV) and chromatin in cells; and Frederick Sanger (b. 1918), a British molecular biologist, who identified the chemical structure of insulin and investigated recombination deoxyribonucleic acid (DNA). Professor Sanger is the first person to win the Nobel Prize in *chemistry* twice.

Two outstanding researchers and teachers in the United States were Roger Adams (1889–1971) of the University of Illinois, an organic chemist, and Gilbert Newton Lewis (1875–1946) of the University of California (Berkeley), a physical chemist. These men not only have made great contributions to our chemical knowledge but also have inspired many of their students to make even greater contributions. Professor Lewis contributed to our knowledge of the chemical bond (6-6) and of acids and bases. The discoverer of nylon, Wallace H. Carothers (1897–1937), was a student of Professor Adams. Linus C. Pauling (b. 1901), greatly influenced by the work of Professor Lewis, was probably the foremost American theoretical chemist at one time. Among his varied interests are theories on chemical bonding (6-4), the structure of proteins, the action of anesthetics, and the treatment of the common cold by taking large doses of ascorbic acid (vitamin C). Professor Pauling won the Nobel Prize in chemistry and also the Nobel Peace Prize for his efforts to obtain an international nuclear test ban.

Other famous American chemists who have made or are now making contributions to the development of chemistry are Herbert C. Brown (b. 1912), who developed the synthesis of organic compounds using boron compounds; Melvin Calvin (b. 1911), who unraveled the chemical process of photosynthesis; Paul J. Flory (b. 1910), who devised ways of analyzing and studying polymers, which led to the development of plastics and synthetic materials; Charles Goodyear (1800–1860), who developed a process for the vulcanization of rubber; Charles Martin Hall (1863–1914), who discovered an inexpensive commercial process for preparing aluminum from its ore; Percy L. Julian (1899–1975), a black chemist, who discovered a method of synthetically mass-producing cortisone; Marshall W. Nirenberg (b. 1927), Robert W. Holley (b. 1922), and Har G. Khorana (b. 1922), the last a native-born Indian, who all helped break the genetic code of heredity; Glenn T. Seaborg (b. 1912), who synthesized nine new elements; Wendell M. Stanley (1904–1971), a student of Professor Adams who first crystallized a virus bringing together the properties of living and nonliving matter; and Robert B. Woodward (1917–1979), who was probably one of the world's greatest synthetic chemists, having synthesized quinine, lysergic acid, strychnine, reserpine, chlorophyll, vitamin B_{12}, and other important natural products.

Chapter Summary

In this chapter the scientific method and its basis in *experimentation* were considered. The various fields of science were described, with chemistry classified as a physical science. *Chemistry* is the study of the composition of matter and the changes it undergoes. The various branches of chemistry were considered. The history of chemistry from ancient to modern times was briefly reviewed and discoveries made by people of *all* nationalities and ethnic groups were described.

EXERCISES

1. Define or explain the following terms:
   (a) science
   (b) hypothesis
   (c) scientific law
   (d) abstract sciences
   (e) physical sciences
   (f) biological sciences
   (g) social sciences
   (h) chemistry
   (i) organic chemistry
   (j) inorganic chemistry
   (k) analytical chemistry
   (l) physical chemistry
   (m) biochemistry

2. Distinguish between
   (a) biological and physical sciences
   (b) hypothesis and scientific law
   (c) inorganic and organic chemistry
   (d) organic chemistry and biochemistry

PROBLEMS

3. Listed below are research interests of international chemists. Based on the definitions of the branches of chemistry, classify these research interests as organic, inorganic, analytical, physical, or biochemical.
   (a) Metal analysis—Samuel E. Q. Ashley
   (b) Chemistry of opium alkaloids, synthesis of morphine substitutes, chemotherapy, antimalarials, antituberculosis drugs, organic phosphorus compounds, antimetabolites and psychopharmacological drugs—Alfred Burger
   (c) Structure of boron hydrides—Hugh Christopher Longuet-Higgins

4. The following is a list of research papers published in various scientific journals. Based on the definitions of the branches of chemistry, classify these research papers as organic, inorganic, analytical, physical, or biochemical.

   (a) "The Structure of Krypton Difluoride"—Felix Schreiner

   (b) "Multiple Forms of the Nerve Growth Factor Protein and Its Subunits"—Andrew P. Smith

   (c) "The Nature of Soluble Copper(I) Hydride [CuH]"—J. A. Dilts

   (d) "The Preparation and Properties of Peroxychromium(III) Species [$Cr_2O_2^{4+}$ and $Cr_3(O_2)_2^{5+}$]"—Edward L. King

   (e) "Thermodynamics of Proton Ionization in Dilute Aqueous Solution"—James L. Christensen

   (f) "Quantitative Analysis of 1,2-Propanediol Monoesters of Long Chain Fatty Acids by Fluorine Magnetic Resonance"—Kazuo Konishi

   (g) "Synthesis of Four Methoxy-Substituted 1,8-Naphthalic Anhydrides [$C_{13}H_8O_4$] and the Three Monomethyl-1,8-Naphthalic Anhydrides [$C_{13}H_8O_3$]"—James Cason

Readings

Fortune, Editors of, *Great American Scientists*. Englewood Cliffs, N.J.: Prentice-Hall, Inc., 1961. Discussion of the role of Americans in the development of science since 1920 and their contributions.

Tarbell, D. Stanley, and Ann Tracy Tarbell, "The Role of Roger Adams in American Science." *J. Chem. Educ.*, 1979, v. 56, p. 163, and the book *Roger Adams: Scientist and Statesman*, Washington, D.C., American Chemical Society, 1981. Both the article and the book review the life of the American researcher and teacher Roger Adams and his role in developing the chemistry department at the University of Illinois.

Ziegler, Gene R., "Eloosis—A Card Game Which Demonstrates the Scientific Method." *J. Chem. Educ.*, 1974, v. 51, p. 532. A very effective method for learning the scientific method through a card game.

Zuckerman, Harriet, "The Sociology of the Nobel Prizes." *Sci. Am.*, Nov. 1967, v. 217, p. 25. Analysis of American Nobel Prize laureates in science with regard to their origins and fate.

2

Measurements

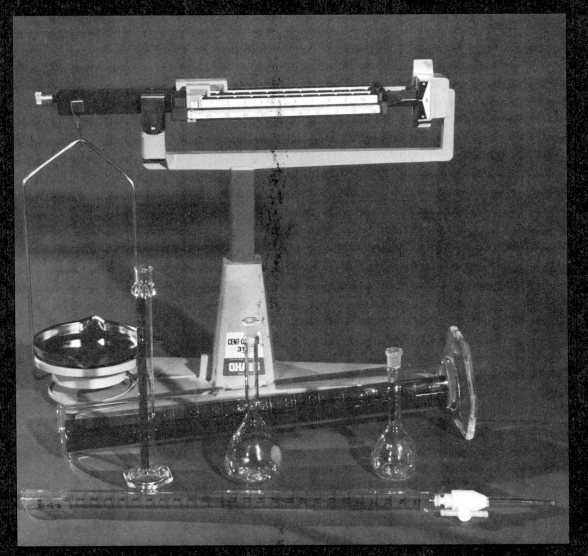

Triple beam balance, volumetric cylinders, volumetric flasks, and a buret represent measuring devices used by the chemist in the laboratory.

TASKS

1. Know the rules for identifying significant digits given in Section 2-1.
2. Know the rules for rounding off the nonsignificant digit(s) given in Section 2-2.
3. Memorize the meaning of the metric unit prefixes given in Table 2-2.
4. Learn the *factor-unit* method of problem solving given in Section 2-9.

OBJECTIVES

1. Given the following terms, define each term and describe the distinguishing characteristic of each:
   (a) significant digits (Section 2-1)
   (b) exponent (Section 2-3)
   (c) exponential notation (Section 2-3)
   (d) scientific notation (Section 2-4)
   (e) matter (Section 2-5)
   (f) mass (Section 2-5)
   (g) weight (Section 2-5)
   (h) energy (Section 2-10)
   (i) calorie (Section 2-12)
   (j) joule (Section 2-12)
   (k) newton (Section 2-12)
   (l) specific heat (Section 2-12)
   (m) density (Section 2-13)
   (n) specific gravity (Section 2-14)

2. Given various numbers, identify the number of significant digits in the numbers (Problem Example 2-1, Problems 5 and 6).

3. Given various numbers, add, subtract, multiply, and divide the numbers, giving your answer to the proper number of significant digits, and round off your answer correctly (Problem Examples 2-2 and 2-3, Problems 7, 8, 9, and 10).

4. Given a number, express the number in exponential notation as requested (Problem Examples 2-4 and 2-5, Problems 11 and 12).

5. Given numbers in exponential notation, add, subtract, multiply, and divide the numbers (Problem Examples 2-6 and 2-7, Problem 13).

6. Given numbers in exponential notation, determine the square root of the numbers (Problem Example 2-8, Problem 14).

7. Given numbers in exponential notation, raise the numbers to the requested power (Problem Example 2-9, Problem 15).

8. Given a number, express the number in scientific notation to three significant digits (Problem Example 2-10, Problems 16 and 17).

9. Given a measurement in the metric system, convert it to any other related unit in the metric system (Problem Examples 2-11, 2-12, 2-13, and 2-14, Problems 18, 19, 20, 21, and 22).

10. Given a temperature measurement in degrees Fahrenheit, convert it to degrees Celsius and kelvin, and the reverse (Problem Examples 2-15, 2-16, and 2-17, Problems 23, 24, 25, 26, 27, and 28).

11. Given calories or joules of heat energy, mass, and temperature change, calculate the specific heat of a substance (Problem Example 2-18, Problem 29).

12. Given the specific heat of a substance (in either cal/g·°C or J/kg·K units) and mass, heat energy, or temperature change, calculate the unknown mass or heat energy (Problem Examples 2-19, 2-20, and 2-21, Problems 30, 31, 32, 33, and 34).

13. Given its mass and volume, calculate the density of a substance (Problem Examples 2-22 and 2-23, Problem 35).

14. Given the density of a substance and its mass or volume, calculate the unknown volume or mass (Problem Examples 2-24, 2-25, 2-26, and 2-27, Problems 36 and 37).

15. Given the specific gravity of a substance and its mass or volume, calculate the unknown volume or mass (Problem Examples 2-28, 2-29, and 2-30, Problems 38 and 39).

Almost daily we use some form of measurement. The distance between two cities, the amount of flour in a cake made from "scratch," and the time it takes you to read this chapter are all examples of measurements.

In the physical sciences, such as chemistry and physics, accurate measurements are required. Suppose a certain piece of metal in a space capsule had to be exactly 16.2 centimeters (cm) long. If a machinist made a mistake and made it 16.1 cm long because he misread a ruler, the entire operation could be lost.

2-1

Significant Digits

Let us consider further the significance of our measurements. The **significant digits** (figures) in a number are the number of digits (figures) that give reasonably *reliable* information. They are important digits. For our 16.2-cm piece of metal, only the last digit, the 2, appears uncertain. From Figure 2-1, ruler *A*, you will note that the length of the metal is approximately $\frac{2}{10}$ or $\frac{1}{5}$ the distance between 16 and 17.

If we had used a ruler with more subdivisions, we would have been able to measure the length of the metal to 16.25 cm; then the 5 would have been

FIGURE 2-1

Measurement of a
piece of metal in a
space capsule. A
portion of the
centimeter scale of
both rulers is
magnified.

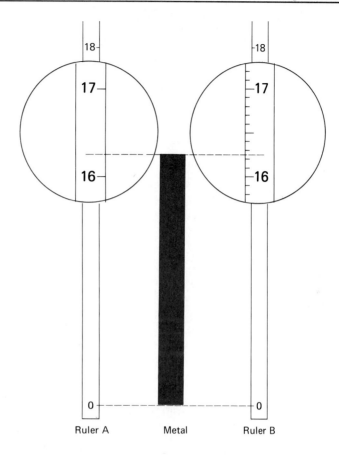

Ruler A   Metal   Ruler B

uncertain. As you will note in Figure 2-1, ruler *B*, the length of the metal is
approximately $\frac{5}{10}$ or $\frac{1}{2}$ the distance between 16.2 and 16.3.

To determine the number of significant digits in a measurement, we follow
certain rules.

1. The digits 1, 2, 3, 4, 5, 6, 7, 8, and 9 are significant. Hence, in the
    preceding example, 16.2 has three significant digits and 16.25 has four
    significant digits.

2. This leaves zero, which presents a peculiar problem.
    a. If zeros appear between nonzero digits, they are considered signifi-
        cant.
        104 contains three significant digits.
        1004 contains four significant digits.
    b. If zeros appear to the right of the decimal in numbers *ending in
        zero(s)*, they are considered significant. This is true for numbers
        greater and less than 1. A measurement has been made and found to
        be zero in that position.

154.00 contains five significant digits.
154.0 contains four significant digits.
15.400 contains five significant digits.
15.40 contains four significant digits.
1.5400 contains five significant digits.
1.540 contains four significant digits.
0.15400 contains five significant digits. (See rule c for the zero in the units place.)
0.1540 contains four significant digits.

c. If a zero appears in a number only to fix the position of the decimal point in a number less than 1, it is not significant, since *no measurement has been made.*

0.564 contains only three significant digits.
0.0564 contains only three significant digits.

d. Terminal zeros (zeros at the end of a number with no decimal point shown) in a number are usually not significant. To avoid confusion in this text, we shall write a terminal zero that is significant with a —— (bar) above it. All others will be considered nonsignificant. Thus, 5600 expressed to two significant digits is written 5600; to three significant digits, 56$\overline{0}$0; to four significant digits, 56$\overline{00}$.[1] Expressed to three significant digits, 56400 is written 56400; to four significant digits, 564$\overline{0}$0; to five significant digits, 564$\overline{00}$. Exponential notation can be used instead of the bar (see 2-3).

Problem Examples 2-1

Determine the number of significant digits in the following numbers:

Number	*Answer* [*rule(s)*]
(a) 747	3 (1)
(b) 1011	4 (1, 2a)
(c) 3.50	3 (1, 2b)
(d) 0.056	2 (1, 2c)
(e) 35$\overline{0}$	3 (1, 2d)
(f) 6.02	3 (1, 2a)
(g) 7065	4 (1, 2a)
(h) 0.604	3 (1, 2a, 2c)
(i) 10.04	4 (1, 2a, 2b)
(j) 122.0	4 (1, 2b)
(k) 7.0200	5 (1, 2a, 2b)

Work Problems 5 and 6.

---

[1]In the case of 5600, to two significant digits, we are saying that the number is between 5500 and 5700. In the case of 56$\overline{0}$0, to three significant digits, we are saying that the number is between 5590 and 5610. In the last case of 56$\overline{00}$, to four significant digits, we are saying that the number is between 5599 and 5601. The number 56$\overline{00}$ to four significant digits may also be written 5600. with a decimal point after the final zero. The bar is also used in algebra above *digits* to indicate a repeating decimal. Either above a zero or above a digit, the bar implies a more significant meaning in both cases.

2-2

Mathematical Operations Involving Measurements and Significant Digits[2]

We shall now use these significant digits rules in some simple calculations.

Addition and Subtraction

In addition and subtraction, *the answer must not contain any more places* (i.e., decimal, units, tens, etc.) *than the number with the least number of places.* The sum of

$$
\begin{array}{r}
25.1 \\
+\ \underline{22.11} \\
\text{is}\quad 47.21
\end{array}
$$

but the answer must be expressed to only the tenths decimal place since only the tenths decimal place is the least place in the number 25.1; hence, the answer is 47.2. The reason becomes obvious if you note that the hundredths decimal place is not measured in the number 25.1 and, thus, could vary widely. The difference of

$$
\begin{array}{r}
4.732 \\
-\ \underline{3.62} \\
\text{is}\quad 1.112
\end{array}
$$

but the answer must be expressed to only the hundredths decimal place, because only the hundredths decimal place is the least place in the number 3.62; hence, the answer is 1.11.

Multiplication and Division

In multiplication and division, *the answer must not contain any more significant digits than the **least** number of significant digits in the numbers used in the multiplication or division.* The product of 17.21×11.1 is 191.031, but the answer must be expressed to only three significant digits since 11.1 has only three significant digits; hence, the answer is 191. The quotient of $\dfrac{26.32}{2.23}$ is 11.80269, but the answer again must be expressed to only three significant digits since 2.23 has only three significant digits; therefore, the answer is 11.8.

---

[2]In chemistry you will be solving many mathematical problems. We suggest that you buy an inexpensive calculator. But, before you do this, ask your instructor what model he or she recommends. The calculator should have the following functions: $+$, $-$, $\times$, $\div$, and $\sqrt{}$. Although it is not essential it is helpful if the calculator also has the functions $\log x$ and y^x. Instructions for operating the calculator come with each model. The instructions vary with each model, so you should follow them very carefully in operating your calculator.

Rounding Off

The next problem facing us is the method of rounding off the nonsignificant digits so as to arrive at the desired significant digits. The following rules apply to rounding off the nonsignificant digits:

1. If the nonsignificant digit is *less* than 5, it is dropped and the last significant digit remains the same. Thus, 47.21 is equal to 47.2 to three significant digits.

2. If the nonsignificant digit is *more* than 5 or *is* 5 followed by *numbers other than zeros*, the nonsignificant digit(s) is (are) dropped and the last significant digit is increased by one. Hence, 47.26 and 47.252 are both equal to 47.3 to three significant digits.

3. If the nonsignificant digit is 5 and is followed by zeros, the 5 is dropped and the last significant digit is *increased by one if it is odd* and *left the same if it is even*. Thus, 47.250 is equal to 47.2 to three significant digits, and 47.350 is equal to 47.4.

Problem Example 2-2

Round off the following numbers to three significant digits.

Number	*Answer (rule)*
(a) 462.2	462 (1)
(b) 453.6	454 (2)
(c) 474.50	474 (3)
(d) 687.54	688 (2)
(e) 687.50	688 (3)
(f) 688.50	688 (3)
(g) 12.750	12.8 (3)
(h) 0.027650	0.0276 (3)
(i) 0.027654	0.0277 (2)
(j) 0.027750	0.0278 (3)

Now let us consider some mathematical operations applying the rules governing addition and subtraction, multiplication and division, and rounding off.

Problem Example 2-3

Perform the indicated mathematical operations and express your answer to the proper number of significant digits.

(a) 17.8 + 14.73 + 16

SOLUTION: The least place is the units place in the number 16; hence, the answer must be expressed to the units place.

$$
\begin{array}{l}
17.8 \\
14.73 \\
\underline{16} \longleftarrow \text{number with the least number of places} \\
48.53 \qquad\qquad 49 \quad \textit{Answer}
\end{array}
$$

Rounded off to the units place 48.53 is 49.

(b) 0.647 + 0.03 + 0.31

SOLUTION: The least place is the hundredths decimal place in the numbers 0.03 and 0.31; and the answer must be expressed to the hundredths decimal place.

$$
\begin{array}{l}
0.647 \\
0.03 \longleftarrow \\
\underline{0.31} \longleftarrow \text{numbers with the least number of places} \\
0.987 \qquad\qquad 0.99 \quad \textit{Answer}
\end{array}
$$

Rounded off to the hundredths decimal place, 0.987 is 0.99.

(c) 14.72 − 6.8

SOLUTION: The least place is the tenths decimal place in the number 6.8; thus, the answer must be expressed to the tenths decimal place.

$$
\begin{array}{l}
14.72 \\
\underline{-\ 6.8} \longleftarrow \text{number with the least number of places} \\
7.92 \qquad\qquad 7.9 \quad \textit{Answer}
\end{array}
$$

Rounded off to the tenths decimal place, 7.92 is 7.9

(d) 24.78 − 0.065

SOLUTION: The least place is the hundredths decimal place in the number 24.78; and, the answer must be expressed to the hundredths decimal place.

$$
\begin{array}{l}
24.78 \longleftarrow \text{number with the least number of places} \\
\underline{-0.065} \\
24.715 \qquad\qquad 24.72 \quad \textit{Answer}
\end{array}
$$

Rounded off to the hundredths decimal place, 24.715 is 24.72.

(e) 13 × 752

SOLUTION: The number with the least number of significant digits is 13, which has only two; thus, the answer must be expressed to no more than two significant digits.

$$752 \times 13 = 9776 \qquad 9800 \quad \textit{Answer}$$

Rounded off to two significant digits, 9776 is 9800.

(f) 0.02×47

SOLUTION: The number with the least number of significant digits is 0.02, which has only one; hence, the answer must be expressed to only one significant digit.

$$47 \times 0.02 = 0.94 \qquad 0.9 \quad \textit{Answer}$$

Rounded off to one significant digit, 0.94 is 0.9.

(g) $\dfrac{181.8}{75}$

SOLUTION: The number with the least number of significant digits is 75, which has two; therefore, the answer must be expressed to no more than two significant digits.

$$\frac{181.8}{75} = 2.424, \text{ from your calculator}$$

Rounded off to two significant digits, 2.424 is 2.4   *Answer*

(h) $\dfrac{13.65}{2.26}$

SOLUTION: The number with the least number of significant digits is 2.26, which has three; and, the answer must be expressed to no more than three significant digits.

$$\frac{13.65}{2.26} = 6.039823, \text{ from your calculator}$$

Rounded off to three significant digits, 6.039823 is 6.04.   *Answer*

(i) $\dfrac{9.74 \times 0.12}{1.28}$

SOLUTION: The number with the least number of significant digits is 0.12, which has two, and the answer must be expressed to two significant digits.

$$\frac{9.74 \times 0.12}{1.28} = 0.913125, \text{ from your calculator}$$

Rounded off to two significant digits, 0.913125 is 0.91. *Answer*

In performing this operation on your calculator, do the entire operation in *one* step; that is, divide, multiply *or* multiply, divide. Both ways give the same answer on your calculator, that is, 0.913125. Do NOT carry out the operation in *two* steps. Do not perform the multiplication or division, *remove* the answer from your calculator, and then put this answer back in your calculator and do the next step. If you do the operation this way, you may make a mistake in writing down the answer from one of the steps.

Work Problems 7, 8, 9, and 10.

2-3

Exponents

In scientific work it is often necessary to use large numbers or extremely small numbers. As a practical example, the world population is estimated at 4,700,000,000 people and the U.S. population at 234,000,000 people. The U.S. population represents a mere 0.050 factor of the total world population. A method whereby these numbers may be written in a more condensed form will be given here.

An **exponent** is a whole number or symbol written as a superscript above another number or symbol, the *base,* denoting the number of times the base must be repeated as a factor. The number of times a base is repeated as a factor is called the "power of the base." For example, in the symbol x^n, *n* is the exponent and *x* is the base. We read "x^n" as "*x* raised to the *n*th power," which is equal to

$$\underbrace{x \cdot x \cdot x \ldots x}_{n \text{ factors}}$$

We read "10^6" as "10 raised to the sixth power," and it equals

$$\underbrace{10 \cdot 10 \cdot 10 \cdot 10 \cdot 10 \cdot 10}_{6 \text{ factors}} \quad \text{or} \quad 1{,}000{,}000$$

As review, let us consider some powers of 10 as given in Table 2-1. We can express 1000 as $10 \cdot 10 \cdot 10$, or 10^3; $\frac{1}{1000}$ can be expressed as $\frac{1}{10} \cdot \frac{1}{10} \cdot \frac{1}{10}$ or $1/10^3$—hence, 10^{-3}.

Now let us consider expressing numbers in exponential notation. **Exponential notation** is a form for expressing a number using a product of two numbers, one of the numbers as a decimal and the other as a power of 10. For example, 24.1×10^4 is in exponential notation with 24.1 being the decimal and 10^4 as a power of 10.

To express a number in exponential notation, you may find the following guidelines helpful:

1. Changing a number by shifting the decimal point to the **left** of its original position involves a factor of "a power of 10" and uses a **positive** exponent;

TABLE 2-1   Powers of 10

$$1000 = 10^3$$
$$100 = 10^2$$
$$10 = 10^1$$
$$1 = 10^{0a}$$
$$\tfrac{1}{10} = 10^{-1b}$$
$$\tfrac{1}{100} = 0.01 = 10^{-2b}$$
$$\tfrac{1}{1000} = 0.001 = 10^{-3b}$$

[a]Any nonzero number raised to the zero power is equal to 1, such as $x^0 = 1$, or $10^0 = 1$.

[b]Any number written with a base and a negative exponent is the inverse of another number using the *same base,* and a corresponding *positive* exponent. For example, x^{-n} is the inverse of x^n (since $x^{-n} = 1/x^n$), and 10^{-6} is the inverse of 10^6 (since $10^{-6} = 1/10^6$).

changing a number by shifting the decimal point to the **right** of its original position involves a factor of "a power of 10" and uses a **negative** exponent.

2. In moving the decimal point to the left or the right, the *exponent* is *equal* numerically to the *number of places* the decimal point has been moved. A positive exponent means a number *larger* than 1, while negative exponent means a number *smaller* than 1.

Problem Example 2-4

Express the estimated world population of 4,700,000,000 people (or 4,700,000,000) in exponential notation as 4.7×10^n.

SOLUTION: Moving the decimal place to the left to obtain 4.7 means that the exponent (n) must be positive, and to obtain the 4.7, the decimal point must be moved *nine* places to the left of its original position—hence, 4.7×10^9 people.   *Answer*

Problem Example 2-5

Express the factor of the U.S. population to that of the world population, 0.050 in exponential notation as $5\bar{0} \times 10^n$ ($50. \times 10^n$).

SOLUTION: Moving the decimal place to the right to obtain 50. means that the exponent (n) must be negative, and to obtain 50., the decimal must be moved three places to the right; and, the answer is $5\bar{0} \times 10^{-3}$.

Addition and Subtraction of Exponential Numbers

To add or subtract exponential numbers, we must express *each* quantity to the *same power of 10.* The decimals are added or subtracted in the usual manner and the powers of 10 are recorded.

Problem Example 2-6

Carry out the operations indicated on the following exponential numbers:

(a) $3.40 \times 10^3 + 2.10 \times 10^3$

SOLUTION: Both numbers have the same power of 10 (10^3); hence, they can be added:

$$\begin{array}{r} 3.40 \times 10^3 \\ \underline{2.10 \times 10^3} \\ 5.50 \times 10^3 \quad Answer \end{array}$$

(b) $4.20 \times 10^{-3} + 1.2 \times 10^{-4}$

SOLUTION: To add these numbers, first convert them to the *same power of 10*. The number 1.2×10^{-4} converts to 0.12×10^{-3}, following the guidelines previously mentioned. These two numbers can now be added.

$$\begin{array}{r} 4.20 \times 10^{-3} \\ \underline{0.12 \times 10^{-3}} \\ 4.32 \times 10^{-3} \quad Answer \end{array}$$

Multiplication and Division of Exponential Numbers

For multiplying or dividing exponential numbers, the only requirement is that the numbers are expressed to the same *base*, which is 10 in exponential notation. In multiplication, the decimals are multiplied in the usual manner, but the *exponents* to the base 10 are **added** algebraically. In division, the decimals are divided in the usual manner, but the *exponents* to the base 10 are **subtracted** algebraically.

Problem Example 2-7

Carry out the operations indicated on the following exponential numbers:

(a) $1.70 \times 10^6 \times 2.40 \times 10^3$

SOLUTION: Multiply the decimals and then add the exponents algebraically, as

$$(1.70 \times 2.40)\,(10^6 \times 10^3) = 4.08 \times 10^{6+3} = 4.08 \times 10^9 \quad Answer$$

(b) $1.70 \times 10^6 \times 2.40 \times 10^{-3}$

SOLUTION

$$(1.70 \times 2.40)\,(10^6 \times 10^{-3}) = 4.08 \times 10^{6-3} = 4.08 \times 10^3 \quad Answer$$

(c) $\dfrac{2.40 \times 10^5}{1.30 \times 10^3}$

SOLUTION: Divide the decimals and then subtract the exponents algebraically, as

$$\frac{2.40 \times 10^5}{1.30 \times 10^3} = \frac{2.40}{1.30} \times \frac{10^5}{10^3} = 1.85 \times 10^{5-3} = 1.85 \times 10^2 \quad Answer$$

(d) $\dfrac{2.40 \times 10^5}{1.30 \times 10^{-3}}$

SOLUTION

$$\frac{2.40 \times 10^5}{1.30 \times 10^{-3}} = \frac{2.40}{1.30} \times \frac{10^5}{10^{-3}} = 1.85 \times 10^{5-(-3)} = 1.85 \times 10^{5+3} = 1.85 \times 10^8$$
$$Answer$$

Note that the exponents are *subtracted algebraically*.

Square Root of Exponential Numbers

To obtain the positive value of the square root of a number, first express the number in exponential notation in which the power of 10 has an **even** exponent. To obtain the positive square root of the exponential number, obtain the square root of the decimal from a calculator, and obtain the square root of the power of 10 by dividing the exponent by 2.

Problem Example 2-8

Determine the value of each of the following numbers:

(a) $\sqrt{4.00 \times 10^{-4}}$

SOLUTION: Take the positive square root of the decimal and then divide the exponent by 2.

$$\sqrt{4.00 \times 10^{-4}} = \sqrt{4.00} \times \sqrt{10^{-4}} = 2.00 \times 10^{-(4/2)}$$
$$= 2.00 \times 10^{-2} \quad Answer$$

(b) $\sqrt{5.60 \times 10^{-5}}$

SOLUTION: Change the number 5.60×10^{-5} to a number with an *even* exponent, following the guidelines previously mentioned, that is, 56.0×10^{-6}. Then, take the positive square root of 56.0×10^{-6} as

$$\sqrt{5.60 \times 10^{-5}} = \sqrt{56.0 \times 10^{-6}} = \sqrt{56.0} \times \sqrt{10^{-6}}$$
$$= 7.48 \times 10^{-(6/2)} = 7.48 \times 10^{-3} \quad Answer$$

(The 7.48 is obtained from a calculator.)

Positive Powers of Exponential Numbers

To raise an exponential number to a given positive power, first raise the decimal part to the power by using it as a factor the number of times indicated by the power and then multiply the exponent of 10 by the indicated power.

Problem Example 2-9

Perform the indicated operations on the following exponential numbers.

(a) Raise 2.45×10^4 to the second power.

SOLUTION: Multiply 2.45×2.45; then multiply the exponent (4) by 2 (the second power).

$$(2.45 \times 10^4)^2 = (2.45)^2 \times (10^4)^2 = 2.45 \times 2.45 \times 10^8 = 6.00 \times 10^8 \quad \textit{Answer}$$

(b) Raise 3.14×10^2 to the third power.

SOLUTION: Multiply $3.14 \times 3.14 \times 3.14$, then multiply the exponent (2) by 3 (the third power).

$$
\begin{aligned}
(3.14 \times 10^2)^3 &= (3.14)^3 \times (10^2)^3 = 3.14 \times 3.14 \times 3.14 \times 10^6 \\
&= 31.0 \times 10^6 \text{ or } 3.10 \times 10^7 \text{ (in scientific notation,}
\end{aligned}
$$

Work Problems 11, 12, 13, 14, and 15.

see next section)   *Answer*

2-4

Scientific Notation

Scientific notation is a more systematic form of exponential notation. In **scientific notation,**[3] the decimal part is from *one* to **less** *than* 10. Hence, in scientific notation, the decimal part must have *exactly one* digit to the left of the decimal point.

Problem Example 2-10

Express the following in scientific notation to three significant digits:

(a) 6,780,000

(b) 2170

(c) 0.0756

(d) 10.7

---

[3]Another form of exponential notation is engineering notation. Engineering notation is used in the engineering sciences and expresses the decimal from 1 to less than 1000 with the exponent of the power of 10, divisible by 3. The number 2.43×10^4 expressed in engineering notation is 24.3×10^3.

SOLUTION: Express the preceding in scientific notation by shifting the decimal place following the guidelines previously mentioned or by a second method[4] of operating *equally* on the numerator and denominator to obtain a number between 1 and *less* than 10, as

(a) $6,780,000 = 6.78 \times 10^6$ *Answer*

or

$$6,780,000 \times \frac{1,000,000}{1,000,000} = \frac{6,780,000}{1,000,000} \times 1,000,000 = 6.78 \times 1,000,000$$
$$= 6.78 \times 10^6 \quad Answer$$

(b) $2170 = 2.17 \times 10^3$ *Answer*

or

$$2170 \times \frac{1000}{1000} = \frac{2170}{1000} \times 1000 = 2.17 \times 1000 = 2.17 \times 10^3 \quad Answer$$

(c) $0.0756 = 7.56 \times 10^{-2}$ *Answer*

or

$$0.0756 \times \frac{100}{100} = 7.56 \times \frac{1}{100} = 7.56 \times \frac{1}{10^2} = 7.56 \times 10^{-2} \quad Answer$$

(d) $10.7 = 1.07 \times 10^1$ *Answer*

or

Work Problems 16 and 17.

$$10.7 \times \frac{10}{10} = 1.07 \times 10^1 \quad Answer$$

2-5

Matter, Mass, and Weight

Basic to the study of chemistry is matter, but exactly what is matter? **Matter** can be defined as anything that has mass and occupies space. Now this presents another problem—what is mass? **Mass** is the quantity of matter in a particular body. *The mass of a body is constant and does not change* whether it is measured in New Mexico or Bombay, or even on the moon. The **weight** of a body is the *gravitational force of attraction* between the body's mass and the mass of the planet or satellite on which it is weighed, whether it be the earth or the moon.

The *weight* of a body *varies*, but the *mass does not*. The earth is not spherical, but is slightly pear-shaped, and since the gravitational attraction between the earth and a body varies with the distance between their centers, the weight

---

[4]This alternate method can also be used to express numbers in exponential notation. It is introduced here only after multiplication and division of exponents have been covered and as an alternative, since some students find it difficult to remember the guidelines. You will note that to express a number in scientific notation, any number **larger** than 1 will have a **positive** exponent, and any number **smaller** than 1 will have a **negative** exponent, with 1 having an exponent of 0.

of the body varies slightly depending on where it is measured. For example, a body that weighs 10 pounds (lb) at the North Pole would weigh approximately 9 lb 15 ounces (oz) at the Equator—a 1-oz difference. Hence, a person can lose weight, but not mass, in flying from the North Pole to the Equator! On the moon, the gravitational attraction is considerably less and an object would weigh less on the moon than on the earth. For example, the same 10-lb body at the North Pole would weight approximately 1 lb 11 oz on the moon.

Figure 2-2 summarizes the relationship of matter, mass, and weight.

FIGURE 2-2

Relations of matter, mass, and weight.

2-6

Measuring Mass and Weight

The mass of a body may be determined on a balance, as shown in Figure 2-3a, b, and c. In the balances shown in a and b, the body to be "weighed" is placed on the left-hand pan and the "weights" of known masses are placed on the right-hand pan and/or on the sliding scale to counterbalance the unknown body. The weight of the unknown body and the weights of known masses are

FIGURE 2-3

Measuring mass: (a) the platform balance, (b) the triple beam balance, (c) a single pan analytical balance. Measuring weight: (d) the spring scale.

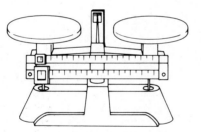

(a) Platform balance

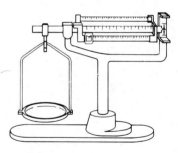

(b) Triple beam balance

(c) A single pan analytical balance

(d) Spring scale

in balance when the pointer is on the center of the scale. A balance operates on the same principle as the balancing of a pencil on your finger. The force of gravity acts equally on the known masses and on the body when balance is achieved; thus, the *mass* of the object would be found to be the same, despite where it was measured.

Spring scales, such as those found in the produce departments of grocery stores, are used to determine the *weight* of a body, as is shown in Figure 2-3d. The body to be weighed is attached to a hook and the spring is stretched. The stretching of the spring is dependent upon the gravitational attraction of the planet or satellite on which the body is weighed. The spring scale is also used by fishermen to measure the weight of their catch.

In chemistry, the balance is used exclusively, and hence we measure the *mass* of a body. The two terms "mass" and "weight" are unfortunately used interchangeably, but if a balance is used the correct term is "mass."

2-7

Quantitative Measurement of Matter

Two basic measures of matter are mass and volume. In addition, linear measurements may also be made on a sample of matter. In chemistry, mass is measured with a chemical balance (Figure 2-3a, b, and c); volume is measured with a graduated cylinder, buret, or pipette (Figure 2-4); and length is commonly measured with a ruler (Figure 2-5).

FIGURE 2-4

Measuring volume: (a) the graduated cylinder, (b) the buret, (c) the pipette (a bulb is used to fill the pipette).

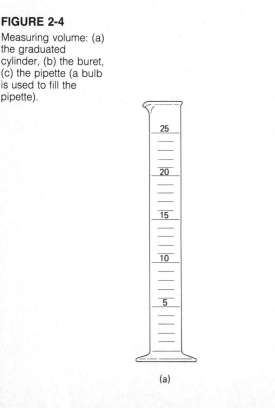

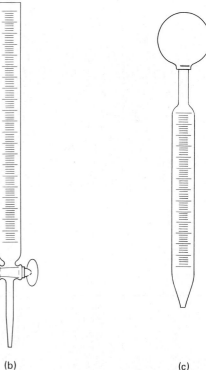

(a)　　　　　　　　(b)　　　　　　　　(c)

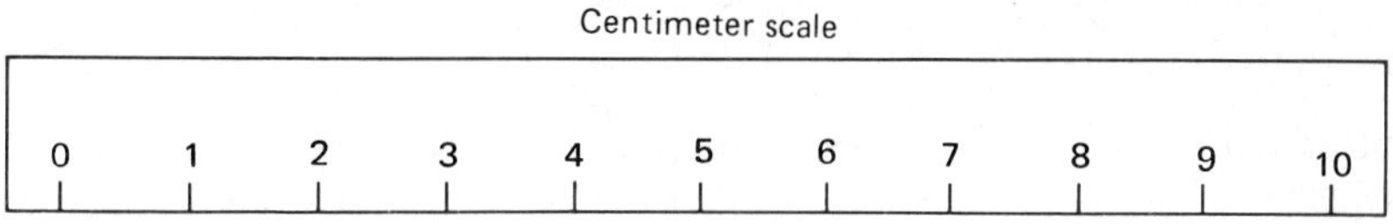

FIGURE 2-5

Measuring length by a ruler.

Centimeter scale

0   1   2   3   4   5   6   7   8   9   10

There are two methods for quantitatively measuring matter: the metric system[5] and the English-based units.[6] The English-based units are used primarily in the United States; the metric system is used throughout almost all of the rest of the world (even in Great Britain) and is gradually being adopted in the United States, as shown in Figure 2-6. The chemical industry is now in the process of "going metric" in the shipping and billing of industrial chemicals.

ALBUQUERQUE N.M.
ELEVATION 1 MILE
5280 FT. 1609 M

(a)

LITER PRICING

67 ¢
PER LITER
EQUALS
$1.41
PER GALLON
3.785 LITERS EQUALS 1 GALLON

(b)

Campbell's

PORK & BEANS

IN TOMATO SAUCE

NET WT. 28 OZ. (1 LB. 12 OZ.) (794 GRAMS)

(c)

FIGURE 2-6

The metric system. Even in the United States the metric system is being adopted: (a) length, (b) volume, and (c) mass.

---

[5]In 1960 an international group of scientists adopted a system of units called the International System of Units, abbreviated SI from the French *Système International*. This system includes the metric system, although it sometimes uses other base units. SI uses seven base units, two supplementary units, a series of derived units, and a set of prefixes to define multiples or fractions of these units. Appendix I summarizes the SI units and provides conversion factors in going from SI units to non-SI units.

[6]The English-based units are discussed in Appendix II.

2-8

The Metric System

The metric system has as its basic units the gram (g, mass), the liter (or litre, L, volume), the meter (or metre, m, length), and the second (s or sec, time).[7] The units for mass, volume, and length in the metric system are related in multiples of 10, 100, 1000, etc., similar to some parts of our monetary system. The prefixes used to define multiples or fractions of the basic units and their relation to parts of our monetary system are shown in Table 2-2.

TABLE 2-2 Metric Units in General vs. Monetary System

kilo	1000 units	vs. $1000	A "grand"
deci	1/10 unit	vs. 1/10 dollar	A **d**ime
centi	1/100 unit	vs. 1/100 dollar	A **c**ent
milli	1/1000 unit	vs. 1/1000 dollar	A **m**ill
micro	1/1,000,000 unit	No analogy	—

TABLE 2-3 Metric Units of Mass, Volume, and Length

Basic Units per Derived Unit			Mass		Volume		Length[a]	
kilo	1000	(10^3)	kilogram	(kg)	kiloliter	(kL)	kilometer	(km)
basic unit	1	(10^0)	gram	(g)	liter	(L)	meter	(m)
deci	0.1	(10^{-1})	decigram	(dg)	deciliter	(dL)	decimeter	(dm)
centi	0.01	(10^{-2})	centigram	(cg)	centiliter	(cL)	centimeter	(cm)
milli	0.001	(10^{-3})	milligram	(mg)	milliliter	(mL)[b]	millimeter	(mm)
micro	0.000001	(10^{-6})	microgram or gamma	(μg) (γ)	microliter	(μL)	micrometer or micron	(μm) (μ)

[a]Smaller units of length are the nanometer (nm, 1/1000 micron, once called a millimicron, mμ) and the **angstrom** (Å, 1/10 nanometer or 1/10,000 micron). We, therefore, have the following two equivalents: 10 Å = 1 nm and 10,000 Å = 1μ. Hence, a nanometer is 0.000000001 meter (10^{-9} m), and an **angstrom** is 0.0000000001 meter (10^{-10} m). The **angstrom** is not used in the SI, but the nanometer is accepted.

[b]The milliliter (mL) and the cubic centimeter (cm^3 or cc) are exactly equivalent, since 1 L = 1 dm^3 according to the SI definition.

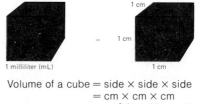

Volume of a cube = side × side × side
= cm × cm × cm
= cm^3 (1 cubic centimeter, cc)

[7]The SI uses the basic units kilogram (kg), meter (m), and second (s) for mass, length, and time, respectively; and the derived unit cubic meter (m^3) as the unit for volume. In the SI one liter is defined as being exactly equal to one cubic decimeter (1 L = 1 dm^3) or one cubic meter equals 1000 liters (1 m^3 = 1000 L).

The specific metric units of mass (gram), volume (liter), and length (meter) are shown in Table 2-3, which is an extension of Table 2-2 using the specific units. You must learn these units and their equivalents in powers of 10 in order to work problems. For example, you should know that 10 dg = 1 g, 100 cm = 1 m, and 1000 mL = 1L, etc.

2-9

The Factor-Unit Method of Problem Solving. Conversion Within the Metric System

Before we consider conversion within the metric system, we shall describe a general method of problem solving called the *factor-unit* method.[8] This method is quite simple and is based on developing a relationship between different units expressing the same physical dimension. A very simple example will serve to illustrate our point. Suppose you decide to have for dinner a salad that uses avocados. You go to the neighborhood grocery store and discover that avocados cost 40¢ each. If you have only 80¢, how many avocados can you buy to put in this salad? We all know that you can buy two avocados, but let us apply our *factor-unit* method of problem solving to this problem. The relationship we have between the cents and the number of avocados can be expressed as

$$40¢ = 1 \text{ avocado}$$

Dividing both sides of the equation by 40¢, we have

$$\frac{40¢}{40¢} = 1 = \frac{1 \text{ avocado}}{40¢}, \text{ which we will call factor } A$$

Now, dividing both sides of the equation 40¢ = 1 avocado by 1 avocado, we have

$$\frac{40¢}{1 \text{ avocado}} = \frac{1 \text{ avocado}}{1 \text{ avocado}} = 1, \text{ which we will call factor } B$$

We have given (in the problem) 80¢, but we wish to express our answer as the number of avocados. We wish to multiply the given quantity, 80¢, by a factor, so our cents will cancel and our answer will be in avocados. The factor which we must use is factor A, which has avocados in the numerator and cents in the denominator. This will cancel out cents and leave avocados as the unit.

$$80¢ \times (\text{factor } A) = \text{number of avocados}$$

We can multiply the 80¢ by 1 (both factors equal 1) and not change the value

---

[8]This method has also been called "dimensional analysis," "unit conversion," and the "factor-label" method for problem solving.

based on the unit multiplicative identity property from mathematics, but only factor A will give the correct unit (label), and hence the correct answer.

$$80¢ \times \frac{1 \text{ avocado}}{40¢} = 2 \text{ avocados} \quad Answer$$

If we had chosen factor B in error, then

$$80¢ \times \frac{40¢}{1 \text{ avocado}} = \frac{3200¢^2}{1 \text{ avocado}}$$

which does not answer our original question, and the units are also meaningless.

Whenever possible in this text, we shall use the *factor-unit* method in problem solving.

Before we consider conversions within the metric system, a few hints in problem solving should be given since you will be solving many problems in this course. The following are some useful hints in problem solving:

1. Read the problem first very carefully to determine what is actually asked for.

2. Organize the data that are given, being sure to include *both* the *units* of the *given* and the *units* of the *unknown*.

3. **Write down the *given* along with the *units*, and at the end of the line the *units* of the *unknown*.**

4. Apply the principles you have learned throughout the course to develop **factors** so that these factors, used properly, will give the correct units in the unknown.

5. Check your answer to see if it is reasonable by checking both the mathematics and the units.

6. Finally, check the number of significant digits.

Now let us consider a number of problem examples for conversions within the metric system.

Problem Example 2-11

Convert 3.85 m to mm.

SOLUTION

$$3.85 \text{ m} \times (\text{factor}) = \text{mm}$$

We know that 1000 mm = 1 m; hence, our factors are as follows:

$$\frac{1000 \text{ mm}}{1 \text{ m}}; \quad \frac{1 \text{ m}}{1000 \text{ mm}}$$

$$A \qquad\qquad B$$

To obtain the correct units (mm), the factor we must use is factor A.

$$3.85 \; \cancel{m} \times \frac{1000 \; mm}{1 \; \cancel{m}} = 3850 \; mm \quad \textit{Answer}$$

Multiplying 3.85×1000, we arrive at the answer 3850 expressed in mm. The number 3.85 is expressed to three significant digits, and hence our answer should be expressed accordingly; the factor $\frac{1000 \; mm}{1 \; m}$ is not considered in significant digits since it is an exact value, and could be expressed as $\frac{10\overline{00} \; mm}{1 \; m}$. (This number has as many significant digits as are necessary to match those in a measurement.)

Problem Example 2-12

Convert 75.2 mg to kg.

SOLUTION

$$75.2 \; mg \times (\text{factor}) = kg$$

We do not know a factor converting mg directly to kg, but we do know factors that convert mg to g and g to kg; that is,

$$1000 \; mg = 1 \; g \quad \text{and} \quad 1000 \; g = 1 \; kg$$

Considering the first factor, $1000 \; mg = 1 \; g$, we have

$$\frac{1000 \; mg}{1 \; g} ; \quad \frac{1 \; g}{1000 mg}$$

$$\quad\quad A \quad\quad\quad\quad B$$

If we use factor B, we shall cancel the mg and our units will be in g.

$$75.2 \; \cancel{mg} \times \frac{1 \; g}{1000 \; \cancel{mg}} \times (\text{factor}) = kg$$

Now, considering the second factor, $1000 \; g = 1 \; kg$, we have

$$\frac{1000 \; g}{1 \; kg} ; \quad \frac{1 \; kg}{1000 \; g}$$

$$\quad\quad C \quad\quad\quad\quad D$$

If we use factor D, we cancel the g and our units will be in kg, which is the answer to the original question

$$75.2 \; \cancel{mg} \times \frac{1 \; \cancel{g}}{1000 \; \cancel{mg}} \times \frac{1 \; kg}{1000 \; \cancel{g}} = 0.0000752 \; kg \text{ or } 7.52 \times 10^{-5} kg \quad \textit{Answer}$$

Dividing 75.2 by (1000 × 1000), we arrive at the answer of 0.0000752 expressed in kg. The number 75.2 is expressed to three significant digits, and hence our answer should also be expressed to three significant digits.

Problem Example 2-13

The following masses were recorded in a laboratory experiment: 2.0000000 kg, 5.0000 g, 650.0 mg, 0.5 mg. What is the total mass in grams?

SOLUTION

$$2.0000000 \text{ kg} \times \frac{1000 \text{ g}}{1 \text{ kg}} = 2000.0000 \text{ g}$$

$$5.0000 \text{ g}$$

$$650.0 \text{ mg} \times \frac{1 \text{ g}}{1000 \text{ mg}} = 0.6500 \text{ g}$$

$$0.5 \text{ mg} \times \frac{1 \text{ g}}{1000 \text{ mg}} = 0.0005 \text{ g}$$

$$\overline{2005.6505 \text{ g}} \quad Answer$$

Note that all the masses in grams are recorded to the ten-thousandth of a gram.

Problem Example 2-14

Convert 0.0076 μ to Å.

SOLUTION: From Table 2-3, we know that 10,000 Å = 1 μ. Therefore, the solution is as follows:

$$0.0076 \text{ } \mu \times \frac{10,000 \text{ Å}}{1 \mu} = 76 \text{ Å} \quad Answer$$

As you become more proficient with the metric system you may wish to make these conversions by merely shifting the decimal point as you do in our monetary system. For example, in one of the preceding cases, 650.0 mg to g involves shifting the decimal three places to the left, giving 0.6500 g, since 1 g = 1000 mg; hence, a smaller value must be obtained.

Work Problems 18, 19, 20, 21, and 22.

2-10

Energy

All changes and transformations in nature are accompanied by changes in energy. **Energy** is defined as the capacity for doing work or to transfer heat. The primary types of energy are mechanical energy, heat energy, electrical energy, chemical energy, and light or radiant energy. In this discussion, we shall be concerned primarily with heat energy.[9] **Heat energy** is energy transferred from one substance to another when there is a temperature difference between them and is associated with the random motion of very small particles of matter. The quantity of heat energy gained or lost by an object is measured in calories (cal) and joules (J),[10] and the degree of hotness of an object (temperature) is measured in degrees. A common heat-measuring device is a calorimeter (a thermally insulated container); a common temperature-measuring device is a thermometer.

2-11

Temperature

There are three common temperature scales that we shall consider in this book. They are

1.   the Fahrenheit scale—°F

2.   the Celsius scale[11]—°C

3.   the Kelvin scale[11]—K

The Fahrenheit scale, named after the German physicist Gabriel Daniel Fahrenheit (1686–1736), is the scale most familiar to us. On this scale, the freezing point of pure water is 32° and the boiling point of water at 1 atmosphere (atm) pressure[12] is 212°. On the Celsius scale, named after Swedish astronomer Anders Celsius (1701–1744), these points correspond to 0° and 100°, respectively.

In Figure 2-7, these two scales are compared. On the Celsius scale there is 100° between the freezing point (fp) and the boiling point (bp) of water; however, on the Fahrenheit scale this difference corresponds to 180°. Thus, 180 divisions Fahrenheit equal 100 divisions Celsius, or there are 1.8°F to 1°C. In addition, the freezing point of water is 0° on the Celsius scale and 32° on

---

[9]Heat energy can cause environmental problems, as the mean temperature of the earth varies from year to year. If the temperature of the earth becomes too hot, the polar ice caps will melt and the level of the oceans will rise, submerging many of our coastal cities. If the temperature of the earth becomes too cold, the growing season will be shortened, possibly creating an agricultural disaster. With the increase in atmospheric carbon dioxide levels due to the burning of fossil fuels (coal, oil, and gas), scientists have predicted that the mean temperature of the earth will probably increase by 3.6°F by 2040 and 9°F by 2100. This phenomenon has been referred to as the "greenhouse effect."

[10]SI recognizes the joule as a unit of heat energy. The non-SI unit, calorie, is widely used. We will use both calories and joules in heat calculations, but a greater emphasis will be placed on the calorie due to its wide use.

[11]The Celsius scale is the same as the old centigrade scale with the same symbol, °C. The Kelvin scale is an absolute scale with the symbol K and is the SI base unit for temperature.

[12]As we shall see later (12-4), the boiling point of water varies with pressure.

FIGURE 2-7

Comparison of the
Celsius and Fahrenheit
scales.

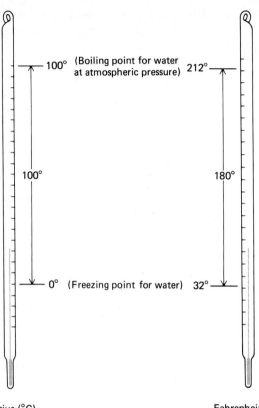

Celsius (°C)   Fahrenheit (°F)

the Fahrenheit scale. To convert a given temperature from °C to °F, we need only to consider the preceding facts.

$$100 \text{ divisions °C} = 180 \text{ divisions °F}$$

or

$$5 \text{ divisions °C} = 9 \text{ divisions °F (by dividing both sides}$$
$$\text{of the equation by 20)}$$

To convert from °C to °F,

$$°C \times \frac{9 \text{ divisions °F}}{5 \text{ divisions °C}} = \text{ divisions °F above or below}$$
$$\text{the freezing point of water}$$

Since the freezing point of water is 32°F, we must add these divisions to 32 to get the temperature in °F:

$$\frac{9}{5}°C + 32 = °F \tag{2-1}$$

$$1.8°C + 32 = °F \tag{2-2}$$

Equation 2-2 may be rearranged as follows to convert °F to °C:

$$1.8° \text{ C} = °\text{F} - 32$$
$$°\text{C} = \frac{(°\text{F} - 32)}{1.8} \qquad (2\text{-}3)^{13}$$

In this case, we must remember to subtract 32 from the given temperature in °F to obtain the number of Fahrenheit divisions above or below the freezing point of water and then convert this to °C by dividing by 1.8. Using the number 1.8 in Equations 2-2 and 2-3 simplifies the calculation if you are using a calculator.

The Kelvin scale, named after British physicist and mathematician William Thomson (1824–1907), who was later titled Lord or Baron Kelvin, consists of a new scale with the zero point equal to −273°C (more accurately, −273.15°). To convert from °C to K we need only to add 273°. (In this text, we shall use 273 instead of 273.15 to simplify calculations.)

$$\text{K} = °\text{C} + 273 \quad \text{or} \quad °\text{C} = \text{K} - 273 \qquad (2\text{-}4)$$

The lower limit of this scale is theoretically zero, with no upper limit.[14] The temperature of some stars is estimated at many millions of degrees kelvin.

Problem Example 2-15

Convert 25°C to °F.

SOLUTION: Substituting into our derived Equation 2-2, we get

$$25°\text{C} = [(1.8 \times 25) + 32]°\text{F} = 77°\text{F} \quad \textit{Answer}$$

In regard to significant digits, express your answer to the smallest place (units, tenths, etc.) as given to you in the number in the problem. In the number above, the degrees C was given to the units place, so your answer in degrees F should be expressed to the units place. Regardless of the thermometer used, you should be able to read each with the same exactness.

Problem Example 2-16

Convert −25.0°F to °C and K.

---

[13]Other formulas are also useful, as $°\text{F} = \frac{9}{5}(°\text{C} + 40) - 40$ and $°\text{C} = \frac{5}{9}(°\text{F} + 40) - 40$. These formulas are based on the fact that Celsius and Fahrenheit are equal at −40°.

[14]William F. Giauque (b. 1895), an American chemist and a student of Professor Lewis (1-3), received the Nobel Prize in chemistry in 1949 for his novel method of reaching a few thousandths of a degree above 0 K and for his experiments on the properties of matter at this extremely low temperature. The study of the behavior of matter at extremely low temperatures has application in space travel, where the temperature is quite low.

SOLUTION TO °C: Substituting into our derived Equation 2-3, we have

$$-25.0°F = \frac{(-25.0 - 32)}{1.8}°C = \frac{(-57.0)°C}{1.8} = -31.7°C \quad \textit{Answer}$$

Note that -25.0 was expressed to the tenths place; hence, the answer is expressed to the tenths place.

SOLUTION TO K: Substituting into Equation 2-4, we get

$$-31.7°C = (-31.7 + 273)\,K = 241.3\,K \quad \textit{Answer}$$

Again we express our answer to the tenths place. The number 273 is not taken into account in considering significant digits.

Problem Example 2-17

Xenon has a freezing point of 133 K. What is its freezing point on the Fahrenheit scale?

SOLUTION TO °C

$$133\,K = (133 - 273)°C = -14\overline{0}°C$$

SOLUTION TO °F

Work Problems 23, 24, 25, 26, 27, and 28.

$$-14\overline{0}°C = [1.8 \times (-14\overline{0}) + 32]°F = -22\overline{0}°F \quad \textit{Answer}$$

2-12

Specific heat

A **calorie** (cal), a unit of measurement for heat energy, is equal to the quantity of heat required to raise the temperature of 1 g of water from 14.5°C to 15.5°C. In nutrition, the large calorie or Calorie (Cal), which is equal to 1000 small calories or 1 kilocalorie, is used. One tablespoon of sugar (12 g), when burned in the body, produces 45 Calories or kilocalories, which is equivalent to 45,000 calories of heat energy.

For an activity such as going to college the Calorie requirement is about 3000 Cal per day for a male and 2200 Cal per day for a female. Depending on the activity, more Calories are required.

A **joule** (J, pronounced $\overline{joo}$l) is the unit of energy defined within the SI. It is the energy equal to a force of one newton (N) acting through a distance of one meter in the direction of the force. A **newton**[15] is a unit of force which gives a mass of one kilogram an acceleration of one meter per second per

---

[15]The newton is named in honor of Sir Isaac Newton (1642–1727), English scientist and philosopher who among other things formulated the laws of motion and invented the calculus. He was also considered to be a biblical scholar.

FIGURE 2-8

Comparison of specific heats of water (liquid) and sodium chloride (salt). Notice the difference in calories of heat energy.

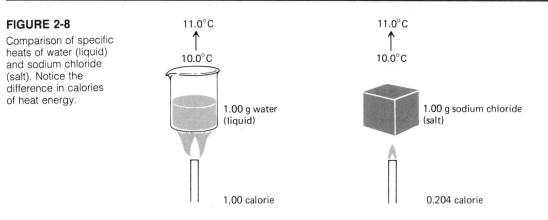

second and has the units kg · m/s^2. Therefore, the units of joules are N · m or kg · m^2/s^2. *One calorie* (cal) is equal to *4.184 joules* (J).

One of the properties of matter is that it requires a certain quantity of heat to produce a given change in temperature per unit mass of a given substance.[16] This is called the specific heat of the substance. **Specific heat** is defined in non-SI units as the number of calories required to raise the temperature of 1.00 g of a substance 1.00°C. Figure 2-8 illustrates the difference in specific heats of water (liquid) and sodium chloride (salt). In SI units, specific heat is the number of joules required to raise the temperature of 1.00 kg of a substance 1.00 K. Table 2-4 lists the specific heats of a few substances in both units. Notice that water has a relatively high specific heat while the other substances are relatively low. Water acts to absorb heat, that is, it is a "heat sponge," whereas the other substances, especially lead and silver, conduct heat.

TABLE 2-4   Specific Heat of Some Substances

Substance	Specific Heat[a]	
	cal/g ·°C	J/kg · K[b]
Water (liquid)	1.00	4.18×10^3
Aluminum	0.217	9.08×10^2
Lead	0.031	1.30×10^2
Sugar	0.299	1.25×10^3
Silver	0.056	2.34×10^2
Sodium chloride (salt)	0.204	8.54×10^2

[a]The notations 1/g and g^{-1} or 1/kg and kg^{-1} have identical meanings; the notation cal/g · °C may also be expressed as cal g^{-1} · °C^{-1} and the notation J/kg · K as J kg^{-1} · K^{-1}.

[b]See 2-3 and 2-4 for a discussion of exponents and scientific notation.

[16]The term "substance" is defined in 3-2.

Problem Example 2-18

Exactly 75.0 cal of heat energy will raise the temperature of 10.0 g of an unknown metal from 25.0°C to 60.0°C. Calculate the specific heat of the metal in cal/g · °C.

SOLUTION

$$\frac{75.0 \text{ cal}}{10.0 \text{ g} \times (60.0 - 25.0)°\text{C}} = 0.214 \text{ cal/g} \cdot °\text{C} \quad \textit{Answer}$$

Problem Example 2-19

Calculate the number of calories that would be required to raise the temperature of 12.0 g of sodium chloride from 25.0°C to 85.0°C.

SOLUTION: From Table 2-4, the specific heat of sodium chloride is 0.204 cal/g · °C.

$$\frac{0.204 \text{ cal}}{1 \text{ g} \cdot 1°\text{C}} \times 12.0 \text{ g} \times (85.0 - 25.0)°\text{C} = 147 \text{ cal} \quad \textit{Answer}$$

Problem Example 2-20

Calculate the amount of aluminum in grams used if $\overline{8}00$ cal of heat is required when the aluminum is heated from $4\overline{0}$°C to 80°C.

SOLUTION: From Table 2-4, the specific heat of aluminum is 0.217 cal/g · °C. We are asked for the amount of aluminum in grams, so grams must appear in the answer; hence, the solution is as follows:

$$\frac{1.00 \text{ g} \cdot °\text{C}}{0.217 \text{ cal}} \times \frac{\overline{8}00 \text{ cal}}{(8\overline{0} - 4\overline{0})°\text{C}} = 92 \text{ g} \quad \textit{Answer}$$

Notice that the number (0.217) and the units (calories) are *both* inverted.

Problem Example 2-21

Calculate the number of joules that would be required to raise the temperature of $25\overline{0}$ g of lead from 373 K to 473 K.

SOLUTION: From Table 2-4, the specific heat of lead is 1.30×10^2 J/kg · K.

Work Problems 29, 30, 31, 32, 33, and 34.

$$\frac{1.30 \times 10^2 \text{J}}{1 \text{ kg} \cdot \text{K}} \times \frac{1 \text{ kg}}{1000 \text{ g}} \times 25\overline{0} \text{ g} \times (473 - 373)\text{K} = 3.25 \times 10^3 \text{ J} \quad \textit{Answer}$$

2-13

Density

Another property of matter is that a given volume of different substances may have different masses; this property is measured by the density. **Density** is defined as the mass of a substance occupying a unit volume, or

$$\text{density} = \frac{\text{mass}}{\text{volume}}$$

We know that certain substances are heavier than other substances, even though the volumes are the same. For example, balsa wood is much lighter than a lead brick of the same volume, illustrated in Figure 2-9.

The unit of density generally used for solids and liquids is g/mL (g/cm^3), while the unit generally used for gases is g/L. In the SI the unit used for the density of solids, liquids, and gases is kg/m^3. Density has units of mass/volume, and whenever the density of a substance is expressed, the particular units of mass and volume must be given. For example, the density of water is 1.00 g/mL in the metric system and 1.00×10^3 kg/m^3 in the SI. Thus, we must realize that it is insufficient to express the density of a substance as a pure number without units.

Density is often expressed as follows: $d^{20^\circ} = 13.55$ g/mL for mercury. The 20° indicates the temperature in °C at which the measurement was taken; hence, mercury at 20°C has a density of 13.55 g/mL. The reason for recording the temperature is that almost all substances expand when heated, and therefore the density would decrease at a higher temperature; for example, $d^{270^\circ} = 12.95$ g/mL for mercury. Thus, the density is dependent on the temperature.

FIGURE 2-9
Balsa wood is much lighter than a lead brick of the same volume.

Problem Example 2-22

Calculate the density of a piece of metal that has a mass of 25 g and occupies a volume of 6.0 mL.

SOLUTION

$$\frac{25 \text{ g}}{6.0 \text{ mL}} = 4.2 \text{ g/mL} \quad \textit{Answer}$$

Problem Example 2-23

A cube of iron measures 2.00 cm on each edge and has a mass of 62.9 g. Calculate its density in kg/m³.

SOLUTION: The volume of the cube of iron in cubic meters is

$$2.00 \text{ cm} \times \frac{1 \text{ m}}{100 \text{ cm}} \times 2.00 \text{ cm} \times \frac{1 \text{ m}}{100 \text{ cm}} \times 2.00 \text{ cm} \times \frac{1 \text{ m}}{100 \text{ cm}} = 8.00 \times 10^{-6} \text{ m}^3$$

The density in kg/m³ is

$$\frac{62.9 \text{ g}}{8.00 \times 10^{-6} \text{ m}^3} \times \frac{1 \text{ kg}}{1000 \text{ g}} = 7.86 \times 10^3 \text{ kg/m}^3 \quad \textit{Answer}$$

Problem Example 2-24

Calculate the volume in liters occupied by 880 g of benzene at 20°C. For benzene, $d^{20°}$ = 0.88 g/mL.

SOLUTION: We have 880 g of benzene and the density at 20°C is 0.88 g/mL. We are asked to calculate the volume in liters. In other words, we wish to convert a given amount of benzene from mass units to volume units. This is readily done by using the density of benzene as a conversion factor, since 1 mL = 0.88 g.

$$880 \text{ g} \times (\text{factor}) = \text{volume units}$$

The choice of factors is as follows:

$$\frac{0.88 \text{ g}}{1 \text{ mL}}; \quad \frac{1 \text{ mL}}{0.88 \text{ g}}$$

$$A \qquad B$$

If we use factor A, our units would be g²/mL, which have no meaning and do not answer our question. But, let us consider factor B:

$$880 \ g \times \frac{1 \ mL}{0.88 \ g}$$

Conversion from mL to L yields the complete setup:

$$880 \ g \times \frac{1 \ mL}{0.88 \ g} \times \frac{1 \ L}{1000 \ mL} = 1.0 \ L \quad \textit{Answer}$$

Problem Example 2-25

An experiment requires 0.156 kg of bromine. How many milliliters (20°C) should the chemist use? For bromine, $d^{20°} = 3.12$ g/mL.

SOLUTION

$$0.156 \ kg \times \frac{1000 \ g}{1 \ kg} \times \frac{1 \ mL}{3.12 \ g} = 50.0 \ mL \quad \textit{Answer}$$

Problem Example 2-26

Calculate the volume in milliliters at 20°C occupied by 1.25 kg of chloroform. For chloroform, $d^{20°} = 1.49 \times 10^3$ kg/m³.

SOLUTION: The density converts the mass to volume—meter cubed—but the unknown volume has the units mL. Therefore, to convert m³ to mL, we must know that 1 mL = 1 cm³ (Table 2-3) and to convert from m³ to cm³ we must cube 100 cm = 1 m (Table 2-3). The complete solution is as follows:

$$1.25 \ kg \times \frac{1 \ m^3}{1.49 \times 10^3 \ kg} \times \frac{(100)^3 \ (cm)^3}{1 \ m^3} \times \frac{1 \ mL}{1 \ cm^3} = 839 \ mL \quad \textit{Answer}$$

[In cubing 100, remember to multiply it by itself three times, that is, 100 × 100 × 100 (see 2-3), or express 100 in exponential notation as 10^2 and to cube 10^2, we would express it as $(10^2)^3 = 10^6$.]

Problem Example 2-27

Calculate the mass in grams of 400 mL (20°C) of glycerine; $d^{20°} = 1.26$ g/mL for glycerine.

SOLUTION

Work Problems 35, 36, and 37.

$$400 \ mL \times \frac{1.26 \ g}{mL} = 504 \ g, \text{ rounded off to one significant digit is 500 g} \quad \textit{Answer}$$

2-14

Specific Gravity

The **specific gravity** of a substance is the density of the substance divided by the density of some substance taken as a standard. For expressing the specific gravity of liquids and solids, water at 4°C is the standard with a density of 1.00 g/mL in the metric system:

$$\text{specific gravity} = \frac{\text{density of substance}}{\text{density of water at 4°C}}$$

$$\text{density of substance} = \text{specific gravity} \times \text{density of water at 4°C}$$

In calculating the specific gravity of a substance, we must express both densities in the **same** *units*. Specific gravity, therefore, has **no** *units*. To convert from specific gravity to density, we merely have to multiply specific gravity by the density of the reference substance (in most cases, water). We may thus find the density of any substance for which we have a reference density. Since the density of water in the metric system is 1.00 g/mL, the density of solids or liquids expressed as g/mL is numerically equal to their specific gravities.

Specific gravity is often expressed as follows:

$$\text{sp gr} = 0.708^{25°/4} \text{ of ether}$$

The 25° refers to the temperature in °C at which the density of ether was measured, and the 4 refers to the temperature in °C at which the density of water was measured. Table 2-5 lists the specific gravity of a few substances.

TABLE 2-5 Specific Gravity of Some Substances

Substances	Specific Gravity
Water	$1.00^{4°/4}$
Ether	$0.708^{25°/4}$
Benzene	$0.880^{20°/4}$
Acetic acid	$1.05^{20°/4}$
Chloroform	$1.49^{20°/4}$
Carbon tetrachloride	$1.60^{20°/4}$
Sulfuric acid (conc)	$1.83^{18°/4}$

Problem Example 2-28

Calculate the mass in grams of $11\overline{0}$ mL (20°C) of chloroform.

SOLUTION: From Table 2-5, the specific gravity of chloroform is 1.49 at 20°C. Therefore, in the metric system the density is 1.00 g/mL × 1.49 = 1.49 g/mL at 20°C.

$$11\overline{0} \text{ mL} \times \frac{1.49 \text{ g}}{1 \text{ mL}} = 163.9 \text{ g, rounded off to three significant digits is 164 g} \quad \textit{Answer}$$

Problem Example 2-29

The specific gravity of a certain organic liquid is 1.20. Calculate the number of liters in $84\overline{0}$ g of the liquid.

SOLUTION: The specific gravity is 1.20. Hence, in the metric system the density is 1.00 g/mL $\times$ 1.20 = 1.20 g/mL.

$$84\overline{0} \ \cancel{g} \times \frac{1 \ \cancel{mL}}{1.20 \ \cancel{g}} \times \frac{1 \ L}{1000 \ \cancel{mL}} = 0.700 \ L \quad \textit{Answer}$$

Problem Example 2-30

The specific gravity of a certain organic liquid is 0.950. Calculate the number of kilograms in 3.75 L of the liquid.

SOLUTION: The specific gravity is 0.950. Therefore, in the metric system the density is 1.00 g/mL $\times$ 0.950 = 0.950 g/mL.

Work Problems 38 and 39.

$$3.75 \ \cancel{L} \times \frac{1000 \ \cancel{mL}}{1 \ \cancel{L}} \times \frac{0.950 \ \cancel{g}}{1 \ \cancel{mL}} \times \frac{1 \ kg}{1000 \ \cancel{g}} = 3.56 \ kg \quad \textit{Answer}$$

Chapter Summary

This chapter reviewed basic mathematics including significant digits, exponents, and exponential and scientific notations.

Matter is anything that has mass and occupies space. The measurement of matter in the metric system, which is based on multiples of 10, 100, 1000, and so forth, was considered. Also, the measurement of temperature using the Fahrenheit, Celsius, and Kelvin scales was considered.

The *factor-unit* method in solving problems was introduced. The basis of this method is that numbers have units and through a logical approach these units can be used to solve quantitative problems. Using the *factor-unit* method, problems involving conversion within the metric system, specific heat, density, and specific gravity were solved.

EXERCISES

1. Define or explain the following terms:
   - (a) significant digit
   - (b) exponent
   - (c) exponential notation
   - (d) scientific notation
   - (e) SI units
   - (f) kilogram
   - (g) millimeter
   - (h) angstrom
   - (i) nanometer
   - (j) micron
   - (k) gamma
   - (l) energy
   - (m) calorie
   - (n) joule
   - (o) newton
   - (p) specific heat
   - (q) heat energy
   - (r) density
   - (s) specific gravity

2. Distinguish between
   (a)   mass and weight
   (b)   a chemical balance and a spring scale
   (c)   calorie and temperature
   (d)   calorie and Calorie
   (e)   calorie and joule
   (f)   the Celsius and Fahrenheit temperature scales
   (g)   the Kelvin and Celsius temperature scales
   (h)   specific heat and specific gravity
   (i)   density and specific gravity

3. Explain how you would determine the mass of an object on
   (a)   a platform balance   (b)   a triple beam balance

4. "The weight of a small medicine bottle on a triple beam balance was found to be 8.53 g." Criticize this statement.

PROBLEMS

Significant Digits   (See Sections 2-1 and 2-2)

5. Determine the number of significant digits in the following numbers:
   (a)   127   (b)   320
   (c)   4,000,020   (d)   1610
   (e)   18,070   (f)   0.2

6. Determine the number of significant digits in the following numbers:
   (a)   0.05   (b)   0.00301
   (c)   60.02   (d)   0.006180
   (e)   170   (f)   7.501

7. Round off the following numbers to three significant digits:
   (a)   1.366   (b)   3.37286
   (c)   16.450   (d)   2.66455
   (e)   16.550   (f)   0.03659

8. Round off the following numbers to three significant digits:
   (a)   7.267   (b)   0.003321
   (c)   0.054350   (d)   0.64650
   (e)   0.53257   (f)   9.742

9. The relation of the pound to the kilogram in the United States is 1 pound = 0.453592428 kilogram. Round this number off to three significant digits.

10. Perform the indicated mathematical operations and express your answer to the proper number of significant digits.
   (a) $3.68 + 7.3654 + 0.5$ (b) $0.243 + 76.720 + 4.6494$
   (c) $14.745 - 1.60$ (d) $0.5642 - 0.260$
   (e) 6.02×2.0 (f) 0.65×427
   (g) 0.022×0.467 (h) $\dfrac{174}{24}$
   (i) $\dfrac{420}{17.5}$

Exponents (See Section 2-3)

11. Express 27,500 as 2.75×10^n.

12. Express 0.0325 as 3.25×10^n.

13. Carry out the operations indicated on the following exponential numbers:
    (a) $3.24 \times 10^3 + 1.50 \times 10^3$
    (b) $4.73 \times 10^2 + 6.6 \times 10^1$
    (c) $3.75 \times 10^3 - 2.74 \times 10^3$
    (d) $6.54 \times 10^5 - 2 \times 10^3$
    (e) $6.45 \times 10^3 \times 1.42 \times 10^2$
    (f) $3.28 \times 10^6 \times 1.24 \times 10^{-2}$
    (g) $\dfrac{7.72 \times 10^6}{2.82 \times 10^2}$
    (h) $\dfrac{6.73 \times 10^{-5}}{2.32 \times 10^{-2}}$

14. Determine the value of each of the following numbers:
    (a) $\sqrt{9.00 \times 10^8}$ (b) $\sqrt{3.60 \times 10^5}$
    (c) $\sqrt{4.90 \times 10^9}$ (d) $\sqrt{2.00 \times 10^{-7}}$

15. Perform the indicated operation on the following exponential numbers:
    (a) Raise 2.11×10^3 to the second power.
    (b) Raise 1.24×10^8 to the second power.
    (c) Raise 1.45×10^2 to the third power.
    (d) Raise 2.01×10^3 to the third power.

Scientific Notation (See Section 2-4)

16. Express the following in scientific notation to three significant digits:
    (a) 8,720,000   (b) 0.0745
    (c) 7272   (d) 0.03275

17. Express the following in scientific notation to three significant digits:
    (a) 0.00764   (b) 725,000
    (c) 9738   (d) 0.006285

The Metric System (See Sections 2-8 and 2-9)

18. Carry out each of the following conversions showing a solution setup:
    (a) 6.5 kg to mg   (b) 12,000 m to km
    (c) 35 mg to kg   (d) 300 mL to L
    (e) 764 dm^3 to L   (f) 35 Å to μ

19. Carry out each of the following conversions showing a solution setup:
    (a) 3.2 km to m   (b) 15,000 g to kg
    (c) 85 mg to kg   (d) 675 mL to L
    (e) 35.0 nm to Å   (f) 6.75 mL to cc

20. Add the following masses: 375 mg, 0.500 g, 0.002000 kg, 200.0 cg, 1.00 dg. What is the total mass in grams?

21. Add the following lengths: 2.0000000 km, 370.00 cm, 7.0000 m, 0.4 mm. What is the total length in meters?

22. The relation of the yard to the meter in the United States is 1 yard = 0.91440183 meter. What is this relation in centimeters to three significant digits?

Temperature (See Section 2-11)

23. Convert each of the following temperatures to °F and K:
    (a) 30.0°C
    (b) $11\overline{0}$°C
    (c) −70.0°C
    (d) −$13\overline{0}$°C

24. Convert each of the following temperatures to °C and K:
    (a) 66.0°F
    (b) −10.0°F
    (c) −35°F
    (d) $45\overline{0}$°F

25. Liquid nitrogen has a boiling point of 77 K at 1 atm pressure. What is its boiling point on the Fahrenheit scale?

26. At what temperature will the Celsius and Fahrenheit scales have the same numerical reading? Carry out your answer to three significant digits.

27. The official coldest temperature recorded in the United States was −79.8°F at Prospect Creek, Alaska, in January 1971. What is this temperature on the Celsius scale?

28. The highest recorded temperature in the world was recorded as 136.4°F at Azizia, Libya, in the Sahara desert on September 13, 1922. What is this temperature in °C and K?

Specific Heat (See Section 2-12)

29. Exactly 80.0 cal of heat energy will raise the temperature of 10.0 g of an unknown metal from 20.0°C to 60.0°C. Calculate the specific heat of the metal.

30. Calculate the number of kilocalories that would be required to raise the temperature of $20\overline{0}$ g of water (liquid) from 10.0°C to 80.0°C (see Table 2-4).

31. Calculate the number of calories that would be required to raise the temperature of 0.200 kg of aluminum from 20.0°C to 90.0°C (see Table 2-4).

32. Calculate the mass of sodium chloride in grams used if 75.0 cal of heat is absorbed when the sodium chloride is heated from 35.0°C to 85.0°C (see Table 2-4).

33. Calculate the mass of lead in grams used if 90.0 cal of heat is absorbed when lead is heated from 25.0°C to 65.0°C (see Table 2-4).

34. Calculate the number of joules that would be required to raise the temperature of 325 g of silver from 315 K to 345 K (see Table 2-4).

Density (See Section 2-13)

35. Calculate the density in g/mL for each of the following:
    (a) a piece of metal of volume $6\overline{0}$ mL and mass 350 g

(b)   a substance occupying a volume of $7\overline{0}$ mL and having a mass of 220 g

(c)   a sample of metal occupying a volume of 5.4 mL and having a mass of 65 g

(d)   a piece of metal measuring 2.0 cm × 0.10 dm × 25 mm and having a mass of 35.0 g

36. Calculate the volume in milliliters at 20°C occupied by each of the following:

(a)   a sample of carbon tetrachloride having a mass of 75.0 g; $d^{20°} = 1.60$ g/mL

(b)   a sample of acetic acid having a mass of 225 g; $d^{20°} = 1.05$ g/mL

(c)   a sample of chloroform having a mass of 38.5 g; $d^{20°} = 1.49$ g/mL

(d)   a sample of benzene having a mass of 1.7 kg; $d^{20°} = 8.8 \times 10^2$ kg/m^3

37. Calculate the mass in grams of each of the following:

(a)   a $25.\underline{0}$-mL volume of ether; $d^{20°} = 0.708$ g/mL

(b)   a $32\overline{0}$-mL volume of glycerine; $d^{20°} = 1.26$ g/mL

(c)   a 4.75-mL volume of carbon tetrachloride; $d^{20°} = 1.60$ g/mL

(d)   a 0.220-L volume of bromine; $d^{20°} = 3.12$ g/mL

Specific Gravity   (See Section 2-14)

38. Calculate the volume in liters occupied by each of the following (see Table 2-5):

(a)   a sample of sulfuric acid (conc) having a mass of 265 g

(b)   a sample of acetic acid having a mass of 3520 g

(c)   a sample of chloroform having a mass of 0.445 kg

(d)   a sample of carbon tetrachloride having a mass of 2.00 kg

39. Calculate the mass in grams of each of the following (see Table 2-5):

(a)   a 25.0-mL volume of benzene

(b)   a 152-mL volume of acetic acid

(c)   a 1.75-L volume of chloroform

(d)   a 3.25-L volume of carbon tetrachloride

General Problems

40. Calculate the number of (a) calories and (b) joules that would be required to raise the temperature of 125 g of sodium chloride from 35.0°F to 65.0°F (see Table 2-4)

41. Calculate the specific heat in (a) cal/g · °C and (b) J/kg · K if 6.00 kcal of heat energy will raise the temperature of 1.00 kg of a certain metal from 25.0°F to 52.0°F.

Readings

Adamson, Arthur W., "SI Units? A Camel Is a Camel." *J. Chem. Educ.*, 1978, v. 55, p. 634. Reviews the historical development of the base units leading to the SI units; lists the advantages and disadvantages of this system with a comprehensive discussion of the latter.

Pinkerton, Richard C., and Chester E. Gleit, "The Significance of Significant Figures." *J. Chem. Educ.*, 1967, v. 44, p. 232. Discusses the concepts of significant figures, precision, resolution, and the meaning of numerical data.

Revelle, Roger, "Carbon Dioxide and World Climate." *Sci. Am.*, Aug. 1982, v. 247, p. 35. Discusses the increase in atmospheric carbon dioxide and the effects of a warmer and wetter climate, which may not all be bad.

Socrates, G., "SI Units." *J. Chem. Educ.*, 1969, v. 46, p. 710. Reviews the historical development of the English, metric, and SI units and gives the accepted SI units and their conversions.

Woodwell, George M., "The Carbon Dioxide Question." *Sci. Am.*, Jan. 1978, v. 238, p. 34. Considers the increase in carbon dioxide in the earth's atmosphere brought about by human activities and what effect this will have on climate.

3

Basic Concepts of Matter

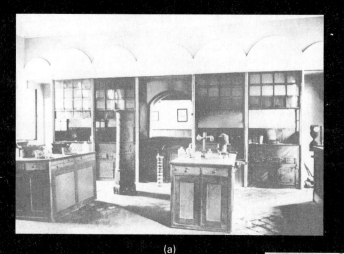

(a)

(b)

Justus von Liebig (1803–1873) was a German chemist and teacher who made many contributions on properties both chemical and physical of inorganic and organic compounds. (a) Liebig's analytical laboratory, (b) Liebig's office in Giessen, Germany.

TASKS

1. Identify the three physical states of matter and give an example of each (Section 3-1).
2. Memorize the symbols for the 47 elements listed in Table 3-1.
3. Memorize the general physical and chemical properties of metals and nonmetals (Section 3-9).

OBJECTIVES

1. Given the following terms, define each term and describe the distinguishing characteristic of each:
   (a) homogeneous matter (Section 3-2)
   (b) heterogeneous matter (Section 3-2)
   (c) pure substance (Section 3-2)
   (d) homogeneous mixture (Section 3-2)
   (e) solution (Section 3-2)
   (f) mixture (Section 3-2)
   (g) compound (Section 3-3)
   (h) element (Section 3-3)
   (i) physical properties (Section 3-4)
   (j) chemical properties (Section 3-4)
   (k) physical changes (Section 3-5)
   (l) chemical changes (Section 3-5)
   (m) potential energy (Section 3-6)
   (n) kinetic energy (Section 3-6)
   (o) Law of Conservation of Energy (Section 3-6)
   (p) Law of Conservation of Mass (Section 3-6)
   (q) atom (Section 3-7)
   (r) formula unit (Section 3-8)
   (s) molecule (Section 3-8)
   (t) molecular formula (Section 3-8)
   (u) Law of Definite Proportions or Constant Composition (Section 3-8)
2. Given the name of a common substance, classify its physical state as one of the three physical states of matter, at room temperature and atmospheric pressure (Section 3-1, Problem 3).
3. Given the name or symbol of a metal or nonmetal, write the corresponding symbol or name (Section 3-3, Problem 4).

50

4. Given the property of a substance, classify it as a physical or chemical property (Section 3-4, Problems 5 and 6).

5. Given a change of a substance, classify it as a physical or chemical change (Section 3-5, Problems 7 and 8).

6. Given the molecular formula of a compound, write the number of atoms, name of each element, and total number of atoms in one unit of the compound (Section 3-8, Problem 9).

7. Given the number of atoms of each element in one molecule of a compound and the name of the element, write the molecular formula of the compound (Section 3-8, Problem 10).

8. Given the name of a common substance, classify it as a compound, element, or mixture (Sections 3-3 and 3-8, Problem 11).

9. Given the Periodic Table, identify the elements called metals, nonmetals, noble gases, and metalloids (Sections 3-3 and 3-9, Problems 12, 13, 14, 15, 16, 17, 18, and 19).

In Chapter 2, we defined matter as anything that has mass and occupies space. In this chapter, we shall consider different types of matter and study the properties of matter and the changes it undergoes.

3-1

Physical States of Matter

The three physical states of matter are solid, liquid, and gas.[1] Matter exists either as a solid, liquid, or gas, depending upon the temperature and pressure and the specific characteristics of the particular type of matter. Some matter exists in all three physical states, whereas other matter decomposes when an attempt is made to change its physical state. Water exists in all three physical states, as shown in Figure 3-1:

Solid:	ice, snow
Liquid:	water
Gas:	water vapor and steam

---

[1]Intermediate between solids and liquids are *liquid crystals*. Liquid crystals are substances (organic compounds) whose properties are intermediate between those of a true crystalline solid and those of a true liquid as they go from the solid to the liquid physical state. One property of liquid crystals is that they may change color with a change in temperature. Liquid crystals are used to display numbers on wristwatches, pocket calculators, and electronic games. They have also been used for detecting cancer of the skin and breasts and abnormal vascular diseases. This is based on their property of changing color with temperature. *Plasma*, sometimes considered a fourth state of matter, is similar to a gas, except that the particles are charged instead of being neutral.

FIGURE 3-1

The three physical states of water.

Solid—ice   Liquid—water   Gas—water vapor

The particular physical state of water is determined by the conditions (temperature and pressure) under which the observation is made. Common table sugar exists under normal conditions in only one physical state—solid. Attempts to change it to a liquid or a gas by heating at atmospheric pressure results in decomposition of the sugar; the sugar turns caramel brown to black in color.

Work Problem 3.

3-2

Homogeneous and Heterogeneous Matter

Matter is further divided into two major subdivisions: homogeneous and heterogeneous matter. **Homogeneous matter** is *uniform* in composition and properties throughout. It is the same throughout. **Heterogeneous matter** is *not uniform* in composition and properties; it consists of two or more physically distinct *portions* or *phases* unevenly distributed. A class consisting of all women would be analogous to homogeneous matter, while a class of both men and women would be analogous to heterogeneous matter.

Homogeneous matter is divided into three categories: pure substances, homogeneous mixtures, and solutions. A **pure substance** is characterized by *definite* and *constant composition;* and a pure substance has *definite* and *constant properties* under a given set of conditions. A pure substance obeys our definition of homogeneous matter, not only in that it is uniform throughout in both composition and properties, but also in that it has the additional requirement of a definite and constant composition and properties. Some examples of pure substances are water, salt (sodium chloride), sugar (sucrose), mercuric or mercury(II) oxide, gold, iron, and aluminum.

A **homogeneous mixture** is homogeneous throughout and is composed of two or more pure substances whose proportions may be *varied* in some cases without limit. One example of a homogeneous mixture is air, which is a mixture of oxygen, nitrogen, and certain other gases. Mixtures of gases are generally called homogeneous mixtures, but homogeneous mixtures composed of

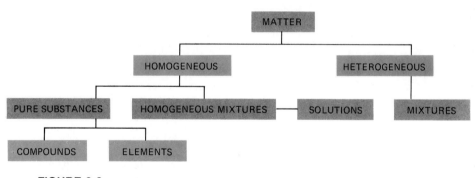

FIGURE 3-2

Classification of matter.

gases, liquids, or solids dissolved in liquids are called *solutions*. A **solution** is homogeneous throughout and is composed of two or more pure substances; its composition can be varied usually *within* certain limits. In some cases solids may dissolve in other solids to form homogeneous mixtures which are called *solid solutions*. Some common examples of solutions are a sugar solution (sugar dissolved in water), a salt solution (salt dissolved in water), carbonated water (carbon dioxide dissolved in water), and an alcohol solution (ethyl alcohol dissolved in water).

Homogeneous mixtures (or solutions) differ from pure substances in that these mixtures consist of two or more pure substances in *variable* proportions, whereas pure substances have *definite* and *constant compositions*.

Heterogenous matter is also commonly called a *mixture. This* type of **mixture** is composed of two or more pure substances each of which retains its identity and specific properties. In many mixtures, the substances can be readily identified by visual observations. For example, in a mixture of salt and sand, the colorless crystals of salt can be distinguished from the tan crystals of sand by the eye or a hand lens. Likewise, in a mixture of iron and sulfur, the yellow sulfur can be identified from the black iron by visual observation. Mixtures can usually be separated by a simple operation that does not change the composition of the several pure substances comprising the mixture. For example, a mixture of salt and sand can be separated by using water. The salt dissolves in water; the sand is insoluble. A mixture of iron and sulfur can be separated by dissolving the sulfur in liquid carbon disulfide (the iron is insoluble), or by attracting the iron to a magnet (the sulfur is not attracted).

3-3

Compounds and Elements

Pure substances are divided into two groups: compounds and elements. Figure 3-2 summarizes classification of matter. A **compound** is a pure substance that *can be broken down* by various chemical means into two or more different simpler substances. As mentioned in 3-2, pure substances that are compounds are water, table salt, sugar, and mercuric or mercury(II) oxide. The action of an electric current (electrolysis; see Figure 3-3) decomposes both water (Equation 3-1) and molten table salt (Equation 3-2) to simpler substances. The action of heat decomposes both sugar (Equation 3-3) and mercuric or mercury(II) oxide (Equation 3-4):

$$\text{water} \xrightarrow[\text{current}]{\overset{\text{direct}}{\overset{\text{electric}}{}}} \text{hydrogen} + \text{oxygen} \qquad (3\text{-}1)$$

$$\text{table salt (molten)} \xrightarrow[\text{current}]{\overset{\text{direct}}{\overset{\text{electric}}{}}} \text{sodium} + \text{chlorine} \qquad (3\text{-}2)$$

$$\text{sugar} \xrightarrow{\text{heat}} \text{carbon} + \text{water} \qquad (3\text{-}3)$$

$$\text{mercuric or mercury(II) oxide} \xrightarrow{\text{heat}} \text{mercury} + \text{oxygen} \qquad (3\text{-}4)$$

FIGURE 3-3

Electrolysis of water to produce hydrogen and oxygen.

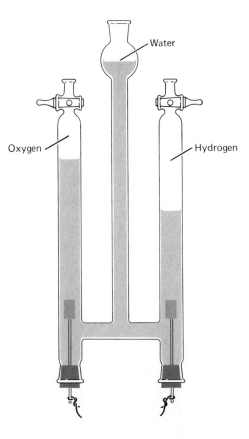

An **element** is a pure substance that *cannot be decomposed* into simpler substances by ordinary chemical means. Such pure substances, previously mentioned in 3-2, are gold, iron, and aluminum; and those pure substances mentioned previously, hydrogen, oxygen, sodium, chlorine, carbon, and mercury, are all examples of elements. (In Equation 3-3, water is produced. Why is water not an element? See Equation 3-1.)

There are 108 elements[2] at this writing, of which 90 have, so far, been found to occur naturally. The remaining 18 have been produced only synthetically by nuclear reactions. Minute amounts of some of these may also exist naturally. The ranking and relative abundance (percent by mass) of the first 10 elements in the earth's crust (upper 10 miles, including the oceans and atmosphere) are given in Table 3-1. The latest element (109) has been reported by a group of West German scientists. Plans are being made to prepare element 116 in the United States and West Germany.

Each of the elements has a symbol which is an abbreviation for the name of the element. On the inside front cover of this book are listed all the elements and their symbols. Since many of the elements are rarely mentioned in a basic chemistry course, Table 3-1 condenses this list of 108 elements to 47 of the most common. You must learn the *names* and *symbols* for the 47 elements in Table 3-1. To do this, we suggest that you make flash cards. For example, the symbol for copper is Cu, so on a small card (cut a 3 × 5-inch file card in half) write "copper" on one side and "Cu" on the other side (see Figure 3-4). Do this for all the elements listed in Table 3-1. Go over the cards until you know them all and then keep going over them so you will not forget them.

Work Problem 4.

(a)   (b)

FIGURE 3-4

A "flash card" for the element copper. (a) side 1: the name of the element.
(b) side 2: the symbol of the element.

---

[2]Two new elements have recently been added. They are elements 107 and 109, reported by a group of West German scientists. Proposed names and symbols for these elements are as follows: element 107 = unnilseptium—Uns, and element 109 = unnilennium—Une. Element 108 has yet to be discovered.

TABLE 3-1 Some Common Elements, Their Symbols, and Ranking of Relative Abundance (Percent by Mass) for the First Ten Elements in the Earth's Crust[a]

Element	Symbol[b]	Ranking (% Mass)	Element	Symbol	Ranking (% Mass)
Aluminum	Al	3 (7.5)	Lithium	Li	
Antimony	Sb		Magnesium	Mg	8 (1.9)
Argon	Ar		Manganese	Mn	
Arsenic	As		Mercury	Hg	
Barium	Ba		Neon	Ne	
Beryllium	Be		Nickel	Ni	
Bismuth	Bi		Nitrogen	N	
Boron	B		Oxygen	O	1 (49.5)
Bromine	Br		Phosphorus	P	
Cadmium	Cd		Platinum	Pt	
Calcium	Ca	5 (3.4)	Potassium	K	7 (2.4)
Carbon	C		Radium	Ra	
Chlorine	Cl		Selenium	Se	
Chromium	Cr		Silicon	Si	2 (25.7)
Cobalt	Co		Silver	Ag	
Copper	Cu		Sodium	Na	6 (2.6)
Fluorine	F		Strontium	Sr	
Gold	Au		Sulfur	S	
Helium	He		Tin	Sn	
Hydrogen	H	9 (0.9)	Titanium	Ti	10 (0.6)[c]
Iodine	I		Uranium	U	
Iron	Fe	4 (4.7)	Xenon	Xe	
Krypton	Kr		Zinc	Zn	
Lead	Pb				

[a]Upper 10 miles, including the oceans and atmosphere. Some other elements that are important for their use but for which you will not need to know their symbols are germanium (transistors), molybdenum (steel), palladium (precious metal), plutonium (nuclear reactors/weapons), tungsten (light bulb filaments), and vanadium (steel and construction).

[b]Some of these symbols do not appear to be related to the names of the elements. In these cases, the symbol used has been obtained from the Latin name, by which the element was known for centuries.

Name of Element	Latin Name (Symbol)
Antimony	*Stibium* (Sb)
Copper	*Cuprum* (Cu)
Gold	*Aurum* (Au)
Iron	*Ferrum* (Fe)
Lead	*Plumbum* (Pb)
Mercury	*Hydrargyrum* (Hg)
Potassium	*Kalium* (K)
Silver	*Argentum* (Ag)
Sodium	*Natrium* (Na)
Tin	*Stannum* (Sn)

[c]It is interesting to note that analysis of lunar rock appeared to be quite high in titanium. Some samples contained 6 percent titanium—10 times that of the earth's crust.

3-4

Properties of Pure Substances

Just as each person has his or her own appearance and personality, each pure substance has its own properties, distinguishing it from other substances. The properties of pure substances are divided into physical and chemical properties.

Physical properties are those properties that can be observed without changing the composition of the substance. These properties include color, odor, taste, solubility, density, specific heat, melting point, and boiling point. Physical properties of a pure substance are analogous to peoples' appearance—the color of their hair and their eyes, their height, and their weight (see Figure 3-5).

FIGURE 3-5

Physical and chemical properties are analogous to an individual's appearance and personality.

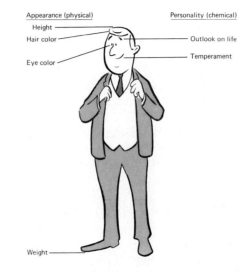

Chemical properties are those properties that can be observed only when a substance undergoes a change in composition. These properties include the fact that iron rusts, that coal or gasoline burns in air, that water undergoes electrolysis, and that chlorine reacts violently with sodium. Chemical properties of a pure substance are analogous to a person's personality, or peoples' outlook on life, or their temperament, or how they react in various situations (see Figure 3-5).

Work Problems 5 and 6.

Table 3-2 lists some physical and chemical properties of water and iron.

TABLE 3-2 Some Physical and Chemical Properties of Water and Iron

		Physical					Chemical
			Specific Heat				
Substance	Color	Density (g/mL, 20°C)	cal/g · °C	J/kg · K	Melting Point (°C)	Boiling Point (°C)[a]	
Water (liquid)	Colorless	0.998	1.000	4.18×10^3	0	100	Undergoes electrolysis; yields hydrogen and oxygen
Iron (solid)	Gray-white	7.874	0.108	4.52×10^2	1535	3000	Rusts; reacts with oxygen in air to form an iron oxide [ferric or iron(III) oxide]

[a]At 1.00 atm pressure.

3-5

Changes of Pure Substances

In determining the properties of pure substances, we shall observe certain changes or conversions from one form to another in these pure substances. These changes are divided into physical and chemical changes.

Physical changes are those changes that can be observed *without* a change in the composition of the substance taking place. The changes in state of water from ice to liquid to water vapor are examples of physical change.

$$\text{ice} \rightleftarrows \text{liquid water} \rightleftarrows \text{water vapor} \tag{3-5}$$

The difference between a property and a change should be noted here; a property distinguishes one *substance* from another substance, whereas a change is a *conversion* from one *form* of substance to another. Physical change in a substance is analogous to the change in a woman's appearance when she puts on eye shadow or lipstick.

Chemical changes are those changes that can be observed only when a *change* in the composition of the substance is occurring. **New substances are formed.** The properties of the new substances are different from those of the old substances. In a chemical change a gas may be produced, heat energy may be given off (the flask gets hot), a color change may occur, or an insoluble substance may appear. We previously mentioned (3-3) that liquid water upon electrolysis yields the gases hydrogen and oxygen. Also, chlorine gas reacts violently with sodium metal. This yields sodium chloride, common table salt:

$$\text{chlorine (Cl)} + \text{sodium (Na)} \rightarrow \text{sodium chloride (NaCl)} \tag{3-6}$$

The change that occurs and that determines chemical properties is a chemical change. These two changes are represented in Equations 3-1 and 3-6, respec-

Work Problems 7
and 8.

tively. Somewhat analogous to a chemical change would be a pugnacious person meeting another person possibly of the opposite sex and becoming more cordial—possibly becoming a "new" person!

Table 3-3 lists various changes and classifies them as chemical or physical.

TABLE 3-3   Classification of Changes as Physical or Chemical

Change	Classification
Boiling of water	Physical
Freezing of water	Physical
Electrolysis of water	Chemical
Reaction of chlorine with sodium	Chemical
Melting of iron	Physical
Rusting of iron	Chemical
Cutting of wood	Physical
Burning of wood	Chemical
Taking a bite of food	Physical
Digestion of food	Chemical

3-6

Energy-Mass Relations

As previously mentioned in 2-10, all changes in nature, whether physical or chemical, are accompanied by changes in energy. Energy may be potential or kinetic. **Potential energy** is the energy possessed by a substance by virtue of its *position* in space or *composition*. **Kinetic energy** is energy possessed by a substance by virtue of its *motion*. A rock high on a cliff has potential energy, but as it falls down the cliff its *potential energy decreases* and its *kinetic energy increases*. This could be analogous to your life, since your "potential" is now great, although your "kinetic" may be nonexistent; but as time passes, your "potential" will decrease and your "kinetic" should increase.

Chemicals in the form of natural gas or gasoline have high potential energy. As they are burned this potential energy is transferred to kinetic energy to produce heat to warm your home and mechanical energy to run your car.

In 2-10 we mentioned that energy takes on various forms and can be transformed from one form to another. These forms of energy are mechanical, heat, electrical, chemical, and light or radiant energy. Starting a car is an example of the transformation of forms of energy. When the ignition key is turned, energy from the lead storage battery (chemical energy) produces an electrical current (electrical energy) that is transmitted to the starting motor (mechanical energy) and to the spark plugs (electrical and heat energy), which ignite the compressed gas in the cylinders (chemical energy), which, in turn, is transferred to the crankshaft (mechanical energy) when the transmission is engaged, and the car moves forward.

In photosynthesis, sunlight (radiant or light energy) initiates the chemical processes for the synthesis of carbohydrates (chemical energy).[3]

Such transformation of energy is expressed in the **Law of Conservation of Energy,**[4] which states that energy can be neither created nor destroyed, but may be transformed from one form to another. The **Law of Conservation of Mass,**[5] which corresponds to the Law of Conservation of Energy, states that mass is neither created nor destroyed, and that the total mass of the substances involved in a physical or chemical change remains unchanged. For example, consider the reaction of hydrogen with oxygen to produce water:

$$\text{hydrogen } + \text{ oxygen} \rightarrow \text{water}$$
$$4.0320 \text{ g } + \text{ } 31.9988 \text{ g } = 36.0308 \text{ g} \tag{3-7}$$

In this reaction, 4.0320 g of hydrogen combines with 31.9988 g of oxygen to yield 36.0308 g of water. The **sum** of the masses of *reactants* (hydrogen and oxygen) is *equal* to the mass of *product* (water), and hence mass is neither created nor destroyed, as is illustrated in Figure 3-6.

FIGURE 3-6

Illustration of the Law of Conservation of Mass, using the reaction of hydrogen with oxygen to produce water. (The mass of the hydrogen and oxygen in the pressure vessels is, for all practical purposes, exactly equal to the mass of the water in the beaker.)

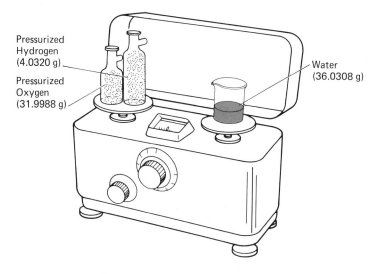

Pressurized Hydrogen (4.0320 g)

Pressurized Oxygen (31.9988 g)

Water (36.0308 g)

---

[3]In 1-3, we mentioned Melvin Calvin, who pioneered the work in the chemical process of photosynthesis, and Robert B. Woodward, who synthesized chlorophyll. Both men have received Nobel Prizes in chemistry for their contributions to science.

[4]James Prescott Joule (1818–1889), English physicist, is credited with formulating this law in the middle of the nineteenth century. The unit of energy, the joule, is named in honor of him (2-12).

[5]Antoine Laurent Lavoisier (lȧ′vwȧ′zyā′) (see 1-3), who published his results in 1789, is credited with formulating this law.

In 1905, the German-American physicist Albert Einstein[6] (1879–1955) concluded that mass and energy are interconvertible, as expressed by the equation $E = mc^2$, where E = energy expressed in units called ergs, m = mass in grams, and c = velocity of light (3.00×10^{10} cm/s). According to this equation, 1.00 g of mass converted *completely* to energy is equivalent to 9.00×10^{20} ergs of energy (9.00×10^{13} J), or, expressed in heat energy (2-10), it is equivalent to 2.15×10^{13} cal. This quantity of heat would raise the temperature of 215,000,000 kg (about 240,000 tons, the size of a small mountain lake) of water from 0°C to 100°C!

In ordinary chemical reactions, the energy changes involved are relatively small (up to 5×10^4 to 5×10^5 cal; or 2×10^5 to 2×10^6 J), and the mass loss (or gain) is of the order of 2×10^{-9} to 2×10^{-8} g, respectively, which is too small to detect on most balances. In the reaction of hydrogen with oxygen to form water (Equation 3-7), the loss of mass would amount to only 5.35×10^{-9} g (0.00535 μg or γ).

In nuclear reactions (fission, fusion), the loss of mass converted to energy is quite significant and is great enough to detect, but we can conclude that the Law of Conservation of Mass is still valid for *ordinary chemical reactions*. To be even more correct, we must combine the Law of Conservation of Mass with the Law of Conservation of Energy and state that the *total* of the *mass* and the *energy* is neither created nor destroyed, but may be transformed.

3-7

Elements and Atoms

In 3-3, we defined an element as a pure substance that cannot be decomposed into simpler substances by ordinary chemical means. *Elements are composed of atoms.* An **atom** is the smallest particle of an element that can undergo chemical changes in a reaction. In 3-3 you were asked to learn the symbols for 47 elements. These symbols represent not only the name of the element but also *one* atom of the element. For example, the symbol Na represents 1 atom of the element sodium, and the symbol H represents 1 atom of the element hydrogen.

3-8

Compounds, Formula Units, and Molecules. Law of Definite Proportions of Compounds

In 3-3 we defined compounds as pure substances that can be broken down into two or more different simpler substances by various chemical means. Compounds may be composed of *charged particles* (ions) or *molecules*.

In the case of an ionic compound, it is convenient to represent the compound with a **formula unit** in which the opposite charges of ions present balance each other so that the formula unit representing the compound has an overall charge of *zero*. Generally, the simplest formula unit possible is used.

---

[6]Einstein received the Nobel Prize in physics in 1921.

For example, sodium chloride is composed of sodium ions (Na^{1+}) and chloride ions (Cl^{1-}) in equal numbers, and a formula unit of this compound is NaCl.

For those compounds existing as molecules, a **molecular formula** is used to represent the compound. The **molecule** is the smallest particle of the compound that can exist and still retain the physical and chemical properties of the compound. Molecules, like atoms, are the particles that undergo chemical changes in a reaction. These molecules are composed of *atoms* of elements, held together by chemical bonds (Chapter 6); *hence these small particles called atoms are fundamental to all compounds.* Molecules may be composed of two or more *nonidentical* atoms (that is, atoms from different elements) which are the smallest particles of a compound. Water molecules are composed of the nonidentical atoms of hydrogen and oxygen. Molecules may also be composed of one or more *identical* atoms (that is, atoms from the same element). Oxygen molecules are composed of two identical atoms of oxygen. For now, we shall consider only molecules of compounds. Later, in Chapter 6 when we discuss bonding, we shall again refer to formula units and molecules from nonidentical and identical atoms.

An atom of an element is represented by a symbol, and a molecule of a compound is represented by a formula—more precisely, a molecular formula. Such a formula is composed of an appropriate number of symbols of elements representing *one* molecule of the given compound. For example, the molecular formula for water is H_2O (read H-two O). The *subscripts* represent the *number of atoms* of the respective elements in *one* molecule of the compound. Where no subscript is given, the number of atoms is one. Hence, in one molecule of water there are 2 atoms of the element hydrogen and 1 atom of the element oxygen, resulting in a total of 3 atoms in one molecule of water.

Consider the following examples:

Formula of Compound (Name)	Atoms of Each Element Present in One Unit	Total Number of Atoms
CO (carbon monoxide)	1 carbon, 1 oxygen	2
H_2S (hydrogen sulfide)	2 hydrogen, 1 sulfur	3
H_2O_2 (hydrogen peroxide)	2 hydrogen, 2 oxygen	4
CH_4O (methyl alcohol)	1 carbon, 4 hydrogen, 1 oxygen	6
$Pb(C_2H_3O_2)_2$ [lead (II) acetate]	1 lead, 4 carbon, 6 hydrogen, 4 oxygen. [*Note:* Clear the ( ) to get the number of atoms of carbon, hydrogen, and oxygen. The 2 (subscript) refers to everything in the ( ).]	15

If the number of atoms of each element in one molecule of the compound is known, we can write the molecular formula of the compound. In the following examples, the left-hand column gives the number of atoms of each element in one molecule of the compound, and the right-hand column gives the molecular formula.

Atoms of Each Element/Molecule of Compound	Molecular Formula
Carbon monoxide (an air pollutant); 1 carbon, 1 oxygen	CO
Ethyl alcohol; 2 carbon, 6 hydrogen, 1 oxygen	C_2H_6O
Ethyl ether; 4 carbon, 10 hydrogen, 1 oxygen	$C_4H_{10}O$
Ethylene glycol (used as an antifreeze); 2 carbon, 6 hydrogen, 2 oxygen	$C_2H_6O_2$
Chlorophyll a; 55 carbon, 72 hydrogen, 1 magnesium, 4 nitrogen, 6 oxygen	$C_{55}H_{72}MgN_4O_6$

Figure 3-7 summarizes the relation of atoms, molecules, formula units, elements, and compounds.

The molecular formula of water (H_2O) does not vary whether the water is Mississippi River water or pure distilled water. [Other impurities may be found in water, such as contaminants in the form of sewage or industrial waste (regrettably found in large rivers), but the formula of water is still H_2O.] This definite formula for water is expressed in the Law of Definite Proportions or Constant Composition. The **Law of Definite Proportions** or **Constant Composition** states that a given pure compound always contains the same elements

FIGURE 3-7

Summary of atoms, molecules, charged particles (formula units), elements, and compounds. Note that atoms are fundamental to compounds and elements.

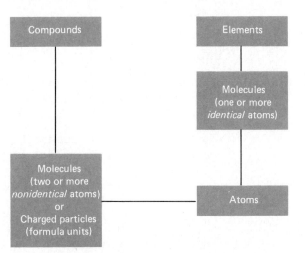

in *exactly* the same proportions by *mass*. For example, exactly 1.0080 parts by mass of hydrogen combine with 7.9997 parts by mass of oxygen to form water. Also, 2.0160 g (2 × 1.0080) of hydrogen will combine with 15.9994 g (2 × 7.9997) of oxygen to form water. How many grams of water would be formed in the latter case? (See **3-6**.) Later (**8-5**), we shall consider the general approach as to how the formula of water as H_2O was obtained, but for the present we can state that the composition by mass for a pure compound is invariable.

The Law of Definite Proportions or Constant Composition readily illustrates one of the chief differences between mixtures and compounds. A compound, which we previously classified as homogeneous matter, has an invariable composition by mass of elements, whereas a mixture, classified as homogeneous and heterogeneous matter, has a variable composition by mass of elements or compounds. Other differences between mixtures and compounds, as we have pointed out in **3-2**, are: (1) elements or compounds in many mixtures can be readily identified, whereas in a compound the elements lose their identity; and (2) the components of mixtures can be separated by simple operations that do not change their respective compositions, whereas compounds cannot be broken down into simpler substances by such procedures.

Work Problems 9, 10 and 11.

3-9

Division of the Elements. Metals and Nonmetals: Physical and Chemical Properties

The elements are divided into **metals** and **nonmetals,** and the differentiating factors are the physical and chemical properties of the elements.

Some general physical properties of the metals are the following:

1. High metallic luster (shine), such as silver.

2. High electrical and thermal conductivity, such as copper.

3. Malleability, that is, ability to be shaped by beating with a hammer, such as tin.

4. Ductility, that is, ability to be drawn out to a thin wire, such as gold.

5. High density, such as lead ($d^{20°}$ = 11.34 g/mL).

6. High melting point, such as iron (mp = 1535°C). Hence, metals are solids at room temperature—the exceptions being mercury (Hg), gallium (Ga), and cesium (Cs), which are liquids at ordinary temperatures.

7. Hardness, such as iron, tungsten, and chromium; however, some metals are soft, such as sodium and lead.

Some general chemical properties of the metals are the following:

1. They do not readily combine with each other.

2. They do combine with nonmetals and hence are normally found in nature in the combined form; iron is found as an iron oxide or sulfide. A few

relatively unreactive metals are found in nature in the free state, such as gold, silver, copper, and platinum.

The physical and chemical properties of the metals just listed are *general* properties, which vary from metal to metal, as we have pointed out with the melting points of mercury, gallium, and cesium.

In the front of this text and in Figure 3-8 is a table of the elements, the **Periodic Table** or **Periodic Chart**.[7] Slightly to the right of the center of the table is a **colored** stair step line which in general separates the metals from the nonmetals. To the *left* of this line lie the *metals* and to the *right* are the *nonmetals*.[8] Elements such as sodium, copper, iron, gold, lead, and platinum are metals. Find them on the table. You need to know their symbols first!

In general, the physical properties of the nonmetals are different from the physical properties of the metals. Nonmetals have little luster and poor electrical and thermal conductivity; they are neither malleable nor ductile; they have low melting points and densities and they are soft. There is at least one of the nonmetals that exists in each category of the three physical states of matter at room temperature and atmospheric pressure: solid—such as sulfur, phosphorus, carbon, and iodine; liquid—such as bromine; and gas—such as fluorine, chlorine, oxygen, and nitrogen.

The general chemical properties of the nonmetals follow:

1. They combine with metals. A few exist in nature in the free state (uncombined)—these are oxygen and nitrogen (both in air); sulfur; and carbon (coal, graphite, and diamond).

2. They may also combine with each other. Carbon dioxide, carbon monoxide, silicon dioxide (sand), sulfur dioxide, and carbon tetrachloride are examples of compounds formed from nonmetals with nonmetals.

Again, refer to the inside front cover of this text or to Figure 3-8. The nonmetals include such elements as carbon, nitrogen, oxygen, fluorine, sulfur, chlorine, and bromine.

The last column of elements is a special group of nonmetals, called the *noble gases*. They exist in nature in the free (uncombined) state. Helium is found in some gas wells, and a small amount is found in air along with argon. The **noble gases** are relatively *unreactive*.

Elements that lie on the color stair step line are called **metalloids** (not aluminum). They have both metallic and nonmetallic properties. Examples of these are boron, silicon, germanium (Ge), arsenic, antimony, tellurium (Te), polonium (Po), and astatine (At), but not aluminum. Find them in the Peri-

---

[7]We shall consider the Periodic Table in Chapter 5.

[8]You will note that hydrogen (H) is to the far left of this line. Hydrogen is not a metal but sometimes acts like a metal. It also sometimes acts like a nonmetal.

FIGURE 3-8

The Periodic Table of the Elements. Note the **colored** stair step line which, in general, separates the metals from the nonmetals. The [ ] indicates that the element has not officially been approved or named.

GROUPS

PERIODS	IA	IIA	IIIB	IVB	VB	VIB	VIIB	VIII			IB	IIB	IIIA	IVA	VA	VIA	VIIA	0
1	1.008 H 1																	4.003 He 2
2	6.941 Li 3	9.012 Be 4											10.811 B 5	12.011 C 6	14.007 N 7	15.999 O 8	18.998 F 9	20.179 Ne 10
3	22.990 Na 11	24.305 Mg 12											26.982 Al 13	28.0855 Si 14	30.9738 P 15	32.06 S 16	35.453 Cl 17	39.948 Ar 18
4	39.0983 K 19	40.08 Ca 20	44.956 Sc 21	47.90 Ti 22	50.9415 V 23	51.996 Cr 24	54.938 Mn 25	55.847 Fe 26	58.933 Co 27	58.71 Ni 28	63.546 Cu 29	65.37 Zn 30	69.72 Ga 31	72.59 Ge 32	74.922 As 33	78.96 Se 34	79.904 Br 35	83.80 Kr 36
5	85.468 Rb 37	87.62 Sr 38	88.906 Y 39	91.22 Zr 40	92.9064 Nb 41	95.94 Mo 42	98.906 Tc 43	101.07 Ru 44	102.906 Rh 45	106.4 Pd 46	107.868 Ag 47	112.41 Cd 48	114.82 In 49	118.69 Sn 50	121.75 Sb 51	127.60 Te 52	126.904 I 53	131.30 Xe 54
6	132.906 Cs 55	137.33 Ba 56	138.906 *La 57	178.49 Hf 72	180.948 Ta 73	183.85 W 74	186.2 Re 75	190.2 Os 76	192.22 Ir 77	195.09 Pt 78	196.967 Au 79	200.59 Hg 80	204.37 Tl 81	207.2 Pb 82	208.981 Bi 83	(209) Po 84	(210) At 85	(222) Rn 86
7	(223) Fr 87	226.025 Ra 88	(227) **Ac 89	(261) [Rf] 104	(262) [Ha] 105	(263) [ ] 106	(262) [ ] 107	(266) [ ] 109										

TRANSITION ELEMENTS

*Lanthanides

140.12 Ce 58	140.908 Pr 59	144.24 Nd 60	(145) Pm 61	150.4 Sm 62	151.96 Eu 63	157.25 Gd 64	158.925 Tb 65	162.50 Dy 66	164.930 Ho 67	167.26 Er 68	168.934 Tm 69	173.04 Yb 70	174.967 Lu 71

**Actinides

232.038 Th 90	231.031 Pa 91	238.029 U 92	237.048 Np 93	(244) Pu 94	(243) Am 95	(247) Cm 96	(247) Bk 97	(251) Cf 98	(254) Es 99	(253) Fm 100	(256) Md 101	(253) No 102	(257) Lr 103

Work Problems 12, 13, 14, 15, 16, 17, 18, and 19.

odic Table. Some metalloids are used as semiconductors of electricity in transistors for pocket calculators and are of recent interest in solar cells. The electrical conduction of the metalloids increases with temperature, whereas with metals the electrical conduction decreases. This is a factor that differentiates metals from metalloids.

Chapter Summary

The three physical states of matter are *solid*, *liquid*, and *gas*. Matter is divided into *homogeneous* and *heterogeneous* matter. Homogeneous matter is further divided into *pure substances*, *homogeneous mixtures*, and *solutions*, also a homogeneous mixture. Heterogeneous matter is called a *mixture*. Pure substances are divided into *compounds* and *elements*. Properties and changes of pure substances are categorized as *physical* or *chemical*.

Energy may be *potential* or *kinetic*. The *Law of Conservation of Energy* governs energy changes in nature and the *Law of Conservation of Mass* governs physical and chemical changes. These laws can be combined due to the interconversion of mass and energy according to Einstein's equation, $E = mc^2$.

Elements are composed of *atoms*. Compounds may be composed of charged particles (ions) and are represented as *formula units*. Those compounds not consisting of charged particles exist as molecules and are represented by a *molecular formula*. The *Law of Definite Proportions* states that a given pure compound always contains the same elements in exactly the same proportions by mass.

The elements are divided into *metals* and *nonmetals* based on the physical and chemical properties of the elements. The metals are separated from the nonmetals on the Periodic Table (see the inside front cover or Figure 3-8) by a **colored** stair step line. To the left of this line lie the metals and to the right are the nonmetals. The last column of nonmetals is called the *noble gases*. Elements that lie on the color stair step line are called *metalloids*.

EXERCISES

1. Define or explain the following terms:

(a)	matter	(b)	heterogeneous matter
(c)	homogeneous matter	(d)	physical states of matter
(e)	pure substance	(f)	solution
(g)	mixture	(h)	compound
(i)	element	(j)	molecule
(k)	atom	(l)	physical properties
(m)	chemical properties	(n)	physical changes
(o)	chemical changes	(p)	formula
(q)	Law of Conservation of Energy	(r)	Law of Conservation of Mass
(s)	$E = mc^2$	(t)	Law of Definite Proportions
(u)	metals	(v)	nonmetals
(w)	malleable	(x)	ductile
(y)	noble gases	(z)	metalloids

2. Distinguish between
   (a) heterogeneous and homogeneous matter
   (b) a pure substance and a homogeneous mixture
   (c) a heterogeneous mixture and a compound
   (d) a compound and an element
   (e) a molecule and an atom
   (f) physical and chemical properties
   (g) properties and changes
   (h) physical and chemical changes
   (i) metal and nonmetal
   (j) metal and metalloid

PROBLEMS

States of Matter   (see Section 3-1)

3. Classify the following as existing in one of the three physical states of matter at room temperature and atmospheric pressure:
   (a) chalk
   (b) alcohol
   (c) antifreeze
   (d) battery acid (dilute sulfuric acid)
   (e) methane (natural gas)
   (f) oxygen

Symbols   (See Section 3-3)

4. List in order the symbols for the 10 most abundant elements in the earth's crust.

Physical and Chemical Properties   (See Section 3-4)

5. Using a reference book such as the *Handbook of Chemistry and Physics* or any suitable reference book found in your library, look up the melting and boiling points of mercury.

6. The following are properties of the element thallium; classify them as physical or chemical properties.
   (a) oxidizes slowly at 25°C   (b) malleable
   (c) bluish white   (d) reacts with nitric acid
   (e) reacts with chlorine   (f) melting point 303.5°C
   (g) poisonous   (h) easily cut with a knife

Physical and Chemical Changes   (See Section 3-5)

7. Classify the following changes as physical or chemical:
   (a) pumping oil out of a well
   (b) separation of components of oil by distillation
   (c) burning gasoline
   (d) flaring of gas from a well
   (e) grinding up beef in a meat grinder
   (f) digestion of the beef

8. Classify the following changes as physical or chemical:
   (a) baking bread
   (b) mixing flour with yeast
   (c) fermentation to produce beer
   (d) smashing a car against a tree
   (e) burning your chemistry book
   (f) burning toast

Formulas (See Section 3-8)

9. Determine the number of atoms of each element; write the name of the element and the total number of atoms in each of the following molecular formulas:
   (a) CH_4 (methane, natural gas)
   (b) $C_6H_{12}O_6$ (glucose)
   (c) CCl_2F_2 (Freon)
   (d) $C_{16}H_{18}N_2O_5S$ (penicillin V)
   (e) $C_{34}H_{32}FeN_4O_4$ (heme from hemoglobin)
   (f) $C_6H_8N_2O_2S$ (sulfanilamide)

10. From the number of atoms of each element in one molecule of the compound, write the molecular formula of the following compounds:
    (a) sulfur dioxide; 1 sulfur, 2 oxygen
    (b) pyrite or fool's gold; 1 iron, 2 sulfur
    (c) argentite; 2 silver, 1 sulfur
    (d) caffeine; 8 carbon, 10 hydrogen, 4 nitrogen, 2 oxygen
    (e) adenosine triphosphate (ATP); 10 carbon, 16 hydrogen, 5 nitrogen, 13 oxygen, 3 phosphorus
    (f) 2,4,6-trinitrotoluene (TNT); 7 carbon, 5 hydrogen, 3 nitrogen, 6 oxygen

Compounds, Elements, and Mixtures (See Sections 3-3 and 3-8)

11. Classify each of the following as a compound, element, or mixture:

(a)	sulfur	(b)	water
(c)	silicon	(d)	salted popcorn
(e)	salt (sodium chloride)	(f)	sugar
(g)	lead	(h)	gasoline

Metals and Nonmetals (See Sections 3-3 and 3-9)

12. Using the Periodic Table on the inside front cover of this text, write the names and symbols for 10 metals.

13. Using the Periodic Table in the front of this text, write the names and symbols for 10 nonmetals.

14. Write the names and symbols for three metals that are found in nature in the free state (uncombined).

15. Write the names and symbols for three nonmetals that are found in nature in the free state (uncombined).

16. For each of the three physical states of matter, list at least one nonmetal (name and symbol) that exists in that state at room temperature and atmospheric pressure.

17. Write the name and symbol of a metal that is a liquid at ordinary temperatures.

18. Write the names and symbols for two noble gases.

19. Write the names and symbols for two metalloids.

General Problems

20. Calculate the volume in liters at 20°C occupied by 2.85 lb of iron (see Table 3-2; ·1 lb = 454 g).

21. Calculate the mass in pounds of 6.35 quarts (20°C) of water (see Table 3-2; 1 lb = 454 g, 1 L = 1.06 quarts).

22. Calculate the amount of heat in (a) joules and (b) kilocalories required to raise the temperature of $12\overline{0}$ g of iron from 535°C to its melting point (see Table 3-2).

23. Calculate the amount of iron in kilograms used if 12.0 kcal of heat is absorbed when the iron is heated from $137\overline{0}$°C to its melting point (see Table 3-2).

Readings

Chalmers, Bruce, "The Photovoltaic Generation of Electricity." *Sci. Am.*, Oct. 1976, v. 235, p. 34. Discusses the conversion of solar energy into electrical energy using solar cells.

Maugh, Thomas H., II, "Element 106: Soviet and American Claim in Muted Conflict." *Science*, Oct. 4, 1974, v. 186, p. 42. Discusses the Soviet and American synthesis of element 106.

Seaborg, Glenn T., and Justin L. Bloom, "The Synthetic Elements: IV." *Sci. Am.*, Apr. 1969, v. 220, p. 56. Describes recently discovered elements and speculates on the possibility of extending the list of elements to 114 and beyond.

Selbin, Joel, "The Origin of the Chemical Elements, 1 and 2." *J. Chem. Educ.*, 1973, v. 50, pp. 306, 380. Examines in detail the origin of the chemical elements, or *nucleosynthesis*, and the "big-bang" theory for the creation of the universe as proposed primarily by the late George Gamow.

Whitaker, Robert D., "An Historical Note on the Conservation of Mass." *J. Chem. Educ.*, 1975, v. 52, p. 658. Discusses the history of the Law of Conservation of Mass, including the contribution of Joseph Black, who preceded Lavoisier by about 30 years.

4

The Structure of the Atom

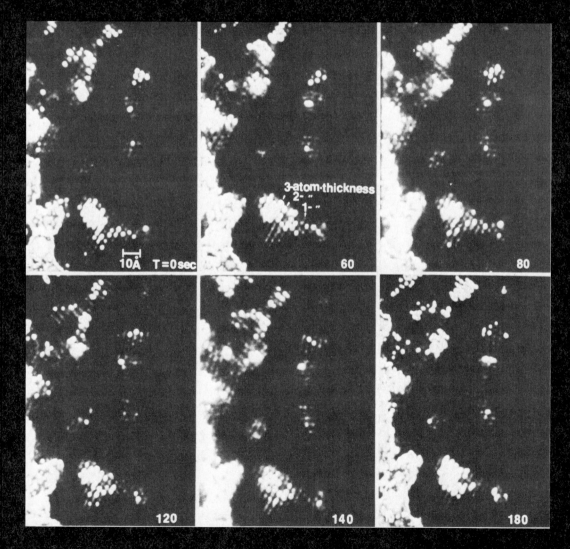

3-atom-thickness
2- "
1- "

10Å T=0sec   60   80

120   140   180

Time-lapse micrographs of uranyl crystals in which uranium atoms can clearly be observed. The pictures were taken with a Scanning Transmission Electron Microscope.

TASKS

1. List and explain in your own words Dalton's basic proposals about atomic theory (Section 4-2).

2. Identify the three basic subatomic particles, their abbreviations, approximate masses in amu, and *relative* charges (Section 4-3).

3. Draw the shape of the *s* and *p* orbitals (Section 4-9).

OBJECTIVES

1. Given the following terms, define each term and describe the distinguishing characteristic of each:
   (a)   atomic mass scale (Section 4-1)
   (b)   electron (Section 4-3)
   (c)   proton (Section 4-3)
   (d)   neutron (Section 4-3)
   (e)   atomic number (Section 4-4)
   (f)   mass number (Section 4-4)
   (g)   isotope (Section 4-5)
   (h)   orbital (Section 4-9)

2. Given the symbol $_Z^A E$, explain the meaning of each letter in the symbol (Section 4-4).

3. Given the atomic number and mass number of any element, calculate the number of protons and neutrons in the nucleus and the number of electrons outside the nucleus (Section 4-4, Problems 5 and 6).

4. Given the properties of an isotope of an element, determine which properties would be the same or different for another isotope of the same element (Section 4-5, Problem 7).

5. Given the exact atomic mass in amu for two or more isotopes and the percent abundance in nature for these isotopes, calculate the atomic mass of the element (Problem Examples 4-1 and 4-2, Problems 8, 9, and 10).

6. Given the principal energy levels, calculate the maximum number of electrons in each level (Problem Example 4-3, Problem 11).

7. Given the atomic number and mass number of any atom of atomic number 1 to 18, diagram the atomic structure of the atom by writing the number of protons and neutrons, arranging electrons in principal

energy levels, and writing the number of valence electrons (Section 4-6, Problems 12 and 13).

8. Given the symbol for an element and its atomic number, write its electron-dot formula [for any atom of atomic number 1 to 18 (Section 4-7, Problems 14 and 15)].

9. Given the atomic number of any element, (1) write the electronic configuration in sublevels, and (2) give the number of valence electrons for each atom (Section 4-8, Problems 16 and 17).

In the third chapter (3-8), we mentioned that the small particles called atoms are fundamental to all compounds. In this chapter, we shall explore further the structure of these minute atoms.

4-1

Atomic Mass

Atoms are very small. The diameter of an atom is in the range of 1 to 5 angstroms (Å; see Table 2-3). If we were to place atoms of a diameter of 1 Å side by side, it would take 10,000,000 of them to occupy a 1-millimeter length, as illustrated in Figure 4-1. That is a lot of atoms!

The mass of an atom is also a very small quantity, too small to be determined on even the most sensitive balance. For example, by indirect methods the mass of a hydrogen atom is found to be 1.67×10^{-24} g, an oxygen atom 2.66×10^{-23} g, and a carbon atom 2.00×10^{-23} g. Since this mass is very small, chemists have devised a scale of relative masses of atoms called the **atomic mass (atomic weight) scale.** The scale is based on an arbitrarily assigned value of exactly 12 atomic mass units (abbreviated amu) for carbon-12. (The nature of carbon-12 will be discussed in 4-5.) Hence, one atomic mass unit (amu) on the atomic mass scale is equal to $\frac{1}{12}$ the mass of a carbon-12 atom. An atom that is twice as heavy as a carbon-12 atom would have a mass of 24 atomic mass units (amu).[1]

FIGURE 4-1

Atoms are very small. If atoms of a diameter of 1 Å were placed side by side, it would take 10,000,000 atoms to occupy a 1-millimeter length.

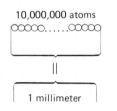

10,000,000 atoms

1 millimeter

---

[1]The equivalence between the mass units, amu, and grams is as follows: $1.00 \text{ g} = 6.02 \times 10^{23}$ amu. Knowing this relationship allows you to interconvert the two units.

A simple analogy will help to illustrate the point. Suppose we wish to relate the "weight" of everyone in your class to one person as our standard. This person we will call John Doe. On *this very day*, John "weighs" 200 pounds, and we assign him a relative mass of 12 units on our arbitrary scale. Since a person's weight changes daily we must pick a particular day. We could have assigned a value of 10, 15, 20 or some other value, but we arbitrarily assigned a value of 12. Now some other person in the room "weighs" 266 pounds or 1.33 ($\frac{266}{200}$) times as much as John Doe; hence, this person would have a relative mass of 16 units (1.33 $\times$ 12). A similar relationship could be worked out for each member of the class and a relative mass could be assigned to each, based on the "weight" of John Doe on *that particular day*.

On the inside front cover of this text, all the elements are listed, and their relative atomic mass units based on carbon-12 are given precisely. As you can see, some of these numbers are very exact and are carried out even to the ten-thousandth place, whereas others are expressed only to the units place. Therefore, for calculations that you will be doing in this course we have developed a Table of Approximate Atomic Masses. It is found on the inside back cover of this text and should be used in all future calculations.

4-2

Dalton's Atomic Theory

In the early part of the nineteenth century, the English scientist John Dalton (1767–1844), to whom we referred in 1-3, proposed an atomic theory based on experimentation and chemical laws known at that time. His proposals, after some modifications due to recent discoveries, still form the framework of our knowledge of the atom. The bases of his proposals are as follows:

1. Elements are composed of tiny, discrete, indivisible, and indestructible particles called atoms. These atoms maintain their identity throughout physical and chemical changes.

2. Atoms of the same element are identical in mass and have the same chemical and physical properties. Atoms of different elements have different masses and different chemical and physical properties.

3. Chemical combinations of two or more elements consist in the uniting of the atoms of these elements in a simple numerical ratio as **one** to **one,** or **one** to **two,** etc., to form a formula unit or molecule of a compound. In 3-8, we mentioned that one molecule of water consists of **2** atoms of hydrogen and **1** atom of oxygen.

4. Atoms of different elements can unite in *different* ratios to form more than one compound. In the preceding case, 2 atoms of hydrogen united with 1 atom of oxygen to form a molecule of water, H_2O. Two atoms of hydrogen can also combine with 2 atoms of oxygen to form a molecule of hydrogen peroxide, H_2O_2. Another example is carbon monoxide, CO, and carbon dioxide, CO_2.

Dalton's first proposal that atoms consist of tiny, discrete particles has been verified in that a single, tiny atom of both uranium and thorium has been photographed using an instrument called the electron microscope. Dalton's first proposal has been modified in that atoms consist of subatomic particles (see 4-3). These atoms can be split and hence are not indestructible (radioactivity, atomic energy), and in such nuclear changes the atoms lose their identity. In his second proposal, Dalton stated that all atoms of the same element have identical masses, but as you will learn in 4-5, isotopes of elements exist. In general, with the minor modifications mentioned above and a few others that we will not cover in this text, Dalton's proposals are valid today.

4-3

Subatomic Particles. Electrons, Protons, and Neutrons

As we mentioned above, the atom is composed of subatomic particles. These subatomic particles are: the **electron,** the **proton,** and the **neutron.** There are other subatomic particles, but these three form the basis of the atom that will be considered in this text.

The **electron,** abbreviated "e−," is a particle having a relative unit negative charge and a mass of 9.109×10^{-28} g or 5.486×10^{-4} (0.0005486) amu.[2] The mass of the electron is thus relatively small in terms of atomic mass units and is considered to be negligible for all practical purposes.

You have encountered electrons every day. When you comb your hair with a hard rubber comb, electrons from your hair collect on the comb and can attract small pieces of paper. When you walk on carpet and then approach certain objects, you get a shock. The electrons from the carpet accumulate in your body and you may be shocked when you touch certain objects. Both phenomena occur best when the humidity and temperature are low, and they are often described as the effects of static electricity.

The **proton,** abbreviated "p or p^+," is a particle having a relative unit positive charge and a mass of 1.6725×10^{-24} g or 1.0073 amu.[3] Since the mass of a proton is very close to 1 amu, it is often rounded off to 1 amu for most calculations.

The **neutron,** abbreviated "n or n^0," is a neutral particle having **no** charge and with a mass of 1.6748×10^{-24} g or 1.0087 amu. Since the mass of a neutron is also very close to 1 amu, it, too, is rounded off to 1 amu for most calculations.

---

[2]The actual charge on an electron is -1.602×10^{-19} coulomb, not −1. The coulomb is a unit used for measuring electrical charge, but as you can see, the value of the charge of an electron in coulombs is quite awkward to handle. Since atomic particles which are charged have charges that are the same or integral multiples of the charge of an electron, the **relative** charge of an electron may be chosen as **−1.**

[3]The actual charge on a proton is $+1.602 \times 10^{-19}$ coulomb. As you may note, this value is exactly the same as that on an electron, but opposite in sign; hence, the **relative** charge is considered **+1.**

Table 4-1 summarizes the data for the subatomic particles. You must be able to identify these particles including their abbreviations, approximate masses in amu, and *relative* charges.

TABLE 4-1 Summary of Subatomic Particles

Particle (abbrev.)	Approximate Mass (amu)	Relative Charge
Electron (e–)	Negligible	– 1
Proton (p or p$^+$)	1	+ 1
Neutron (n or n^0)	1	0

4-4

General Arrangement of Electrons, Protons, and Neutrons. Atomic Number

Now, how are these three subatomic particles arranged in an atom? To answer this question, we must consider a few fundamental facts about the atoms:

1. *All the protons and neutrons are found in the nucleus.* Since the mass of the atom is concentrated in the small volume of the nucleus, the nucleus has a high density (1.0×10^{14} g/mL). One milliliter of nuclear matter would have a mass of 1.1×10^8 tons! Also, since the protons are positively charged and the neutrons are neutral, the relative *charge* on the *nucleus* must be *positive* and *equal* to the *number of protons*.

2. *The number of protons* (mass of a proton, 1 amu) *plus the number of neutrons* (mass of a neutron, 1 amu) *equals the mass number of the atom* since the mass of the electron is negligible. Hence, the number of neutrons present is equal to the mass number minus the number of protons (neutrons = mass number − protons).

3. *An atom is electrically neutral.* From this statement follows the fact that if the number of electrons in an atom does not equal the number of protons, the atom is positively or negatively charged. Hence, in the *neutral atoms, the number of protons equals the number of electrons.*

4. *Electrons are found outside the nucleus in "shells" of certain energy levels.* In these "shells," the electrons are dispersed at a relatively great distance from the nucleus. The nucleus has a diameter of approximately 1×10^{-5} Å, whereas the diameter of the entire atom is in the range of 1 to 5 Å (4-1). Therefore, these electrons are dispersed at distances that extend up to 100,000 times the diameter of the nucleus. Suppose that as you sit at your chair reading this book you represent the size of the nucleus of the atom; the electrons would be dispersed at a distance up to 38 miles away, as illustrated in Figure 4-2.

Before we look at some examples of the general arrangement of the subatomic particles in the atom of some elements, we must consider symbols used

FIGURE 4-2

Distance between the nucleus and the electrons. If you are the size of the nucleus, the electrons would be spread out at distances as great as 38 miles away in proportion.

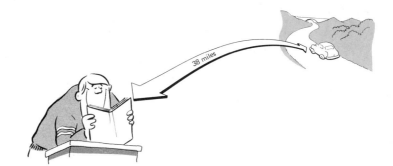

to describe the atom. The following is a general symbol for an atom of an element giving its mass number and atomic number:

$$A = \text{mass number}$$

$$_{Z}^{A}E \qquad E = \text{symbol of the element}$$

$$Z = \text{atomic number}$$

The **atomic number** is *equal* to the number of *protons* found in the nucleus. The **mass number** is *equal* to the *sum* of the *protons* and *neutrons* in the nucleus.

Now let us consider the general arrangement of the subatomic particles in the atom of some elements.

1. $_{1}^{1}$H   1 = atomic number = number of protons in nucleus.
    1 = mass number = sum of protons + neutrons. Hence, number of neutrons = 1 − 1 = 0 neutrons in nucleus.
    Number of electrons = number of protons = 1 electron outside the nucleus.

1e⁻

Nucleus   Outside nucleus

2. $_{2}^{4}$He   2 = atomic number = number of protons in nucleus.
    4 = mass number = sum of protons + neutrons.
    Neutrons = 4 − 2 = 2 neutrons in nucleus.
    Number of electrons = number of protons = 2 electrons outside the nucleus.

2e⁻

Nucleus   Outside nucleus

3. $^{11}_{5}$B   5 = atomic number = number of protons in nucleus.

        11 = mass number = sum of protons + neutrons.

Neutrons = 11 − 5 = 6 neutrons in nucleus.

Number of electrons = number of protons = 5 electrons outside the nucleus.

Nucleus   Outside nucleus

4. $^{27}_{13}$Al   13 = atomic number = number of protons in nucleus.

        27 = mass number = sum of protons + neutrons.

Neutrons = 27 − 13 = 14 neutrons in nucleus.

Number of electrons = number of protons = 13 electrons outside the nucleus.

Nucleus   Outside nucleus

5. $^{260}_{105}$Ha   105 = atomic number = number of protons in nucleus.

        260 = mass number = sum of protons + neutrons.

Neutrons = 260 − 105 = 155 neutrons in nucleus.

Number of electrons = number of protons = 105 electrons outside the nucleus.

Nucleus   Outside nucleus

With the highly positively charged nucleus due to the protons and the negatively charged electrons outside the nucleus, you might wonder why the positive center does not draw in and unite with the negative charges and neutralize them. (Remember, a positive charge attracts a negative charge.) To explain this, the Danish physicist Niels Bohr[4] (1885–1962) proposed a theory in 1913. A description follows. The electrons in an atom have their energy restricted to *certain* energy values called energy levels, which we shall consider

---

[4]Bohr received the Nobel Prize in physics in 1922. The Soviet scientists who also discovered element 105, named *hahnium* by the Americans, wish to name this element after Niels Bohr. Their proposed name for element 105 is *niels bohrium*. The International Union of Pure and Applied Chemistry Commission (7-1) has proposed the name unnilpentium with the symbol Unp.

in 4-8. For an electron to change its energy, it must shift from one energy level to another. To go to a higher energy level, a definite amount of energy is required equal to the energy difference between the two levels. But to go to a lower energy level, *a lower energy level must be available,* and if so, energy equal to the difference between the two levels is given off. The electrons are arranged in their lowest energy levels, and *no* lower energy level is available; therefore, the electrons remain in their low energy levels. This is analogous to a person walking up a flight of stairs. In order to make progress in walking up the stairs, you are restricted to certain levels of progress (the steps). You cannot raise yourself up between the steps! Climbing a ladder is a similar analogy.

Work Problems 5 and 6.

4-5

Isotopes

On close examination of the atomic masses of the elements (inside front cover of this book), you will note that the atomic masses of the elements are not whole numbers (carbon = 12.01115 amu and chlorine = 35.453 amu). Since the mass of the proton and neutron are nearly equal to 1, and since the mass of the electron is very slight, we would expect the atomic mass of an element to be very nearly a whole number—certainly not halfway between, as is the case with chlorine.[5] The reason that many atomic masses are not even close to whole numbers is that all atoms of the *same* element do not necessarily have the same mass, a contradiction to Dalton's second proposal (4-2). Atoms having different atomic masses or mass numbers, but the same atomic numbers, are called **isotopes.**

Carbon exists in nature as two isotopes: carbon-12 ($^{12}_{6}C$, exact atomic mass = 12.00000 amu, the atomic mass unit standard), and carbon-13 ($^{13}_{6}C$, exact atomic mass = 13.00335 amu). Structurally, the difference between these two isotopes is *one* neutron. Carbon-12 has 6 neutrons, while carbon-13 has 7 neu-

---

[5]Even if we consider the exact masses of the protons, the neutrons, and the electrons, the total mass does not equal the mass found for a particular atom of an isotope, but is greater. For example, carbon-12 has an exact atomic mass of 12.0000 amu. Calculations based on the number of protons, neutrons, and electrons, and their respective masses in amu give the following for carbon-12:

$$6 \text{ protons} \times \frac{1.0073 \text{ amu}}{1 \text{ proton}} = 6.0438 \text{ amu}$$

$$6 \text{ neutrons} \times \frac{1.0087 \text{ amu}}{1 \text{ neutron}} = 6.0522 \text{ amu}$$

$$6 \text{ electrons} \times \frac{0.0005486 \text{ amu}}{1 \text{ electron}} = 0.0033 \text{ amu}$$

$$\overline{12.0993 \text{ amu}}$$

This difference in mass (in this case, 0.0993 amu) is considered to be converted to energy, according to Einstein's equation (3-6), and its liberation is required to hold the positive protons and neutrons together in the nucleus. This mass equivalent of energy is called the *binding energy* of the particular atom.

trons. Chlorine also exists in nature as two isotopes: chlorine-35 ($^{35}_{17}Cl$, exact atomic mass = 34.96885 amu) and chlorine-37 ($^{37}_{17}Cl$, exact atomic mass = 36.96590 amu). Structurally, what is the difference between these two isotopes? *Isotopes of the same element have the same chemical properties, but slightly different physical properties.* Hence, both chlorine-35 and chlorine-37 have the same chemical properties but slightly different physical properties.

The atomic mass in amu for the elements C = 12.01115 and Cl = 35.453 is an *average mass* based on the *abundance of the isotopes in nature.* The atomic mass for the element may be obtained by multiplying the exact atomic mass of each isotope by the decimal of its percent abundance in nature and then taking the sum of the values obtained. This is similar to the calculation of your grade in a particular course. For example, if you get a score of 75 on the exam that counts 25 percent of your final grade and a score of 85 on an exam that counts 75 percent, your final average based on the "weight" of each exam is 82.5 (to three significant digits), *not* 80. The calculation is as follows:

$$75(0.25) + 85(0.75) = 18.75 + 63.75 = 82.5.$$

Note that the percent, meaning parts per 100, is converted to a decimal, meaning parts per *one* by dividing by 100.

The following problem examples illustrate the calculation of atomic masses for elements.

Problem Example 4-1

Calculate the atomic mass to four significant digits for carbon, given the following data:

Isotope	Exact Atomic Mass (amu)	Abundance in Nature (%)
^{12}C	12.00000	98.89
^{13}C	13.00335	1.110

SOLUTION: Convert the percents (98.89 and 1.110) to decimal form by dividing by 100 to give 0.9889 and 0.01110, respectively. Therefore,

$$12.00000 \text{ amu } (0.9889) + 13.00335 \text{ amu } (0.01110) = 12.01 \text{ amu}   Answer$$

Problem Example 4-2

Calculate the atomic mass to four significant digits for chlorine, given the following data:

Isotope	Exact Atomic Mass (amu)	Abundance in Nature (%)
^{35}Cl	34.96885	75.53
^{37}Cl	36.96590	24.47

SOLUTION

$$34.96885 \text{ amu } (0.7553) + 36.96590 \text{ amu } (0.2447) = 35.46 \text{ amu} \quad \textit{Answer}$$

In Appendix III is a table of some naturally occurring stable isotopes of the elements with their mass numbers and percent abundance in nature.

In our analogy in 4-1, we pointed out that we compared the "weight" of members of the class to the weight of John Doe on a *particular day*. The weight of a person does vary from day to day, so to be correct we must define when this weight was taken. For example, on the afternoon of October 15, John may "weigh" 200 pounds, but on the afternoon of October 18 he may weigh just 197 pounds. Therefore, with atomic masses of the elements we must use as a standard a particular isotope. Hence, the carbon-12 isotope is used and has been assigned a value of exactly 12 amu.

Based on the average mass, the atomic mass of carbon was found to be 12.01115 amu, but we would never find an atom of carbon that would have a relative mass of 12.01115 amu; it would have a relative mass of 12.00000 or 13.00335 amu, depending on the isotope with which we were working. But in general, for an ordinary size sample of carbon atoms containing the isotopes in the proportions given, we find it convenient to use the average mass, 12.01115 amu. The same reasoning applies to all the other elements and their atomic mass units, which are given inside the front cover of this text and are the average masses of the naturally occurring isotopes of the elements.

Work Problems 7, 8, 9, and 10.

4-6

Arrangement of Electrons in Principal Energy Levels

In 4-4 we did not specify how the electrons are arranged. We just said that they were outside the nucleus. In this section, we shall be more specific.

The electrons can exist at *principal energy levels* and are arranged in *shells*, which increase in energy as they increase in distance from the nucleus. That is, the nearer the electron is to the nucleus the less energy the electron has; the farther away it is, the more energy it has. An analogy to this is that it appears that the farther away from home (nucleus) a young college student gets, the more freedom he has! The principal energy levels are designated by the whole numbers 1, 2, 3, 4, 5, 6, and 7. There is a maximum number of electrons that can exist in a given energy level. This number is found from the following equation:

$$\text{Maximum number of electrons at principal energy levels} = 2n^2$$

$$n = \text{integers 1 to 7 of the principal energy levels}$$

Problem Example 4-3

Calculate the maximum number of electrons at the 1 and 2 principal energy levels.

SOLUTION

For 1: Maximum number of electrons $= 2 \times 1^2 = 2 \times 1 = 2$ *Answer*
For 2: Maximum number of electrons $= 2 \times 2^2 = 2 \times 4 = 8$ *Answer*

Table 4-2 lists the principal energy levels and the maximum number of electrons at that level. These are the maximum numbers of electrons that can be accommodated at a given energy level, but an energy level *may have less* than the maximum.

TABLE 4-2 Maximum Number of Electrons in Principal Energy Levels

Principal Energy Level	Maximum Number of Electrons
1	2
2	8
3	18
4	32
5	50
6	72
7	98

Increasing energy (applies to levels 1 through 7)

Now let us consider the arrangement of the electrons in principal energy levels. Consider the following atoms:

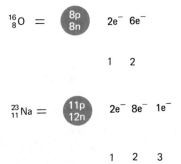

$^{4}_{2}\text{He}$ = 2p 2n   2e⁻
                          1

$^{11}_{5}\text{B}$ = 5p 6n   2e⁻ 3e⁻
                          1   2

Note here that the maximum number of electrons in energy level 1 is 2, so to place 5 electrons outside the nucleus we must go to a higher energy level—level 2.

$^{16}_{8}\text{O}$ = 8p 8n   2e⁻ 6e⁻
                          1   2

$^{23}_{11}\text{Na}$ = 11p 12n   2e⁻ 8e⁻ 1e⁻
                                1   2   3

Level 2 can accommodate a maximum of 8 electrons, so to place 11 electrons outside the nucleus we must use not only levels 1 and 2 but also a higher energy level—level 3.

$$^{40}_{18}Ar \;=\; \boxed{\begin{array}{c}18p\\22n\end{array}} \quad 2e^-\ \ 8e^-\ \ 8e^-$$

$$1\qquad 2\qquad 3$$

Work Problems 11, 12, and 13.

Argon is a noble gas (3-9) and has *eight* electrons in its highest principal energy level.

4-7

Electron-Dot Formulas of Elements

The electrons in the highest *principal* energy level in the preceding diagrams of the atom are usually called valence energy level (shell) of electrons—**valence electrons.** The remainder of the atom (nucleus and other electrons) is called the **core** (kernel). These other electrons usually have a noble gas arrangement. The electrons in the valence energy level are of higher energy than the inner electrons and are *gained, lost,* or *shared* when an atom of one element unites with an atom of another element to form a molecule or ion. These valence electrons, due to their activity, are depicted in electron-dot formulas.

To write electron-dot formulas of elements, we need to follow a few simple rules:

1. The symbol for the element is written to represent the *core.*

2. We consider four sides to the symbol of the element with a maximum of *two* electrons on each side up to a maximum of *eight* electrons around the symbol.

3. The valence *electrons* (highest principal energy level) are shown on each of the four sides of the symbol, with *one* electron on *each side to a maximum of four,* and then the electrons are *paired up to a maximum of eight.* (Helium is an exception, in that both of its electrons are on the same side, since it has a completed first principal energy level.)

Consider the electron-dot formulas for the following atoms (in each, be sure to determine the number of valence electrons):

1. $^{1}_{1}H$ = H $\cdot$ or $\underset{\cdot}{H}$, etc. (1 valence electron; the four sides are equivalent)

2. $^{4}_{2}He$ = He $\vcentcolon$ (exception—see rule 3)

3. $^{7}_{3}Li$ = Li $\cdot$ (1 valence electron)

4. $^{11}_{5}B$ = $\cdot\overset{\cdot}{B}$ $\cdot$ or $\overset{\cdot}{B}$ $\vcentcolon$; etc. (3 valence electrons)

5. $^{12}_{6}C$ = $\cdot\overset{\cdot}{C}$ $\cdot$ (4 valence electrons)

6. $^{15}_{7}$N $=$ $\cdot\dot{\text{N}}\vdots$ (5 valence electrons, 2 paired up)

7. $^{20}_{10}$Ne $=$ $\vdots\ddot{\text{Ne}}\vdots$ (8 valence electrons, all sides filled)

8. $^{24}_{12}$Mg $=$ Ṃg $\cdot$ (2 valence electrons)

9. $^{32}_{16}$S $=$ $\cdot\dot{\text{S}}\vdots$ (6 valence electrons)

10. $^{35}_{17}$Cl $=$ $\vdots\ddot{\text{Cl}}\cdot$ (7 valence electrons)

In all the preceding examples, you may have noted that eight electrons filled all four sides, as in the case of neon (Ne, example 7). There is a specific rule governing this, the **"rule of eight"** or "octet rule." In the formation of molecules from atoms, most atoms attempt to obtain this stable configuration of eight electrons around each atom. The elements helium (He), neon (Ne), argon (Ar), krypton (Kr), xenon (Xe), and radon (Rn) are called the *noble gases*. All of them except helium have eight valence electrons and all are relatively unreactive, including helium.[6] In fact, they were once called the inert gases due to their lack of reactivity, but compounds containing the inert gases have now been prepared. We shall refer to this "rule of eight" again in 6-2.

Work Problems 14 and 15.

4-8

Arrangement of the Electrons in Sublevels

The electrons in the principal energy levels are further divided into sublevels (subshells). These sublevels are labeled *s*, *p*, *d*, and *f*; they also have a maximum number of electrons they can contain, which are 2, 6, 10, and 14, respectively, as shown in Table 4-3. (Note that 4 electrons are added each time.)[7]

As you see in Table 4-3, the number of sublevels equals the number of the principal energy level. For example, the first level has one (*s*), the second has two (*s* and *p*), the third has three (*s*, *p*, and *d*), etc. Each of these sublevels, with its respective principal energy level, has an order of increasing en-

---

[6]Helium has two valence electrons that complete its principal energy level 1; hence, it, too, is relatively unreactive.

[7]This arrangement of electrons in sublevels of principal energy levels is analogous to the arrangement of students in rooms in the various floors of the Electra Hostel. In the Electra Hostel, on the *first* (1) floor there is just one (1) room for guests which can accommodate a maximum of two (2) students. On the *second* (2) floor there are two (2) rooms for guests. One room can accommodate a maximum of two (2) students and the other room can accommodate a maximum of six (6) students with a total maximum of eight (8) students on the second floor. On the *third* (3) floor, there are three (3) rooms for guests. One room can accommodate a maximum of two (2) students, the other room a maximum of six (6) students, and the third room a maximum of ten (10) students with a total maximum of eighteen (18) students on the third floor. On the *fourth* (4) floor, there are four (4) rooms for guests. One room can accommodate a maximum of two (2) students, the second room a maximum of six (6) students, the third room a maximum of ten (10) students, and the fourth room a maximum of fourteen (14) students, with a total maximum of thirty-two (32) students on the fourth floor.

TABLE 4-3 Maximum Number of Electrons in Principal Energy Levels 1 to 7 and Their Respective Sublevels

		Maximum Number of Electrons	
Principal Energy Level (Increasing energy ↓)	Sublevel	Sublevel	Principal Energy Level
1	*s*	2	2
2	*s*	2	8
	p	6	
3	*s*	2	18
	p	6	
	d	10	
4	*s*	2	32
	p	6	
	d	10	
	f	14	
5	*s*	2	50 (actually 32[a])
	p	6	
	d	10	
	f	14	
	g	18	
6	*s*	2	72 (actually 14[a])
	p	6	
	d	10	
	f	14	
	g	18	
	h	22	
7	*s*	2	98 (actually 2[a])
	p	6	
	d	10	
	f	14	
	g	18	
	h	22	
	i	26	

[a]This is the actual maximum number of electrons found for the elements known at present; hence, these principal energy levels are incomplete.

ergy. This increasing energy level is as follows: $1s < 2s < 2p < 3s < 3p < 4s < 3d < 4p < 5s < 4d < 5p < 6s < (4f < 5d) < 6p < 7s\ (5f < 6d)$. (The $<$ is read "less than.") In filling the sublevels, the *lower energy sublevels are filled first*, as are the principal energy levels. Figure 4-3 is a simplified way of remembering the order of filling. From either the preceding order or the diagram in Figure 4-3, you will note that the 4*s* fills before the 3*d*. Also from the preceding order, you will note that the 4*f* and 5*d*, and the 5*f* and 6*d*, sublevels have been placed in parentheses, since the energy of these sublevels is very

FIGURE 4-3

Order of filling the sublevels. Write down the principal energy levels with their sublevels to the *f* sublevel and then draw diagonal lines which follow the order of filling. (The diagonal lines need not be extended beyond the 6*d* sublevel since no elements at present have been discovered that have electronic configurations beyond the 6*d* sublevel.) The 5*d* and 6*d* sublevels are circled since only *one* electron is placed in each of these sublevels *before* filling the 4*f* or 5*f*, respectively. After filling the *f* sublevels, the *d* sublevels fill to their maximum of 10 electrons.

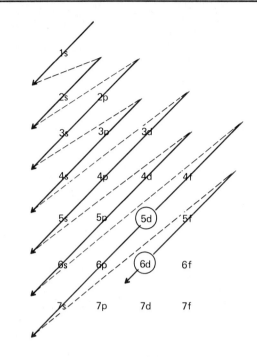

close. One electron is placed in the 5*d* before filling the 4*f*. The same is true for the 6*d*, before filling the 5*f*. After the 4*f* is filled, the 5*d* is then filled to its maximum of 10 electrons. The same is done for the 6*d* after filling the 5*f*. Exceptions to the order of filling sometimes occur in these sublevels, but in this text we shall not consider them.

When writing the sublevel electron configuration of an atom, write the principal energy level *number* and the sublevel *letter*, followed by the number of electrons in the sublevel written as a superscript. The sublevels of a given principal energy level may be *grouped together*, or as *they are filled*.

Consider the following atoms:

1. $_1^1H = 1s^1$ ← sublevel, number of electrons in that sublevel, principal energy level
   (**1** valence electron)

2. $_2^4He = 1s^2$: principal energy level 1 is now filled (**2** valence electrons)

3. $_3^7Li = 1s^2, 2s^1$: (**1** valence electron)

4. $_5^{11}B = 1s^2, 2s^2 2p^1$: the maximum in the 2*s* is 2, so we next fill the 2*p*
   (**3** valence electrons; see **4-7**)

5. $^{14}_{7}N = 1s^2, 2s^22p^3$: (**5** valence electrons)

6. $^{20}_{10}Ne = 1s^2, 2s^22p^6$: principal energy level 2 is now complete (**8** valence electrons)

7. $^{24}_{12}Mg = 1s^2, 2s^22p^6, 3s^2$: (**2** valence electrons)

8. $^{29}_{14}Si = 1s^2, 2s^22p^6, 3s^23p^2$: (**4** valence electrons)

9. $^{37}_{17}Cl = 1s^2, 2s^22p^6, 3s^23p^5$: (**7** valence electrons)

10. $^{39}_{19}K = 1s^2, 2s^22p^6, 3s^23p^6, 4s^1$: (**1** valence electron)

The next energy level after the $3p$ is the $4s$, so we go to that level before the $3d$.

11. $^{64}_{30}Zn = 1s^2, 2s^22p^6, 3s^23p^63d^{10}, 4s^2$ (**2** valence electrons); the $3d$ sublevel fills after the $4s$. We can group the $3d$ sublevel with the other sublevels of principal energy level 3, regardless of the order of filling, or equally acceptable is the electronic configuration $1s^2, 2s^22p^6, 3s^23p^6, 4s^2, 3d^{10}$, following the order of filling.

12. $^{75}_{33}As = 1s^2, 2s^22p^6, 3s^23p^63d^{10}, 4s^24p^3$ (**5** valence electrons: the $4p$ sublevels fills after the $3d$. Or, $1s^2, 2s^22p^6, 3s^2\,3p^6, 4s^2, 3d^{10}, 4p^3$.

13. $^{138}_{56}Ba = 1s^2, 2s^22p^6, 3s^23p^63d^{10}, 4s^24p^64d^{10}, 5s^25p^6, 6s^2$ (**2** valence electrons); the $5s$ sublevel fills after the $4p$, then the $4d$, $5p$ fills, and last the $6s$ fills. Or, $1s^2, 2s^2\,2p^6, 3s^2\,3p^6, 4s^2, 3d^{10}, 4p^6, 5s^2, 4d^{10}, 5p^6, 6s^2$.

14. $^{153}_{63}Eu = 1s^2, 2s^22p^6, 3s^23p^63d^{10}, 4s^24p^64d^{10}4f^6, 5s^25p^65d^1, 6s^2$ (*usually* **2** valence electrons); after filling the $6s$, *one* electron is placed in the $5d$ and then *six* electrons are placed in the $4f$. All sublevels of the same principal energy level can be grouped together. Or, $1s^2, 2s^2\,2p^6, 3s^2\,3p^6, 4s^2, 3d^{10}, 4p^6, 5s^2, 4d^{10}, 5p^6, 6s^2, 5d^1, 4f^6$.

The electronic configuration in sublevels for all the elements is given in Appendix IV, including the exceptions to the order of filling.

The filling of the sublevels correlates with the Periodic Table (3-9). The Periodic Table will be covered in the next chapter. This correlation is shown in Figure 4-4. Note that there are blocks of elements that fill just the s sublevels, those that fill just the p sublevels, those that fill just the d sublevel, and finally at the bottom of the table those that fill just the f sublevel. Using these blocks of elements, this means that the last electron will go into a specific sublevel corresponding to that block. For example, vanadium ($^{51}_{23}V$) has the electronic configuration $1s^2, 2s^2\,2p^6, 3s^2\,3p^6, 4s^2, 3d^3$. Vanadium is the third element in the d block filling the $3d$ sublevel; therefore, its last electron would be $3d^3$. Using the Periodic Table, you can check to see that you filled the sublevels properly.

Work Problems 16 and 17.

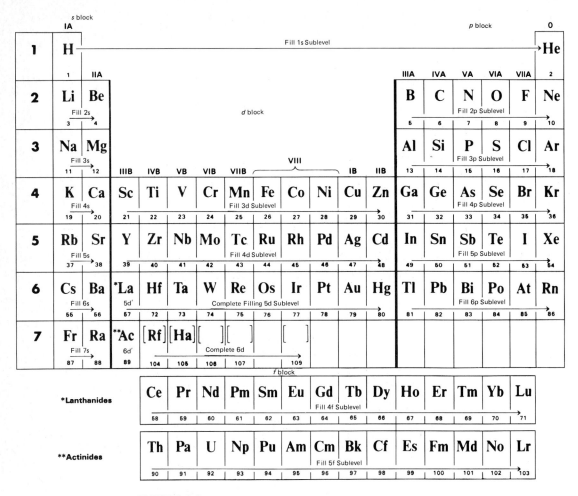

FIGURE 4-4

The corelation of the filling of the sublevels with the Periodic Table. Note the groups of elements that fill the *s, p, d,* and *f* sublevels. These groups of elements are indicated in the Periodic Table as follows: ☐ *s* block, ☐*p* block, ☐ *d* block, and ☐ *f* block. The numbers at the bottom of each box refer to the atomic number of each element.

4-9

Orbitals

In the preceding discussions, we considered the electrons of definite energy levels in discrete, definite orbital paths. This approach was formulated by the Danish physicist Niels Bohr (see 4-4) from 1913 to 1915, and is called the *Bohr model* of the atom. Since then, our concept of the arrangement of electrons in atoms has been modified. We do not now consider the electrons as traveling fixed orbital paths but consider them to be occupying orbital *volumes* of space. An **orbital** is a region of space within an atom in which there can be *no more than two* electrons. Orbitals have *shape*, which is defined by a 95 percent probability of the two electrons being found in a certain region. The *s, p, d, f* sublevel electrons occupy the *s, p, d, f* orbitals. Hence, we are only giving a shape to the *s, p, d, f* sublevel electrons when we call them orbitals. These electrons are found *somewhere* within the shape of the orbitals. The orbitals are *not* hollow. The shape of the *s* and *p* orbitals are shown in Figure 4-5. The shape of an *s* orbital is a sphere with the electrons—a maximum of two—found somewhere within the sphere. The *p* orbitals, of which there are three —p_x, p_y, p_z—resemble dumbbells arranged on the *x, y,* and *z* axes. In each of these orbitals are found no more than two electrons; for example, p_x = 2, p_y = 2, p_z = 2, to give a total of six *p* electrons—the maximum for this sublevel.

To clarify the idea of 95 percent probability of finding an electron somewhere in an orbital, let us consider an analogy. Suppose that every Monday afternoon you have chemistry laboratory between 1:00 and 4:00 in room 100. The probability of finding you in room 100 from 1:00 to 4:00 on Monday afternoon is very high, say 95 percent probability, allowing a 5 percent probability in case you are ill! Now you may be assigned a certain desk where you are to perform your experiments, but you may not always be at that desk for the entire time. You may leave your desk to get chemicals, or to talk to a friend, or even to ask the laboratory instructor some questions, but you are still in room 100.

Now, suppose we checked on your exact position every five minutes for the three-hour period and placed pins in a drawing of the room wherever we found you. Assuming you completed your experiment, we would probably find most of the pins around your desk. That is, the *region of highest probability* of finding you would be somewhere near your desk. If someone were looking for you, his chances of finding you somewhere in room 100 would be ex-

FIGURE 4-5

The shape of s, p_x, p_y, and p_z orbitals. The shape of the orbital is defined by a 95 percent probability of finding a maximum of two electrons somewhere within these orbitals. The arrows point to the nuclei. The p orbitals have two lobes with the electrons found somewhere within those lobes.

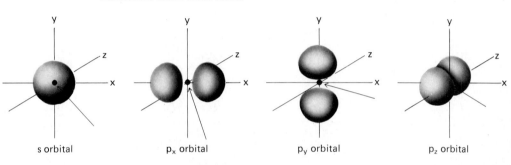

s orbital p_x orbital p_y orbital p_z orbital

tremely high and if he looked in the general region of your desk he would probably find you! The same reasoning applies to the electrons, in that there is a 95 percent probability of finding the electrons somewhere within these orbitals.

Let us now consider the relation of the *s* orbitals to one another in the principal energy levels—1, 2, 3, etc. The 1*s* orbital is analogous to a tennis ball suspended in the center of a volley ball—2*s*, which, in turn, is suspended in a basket ball—3*s* (see Figure 4-6).

FIGURE 4-6

Relation of the 1s, 2s, and 3s orbitals to each other.

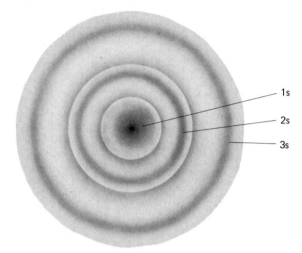

The *p* orbitals are related to each other as are *s* orbitals, but here we would have small dumbbells—2*p*, suspended in larger dumbbells—3*p* (see Figure 4-7).

We shall not cover the shape of the *d* and *f* orbitals in this text but leave that to more advanced chemistry courses.

FIGURE 4-7

Relation of the 2p$_y$ and 3p$_y$ orbitals to each other.

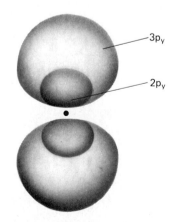

4-10

Historical Development of Modern Atomic Structure

In this section, we will describe briefly the contributions of several scientists whose experiments led to the discovery of subatomic particles such as the electron, proton, and neutron that we studied in 4-3, and the concept of the nuclear atom mentioned in 4-4.

The Electron

The discovery of the electron in the latter part of the nineteenth century was one of the most important steps in the development of modern atomic theory, since it stimulated further research which eventually led to the present theory of atomic structure.

In 1879, an English physicist and chemist, Sir William Crookes (1832–1911), discovered that when an electric discharge was passed through a tube containing a gas at low pressures (about 0.001 mm Hg), invisible rays (cathode rays) emanated from the cathode (negative electrode). At ordinary pressures, gases are very poor conductors of electricity, and a high voltage is required to produce a discharge; however, as the pressure is decreased the conductivity rises rapidly and cathode rays are produced. These rays are deflected by magnetic or electrical fields (see Figure 4-8), are independent of the cathode material and the gas present in the tube, and are composed of small, negatively charged particles of definite mass (electrons).

In 1897, another English scientist, Sir J. J. Thomson (1856–1940), deter-

FIGURE 4-8

The cathode ray tube. (a) Cathode rays are deflected by an electrical field; (b) cathode rays are deflected by a magnetic field.

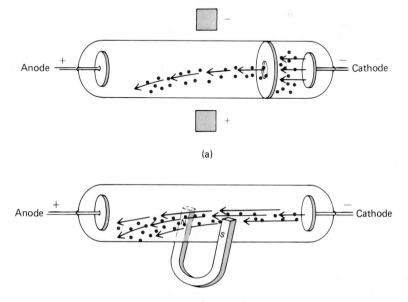

Anode

Cathode

(a)

Anode

Cathode

(b)

mined the charge-to-mass ratio for the electron by studying the simultaneous effect of electrical and magnetic fields on the electron beam (see Figure 4-9).

FIGURE 4-9

Thomson's experiment on the charge/mass ratio of the electron. Application of the magnetic field alone caused deflection of cathode rays along path M. Application of the electrical field alone caused deflection along path E. The rays could be followed by the appearance of flashes of light on the fluorescent zinc sulfide screen.

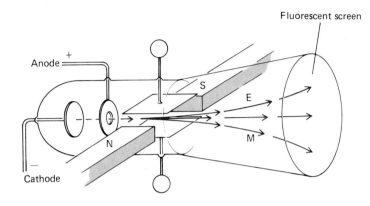

From the data collected in his experiments he calculated the charge-to-mass ratio for the electron to be 1.76×10^8 coulombs (C) of charge per gram of electron mass.

Later, in 1909, the American physicist Robert A. Millikan (1868–1953) measured the actual charge of a single electron in his classic oil drop experiment. He studied the behavior of very tiny oil droplets which had acquired a charge by picking up a stray electron or electrons (produced by slight, natural ionization of air); see Figure 4-10. Millikan observed these charged droplets as they fell through the space between two charged horizontal plates whose potential could be varied and opposed the force of gravity on the droplets. From the rate of fall of a droplet, its radius, and the electrical potential applied to

FIGURE 4-10

Millikan's oil drop experiment. Charge on plates A and C was adjusted to keep the droplet suspended. This experiment allowed Millikan to calculate the charge on the electron.

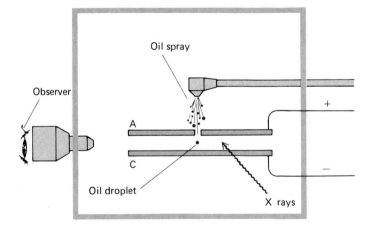

the plates, Millikan calculated that the total charge on a droplet was always some integer multiple of 1.60×10^{-19} C, depending on the number of electrons it had acquired in his experiment. He thus deduced that the charge on a single electron was indeed—1.60×10^{-19} C, and using Thomson's value for the charge-to-mass ratio for the electron, he calculated the mass of the electron to be 9.1×10^{-28} g or $\frac{1}{1837}$ that of the hydrogen atom. For convenience, the relative charge on the electron is referred to as $1-$.

The Proton

Eugen Goldstein (1850–1930), a German physicist, conducted experiments in 1886 in which he showed that positively charged particles are also formed in a cathode ray tube. If the cathode is perforated, rays of positively charged particles can be observed moving in the opposite direction of the cathode rays (see Figure 4-11). These rays originate in the space between the anode and cathode

FIGURE 4-11

Positive rays. Perforation of the cathode in a cathode ray tube allows the observation of positive rays which pass through the holes in the cathode. When the gas in the cathode ray tube is hydrogen the positive rays can be identified as protons.

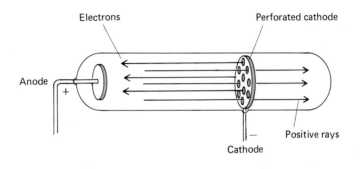

and are gas molecules or atoms which have had one or more electrons removed by bombardment with the cathode rays.

Later, J. J. Thomson (1856–1940) studied these positive rays in order to determine the charge-to-mass ratio of the positive particles as he had done earlier for the electron. The smallest positive particle which he could observe was obtained when hydrogen gas was present in the tube; this particle (the proton) had a charge-to-mass ratio of 9.57×10^4 C per gram. Since the charge on the proton was known to be positive and of equal magnitude to that of the electron (1.60×10^{-19} C per proton), the mass of the proton was calculated as 1.67×10^{-24} g per proton.

The Nuclear Atom

J. J. Thomson pictured the atoms as a particle having protons and electrons mixed together like raisins (electrons) in a plum pudding (protons); however, in 1911, Ernest Rutherford (1871–1937), a British physicist and former student of Thomson, modified this idea based on his experiments. Rutherford studied the scattering of alpha particles ($_2^4\text{He}^{2+}$) by a very thin gold foil, and he found that although 99 percent of the alpha particles went through, some were strongly deflected. About one in every 20,000 was deflected more than 90° (see Figure 4-12). These results could not be explained using the Thomson

FIGURE 4-12

Rutherford's gold foil experiment. Alpha particles (positively charged) are allowed to bombard a thin gold foil. Most of the alpha particles pass straight through the foil, striking the fluorescent screen at point 1. Some of the particles are deflected by the gold nuclei as indicated by the colored arrows (→).

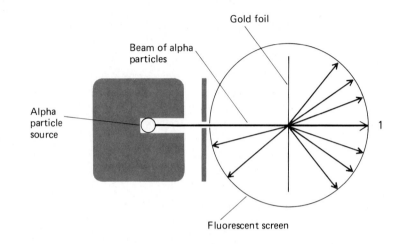

model of the atom, and Rutherford therefore described the nuclear atom, which was consistent with his observations. Since most of the alpha particles passed through undeflected, the volume of an atom must be mostly empty space, and there must be a heavy positively charged body present to account for the fact that some of the alpha particles were strongly deflected. He described the nuclear atom as one having a very small positive body (the *nucleus*) located at its center; about this nucleus are electrons whose number exactly balance the nuclear charge. The mass of an atom is concentrated in the nucleus, and atomic radii are in the order of 10^{-8} cm, with nuclear radii being in the order of 10^{-12} to 10^{-13} cm.

The Neutron

In 1932, Sir James Chadwick (1891–1974), an English physicist, discovered the third subatomic particle whose existence Rutherford had predicted many years earlier. In studying the bombardment of the element beryllium with alpha particles, Chadwick found that a new particle was emitted. This particle had a mass of approximately 1 amu and had no charge; Chadwick named it the *neutron*. The reaction that occurred in Chadwick's experiment is an example of artificial transmutation, where an atom of beryllium is converted to an atom of carbon through a nuclear reaction as simplified below:

$$\text{alpha particle} + {}_4^9\text{Be} \longrightarrow {}_6^{12}\text{C} + \text{neutron}$$

Chapter Summary

The *atomic mass* of the elements is based on carbon-12, which is assigned a value of exactly 12 atomic mass units (amu). The atom is composed of three basic subatomic particles: the *electron*, the *proton*, and the *neutron*. The protons and neutrons are located in the nucleus of the atom, and the electrons are located outside the nucleus and are equal to the number of protons. The number of protons in an atom is equal to the *atomic number* of the element. The sum of the number of protons and neutrons is called the *mass number*.

Isotopes of atoms have different mass numbers but the same atomic number. Using the exact atomic mass and the percent abundance in nature of isotopes of an element, the average atomic mass of the element can be calculated.

The electrons are arranged in both *principal energy levels* and in *sublevels* with a maximum number of electrons for each of the principal energy levels and sublevels. Using the number of *valence electrons* of an element, the *electron-dot* formulas of the element can be drawn. An *orbital* is a region of space within an atom in which there can be no more than two electrons. Orbitals have the same letter designation as those given to the sublevels.

The historical development of modern atomic structure was briefly considered. The discovery of the electron, proton, and neutron and how these particles were put together to form an atom was discussed. Many scientists with diverse nationalities contributed to these discoveries.

EXERCISES

1. Define or explain the following terms:
   - (a)   mass number
   - (b)   atomic mass units (amu)
   - (c)   nucleus of an atom
   - (d)   electron
   - (e)   proton
   - (f)   neutron
   - (g)   atomic number
   - (h)   isotope
   - (i)   principal energy level of electrons
   - (j)   valence electrons
   - (k)   core
   - (l)   rule of eight
   - (m)   electron-dot formula of an element
   - (n)   sublevels of electrons
   - (o)   orbitals
   - (p)   95 percent probability

2. Distinguish between
   - (a)   mass number and atomic number
   - (b)   electron and proton
   - (c)   neutron and proton
   - (d)   neutron and electron
   - (e)   subatomic particles in the nucleus and those outside the nucleus
   - (f)   valence electrons and core of an atom
   - (g)   shape of s and p orbitals
   - (h) p_x and p_z orbitals

3. Explain the meaning of the following symbols or numbers:
    (a) $_Z^A E$ (b) $1s^2$

4. Draw the shape of
    (a)   an s orbital   (b)   a p_x orbital
    (c)   a p_y orbital   (d)   a p_z orbital

PROBLEMS

General Arrangements of Subatomic Particles   (See Section 4-4)

5. For each of the following atoms, calculate the number of protons and neutrons in the nucleus and the number of electrons outside the nucleus:
    (a) $_5^{11}B$ (b) $_{18}^{36}Ar$
    (c) $_{22}^{46}Ti$ (d) $_{19}^{41}K$
    (e) $_{27}^{59}Co$ (f) $_{44}^{96}Ru$

    (All of the atoms above exist, although they may not be the most abundant isotope in nature.)

6. For each of the following atoms, calculate the number of protons and neutrons in the nucleus and the number of electrons outside the nucleus:
    (a) $_{11}^{23}Na$ (b) $_{34}^{74}Se$
    (c) $_{46}^{102}Pd$ (d) $_{40}^{90}Zr$
    (e) $_{58}^{142}Ce$ (f) $_{92}^{235}U$

    (All the atoms above exist, although they may not be the most abundant isotope in nature.)

Isotopes   (See Section 4-5)

7. The following are properties of uranium-238. Which of these properties would be the same for uranium-235?
    (a)   atomic mass = 238.0508 amu
    (b)   reacts rapidly with oxygen to form U_3O_8

8. Boron (10.811 amu) consists of two isotopes: boron-10 (10.013 amu) and boron-11 (11.009 amu). Based on the average atomic mass of boron, which one of the two isotopes is most abundant in nature?

9. Calculate the atomic mass to four significant digits for gallium, given the following data:

Isotope	Exact Atomic Mass (amu)	Abundance in Nature (%)
^{69}Ga	68.9257	60.40
^{71}Ga	70.9249	39.60

10. Calculate the atomic mass to four significant digits for antimony, given the following data:

Isotope	Exact Atomic Mass (amu)	Abundance in Nature (%)
^{121}Sb	120.9038	57.25
^{123}Sb	122.9041	42.75

Arrangement of Electrons in Principal Energy Levels   (See Section 4-6)

11. Calculate the maximum number of electrons that can exist in the following principal energy levels:
    (a)   1   (b)   2
    (c)   4   (d)   3
    (e)   6   (f)   7

12. (1) Diagram the atomic structure for each of the following atoms. Indicate the number of protons and neutrons, and arrange the electrons in principal energy levels (see 4-6). (2) Give the number of valence electrons for each atom.
    (a) $^{7}_{3}Li$ (b) $^{13}_{6}C$
    (c) $^{18}_{8}O$ (d) $^{23}_{11}Na$
    (e) $^{31}_{15}P$ (f) $^{36}_{18}Ar$

13. (1) Diagram the atomic structure for each of the following atoms. Indicate the number of protons and neutrons, and arrange the electrons in principal energy levels (see 4-6). (2) Give the number of valence electrons for each atom.
    (a) $^{11}_{4}Be$ (b) $^{15}_{7}N$
    (c) $^{19}_{9}F$ (d) $^{22}_{10}Ne$
    (e) $^{26}_{12}Mg$ (f) $^{32}_{16}S$

Electron-Dot Formulas of Elements   (See Section 4-7)

14. Write the electron-dot formulas for the following atoms:
    (a) $^{4}_{2}He$ (b) $^{7}_{3}Li$
    (c) $^{9}_{4}Be$ (d) $^{16}_{8}O$
    (e) $^{19}_{9}F$ (f) $^{38}_{18}Ar$

15. Write the electron-dot formulas for the following atoms:
    (a) $^{12}_{6}C$ (b) $^{22}_{10}Ne$
    (c) $^{23}_{11}Na$ (d) $^{24}_{12}Mg$
    (e) $^{31}_{15}P$ (f) $^{32}_{16}S$

Arrangement of the Electrons in Sublevels   (See Section 4-8)

16. (1) Write the electronic configuration in sublevels for the following atoms. (2) Give the number of valence electrons for each.
    (a) $^{7}_{3}Li$ (b) $^{9}_{4}Be$
    (c) $^{12}_{6}C$ (d) $^{16}_{8}O$
    (e) $^{31}_{15}P$ (f) $^{78}_{34}Se$

17. (1) Write the electronic configuration in sublevels for the following atoms.
    (2) Give the number of valence electrons for each.

    (a) $^{27}_{13}$Al

    (b) $^{35}_{17}$Cl

    (c) $^{81}_{35}$Br

    (d) $^{88}_{38}$Sr

    (e) $^{115}_{49}$In (Indium)

    (f) $^{158}_{64}$Gd (Gadolinium)

General Problems

18. The element osmium (Os) has the following physical properties:

$$mp = 3045°C, \quad bp = 5027°C, \quad density = 22.57 \text{ g/cm}^3$$

    (a) Calculate its melting point in °F.
    (b) Calculate its density in kg/m³.
    (c) Calculate its density in lb/ft³. (1 lb = 454 g; 1 in. = 2.54 cm.)

19. Write the electronic configuration in sublevels for an isotope of osmium (Os, atomic number = 76) having a mass number of 192.

Readings

Fermi, Laura, *The Story of Atomic Energy*. New York: Random House, Inc., 1961. A fascinating story of the history of atomic energy with personal anecdotes by the wife of Enrico Fermi, a pioneer in nuclear fission.

Gamow, G., *Mr. Tompkins Explores the Atom*. New York: The Macmillan Company, 1945. A simplified approach to the atom, some of which you should now be able to understand and enjoy.

——, *The Atom and Its Nucleus*. Englewood Cliffs, N.J.: Prentice-Hall, Inc., 1961. A more detailed discussion of the atom than *Mr. Tompkins Explores the Atom*. The two books by the late Professor Gamow are considered classics in chemical literature.

Holliday, Leslie, "Early Views on Forces Between Atoms," *Sci. Am.*, May 1970, v. 222, p. 116. Discusses the atomic theory of matter from the Greek philosophers to the scientists in the early part of the nineteenth century.

5

The Periodic Classification of the Elements

Section of a page of a manuscript of "Essay on the System of Elements" by Mendeleev dated February 17, 1869. In this preliminary version the periods were vertical and families were horizontal. Note the absence of the noble gases since they were unknown at that time. Elements were placed in order of atomic masses, but in cases of discrepencies similarities in chemical properties were used to place the elements (note Te and I). Note the various "question marks" entered.

OBJECTIVES

1. Given the following terms, define each term and give the distinguishing characteristic of each:
   (a) Periodic Law (Section 5-1)
   (b) periods (Section 5-2)
   (c) groups (Section 5-2)
   (d) representative elements (Section 5-2)
   (e) transition elements (Section 5-2)

2. Given the Periodic Table, point to the elements in a period and in a group (Section 5-2).

3. Given the Periodic Table and any element in the table, determine the following:
   (a) whether the element is a metal, nonmetal, or metalloid (Section 5-3, Problems 3 and 4)
   (b) the number of valence electrons for any A group elements and Group 0 (Section 5-3, Problems 5 and 6)

4. Given the electronic configurations of a number of elements, group the elements together according to those you would expect to show similar chemical properties (Section 5-3, Problems 7, 8, 9, and 10).

5. Given the Periodic Table, determine the trend in metallic and nonmetallic properties in the A group elements as you move down a given group (Section 5-3, Problems 11 and 12).

6. Given the Periodic Table, determine the trend in atomic radii as you move down a given group (Section 5-3, Problems 13 and 14).

In Chapter 3 (3-9), we mentioned the Periodic Table in connection with the separation of the metals from the nonmetals. Also, in Chapter 4 (4-8), we showed how the Periodic Table correlates to the filling of the sublevels. In this chapter, we shall consider the classification of the elements in the Periodic Table and some general characteristics of groups of elements.

5-1

The Periodic Law

As more elements were being discovered, chemists in the early 1800s attempted to *classify* the *elements* that had similar properties in groups or families, the way the various mammals may be classified. For example, one of the classifications of mammals is the cat family, whose characteristics are a round head, 28 to 30 teeth, eyes with vertically slit pupils, retractable claws, etc. The cat family includes not only domestic house cats but also lions, tigers, leopards, jaguars, and bobcats, to name just a few. All have the *same* general characteristics mentioned. Many of the elements also have general characteristics that can be used to classify them as belonging in a particular group or family.

In the nineteenth century, two chemists working independently of each other classified the elements known at that time, and their classification is the basis of the present one. Lothar Meyer (1830–1895), a German chemist, in 1864 devised an incomplete periodic table and published it in a book; in 1869, he extended it to include a total of 56 elements. Also in 1869, a Russian chemist, Dmitri Mendeleev (1834–1906), presented a paper describing a periodic table. Mendeleev went further than Meyer in that he left gaps in his table and predicted that new elements would be discovered to fill them. He also predicted the properties of these yet undiscovered new elements—truly a bold undertaking in science. Mendeleev lived to see the discovery of some of the elements he predicted, with properties similar to those he forecast.

Since both Meyer's and Mendeleev's periodic tables were based on *increasing atomic* **masses,** several discrepancies occurred in their tables. After the discovery of the proton, Henry G. J. Moseley (1888–1915), a British physicist, studied and determined the nuclear charge on the atoms of the elements and concluded that elements should be arranged by *increasing atomic* **number.** Thus, he corrected the discrepancies of the periodic table.

The elements are arranged in order of **increasing atomic number,** and elements with similar chemical properties recur at definite intervals (see Figure 5-1). In Figure 5-1, you will note that all the elements with the same number

FIGURE 5-1

A abbreviated periodic classification of the elements, based on atomic number. Similar chemical properties recur at definite intervals. (The numbers in color represent the atomic numbers of the elements.

H							He
1							2
Li	Be	B	C	N	O	F	Ne
3	4	5	6	7	8	9	10
Na	Mg	Al	Si	P	S	Cl	Ar
11	12	13	14	15	16	17	18

and kind of valence electrons are located in the same vertical column. For example, Be and Mg, both have 2 valence electrons in an *s* sublevel (4-8). The noble gases (He, Ne, Ar; see 4-7) all appear in the same vertical column and all have 8 electrons in their highest energy level ("rule of eight"), except helium, with 2 (a completed first energy level). The basis of the Periodic Law is the classification of elements by increasing atomic number. Therefore, the **Pe-**

riodic Law states that the chemical properties of the elements are periodic functions of their *atomic numbers*.

5-2

The Periodic Table. Periods and Groups

Following the Periodic Law and completing our abbreviated classification of the elements begun in Figure 5-1, we obtain a complete Periodic Table, as shown in Figure 5-2 and inside the front cover of this text. This Periodic Table, the one we now use, was first proposed in 1895 by Julius Thomsen (1826–1909), a Danish chemist.

The Periodic Table is arranged in 7 horizontal rows called **periods** or **series,** and 18 vertical columns called **groups** or **families.** The elements from left to right in a given period vary gradually from very metallic properties, such as sodium (Na), to nonmetallic properties, such as chlorine (Cl). At the end of each period is group 0, the noble gases, which are relatively inert (unreactive). The elements in a given group resemble each other in that they have similar chemical properties.

Now, let us consider in detail each of the seven periods (horizontal rows). Follow this discussion by studying the Periodic Table (Figure 5-2). Follow the addition of electrons by studying Figure 4-4 in Chapter 4.

Period 1 contains only two elements—hydrogen (H) and helium (He). In this period, the first principal energy level is being filled ($1s$ sublevel). The first energy level is filled with two electrons and helium is placed in group 0, the noble gases. The number of the period gives the principal energy level number that the electrons *begin* to fill.

Period 2 contains eight elements from lithium (Li) to neon (Ne). In this period, the second principal energy level is being filled ($2s$ and $2p$ sublevels), resulting in a completely filled second energy level in neon.

Period 3 also contains eight elements—from sodium (Na) to argon (Ar), with the third principal energy level being filled ($3s$ and $3p$ sublevels *only*). Argon, the last element in the period, has eight electrons in its third energy level. Periods 2 and 3, since they contain only eight elements each, are called the *short periods*.

Period 4 contains 18 elements—from potassium (K) to krypton (Kr). In this period, the $4s$ and $4p$ energy levels are filling and the $3d$ sublevel is being filled from scandium (Sc) to zinc (Zn).

Period 5 contains 18 elements—from rubidium (Rb) to xenon (Xe). In this period, the $5s$ and $5p$ energy levels are being filled and the $4d$ sublevel is being filled from yttrium (Y) to cadmium (Cd).

Period 6 consists of 32 elements—from cesium (Cs) to radon (Rn). In this period, the $6s$ and $6p$ energy levels are being filled. At the same time, the $5d$ and $4f$ sublevels are also being filled. Elements 58 to 71 , cerium (Ce) to lutetium (Lu), are called the *lanthanides* (like lanthanum) and correspond to the filling of the $4f$ sublevel. These elements are placed at the bottom of the

PERIODS	IA	IIA	IIIB	IVB	VB	VIB	VIIB	VIII			IB	IIB	IIIA	IVA	VA	VIA	VIIA	0
1	H 1																	He 2
2	Li 3	Be 4											B 5	C 6	N 7	O 8	F 9	Ne 10
3	Na 11	Mg 12											Al 13	Si 14	P 15	S 16	Cl 17	Ar 18
4	K 19	Ca 20	Sc 21	Ti 22	V 23	Cr 24	Mn 25	Fe 26	Co 27	Ni 28	Cu 29	Zn 30	Ga 31	Ge 32	As 33	Se 34	Br 35	Kr 36
5	Rb 37	Sr 38	Y 39	Zr 40	Nb 41	Mo 42	Tc 43	Ru 44	Rh 45	Pd 46	Ag 47	Cd 48	In 49	Sn 50	Sb 51	Te 52	I 53	Xe 54
6	Cs 55	Ba 56	*La 57	Hf 72	Ta 73	W 74	Re 76	Os 76	Ir 77	Pt 78	Au 79	Hg 80	Tl 81	Pb 82	Bi 83	Po 84	At 85	Rn 86
7	Fr 87	Ra 88	**Ac 89	[Rf] 104	[Ha] 105	106	107		109									

GROUPS

TRANSITION ELEMENTS

*Lanthanides	Ce 58	Pr 59	Nd 60	Pm 61	Sm 62	Eu 63	Gd 64	Tb 65	Dy 66	Ho 67	Er 68	Tm 69	Yb 70	Lu 71
**Actinides	Th 90	Pa 91	U 92	Np 93	Pu 94	Am 95	Cm 96	Bk 97	Cf 98	Es 99	Fm 100	Md 101	No 102	Lr 103

FIGURE 5-2

The Periodic Table of the Elements. (The numbers below the symbol of the elements represent the atomic numbers of the elements. The [ ] indicates that the element has not officially been approved or named.)

table for convenience, since if they were placed in the main body, the table would be extremely long and cumbersome.

Period 7 consists at present of 22 elements[1]—from francium (Fr) to the newly discovered element, atomic number 109. In this period, the 7s energy level is filled and the 6d and 5f sublevels are being filled. Elements **90 to 103,** thorium (Th) to lawrencium (Lr), are called the *actinides* (like actinium) and correspond to the filling of the 5f sublevel. Again, for convenience these elements are placed at the bottom of the table. This period is incomplete and could end with element 118, which would be one of the noble gases and which should have properties like radon (Rn). Periods 4, 5, 6 and 7 are called the *long periods* because they contain more elements than the other periods.

Most of the 16 groups or families (vertical columns) are classed as **A** or **B** groups. The **representative elements** consist of the **A** group elements and group **0** elements. In this text, we shall include all the **B** group elements and the group **VIII** elements (three vertical columns in this group) in the **transition elements.** Lanthanum (La) plus the lanthanides and actinium (Ac) plus the actinides are classed as transition elements in group IIIB. The gradual change from metallic to nonmetallic properties from left to right within a given period is more evident in the representative elements than in the transition elements. The transition elements are all metals and have one or two electrons in their outermost level. In addition, they also have valence electrons in the next lower d sublevel or the f sublevel which lies below that. In this respect they differ markedly from the representative elements which have all of their valence electrons in their outermost level. There are three transition series corresponding each to the filling of the 3d, 4d, and 5d sublevels and two series involving the filling of the 4f and 5f sublevels, respectively. Point to them in the Periodic Table. Check your answer by referring to Figure 4-4.

Since the groups or families have similar properties, they also have special names (remember the cat family in our classification of the mammals). Group IA elements (except hydrogen) are called the *alkali metals*. Hydrogen, although present in group IA, is not considered with the alkali metals, because not all its properties resemble those of the alkali metals. The elements in group IIA are called the *alkaline earth metals;* those in group VIIA are called the *halogens;* and those in group 0 are called the *noble gases.*

5-3

General Characteristics of the Groups

The use of the Periodic Table to correlate general characteristics of the elements is one of the fundamental principles of chemistry. There are five general characteristics of groups that we shall consider here.

1. The Periodic Table separates the **metals** from the **nonmetals,** as shown in Figure 5-2, with a solid **colored** stair step line. To the right of this line

---

[1]This includes the two new elements 107 and 109.

are the nonmetals and to the left are the metals, with the more metallic metals on the *extreme left*. As you can see, most of the elements are considered to be metals, and even some of the so-called nonmetals, such as silicon (Si), phosphorus (P), arsenic (As), and selenium (Se), have considerable metallic properties. Elements that lie on the **colored** stair step line are called **metalloids** (not aluminum). They have both metallic and nonmetallic properties. Examples are boron, silicon, germanium (Ge), arsenic, antimony, tellurium (Te), polonium (Po), and astatine (At), but not aluminum. The elements in group 0 consist of a special group of nonmetals, called the *noble gases*.

Work Problems 3 and 4.

2. In the **A group** elements, the number of *valence electrons* (see 4-8) is given by the *Roman group numeral*. For example, sodium is in group IA; hence, it has 1 valence electron $(1s^2, 2s^2 2p^6, 3s^1)$. Sulfur is in group VIA; hence, it has 6 valence electrons $(1s^2, 2s^2, 2p^6, 3s^2 3p^4)$. The number of valence electrons is 8 for all the elements in group 0, except helium which has only 2. This general characteristic does not hold for the transition elements (B group elements and group VIII elements) since they have *usually* 1 or 2 valence electrons, but the number of valence electrons for the transition elements varies considerably.

Work Problems 5 and 6.

3. Elements in the same group have *similar chemical properties* and *similar electronic configurations*. For example, all the alkali metals (group IA) react rapidly with chlorine to form the metal chloride, MCi (see 3-5). All members of the alkali metals have the same electronic configuration in the valence energy level, with the difference being the addition of principal energy levels.

Li $1s^2, 2s^1$

Na $1s^2, 2s^2 2p^6, 3s^1$

K $1s^2, 2s^2 2p^6, 3s^2 3p^6, 4s^1$

Rb $1s^2, 2s^2 2p^6, 3s^2 3p^6 3d^{10}, 4s^2 4p^6, 5s^1$

Cs $1s^2, 2s^2 2p^6, 3s^2 3p^6 3d^{10}, 4s^2 4p^6 4d^{10}, 5s^2 5p^6, 6s^1$

Fr $1s^2, 2s^2 2p^6, 3s^2 3p^6 3d^{10}, 4s^2 4p^6 4d^{10} 4f^{14},$
$5s^2 5p^6 5d^{10}, 6s^2 6p^6, 7s^1$

Since the electronic configurations of the elements in a group are similar, the formulas of compounds of elements in that group are also similar. Sodium hydroxide has the formula NaOH; hence, the formula for cesium (Cs) hydroxide is CsOH, because cesium is in the same group as is sodium. If there is any exception to this similarity of chemical properties in a given group, it is usually in the first element of the group. For example, lithium

Work Problems 7, 8, 9, and 10.

is not as similar to sodium in chemical properties as sodium is to potassium. Also, boron is not as similar to aluminum as aluminum is to gallium (Ga). In other words, if one of the elements in a group is "out of step," it is usually the first element in the group.

4. In the *A group* elements, the *metallic properties increase* within a given group with *increasing atomic numbers*, and the *nonmetallic properties decrease*. (See 3-9 for properties of metals and nonmetals.) In group VA, the first member of the group is nitrogen, considered to be a nonmetal; the last member of the group is bismuth, with very definite metallic properties. Since the more metallic metals are on the extreme left of the table, and the metallic properties increase with increasing atomic number in a given A group, the most metallic stable (nonradioactive) element would be found in the lower left-hand corner and would be cesium (Cs).[2] The most nonmetallic element (excluding the relatively unreactive group 0, the noble gases) would be found in the upper right-hand corner and would be fluorine.

Work Problems 11 and 12.

5. There is a somewhat *uniform gradation of many physical and most chemical properties* within a given group with increasing atomic number. In group VIIA elements, the halogens (see Table 5-1), the melting and boiling

TABLE 5-1 Some Physical Properties of the Halogens[a]

Element	Melting Point (°C)	Boiling Point (°C)[b]	Density (g/mL)[c]	Radius (Å)[d]
F	−219.6	−188.1	1.11 at bp	0.72
Cl	−101.0	−34.6	1.56 at bp	0.99
Br	−7.2	58.8	2.93 at bp	1.14
I	113.5	184.4	4.93 at 20°C	1.33

[a]Although astatine (At) is a halogen, it is not considered in this table because it is radioactive and so unstable that it is not found in nature. Hence, an insufficient amount of it is present at any one time to allow study of its properties in detail.

[b]At 1.00 atm pressure.

[c]All densities are for the liquid state, except iodine, which is given for the solid state.

[d]Determined by dividing the observed distance between centers of identical adjacent atoms.

points, the densities, and the radii of the elements increase as the atomic number increases. The increase in radii with an increase in atomic number within a given group is true for all the elements, since a new principal energy level is being added as you go down the group to the next period.

[2]Francium (Fr) is radioactive and is unstable, decomposing to other elements. It is not considered here because of its instability.

Work Problems 13
and 14.

Thus, the radius of the atom is increased, as shown in Figure 5-3. Regarding chemical reactivity of the halogens, fluorine is the most reactive, then chlorine, followed by bromine and iodine in that order.

FIGURE 5-3

Radii of group VIIA elements (except astatine). As the atomic number increases in a given group, the radii of the atoms increase.

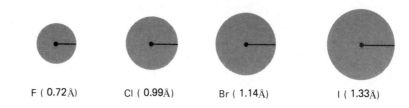

F (0.72Å)   Cl (0.99Å)   Br (1.14Å)   I (1.33Å)

Chapter Summary

The **Periodic Table** is based on the **Periodic Law,** which states that the chemical properties of the elements are periodic functions of their *atomic numbers*. The Periodic Table is arranged in 7 horizontal rows called *periods* and 18 vertical columns called *groups*. The *representative elements* consist of the *A* group elements and group *0* elements. The *transition elements* consist of the *B* group elements and the group *VIII* elements. Special names given to groups of elements are: group IA—the *alkali metals;* group IIA—the *alkaline earth* metals; group VIIA—the *halogens;* group 0—the *noble gases.*

Using the Periodic Table, general characteristics of the elements can be estimated. The **colored** stair step line separates the metals from the nonmetals, with the *metals* to the left and the *nonmetals* to the right. Elements that lie on this stair step line are called *metalloids*. The last column of nonmetals is called the *noble gases*. The number of *valence electrons* for the A group elements can be determined from the Periodic Table and is equal to the *Roman numeral*. For group 0 elements, the number of valence electrons is 8, except for helium which has only 2. Elements in the same group have similar chemical properties and similar electronic configurations. The metallic properties increase within a given group with increasing atomic number, and nonmetallic properties decrease. There is a uniform change of many physical and most chemical properties within a given group with increasing atomic number. Examples of these properties are boiling point, melting point, density, and radii of atoms.

EXERCISES

1. Define or explain the following terms:
   (a)   the Periodic Law
   (b)   periods
   (c)   groups
   (d)   lanthanides
   (e)   actinides
   (f)   short periods
   (g)   long periods
   (h)   representative elements
   (i)   transition elements
   (j)   alkali metals
   (k)   alkaline earth metals
   (l)   halogen elements
   (m)   noble gases

2. Distinguish between
   (a) a period and a group
   (b) lanthanides and actinides
   (c) short and long periods
   (d) representative and transition elements
   (e) alkali and alkaline earth metals

PROBLEMS

If in some of the following problems you are not familiar with the symbols for the elements, look them up inside the front cover of this text.

Metals, Nonmetals, or Metalloids   (See Section 5-3, number 1)

3. Using the Periodic Table, classify the following elements as metals, nonmetals, or metalloids:
   (a) rubidium   (b) iridium
   (c) tellurium   (d) selenium

4. Using the Periodic Table, classify the following elements as metals, nonmetals, or metalloids:
   (a) bromine   (b) thallium
   (c) germanium   (d) tin

Valence Electrons   (See Section 5-3, number 2)

5. Using the Periodic Table, indicate the number of valence electrons for the following elements:
   (a) cesium   (b) germanium
   (c) tellurium   (d) neon

6. Using the Periodic Table, indicate the number of valence electrons for the following elements:
   (a) krypton   (b) astatine
   (c) gallium   (d) arsenic

Electronic Configuration   (See Section 5-3, number 3)

7. Group the following electronic configurations of elements in pairs according to those you would expect to show similar chemical properties:
   (a) $1s^2, 2s^2 2p^6$, $3s^2 3p^6 3d^{10}$, $4s^2 4p^6 4d^{10}, 5s^2 5p^4$
   (b) $1s$, $2s^2 2p^6$, $3s^2 3p^6$, $4s^2$
   (c) $1s^2$, $2s^2 2p^6$, $3s^2\ 3p^4$
   (d) $1s^2$, $2s^2 2p^6$, $3s^2$

8. Using Appendix IV, determine the chemical symbols for the electronic configurations in Problem 7, and then check your answer by referring to the Periodic Table to see if you placed the elements of similar chemical properties in the same group.

9. Group the following electronic configurations of elements in pairs according to those you would expect to show similar chemical properties:
   (a) $1s^2$, $2s^22p^6$, $3s^23p^3$
   (b) $1s^2$, $2s^22p^6$, $3s^23p^63d^{10}$, $4s^24p^64d^{10}$, $5s^25p^6$, $6s^1$
   (c) $1s^2$, $2s^22p^6$, $3s^23p^6$, $4s^1$
   (d) $1s^2$, $2s^22p^6$, $3s^23p^63d^{10}$, $4s^24p^64d^{10}4f^{14}$, $5s^25p^65d^{10}$, $6s^26p^3$

10. Using Appendix IV, determine the chemical symbols for the electronic configurations in Problem 9, and then check your answer by referring to the Periodic Table to see if you placed the elements of similar chemical properties in the same group.

Metallic Properties   (See Section 5-3, number 4)

11. Using the Periodic Table, indicate which one of the following pairs of elements is the most metallic:
    (a) phosphorus and arsenic
    (b) cesium and sodium
    (c) silicon and aluminum
    (d) lead and germanium

12. Using the Periodic Table, indicate which one of the following pairs of elements is the most metallic:
    (a) barium and calcium
    (b) magnesium and phosphorus
    (c) silicon and lead
    (d) oxygen and polonium

Physical and Chemical Properties   (See Section 5-3, number 5)

13. Using the Periodic Table, indicate which one of the following pairs of elements has the greater atomic radius:
    (a) fluorine and chlorine
    (b) sulfur and oxygen
    (c) barium and magnesium
    (d) copper and silver

14. Using the Periodic Table, indicate which one of the following pairs of elements has the greater atomic radius:
    (a) nitrogen and phosphorus
    (b) lead and tin
    (c) barium and strontium
    (d) zinc and mercury

General Problem

15. Prior to the discovery of germanium in 1886, Mendeleev predicted in 1869 the properties of this element. Using the Periodic Table, determine the following for germanium (atomic number 32):
    (a) Would this element be classified as a metal, a nonmetal, or a metalloid?

    (b)   How many valence electrons would it have?

    (c)   Write the electronic configuration in sublevels for both germanium and its group precursor, silicon.

    (d)   Would it be more metallic or nonmetallic than its precursor, silicon?

    (e)   Mendeleev predicted a density for what he called "eka-silicon," now called germanium, of 5.5 g/mL. The actual density for germanium was found to be 5.3 g/mL. Convert the predicted and actual values into SI units. (*Hint:* See 2-13)

Readings

Sanderson, R. T., *Chemical Periodicity*. New York: Reinhold Publishing Corp., 1960. A comprehensive discussion of the Periodic Table, including trends in the properties of elements and compounds.

Sisler, Harry H., *Electronic Structure, Properties, and the Periodic Law*. New York: Reinhold Publishing Corp., 1963. An abbreviated discussion of the electronic structure of the atom, bonding, the Periodic Law, and trends in the properties of elements and compounds using the Periodic Table. Includes a discussion of the compounds formed with the noble gases.

Ternstrom Torolf, "A Periodic Table." *J. Chem. Educ.*, 1964, v. 41, p. 190. A slightly different Periodic Table. Editor's note proposes a clockwise spiral Periodic Table.

The Structure
of Compounds

A time exposure photograph of a three-dimensional model of a water molecule. Courtesy of Disneyland. © Walt Disney Productions.

TASKS

1. Learn the rules for calculating oxidation numbers (Section 6-1).

2. Memorize the seven elements that exist as diatomic molecules (Section 6-4).

3. Begin memorizing the names and formulas of ions in Table 6-1 (cations), Table 6-2 (anions), and Table 6-4 (polyatomic ions), unless instructed otherwise by your instructor. You must know these for Chapter 7, Chemical Nomenclature of Inorganic Compounds.

4. Memorize the partial order of electronegativity, unless instructed otherwise by your instructor (Section 6-4).

OBJECTIVES

1. Given the following terms, define each term and give the distinguishing characteristic of each:
   (a) valence (Section 6-1)
   (b) oxidation number (Section 6-1)
   (c) ions (Section 6-1)
   (d) cations (Section 6-1)
   (e) anions (Section 6-1)
   (f) electrovalent or ionic bond (Section 6-3)
   (g) covalent bond (Section 6-4)
   (h) bond length (Section 6-4)
   (i) coordinate covalent bond (Section 6-5)
   (j) structural formula (Section 6-6)
   (k) bond angle (Section 6-6)
   (l) polyatomic ion (Section 6-6)

2. Given the rules for calculating oxidation numbers, calculate the oxidation numbers for any element in any compound or any ion (Problem Examples 6-1, 6-2, 6-3, and 6-4, Problems 4 and 5).

3. Given the radius of an atom of an element and an ion, explain the change in size (Section 6-3, Problems 6 and 7).

4. Given the atomic number and mass number of an ion of an element of atomic number 1 to 18, diagram the atomic structure of the ion by writing the number of protons and neutrons and arranging the electrons in principal energy levels (Section 6-3, Problem 8).

5. Given the atomic number of an ion of any element, write the electronic configuration in sublevels for the ion (Section 6-3, Problem 9).

6. Given the energy evolved in bond formation or the energy required to break a bond, calculate the energy required to break a bond or evolved in bond formation (Section 6-4, Problem 10).

7. Given the formula of a compound containing two nonmetals and the order of electronegativity, mark above the atom the one that is the most positive with a $\delta^{(+)}$ and the one that is the most negative with a $\delta^{(-)}$ (Section 6-4, Problem 11).

8. Given the formula of a molecule or ion and the atomic number and symbols of each atom in the compound or ion, write the electron-dot and structural formula for the molecule or ion (Section 6-6, Problems 12 and 13).

9. Given the formulas of the ions, write the correct formula for a compound containing these ions (Section 6-7, Problems 14 and 15).

10. Given the Periodic Table, determine the maximum positive oxidation number of any element and the maximum negative oxidation number of any nonmetal (Section 6-8, Problem 16).

11. Given the Periodic Table and any A group elements, determine the correct formula of a binary compound (Section 6-8, Problem 17).

12. Given the Periodic Table and numerical values of various properties of some elements in a group, predict the value of the comparable property of another element in the same group (Section 6-8, Problem 18).

13. Given the Periodic Table and the formula of a compound, predict the formula of a second compound containing an element that differs from, but is in the same group as, the element in the first compound (Section 6-8, Problem 19).

14. Given the Periodic Table and the formula of a compound, predict the type of bonding in binary and ternary compounds (Section 6-8, Problem 20).

In Chapter 4, we considered the structure of the atoms of the elements. In this chapter, we shall consider how these atoms are put together to form compounds. We will also examine again the Periodic Table and use it to make predictions about elements.

6-1

Valence and Oxidation Numbers. Calculating Oxidation Numbers

Before we can consider the structure of compounds, we must know the meanings of the fundamental terms "valence" and "oxidation number."

Valence is a whole number used to describe the *combining capacity* of an element in a compound. Since a hydrogen atom never holds in combination more than *one* atom of another element in a binary compound (compounds containing only two different elements), the valence of hydrogen is arbitrarily assigned the value of 1. The valence of other elements compared with the value of 1 for hydrogen are 1, 2, 3, 4, etc., depending upon the number of hydrogen atoms the other atom could hold in combination. In hydrogen chloride, the elements are combined in a ratio of one atom of hydrogen to one of chlorine. The valence of chlorine in hydrogen chloride is therefore 1. In water, 2 atoms of hydrogen are combined with one atom of oxygen, and the valence of oxygen is 2. By relating the valences of certain elements, which have previously been compared with hydrogen, to other elements, we can determine the valences of all the elements. The formulas for the compounds of some metals with chlorine are $NaCl$, $MgCl_2$, $AlCl_3$, and $SiCl_4$. Since chlorine has a valence of 1, as shown, the valences of Na, Mg, Al, and Si are 1, 2, 3, and 4, respectively.

Some of the elements have only one valence, or a *fixed valence*, as 1, 2, or 3, etc., whereas a large number of them have more than one valence, or *variable valence*, as 1 *and* 2, or 2 *and* 3. We have previously encountered some of these variable valence compounds (4-2) in CO (carbon monoxide) and CO_2 (carbon dioxide). In CO, the carbon has a valence of 2, since oxygen has a valence of 2, as previously determined from the composition of water; and in CO_2, the carbon has a valence of 4, since 2 atoms of oxygen are combined with one of carbon. Another example of variable valence is found in the oxides of nitrogen; in N_2O, NO, N_2O_3, $N_2O_4(NO_2)$, and N_2O_5, the respective valences of nitrogen are 1, 2, 3, 4, and 5.

To define more precisely the *valence* as either *positive* or *negative* for the atoms in a compound, we use the term "oxidation number." **Oxidation number** (ox no) is a *positive or negative whole number* used to describe the combining capacity of an element in a compound.[1] The oxidation number is an arbitrary assignment based on certain rules (see below). The actual charge on an ion is called the **ionic charge**. The ionic charge of an element in the combined state implies the number of electrons lost (positive) or gained (negative) compared with the free state (zero oxidation number for the element). For ions consisting of a single atom, the oxidation number will be equal to the ionic charge. As a general rule, the *sum of the oxidation numbers of all the atoms in a*

---

[1] Fractional oxidation numbers of atoms in compounds do exist, but they are not too common for inorganic compounds. An example is $Na_2S_4O_6$ (sodium tetrathionate), where the sulfur atom has an average oxidation number of $2\frac{1}{2}^+$.

compound is zero. This principle applies to both electrovalent compounds (6-3) and covalent compounds (6-4). In the case of covalent compounds, each of certain key elements will be assigned a constant oxidation number.

When these positive or negative oxidation numbers of the atoms actually exist as charges, the charged particles are called **ions**. Ions with a positive charge are called **cations** (pronounced kat′ī·ons), and those having a negative charge are **anions** (pronounced an′ī·ons). In general, metals will have positive oxidation numbers, and *nonmetals* will have *negative* oxidation numbers when combined with metals. In compounds formed by the combination of two non-metals, one will be assigned a positive oxidation number, whereas the other will be negative. This will be determined by consideration of the relative electronegativities (6-4) of the two nonmetals.

The following are rules for assigning or determining oxidation numbers:

1. The algebraic sum of the oxidation numbers of all the atoms in the formula for a compound is *zero.*

2. The oxidation number of an element in the *free* or *uncombined state* is always *zero.*

3. The oxidation number of a monatomic ion (one atom ion) is considered the same as its ionic charge. The algebraic sum of the oxidation numbers of all the atoms in a polyatomic ion (many atoms ions) is equal to the oxidation number of the ion, which is the same as its ionic charge.

4. Negative oxidation numbers in compounds of two unlike atoms are assigned to the more electronegative atom (see Figure 6-9). For example, in hydrogen chloride (HCl) the oxidation number of hydrogen is 1^+, since chlorine is more electronegative than hydrogen (see Figures 6-10 and 6-11). In water (H_2O) the oxidation number of the hydrogen is 1^+ and that of oxygen is 2^-.

5. In most compounds containing hydrogen, the oxidation number of hydrogen is 1^+. The exceptions to this rule are the hydrides of metals, where hydrogen has an oxidation number of 1^- (NaH, LiH, CaH_2, AlH_3, etc.). Note that here the hydrogen atom is written second. In forming hydrides, hydrogen has acted as a nonmetal.

6. In most oxygen compounds, the oxidation number of oxygen is 2^-. The exceptions to this rule include peroxides, in which oxygen has an oxidation number of 1^- (Na_2O_2, H_2O_2, BaO_2, etc.).[2]

---

[2]Other exceptions do exist. Some of these exceptions are as follows: OF_2 in which oxygen has an oxidation number of 2^+, since fluorine is more electronegative than oxygen (see Figure 6-9); O_2F_2 in which oxygen has an oxidation number of 1^+; and the superoxides, such as KO_2 in which oxygen has an oxidation number of $\frac{1}{2}^-$. In this text the exceptions noted above will not be considered.

TABLE 6-1 Some Common Metals with the Formula of the Cations and Their Names

Metal (Symbol)	Cation	Name of Cation[a]
Aluminum (Al)	$*Al^{3+}$	Aluminum
Barium (Ba)	$*Ba^{2+}$	Barium
Bismuth (Bi)	Bi^{3+}	Bismuth
Cadmium (Cd)	$*Cd^{2+}$	Cadmium
Calcium (Ca)	$*Ca^{2+}$	Calcium
Copper (Cu)	Cu^{1+}	Copper(I) or cuprous
	Cu^{2+}	Copper(II) or cupric
Gold (Au)	Au^{3+}	Gold(III) or auric
Hydrogen[b] (H)	$*H^{1+}$	Hydrogen
Iron (Fe)	Fe^{2+}	Iron(II) or ferrous
	Fe^{3+}	Iron(III) or ferric
Lead (Pb)	Pb^{2+}	Lead(II) or plumbous
	Pb^{4+}	Lead(IV) or plumbic
Lithium (Li)	$*Li^{1+}$	Lithium
Magnesium (Mg)	$*Mg^{2+}$	Magnesium
Mercury (Hg)	Hg_2^{2+} [c]	Mercury(I) or mercurous
	Hg^{2+}	Mercury(II) or mercuric
Nickel (Ni)	Ni^{2+}	Nickel(II)
Potassium (K)	$*K^{1+}$	Potassium
Silver (Ag)	$*Ag^{1+}$	Silver
Sodium (Na)	$*Na^{1+}$	Sodium
Strontium (Sr)	$*Sr^{2+}$	Strontium
Tin (Sn)	Sn^{2+}	Tin(II) or stannous
	Sn^{4+}	Tin(IV) or stannic
Zinc (Zn)	$*Zn^{2+}$	Zinc

[a]The Roman numeral written in parentheses indicates the ionic charge for each atom of the ion. In the cations marked with an asterisk (*), the ionic charge can be determined using the Periodic Table; see 6-8. The ionic charge on all other cations must be memorized.

[b]Not a metal, but often reacts as a metal.

[c]Experimental evidence indicates that this ion exists as a dimer (two units) with an ionic charge of 1^+ on *each* atom $[Hg^{1+}]_2 = Hg_2^{2+}$.

Table 6-1 lists some common metals, their symbols, the symbols and charges of their cations, and the names of the cations. Table 6-2 lists common nonmetals, their symbols, the symbols and charges of their anions, and the names of the anions. You must learn these names, symbols, and ionic charges of both metals and nonmetals so you can use them to write formulas of compounds (6-7). To do this, we again suggest that you make flash cards (see Figure 3-4), placing the name of the ion on one side and the symbol with its ionic charge on the other.

TABLE 6-2   Some Common Nonmetals with the Formulas of the Anions and Their Names

Nonmetal (Symbol)	Anion[a]	Name of Anion
Bromine (Br)	Br^{1-}	Brom*ide*
Chlorine (Cl)	Cl^{1-}	Chlor*ide*
Fluorine (F)	F^{1-}	Fluor*ide*
Hydrogen (H)	H^{1-}	Hydr*ide*
Iodine (I)	I^{1-}	Iod*ide*
Nitrogen (N)	N^{3-}	Nitr*ide*
Oxygen (O)	O^{2-}	Ox*ide*
Phosphorus (P)	P^{3-}	Phosph*ide*
Sulfur (S)	S^{2-}	Sulf*ide*

[a]The ionic charge on all these anions, except hydride (H^{1-}) ion, can be determined by subtracting 8 from the Roman group number; see 6-8.

Consider the following problem examples:

Problem Example 6-1

Calculate the oxidation number of P in H_3PO_4.

SOLUTION: The oxidation numbers (ox no) of H and O in the compound are 1^+ and 2^- (see rules 5 and 6), respectively. The sum of the oxidation numbers of all the elements in the compound must equal zero. Therefore,[3]

$$3(+1) + \text{ox no of P} + 4(-2) = 0$$
$$+3 + \text{ox no of P} - 8 = 0$$
$$\text{ox no of P} - 5 = 0$$
$$\text{ox no of P} = +5 \text{ or } 5^+ \quad \textit{Answer}$$

---

[3]In solving for the oxidation numbers of elements, x may be substituted for "ox no of the element." The equation is then solved as a linear equation (see Appendix VII). Hence, Problem Example 6-1 could be solved as follows:

$$3(+1) + x + 4(-2) = 0$$
$$+3 + x - 8 = 0$$
$$x - 5 = 0$$
$$x = +5 \text{ or } 5^+ \quad \textit{Answer}$$

Problem Example 6-2

Calculate the oxidation number of Cr in H_2CrO_4.

SOLUTION: The oxidation numbers of H and O in the compound are 1^+ and 2^-. The sum of the oxidation numbers of all the elements in the compound must equal zero. Therefore,

$$2(+1) + \text{ox no Cr} + 4(-2) = 0$$
$$+2 + \text{ox no Cr} - 8 = 0$$
$$\text{ox no Cr} - 6 = 0$$
$$\text{ox no Cr} = +6 \text{ or } 6^+ \quad \textit{Answer}$$

Problem Example 6-3

Calculate the oxidation number of Cl in the $ClO_3{}^{1-}$ ion.

SOLUTION: The oxidation number of oxygen is 2^-, and the sum of the oxidation numbers of all the elements in the ion *must equal the charge on the ion or 1^-* (see rule 3). Therefore,

$$\text{ox no Cl} + 3(-2) = -1$$
$$\text{ox no Cl} - 6 = -1$$
$$\text{ox no Cl} = +6 - 1$$
$$\text{ox no Cl} = +5 \text{ or } 5^+ \quad \textit{Answer}$$

Problem Example 6-4

Calculate the oxidation number of S in the $S_2O_3{}^{2-}$ ion.

SOLUTION; The oxidation number of oxygen is 2^-, and the sum of the oxidation numbers of all the elements in the ion must equal the charge on the ion or 2^-. Therefore,

$$2 \, (\text{ox no S}) + 3(-2) = -2 \quad (\textit{Note: } 2 \text{ S atoms})$$
$$2 \, (\text{ox no S}) - 6 = -2$$
$$2 \, (\text{ox no S}) = +6 - 2 = +4$$
$$\text{ox no S} = +\frac{4}{2}$$
$$\text{ox no S} = +2 \text{ or } 2^+ \quad \textit{Answer}$$

Work problems 4 and 5.

6-2

Chemical Bonds

There are three general types of bonds between atoms in a compound: (1) electrovalent or ionic, (2) covalent, and (3) coordinate covalent. *These bonds are formed through use of the valence electrons of the atoms.* To understand the types of bonding, refer to the "rule of eight" (4-7). In the **"rule of eight,"** a stable configuration is achieved in many cases if 8 electrons are present in the valence energy level surrounding each atom by *gaining, losing, or sharing electrons.* One exception to the "rule of eight" is helium, whose first principal energy level is complete with just 2 electrons; *a completed first principal energy level is also a stable configuration*—the **"rule of two."** Therefore, in chemical bonding the valence electrons determine the bonding in a compound.

In general, atoms having 1, 2, or 3 valence electrons tend to *lose* these electrons to become positively charged ions (cations), as the metals; atoms with 5, 6, or 7 valence electrons may *gain* electrons to obtain *8* electrons in their highest energy level and become negatively charged ions (anions), as the nonmetals. These nonmetals may also share electrons to obtain a filled valence energy level; in such cases, the atom involved attains a positive oxidation number as high as 5^+, 6^+, or 7^+, and even 8^+. Those elements with 4 valence electrons tend to share their valence electrons in an attempt to obtain 8 electrons in their highest energy level. Therefore, the "rule of eight" is most important in chemical bonding.

We shall now consider the three general types of bonding—the electrovalent or ionic, the covalent, and the coordinate covalent bonds—and the compounds in which they are involved.

6-3

The Electrovalent or Ionic Bond

The **electrovalent** or **ionic bond** is formed by the *transfer* of one or more electrons from one atom to another. The bond formed between the two oppositely charged particles is based on the attraction of a positively charged particle for a negatively charged particle. *Unlike particles attract each other and like particles repel each other.* This results in a weak bond. But in the crystal, there are many such bonds holding each ion in place, therefore making the weak bond a *strong* bonding force. Compounds formed by the transfer of electrons from one atom to another atom are called *ionic compounds.*

Let us consider some examples of ionic compounds. Sodium chloride, NaCl, is formed when a sodium atom combines with a chlorine atom, as Figure 6-1 shows. In the sodium atom there is 1 valence electron, and in the chlorine atom there are 7 valence electrons. The 1 valence electron from sodium is lost to the chlorine atom, giving 8 electrons in the highest energy level of the sodium ion (**neon** noble gas configuration) and 8 electrons in the highest energy level of the chloride ion (**argon** noble gas configuration). The "rule of eight" is complete for both the sodium and chloride ions. The bond now

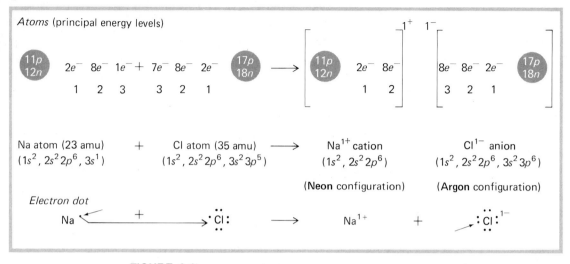

FIGURE 6-1

The formation of sodium chloride, NaCl, from a sodium atom and a chlorine atom is an example of a compound formed by electrovalent or ionic bonding. The positive sodium ion is attracted to the negative chloride ion.

formed between the positive sodium ion and the negative chloride ion is an *electrovalent* or *ionic* bond.

There are five important points to consider regarding the formation of all ionic compounds. **First,** the transfer of electrons can result in great changes in properties. For example, the sodium atoms and the chlorine atoms differ considerably from the sodium chloride (sodium ions and chloride ions). Sodium, composed of sodium atoms, is a soft metallic solid and can be cut with a knife, whereas chlorine, composed of chlorine molecules (Cl_2), is a greenish gas with a strong, irritating odor. Sodium chloride, common table salt, is a colorless crystalline solid. Sodium chloride is edible, but both sodium metal and chlorine gas are poisonous. Sodium reacts with water to give an explosive reaction, while sodium chloride dissolves in water. The transfer of an electron from one atom to another produces this drastic change in properties in the newly formed compound. (Table 6-3 lists some physical properties of sodium, chlorine, and sodium chloride.)

Second, the charge of the ion is related to the numbers of *protons* and *electrons* in the ion. In the sodium atom, there are 11 protons in the nucleus and 11 electrons about the nucleus; hence, the atom is neutral. There are still 11 nuclear protons in the ion but only 10 electrons, since one electron was lost to the chlorine atom. The result is a *net of 1 proton* or *1 positive charge* in excess, giving a charge or oxidation number on the sodium ion of 1^+. In the chlorine atom there are 17 nuclear protons and 17 orbital electrons; thus, the atom is neutral. After an electron is received from the sodium atom, there are 18 electrons and only 17 nuclear protons, resulting in a *net of 1 electron* or *1*

TABLE 6-3 Properties of Sodium, Chlorine, and Sodium Chloride

Element or Compound	Appearance at Room Temperature	Melting Point (°C)	Boiling Point (°C)[a]
Sodium	Soft, silvery, solid; cut with a knife	98	892
Chlorine	Greenish gas; strong irritating order	−101	−35
Sodium chloride	Colorless crystalline solid	801	1413

[a]At 1.00 atm pressure.

negative charge in excess and giving a charge or oxidation number on the chloride ion of 1^-. Therefore, the charges on the ions are directly related to their atomic structures.

Third, the radii of the ions differ from those of the atoms, as shown in Figure 6-2. The radius of the sodium atom is 1.57 Å, whereas the radius of the sodium ion is only 0.95 Å. This decrease in radius results from (1) the loss of an energy level, for the third principal energy level in the sodium atom has been lost in the transfer of the electron to the chlorine atom, and (2) a further decrease in size because of a greater nuclear attraction of the 11 positively charged protons on the remaining 10 electrons. The radius of the chlorine atom is 0.99 Å, whereas the radius of the chloride ion has increased to 1.81 Å. This increase in radius of the chloride ion over that of the chlorine atom is partly due to a smaller nuclear attraction (17 protons) on the 18 orbital electrons, causing an expansion of the radius of the energy level.

Fourth, energy is *given off in bond formation.* In the formation of 1.00 g of sodium chloride, 1.69 kcal or 7.06 kJ of energy is evolved. Therefore, to "break" these ionic bonds in 1.00 g of solid sodium chloride and to form the sodium and chlorine atoms, 1.69 kcal or 7.06 kJ of energy would be required. This energy is important and is often measured in laboratory experiments.

Fifth, the smallest unit of an *ionic compound* is called a **formula unit** (empirical formula unit), since it is a combination of *ions* and *not* discrete *molecules* (see 3-8). Hence, one formula unit of NaCl consists of *1* sodium ion and *1* chloride ion.

FIGURE 6-2

The radii of ions differ from those of the atoms as shown by a sodium atom and ion, and by a chlorine atom and ion.

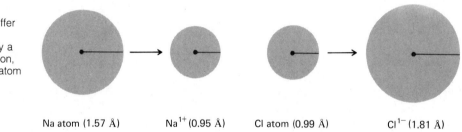

Na atom (1.57 Å) Na^{1+} (0.95 Å) Cl atom (0.99 Å) Cl^{1-} (1.81 Å)

The formula unit of sodium oxide—Na_2O—is formed when 2 atoms of sodium transfer an electron each to 1 atom of oxygen, as shown in Figure 6-3. For the oxygen atom to acquire 8 electrons in its highest energy level, it must gain 2 electrons, since it already has 6 valence electrons. This requires 2 sodium atoms, because each sodium atom has only 1 valence electron to donate. Therefore, each of the 2 sodium atoms *transfer* their valence electrons to 1 negative oxygen to form the formula units, Na_2O, consisting of 2 positive sodium ions and 1 oxide ion in a crystal of many interlaced ions. The loss of an electron from the sodium atom results in a 1^+ ionic charge for each sodium ion, and the gain of 2 electrons by the oxygen atom results in a 2^- ionic charge for the oxide ion.

Some properties of ionic compounds are as follows:

Work Problems 6, 7, 8, and 9.

1.   They have relatively high melting points (above 300°C).

2.   They conduct an electric current in the liquid state or in aqueous solution.

FIGURE 6-3

The formation of sodium oxide, Na_2O, from two sodium atoms and an oxygen atom is an example of a compound formed by electrovalent or ionic bonding. The positive sodium ions are attracted to the negative oxide ion.

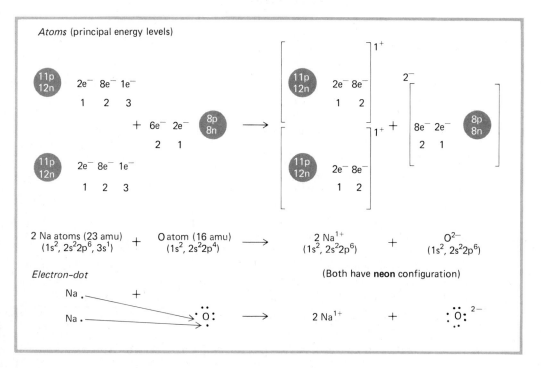

6-4

The Covalent Bond

The **covalent bond** is formed by the *sharing* of electrons between atoms. Compounds formed by the sharing of electrons are called *covalent compounds*. The smallest unit of a *covalent compound* is called a **molecule** (see 3-8); in an *ionic compound*, the smallest unit is a *formula unit* (see 6-3). The term "molecule" is used for compounds consisting of primarily covalent bonds, whereas "formula unit" is used for compounds consisting primarily of ionic bonds. A formula unit is *not* a molecule, since a formula unit does not really exist as a discrete entity but as ions. In 6-8, we shall show how to predict the type of predominant bonding in compounds by using the Periodic Table. Let us consider some examples of covalent compounds.

The hydrogen molecule, H_2, is a simple example of a *covalent* substance, as shown in Figure 6-4. The hydrogen atom as such is relatively unstable since

FIGURE 6-4

The formation of hydrogen, H_2, from two hydrogen atoms is an example of covalent bonding.

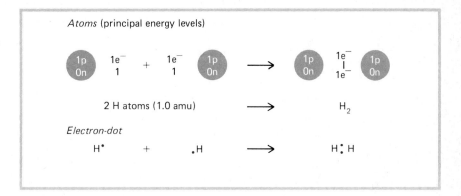

it has only one valence electron, but by sharing its valence electron with another hydrogen atom, it completes the first principal energy level and gives a stable configuration to the molecule. The molecular orbital representation of the H_2 molecule appears as a peanut shell, with the two $1s$ orbitals of the hydrogen atom pushed together or overlapping, as shown in Figure 6-5.

In the hydrogen molecule, as in all covalent substances, there are four important facts to remember. **First,** as with ionic compounds, the individual uncombined atoms differ markedly from the molecules. In fact, individual hydrogen atoms are so unstable that they exist for only a very short time. Thus, when we write the formula for *hydrogen* we must write it as H_2 (2 atoms of hydrogen—a diatomic molecule) and not as H.

Second, the two positive nuclei attract each of the two electrons to produce a molecule more stable than the separate atoms. This attraction by the nuclei for the two electrons counterbalances the repulsion of the two positive nuclei for each other; the greatest probability of finding the electrons is some-

FIGURE 6-5

Molecular orbital representation of a hydrogen, H_2, molecule. (The dot represents the nucleus of the atom.)

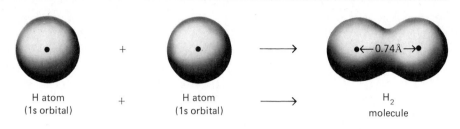

H atom (1s orbital) + H atom (1s orbital) ⟶ H_2 molecule

where *between* the two nuclei. A simple analogy suggested by the late Professor Henry Eyring of the University of Utah, and slightly modified here, may help to illustrate this point. Suppose we consider the nuclei of the two hydrogen atoms as "old potbellied stoves" and the two electrons as children running around each of these "stoves" trying to keep warm (see Figure 6-5). When two atoms come together, the children (electrons) now have two sources of heat (nuclei), and these children can now run between the "stoves" and keep all parts of their body, front and back, warm. Hence, the children (electrons) are now warmer and happier than they were when they had just one "stove" (nucleus), and a stable molecule results.

Third, the distance between the nuclei is such that the 1s orbitals of the hydrogen atoms have the maximum overlap, without having the nuclei so close to each other that they repel each other (causing the molecule to fly apart). In the hydrogen molecule, the distance between the nuclei is 0.74 Å, as shown in Figure 6-5. The distance between the nuclei of covalently bonded atoms is called the **bond length.**

Fourth, during the process of covalent bond formation, energy is evolved. In this case, 52.0 kcal or 2.18×10^5 J of energy is evolved in the formation of 1.0 g of gaseous hydrogen, H_2. Therefore, to "break" these covalent bonds in 1.0 g of gaseous hydrogen and to form the hydrogen atoms, 52.0 kcal or 2.18×10^5 J of energy would be required.

The molecule Cl_2 is formed when two atoms of chlorine share their electrons, as shown in Figure 6-6. For each chlorine atom to raise its third principal energy level to a total of 8 electrons, it must share one unpaired electron from each atom with the other atom. This sharing of electrons obeys the "rule of eight" for both atoms. Figures 6-7 and 6-8 depict the chlorine molecule. In Figure 6-8, the Prentice-Hall model shows the electrons between the two atoms—the shared pair, the electrons not involved in the bond—the *unshared* pairs; and the relative bond distances between the nuclei. The Stuart-Briegleb model shows the relative size of the atoms in the molecule.

H_2
Cl_2
F_2
Br_2
I_2
O_2
N_2

Besides H_2 and Cl_2, other elements exist as *diatomic* molecules (at room temperature); that is, they are not stable as single atoms. These molecules are F_2, Br_2, I_2, O_2, and N_2. Hence, when we write the *formulas of these elements, we do not write them as single atoms but as diatomic molecules.* On the Periodic Table point to the symbols of the elements that form these *seven* diatomic molecules. Notice that N_2, O_2, F_2, Cl_2, Br_2, and I_2 trace the numeral 7, with the top of the seven pointing in the direction of H_2!

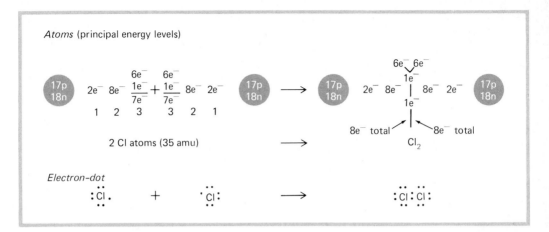

Atoms (principal energy levels)

2 Cl atoms (35 amu)

Electron-dot

FIGURE 6-6

The formation of chlorine, Cl_2, from two chlorine atoms is an example of covalent bonding. (The shared electrons are counted so that each atom achieves the "rule of eight".)

Cl atom
(*3p* orbital)

+

Cl atom
(*3p* orbital)

Molecular orbital

Cl_2
molecule

FIGURE 6-7

Molecular orbital representation of a chlorine, Cl_2, molecule. (The dots represent the nuclei of the chlorine atoms. Note the overlap of the two *p* orbitals to form the molecular orbital.)

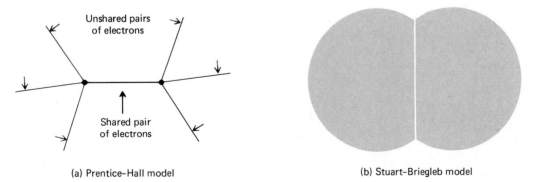

(a) Prentice–Hall model

(b) Stuart–Briegleb model

FIGURE 6-8

Molecular models of the chlorine molecule, Cl_2. (a) The Prentice-Hall model shows the electrons between the two atoms—the shared pair, the electrons not involved in the bond—the unshared pairs ($\downarrow$); and the relative bond distances between the nuclei. (b) The Stuart-Briegleb model shows the relative size of the atoms in the molecule.

F
O
Cl, N
Br
I, C, S
P, H
B
Si

In the preceding examples, the electrons have been shared *equally* by both atoms. This principle of equal sharing is not generally found in molecules that contain different atoms, because some atoms have a greater attraction for electrons than others. The tendency for an atom to attract a pair of electrons in a covalent bond is defined as **electronegativity**. Professor Linus C. Pauling (1-3) has developed a series of electronegativities for the elements. A partial series of decreasing electronegativities is F > O > Cl, N > Br > I, C, S > P, H > B > Si. The assigned values of the electronegativities of a number of the elements are given in the Periodic Table in Figure 6-9. Note that the *metals*

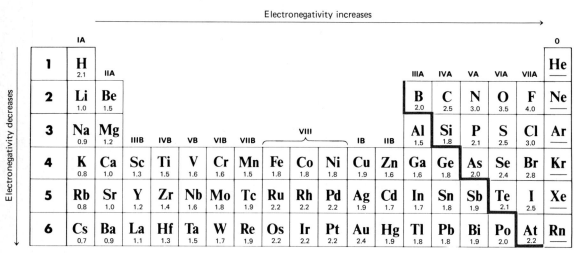

FIGURE 6-9

The electronegativities of a number of the elements.

have *low* electronegativities and the *nonmetals* have *high* electronegativities.

The following are reasons why some elements are more electronegative than others:

1. The smaller the radius of the atom, the greater the attraction for the outermost electrons. The smaller atom often has fewer energy levels, also, and consequently has a greater attraction for the bonding electrons than has a larger atom with more energy levels and hence less attraction. The nitrogen atom has a smaller radius than the carbon atom; thus, nitrogen has a greater attraction for its outermost electrons than does carbon, and hence a greater electronegativity (see Figure 6-9).

2. Atoms having fewer energy levels of electrons between the nucleus and the outermost energy level are more electronegative than those with intervening energy levels. The intervening levels of electrons *shield* the outer electrons from the full electrostatic effect of the positively charged nu-

cleus. This is called the *shielding* effect. For this reason, fluorine is more electronegative than chlorine and chlorine is more electronegative than bromine (see Figure 6-9). Compare the electronic structures of these atoms.

3. The more electrons in the unfilled valence energy level, the greater the attraction for electrons. Therefore, fluorine is more electronegative than oxygen (see Figure 6-9).

Let us consider an example of a molecule in which there is an *unequal* sharing of electrons in the covalent bond due to the difference in electronegativity of the atoms in the molecule. A typical example is hydrogen chloride gas, shown in Figure 6-10. The electronegativity of hydrogen is 2.1, whereas

FIGURE 6-10

The formation of hydrogen chloride, HCl, from one hydrogen atom and one chlorine atom is an example of an unequal sharing of electrons in a covalent bond.

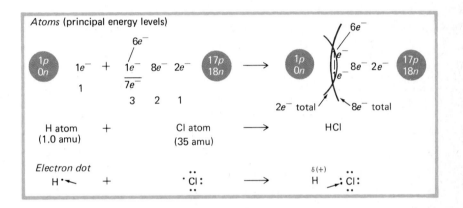

that of chlorine is 3.0 (see Figure 6-9). Hence, in the molecule of hydrogen chloride gas, the more electronegative chlorine would have a greater attraction for the pair of electrons in the covalent bond than would the hydrogen atom. The molecule would appear as shown in Figure 6-11. This unequal sharing of electrons in a covalent bond is often shown by placing a $\delta^{(-)}$ (lowercase Greek

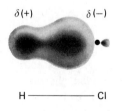

FIGURE 6-11

The hydrogen chloride molecule, showing the greater attraction for the electron pair in the covalent bond by the electronegative chlorine atom. Compare this unequal sharing of electrons with the equal sharing of electrons shown in Figure 6-5 with hydrogen. (The dots represent the nuclei of the atoms.)

letter delta, δ, meaning partially charged) above the relatively negative atom and a $\delta^{(+)}$ above the partially positive atom. Hydrogen chloride gas would be depicted as $\overset{\delta(+)}{H} : \overset{..}{\underset{..}{Cl}} \overset{\delta(-)}{}$. *Unequal sharing of electrons in a covalent bond occurs whenever the atoms differ in electronegativity.* The greater the differences in electronegativities, the more the unequal sharing of electrons in the covalent bond. This type of covalent bond is often referred to as a **polar covalent bond** or **polar bond.**

Unequal sharing of electrons in a covalent bond acts as a transition from equal sharing of electrons in covalent bonding to purely ionic or electrovalent bonding when the difference in electronegativities is great enough, as shown in Figure 6-12. Some compounds that we consider to be purely ionic have

Equal sharing in covalent bonding   Unequal sharing in covalent bonding   Ionic or electrovalent bonding

FIGURE 6-12

The transition from equal sharing in covalent bonding to ionic or electrovalent bonding is bridged by unequal sharing of electrons in a covalent bond. When the difference in electronegativities is sufficiently large, the more electronegative atom gains essentially full possession of the shared pair and ions result.

some covalent bonding. For example, cesium fluoride (CsF), a strongly ionic compound, is considered to have approximately 6 percent covalent bonding.

Covalent bonded compounds have different properties than ionic bonded compounds. Covalent compounds have relatively lower melting points (less than 300°C) and do not conduct an electric current when liquid or in aqueous solution as ionic compounds do.

Work Problems 10 and 11.

6-5

The Coordinate Covalent Bond

In a covalent bond, one electron was contributed by *each* atom to form an electron pair between the *two* atoms. In **coordinate covalent** bonding, also called *coordinate* or *dative* bonding, **both** the electrons of the electron-pair bond are supplied by **one** atom.

An example of a species containing a coordinate covalent bond is the ammonium ion (NH_4^{1+}). Ammonium ion is formed from a proton or hydrogen ion (H atom without an electron, $\left(\overset{1p}{\underset{0n}{}}\right)^{1+}$) and ammonia. First let us consider the formation of ammonia. The ammonia molecule (NH_3) is formed from 3 hydrogen atoms and 1 nitrogen atom, as shown by Figure 6-13. The nitrogen atom with 5 electrons in its second principal energy level shares 1 electron from each of the 3 hydrogen atoms to give a total of 8 electrons around the nitrogen atom. Each hydrogen atom with its 1 electron shares 1 electron with the nitrogen atom to give 2 electrons around the hydrogen atom, completing its first principal energy level. In the electron-dot diagram of Figure 6-13, the pair of electrons not involved in the bonding to the hydrogen atoms (no atom attached to this pair of electrons) is called an **unshared pair of electrons.** Figure 6-14 shows models of the ammonia molecule; the unshared pair of electrons is represented in the Prentice-Hall model in Figure 6-14a.

When a proton is added to ammonia, the proton becomes attached to this unshared pair of electrons to form a coordinate covalent bond, as shown in Figure 6-15.[4] The hydrogen now shares 2 electrons and is stabilized. The un-

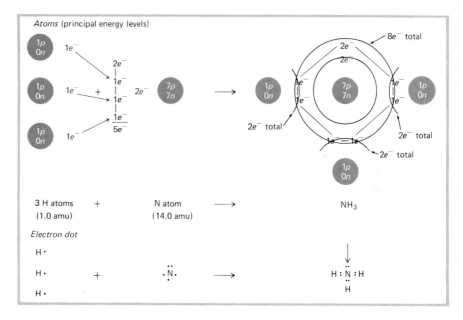

FIGURE 6-13

The formation of ammonia, NH_3, from 3 hydrogen atoms and 1 nitrogen atom. (Note in the electron-dot formula the unshared pair of electrons as shown by the arrow, ↓).

---

[4]The covalent bonds and the coordinate covalent bond in NH_4^{1+} are formed in a different manner, but once they are formed there is no difference between them. A coordinate covalent bond is identical to a covalent bond; only its history is different.

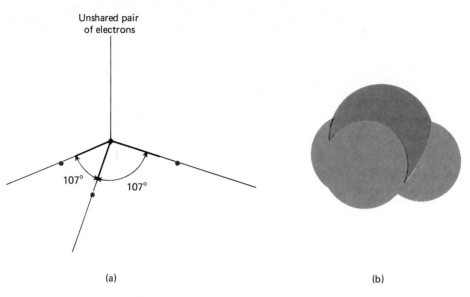

(a)

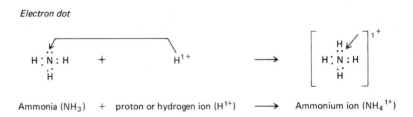

(b)

FIGURE 6-14
Molecular models of ammonia, NH_3. (a) Prentice-Hall model, (b) Stuart-Briegleb model.

shared pair of electrons acts like glue to form the coordinate covalent bond with the proton. The new *ion* that is formed, the ammonium ion—NH_4^{1+}, is charged, since the proton had a charge and the charge is dispersed over the *entire* ion.

Electron dot

Ammonia (NH_3) + proton or hydrogen ion (H^{1+}) $\longrightarrow$ Ammonium ion (NH_4^{1+})

FIGURE 6-15
The formation of an ammonium ion, NH_4^{1+}, from one molecule of ammonia, NH_3, and a proton or hydrogen ion, H^{1+}, is an example of coordinate covalent bonding. The coordinate covalent bond is shown by the arrow, and the positive ionic charge is dispersed over the entire ion. All four N—H bonds are equivalent once NH_4^{1+} is formed.

6-6

Electron-Dot and Structural Formulas of Molecules and Polyatomic Ions

In the discussion of bonding, we have used electron-dot formulas to depict the various types of bonding. All the elements in these formulas had a total of 8 electrons ("rule of eight") in their last energy levels, except the element hydrogen ("rule of two"). Hydrogen, because 2 electrons complete its first principal energy level, is satisfied with just 2 electrons, as is elemental helium. Hence, in writing electron-dot formulas of compounds, we shall follow the "rule of eight" or "octet rule" for all atoms except hydrogen. Compounds do exist in which the elements in these compounds do *not* obey the "rule of eight," but we shall not consider these compounds in this text.

Electron-dot formulas of molecules are of extreme importance in depicting the reactions of molecules to form new compounds.[5] Therefore, we should consider them in detail.

Before we consider some examples, we shall give you a few guidelines in writing electron-dot formulas of molecules.

1.   Write the electron-dot formulas for the elements that occur in the molecule. (Review 4-7).

2.   Arrange the atoms so that each atom obeys the "rule of eight," with hydrogen obeying the "rule of two."

3.   In molecules containing three different atoms, the "central atom" acts as the starting point with the other atoms arranged around this central atom.

Consider the following examples of electron-dot formulas:

1. H_2O

$$H\cdot \qquad \qquad \overset{..}{\underset{.}{\cdot O :}}$$

$$H\cdot$$

Electron-dot formula of elements. Valence electrons are 1 and 6 for H and O, respectively—see 4-7 and 5-3, number 2.

$$H : \overset{..}{\underset{..}{O}} :$$

$$H$$

Electron-dot formula of the molecule. Arranging the hydrogen atoms around the oxygen atom gives 8 electrons about the oxygen and 2 about each of the hydrogens. (All the bonding electrons about oxygen and hydrogen are equivalent, and although we use different colors to identify the electrons from the various atoms, all the bonding electrons in this molecule are equivalent.)

To simplify an electron-dot formula, draw a dash (—) for *each pair of electrons shared between atoms denoting a single covalent bond.* These formulas are called structural formulas. A **structural formula** is a formula showing the arrangement of atoms within the molecule, using a dash for each pair of electrons shared between atoms. Hence, the **structural formula** for water is as follows:

---

[5]Electron-dot formulas of molecules are also called *Lewis structures,* named after Professor Gilbert N. Lewis (1-3), who proposed the theory of covalent bonding in 1916.

The angle between the hydrogen atoms is called the bond angle. In water, this angle has been found to be 105°. A **bond angle** is an angle formed between 3 atoms in a molecule. The reason for the size of the bond angle will be covered in Chapter 13. The unshared pairs of electrons are usually not shown in a structural formula. Figure 6-16 shows models of the water molecule.

FIGURE 6-16

Molecular models of water, H_2O. (a) Prentice-Hall model, (b) Stuart-Briegleb model.

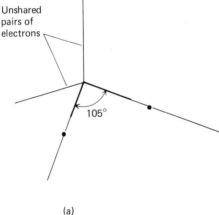

Unshared pairs of electrons

105°

(a)

(b)

2. CH_4, methane—natural gas

H·
H·
H·   ·C·
H·

Electron-dot formula of elements. Valence electrons are 4 and 1 for C and H, respectively, using the Periodic Table (see 5-3, number 2).

H
H:C:H
H

Electron-dot formula of the molecule. Arranging the hydrogen atoms around the carbon gives 8 electrons about the carbon and 2 about each of the hydrogens.

```
    H
    |
H—C—H
    |
    H
```

Structural formula. [Not planar, but *tetrahedral* (similar to a three-sided-base pyramid), with a bond angle of 109.5°.] Figure 6–17 shows models of the methane molecule.

FIGURE 6-17

Molecular models of
methane, CH_4. (a)
Prentice-Hall model,
(b) Stuart-Briegleb
model.

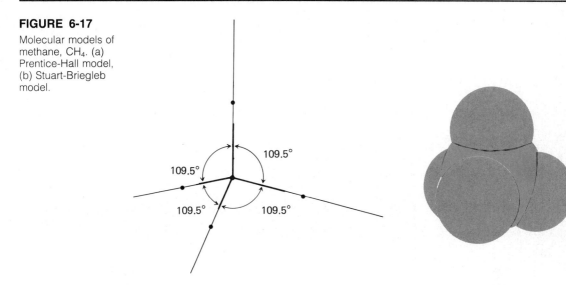

3. CO_2 carbon dixoide

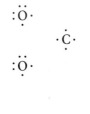

Electron-dot formula of elements.
Carbon must gain 4 electrons to
reach a filled valence energy level,
and each oxygen must gain 2 electrons to complete its valence energy
level.

:Ö: :C: :Ö:
 ̄   ̄   ̄
8   8   8

**Electron-dot formula of the mole-
cule.** If carbon accepts 2 electrons
from each oxygen (by a sharing process), and at the same time shares 2
of its valence electrons with each of
the 2 oxygen atoms, all the atoms in
the molecule will have achieved the
completed valence energy level of
"eight." This will, therefore, require the moving of the single oxygen and carbon electrons to form
pairs of electrons. In each bond between carbon and oxygen, 4 electrons are being shared, 2 having
been donated by the carbon and 2
by the oxygen. Such a bond of 4
electrons being shared is called a
double bond.

$$O{=}C{=}O$$

Structural formula. Draw a dash for each pair of shared electrons; hence, there are two dashes connecting the carbon atom to each of the oxygen atoms—*a double bond.*

4.   HCN, hydrogen cyanide[6]

H·   ·C·   ·N·

Electron-dot formulas of elements. There are 5 valence electrons for N.

H:C·   ·N·

H:C: : :N:

Electron-dot formulas of the molecule. Bonding the hydrogen with the carbon by a covalent bond gives 2 electrons about the hydrogen. To get a total of 8 electrons about the nitrogen, 3 more electrons are needed about the nitrogen. These must come from the carbon atom by covalent sharing. Thus, moving the remaining 3 electrons from carbon and 3 electrons from nitrogen between the carbon and nitrogen atoms gives 8 electrons each about carbon and nitrogen. There are now 6 electrons between the carbon and nitrogen atoms. Such an arrangement is called a *triple bond.*

H—C≡N

Structural formula. For each pair of electrons, draw a dash to the other atom; hence, three dashes connect the carbon and nitrogen atoms—*a triple bond.*

---

[6]Other electron-dot formulas can be drawn for hydrogen cyanide and for the sulfuric acid molecules, but based on the observed properties of hydrogen cyanide and sulfuric acid, the electron-dot formulas given here account for most of these properties.

5. H_2SO_4, sulfuric acid—a molecule

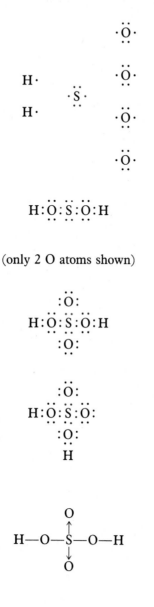

Electron-dot formulas of elements. There are 6 valence electrons for S. In general, *all oxygen atoms* are bound to a *"central"* atom, which in this case is sulfur, and this "central" atom should be our starting point.

H:Ö:S:Ö:H

(only 2 O atoms shown)

Electron-dot formula of the molecule. Bonding the 2 hydrogens to 2 of the more *electronegative* oxygens by covalent bonds and then bonding these oxygens to the central sulfur atom give 8 electrons about the oxygen and sulfur and 2 about the hydrogen. The two other oxygen atoms must also be accounted for. They can be placed on the sulfur atom by *coordinate covalent bonds* and still obey the "rule of eight" for both oxygen and sulfur. The oxygens are arranged about the sulfur in a *tetrahedral* configuration, and the structures at the left are projections of such a three-dimensional structure on a plane surface.

Structural formula. Draw the coordinate covalent bond with an arrow ($\rightarrow$) pointing to the atom that *did not* contribute electrons to form the bond. The O atoms are tetrahedrally located around the S atom as are the H atoms around the C atoms, as in CH_4 (see Figure 6-17).

The electron-dot formulas previously considered have all been for molecules. Following the general guidelines previously established for molecules, we can draw electron-dot formulas and structural formulas for ions containing more than one atom. For negative ions, depending upon the net negative charge on the ion, an excess of an electron or electrons must be present. Ions consisting of two or more atoms with a net negative or positive charge on the ion are called **polyatomic ions.** The charge on the polyatomic ion is called the *ionic charge* and is equal to the *oxidation number* of the *polyatomic ion.* The term "oxidation number" is also used to describe the oxidation state of *each* of the *atoms* comprising the polyatomic ion.

Consider the following electron-dot formulas of polyatomic ions:

1. OH^{1-}, hydroxide ion

<table>
<tr><td>H·   ·Ö·</td><td>**Electron-dot formulas of elements.**</td></tr>
<tr><td>⁻ₓÖ:H</td><td>**Electron-dot formulas of the polyatomic ion.** Arranging the hydrogen atom with the oxygen atom and adding 1 electron (x) for the 1^- ionic charge gives 8 electrons about oxygen and 2 about hydrogen.</td></tr>
<tr><td>O—H¹⁻</td><td>**Structural formula** with the ionic charge (1^-) by convention placed above and to the right of the ion.</td></tr>
</table>

2. SO_4^{2-}, sulfate ion

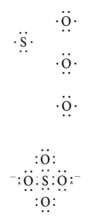

Electron-dot formulas of elements.

:Ö:
⁻ₓÖ:S:Ö:ₓ⁻
:Ö:

Electron-dot formula of the polyatomic ion. Arranging the oxygen atoms around the central sulfur atom as we did for H_2SO_4 and adding 2 electrons (x) for the 2^- ionic charge give 8 electrons about each of the oxygen atoms and the sulfur atom.

$$\left[\begin{array}{c} O \\ \uparrow \\ O - S - O \\ \downarrow \\ O \end{array} \right]^{2-}$$

Structural formula with the ionic charge (2^-) dispersed over the entire ion. Again, the O atoms are tetrahedrally located around the S atom.

3. NO_3^{1-}, nitrate ion

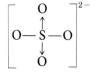

Electron-dot formulas for the elements.

$$\begin{array}{c} :\ddot{O}: \\ \ddot{N}:\ :\ddot{O} \\ :O: \\ \overset{x\cdot}{-} \end{array}$$

Electron-dot formula of the polyatomic ion. Arranging the oxygen atoms around the central nitrogen atom by forming one double bond and a coordinate covalent bond, and adding an electron (x) for the 1^- ionic charge, give 8 electrons about each of the oxygen atoms and the nitrogen atom.

$$\left[\begin{array}{c} O \\ \nwarrow \\ N = O \\ \swarrow \\ O \end{array} \right]^{1-}$$

Structural formula with the ionic charge (1^-) dispersed over the entire ion. The bond angle between the atoms in the nitrate ion is 120°.

The ammonium ion NH_4^{1+}, previously considered in (6-5), is another polyatomic ion. The various polyatomic ions are listed in Table 6-4. The first polyatomic ion in the table is acetate. The formula for the acetate ion is $C_2H_3O_2^{1-}$ and it is read "C-two, H-three, O-two one minus." You must learn the names and formulas of these polyatomic ions so that you can use them to write formulas of compounds. Again, we suggest that you make flash cards.

Work Problems 12 and 13.

TABLE 6-4 Some Common Polyatomic Ions and Their Formulas

$C_2H_3O_2^{1-}$	Acetate
NH_4^{1+}	Ammonium
CO_3^{2-}	Carbonate
ClO_3^{1-}	Chlorate
ClO_2^{1-}	Chlorite
CrO_4^{2-}	Chromate
CN^{1-}	Cyanide
$Cr_2O_7^{2-}$	Dichromate
HCO_3^{1-}	Hydrogen carbonate or bicarbonate
HSO_4^{1-}	Hydrogen sulfate or bisulfate
HSO_3^{1-}	Hydrogen sulfite or bisulfite
OH^{1-}	Hydroxide
ClO^{1-}	Hypochlorite
NO_3^{1-}	Nitrate
NO_2^{1-}	Nitrite
$C_2O_4^{2-}$	Oxalate
ClO_4^{1-}	Perchlorate
MnO_4^{1-}	Permanganate
PO_4^{3-}	Phosphate
SO_4^{2-}	Sulfate
SO_3^{2-}	Sulfite

6-7

Writing Formulas

We shall now use the names and formulas of the cations (Table 6-1), anions (Table 6-2), and polyatomic ions (Table 6-4) to write formulas of compounds. To write the correct formula for a compound, you *must know* or have given to you the ionic charges of the cations and anions. In writing these formulas, *the sum of the total positive charges must be equal to the sum of the total negative charges; that is, the compound must **not** possess a net charge.* When the charge on the positive ion in not equal to the charge on the negative ion, subscripts must be used to balance the positive charges with the negative charges. In most cases, the positive ion is written first, followed by the negative ion. For example, iron(II) bromide, consists of iron(II) ions, Fe^{2+}, and bromide ions, Br^{1-}. For the total positive charges to be equal to the total negative charges, we need to write one Fe^{2+} and two Br^{1-} as $Fe^{2+}Br^{1-}Br^{1-}$, using subscripts, $Fe^{2+}(Br^{1-})_2$. The $2+$ charge of the iron is just balanced by the $2-$ charge of the two bromides. To simplify the formula, delete the charges and the formula is written $FeBr_2$.

Let us now consider more examples to illustrate writing formulas. The names and formulas of the ions in Tables 6-1, 6-2, and 6-4 will be given, but by the time nomenclature is covered (Chapter 7), you should have learned these names and formulas.[7]

1. sodium (Na^{1+}) and chloride (Cl^{1-})

   (Na^{1+}) (Cl^{1-}), NaCl

   $1^+ + 1^- = 0$

2. barium (Ba^{2+}) and fluoride (F^{1-})

   (Ba^{2+}) (F^{1-})$_2$, BaF_2

   $2^+ + 2(1^-) = 0$

3. aluminum (Al^{3+}) and bromide (Br^{1-})

   (Al^{3+}) (Br^{1-})$_3$, $AlBr_3$

   $3^+ + 3(1^-) = 0$

4. ferric (Fe^{3+}) and sulfide (S^{2-})

   (Fe^{3+})$_2$(S^{2-})$_3$, Fe_2S_3

   $2(3^+) + 3(2^-) = 0$

Note: The least common multiple of 6—hence, 2 (3) and 3 (2).

5. cupric (Cu^{2+}) and nitrate (NO_3^{1-})

   (Cu^{2+}) (NO_3^{1-})$_2$, Cu (NO_3)$_2$

   $2^+ + 2(1^-) = 0$

Note: There are two nitrate ions; thus, the ( ) must be used. This ( ) means that there are 2 atoms of nitrogen, 6 atoms of oxygen, and 1 atom of copper in *one* formula unit of cupric nitrate.

6. lithium (Li^{1+}) and sulfate (SO_4^{2-})

   (Li^{1+})$_2$(SO_4^{2-}), Li_2SO_4

   $2(1^+) + 2^- = 0$

7. mercury (I) (Hg_2^{2+}) and acetate ($C_2H_3O_2^{1-}$)

   (Hg_2^{2+}) ($C_2H_3O_2^{1-}$)$_2$, $Hg_2(C_2H_3O_2)_2$

   $2^+ + 2(1^-) = 0$

8. ammonium (NH_4^{1+}) and sulfite (SO_3^{2-})

   (NH_4^{1+})$_2$(SO_3^{2-}), (NH_4)$_2SO_3$

   $2(1^+) + 2^- = 0$

9. strontium (Sr^{2+}) and phosphate (PO_4^{3-})

   (Sr^{2+})$_3$(PO_4^{3-})$_2$, $Sr_3(PO_4)_2$

   $3(2^+) + 2(3^-) = 0$

10. magnesium (Mg^{2+}) and hydrogen carbonate (HCO_3^{1-})

    (Mg^{2+}) (HCO_3^{1-})$_2$, $Mg(HCO_3)_2$

    $2^+ + 2(1^-) = 0$

---

[7]After nomenclature is mastered, you will be asked to write chemical equations (Chapter 9) using these formulas. Using these equations, you will then be asked to determine quantities used or obtained in a given chemical equation (Chapter 10). Therefore, in order to make these calculations and write chemical equations, you must write correct formulas. This means that you must memorize the formulas of the ions given in Tables 6-1, 6-2, and 6-4 (Task 3). Learn them now. Don't delay!

11. calcium (Ca^{2+}) and permanganate (MnO_4^{1-})

$(Ca^{2+})(MnO_4^{1-})_2$, $Ca(MnO_4)_2$

$$2^+ + 2(1^-) = 0$$

12. potassium (K^{1+}) and dichromate $(Cr_2O_7^{2-})$

$(K^{1+})_2(Cr_2O_7^{2-})$, $K_2Cr_2O_7$

$$2(1^+) + 2^- = 0$$

13. aluminum (Al^{3+}) and oxalate $(C_2O_4^{2-})$

$(Al^{3+})_2(C_2O_4^{2-})_3$, $Al_2(C_2O_4)_3$

$$2(3^+) + 3(2^-) = 0$$

Work Problems 14 and 15.

As you should recognize by now, a knowledge of the ionic charges of the cations and anions and their formulas is mandatory in writing the correct formulas for compounds.

6-8

The Use of the Periodic Table for Predicting Oxidation Numbers, Properties, Formulas, and Types of Bonding in Compounds

The Periodic Table can be most helpful to you in learning the ionic charges on the cations (Table 6-1) and anions (Table 6-2).

In general, the *Roman group numeral* represents the **maximum** *positive oxidation number* for the elements in that group.[8] For example, aluminum is in group IIIA and hence has a 3^+ oxidation number or ionic charge (see Table 6-1). For the nonmetals, the Roman numeral represents the maximum positive oxidation number. Also, for the nonmetals the *maximum negative oxidation number* can be calculated by *subtracting 8* from the Roman group number. For example, chlorine, in group VIIA, has a maximum positive oxidation number of 7^+ (group VII) in $KClO_4$ and a maximum negative oxidation number of 1^- (VII − 8 = −1) in KCl. See Table 6-2 for the ionic charge on the chloride ion. Sulfur, in group VIA, has a maximum positive oxidation number of 6^+ (group VI) in H_2SO_4, and a maximum negative oxidation number of 2^- (VI − 8 = −2) in H_2S. Review 6-1 for calculating oxidation numbers of elements. Using the maximum positive and negative oxidation numbers, we can also predict the formulas of some compounds containing two different elements (binary compounds). When barium and iodine form a binary compound, the formula is BaI_2. Barium is in group IIA and has a 2^+ oxidation number or ionic charge, while iodine is in group VIIA and has a 1^- negative oxidation number or ionic charge (VII − 8 = −1). The correct formula (see 6-7) is $(Ba^{2+})(I^{1-})_2$, BaI_2. This prediction of formulas applies primarily to the A group elements. The positive oxidation numbers or ionic charges on *some* of the cations given in Table 6-1 can be determined by using the Periodic Table. The negative oxidation numbers or ionic charges on all the anions given in Table 6-2, except the hydride ion (H^{1-}), can also be determined by using the Periodic Table.

Work Problems 16 and 17.

---

[8]The maximum positive oxidation number is not always the most common oxidation number. In all cases, the oxidation number is also the ionic charge on the monatomic ion.

We can now use the general characteristics outlined in 5-3 for predicting properties of elements, formulas of compounds, and types of bonding in compounds.

We mentioned that there was a somewhat uniform gradation of properties within a given group with increasing atomic number. As an example, let us consider the atomic radii of three elements in group VIA to determine if we can predict the radius of the fourth element in the group, tellurium (Te).

Element	Radius (Å)
O	0.74
S	1.04
Se	1.17
Te	?

The radii increase because of the addition of a new principal energy level; hence, we would expect the radius of tellurium also to increase. A prediction of the value of this radius can be made by taking the difference between the radii of sulfur and selenium and adding it to that of selenium. Hence, we would predict that the radius of tellurium would be 1.30 Å [1.17 + (1.17 − 1.04)]. It has been found to be 1.37 Å. This same general procedure can be applied to many of the properties of the elements with reasonable reliability.[9]

Another property of elements is that of ionization potential. The *ionization potential* of an atom or ion is the amount of energy required to remove the most loosely held electron from the atom or ion. Consider the energy needed to remove the most loosely bound electron from each of the three elements in group VIIA, and then determine if we can predict the ionization potential of the fourth element in the group, iodine (I).

Element	First Ionization Potential (kcal/mol)
F	402
Cl	299
Br	272
I	?

The ionization potential decreases as we move down the group, and we would expect the ionization potential of I to be less than Br. A prediction of this value is made by taking the difference between the ionization potentials of Cl

---

[9]A more accurate prediction can be obtained by graphing the values of the property of the elements vs. the atomic numbers of the elements.

Work Problem 18.

and Br and subtracting it from the value for Br. Hence, we would predict that the ionization potential for iodine (I) would be 245 kcal/mol [272 − (299 − 272)]. It has been found to be 245 kcal/mol.

In 5-3, we mentioned that since the electronic configuration of all elements in a group are similar, the formulas of compounds of elements in that group would be similar. Consider the following examples:

1. The formula for calcium bromide is $CaBr_2$; hence, the formula for radium (Ra) bromide is $RaBr_2$, because radium is in the same group (IIA) as calcium.

2. The formula for water is H_2O; thus, the formula for hydrogen telluride (*Te*) is H_2Te, because tellurium is in the same group (VIA) as oxygen.

3. The formula for magnesium sulfate is $MgSO_4$; and, the formula for the strontium selenate (*Sr* is strontium and **Se** is selenium) is $SrSeO_4$, because strontium is in the same group (IIA) as magnesium, and selenium is in the same group (VIA) as sulfur.

Work Problem 19.

In 6-4, we stated that the term "molecule" was reserved for compounds bonded primarily by covalent bonds and the term "formula unit" for compounds bonded primarily by ionic bonds. The greater the difference in electronegativities (6-4), the greater is the percent of ionic character in a compound. If the *ionic character is greater* than 50 *percent*, the compound is usually considered to be an *ionic compound;* hence, the smallest unit in this compound would be called a *formula unit* and not a molecule. In compounds consisting of only two *different* elements (binary compounds), the greater the difference in electronegativity of the elements, the greater is the ionic character of the compound. Figure 6-9 shows that the elements in a single group with a great degree of electronegativity are the halogens (group VIIA). Therefore, if the halogens combine with elements having relatively low electronegativities, an ionic compound is formed. The low electronegative elements from Figure 6-9 are found in the alkali metals (group IA, *except* hydrogen) and alkaline earth metals (group IIA). Therefore, we can make a general statement that if *binary compounds* are formed between elements in **group IA (except hydrogen)** or **group IIA** with elements in **group VIIA** or **group VIA (oxygen** and **sulfur** only), ionic compounds result. Since both fluorine and oxygen have high electronegativities, any compound formed with *fluorine* or *oxygen* and a **metal** is also classified as an ionic compound. Hence, the smallest unit in these ionic compounds would be a *formula unit*. Consider some examples:

1. Strontium chloride ($SrCl_2$) is an ionic compound, since strontium is in group IIA and chlorine is in group VIIA.

2. Potassium oxide (K_2O) is an ionic compound, since potassium is in group IA and oxygen is in group VIA, and since a compound of any *metal* with *oxygen* is considered ionic.

3. Iron(III) fluoride (FeF_3) is an ionic compound, since any compound formed with *fluorine* and a *metal* is considered ionic.

Other combinations in binary compounds are considered to be covalent with either equal or unequal sharing of electrons, and hence are referred to as molecules. Examples are carbon dioxide (CO_2—carbon is a nonmetal), sulfur dioxide (SO_2—sulfur is a nonmetal), water (H_2O), and methane (CH_4). As with all general statements, there are exceptions, but knowing that a compound is considered ionic if its ionic character is greater than 50 percent and that these binary compounds are formed from certain groups and elements in the Periodic Table will be helpful in further studying the properties of compounds.

The preceding general statements regarding the prediction of the type of bonding were applied only to *binary* compounds, but now let us consider *ternary and higher* compounds (three or more different elements), which involve polyatomic ions, as mentioned in 6-6. The combination of **any element** (*hydrogen* being the sole exception) with any **polyatomic ion** to form a ternary or higher compound results in an *ionic compound*, since the polyatomic *ion* can readily accommodate the positive or negative ionic charge over its many atoms; the smallest unit in these compounds would be a *formula unit* and not a molecule. Consider some examples:

1. Sodium sulfate (Na_2SO_4) is an ionic compound, since sulfate (SO_4^{2-}) is a polyatomic ion.

2. Silver nitrate ($AgNO_3$) is an ionic compound, since nitrate (NO_3^{1-}) is a polyatomic ion.

Work Problem 20.

3. Ammonium chlorate (NH_4ClO_3) is an ionic compound, since both ammonium (NH_4^{1+}) and chlorate (ClO_3^{1-}) are polyatomic ions.

Chapter Summary

In this chapter the calculation of the oxidation numbers of elements in compounds and polyatomic ions was discussed. The three general types of chemical bonds between atoms are: (1) *electrovalent* or *ionic* — formed by the transfer of one or more electrons from one atom to another, (2) *covalent*—formed by the sharing of electrons between atoms, and (3) *coordinate covalent*—formed when both electrons of the electron-pair shared between two atoms are supplied by one atom. Electron-dot and structural formulas of both molecules and polyatomic ions were written, obeying the "rules of two and eight." Correct formulas of compounds were written given the formulas of the ions. The Periodic Table was used to predict oxidation numbers of elements, properties of compounds and elements, formulas of compounds, and types of bonding found in compounds. Figure 6-18 is a Periodic Table that summarizes most of the generalizations of the elements discussed in this chapter and previous

PERIODIC TABLE OF THE ELEMENTS

	alkali metals	alkaline earth metals						GROUPS VIII										noble gases
maximum positive ox. no.	1+	2+	3+	4+	5+	6+	7+	8+		1+ [a]	2+	3+	4+	5+	6+	7+		
maximum negative ox. no.														3-	2-	1-		
PERIODS	IA	IIA	IIIA	IVB	VB	VIB	VIIB	VIII	IB	IIB	IIIA	IVA	VA	VIA	VIIA	0		
1	1.008 H 1 hydrogen																	4.003 He 2 helium
2	6.941 Li 3 lithium	9.012 Be 4 beryllium									10.811 B 5 boron	12.011 C 6 carbon	14.007 N 7 nitrogen	15.999 O 8 oxygen	18.998 F 9 fluorine	20.179 Ne 10 neon		
3	22.990 Na 11 sodium	24.305 Mg 12 magnesium									26.982 Al 13 aluminum	28.0855 Si 14 silicon	30.9738 P 15 phosphorus	32.06 S 16 sulfur	35.453 Cl 17 chlorine	39.948 Ar 18 argon		
4	39.0983 K 19 potassium	40.08 Ca 20 calcium	44.956 Sc 21 scandium	47.90 Ti 22 titanium	50.9415 V 23 vanadium	51.996 Cr 24 chromium	54.938 Mn 25 manganese	55.847 Fe 26 iron / 58.933 Co 27 cobalt / 58.71 Ni 28 nickel	63.546 Cu 29 copper	65.37 Zn 30 zinc	69.72 Ga 31 gallium	72.59 Ge 32 germanium	74.922 As 33 arsenic	78.96 Se 34 selenium	79.904 Br 35 bromine	83.80 Kr 36 krypton		
5	85.468 Rb 37 rubidium	87.62 Sr 38 strontium	88.906 Y 39 yttrium	91.22 Zr 40 zirconium	92.9064 Nb 41 niobium	95.94 Mo 42 molybdenum	98.906 Tc 43 technetium	101.07 Ru 44 ruthenium / 102.906 Rh 45 rhodium / 106.4 Pd 46 palladium	107.868 Ag 47 silver	112.41 Cd 48 cadmium	114.82 In 49 indium	118.69 Sn 50 tin	121.75 Sb 51 antimony	127.60 Te 52 tellurium	126.90 I 53 iodine	131.30 Xe 54 xenon		
6	132.906 Cs 55 cesium	137.33 Ba 56 barium	138.906 *La 57 lanthanum	178.49 Hf 72 hafnium	180.948 Ta 73 tantalum	183.85 W 74 tungsten	186.2 Re 75 rhenium	190.2 Os 76 osmium / 192.22 Ir 77 iridium / 195.09 Pt 78 platinum	196.967 Au 79 gold	200.59 Hg 80 mercury	204.37 Tl 81 thallium	207.2 Pb 82 lead	208.981 Bi 83 bismuth	(209) Po 84 polonium	(210) At 85 astatine	(222) Rn 86 radon		
7	(223) Fr 87 francium	226.025 Ra 88 radium	(227) **Ac 89 actinium	(261) [Rf] 104 rutherfordium	(262) [Ha] 105 hahnium	(263) [ ] 106	(262) [ ] 107	(266) [ ] 109										

*Lanthanides

140.12 Ce 58 cerium	140.908 Pr 59 praseodymium	144.24 Nd 60 neodymium	(145) Pm 61 promethium	150.4 Sm 62 samarium	151.96 Eu 63 europium	157.25 Gd 64 gadolinium	158.925 Tb 65 terbium	162.50 Dy 66 dysprosium	164.930 Ho 67 holmium	167.26 Er 68 erbium	168.934 Tm 69 thulium	173.04 Yb 70 ytterbium	174.967 Lu 71 lutetium

**Actinides

232.038 Th 90 thorium	231.031 Pa 91 protactinium	238.029 U 92 uranium	237.048 Np 93 neptunium	(244) Pu 94 plutonium	(243) Am 95 americium	(247) Cm 96 curium	(247) Bk 97 berkelium	(251) Cf 98 californium	(254) Es 99 einsteinium	(253) Fm 100 fermium	(256) Md 101 mendelevium	(253) No 102 nobelium	(257) Lr 103 lawrencium

| metals; | metalloids; | nonmetals; | noble gases |

TRANSITION ELEMENTS

Key:
1.008 — atomic mass
H — symbol
1 — atomic number
hydrogen — name

Numbers below the symbol of the element indicate the atomic numbers. Atomic masses, above the symbol of the element, are based on the assigned relative atomic mass of ^{12}C = exactly 12. ( ) indicates the mass number of the isotope with the longest half-life. [ ] indicates not officially approved or named.

[a]Certain elements in Group IB also form 2^+ and 3^+ oxidation numbers.

FIGURE 6-18

Periodic Table summarizing most of the generalizations of the elements discussed in this chapter and previous chapters.

chapters. You may use it as a summary, but you will not be permitted to use it on quizzes or exams.

EXERCISES

1. Define or explain the following terms:
   - (a) valence
   - (b) oxidation number
   - (c) ions
   - (d) cations
   - (e) anion
   - (f) electrovalent or ionic bond
   - (g) covalent bond
   - (h) coordinate covalent bond
   - (i) rule of eight
   - (j) ionic compounds
   - (k) formula unit
   - (l) bond length
   - (m) unshared pair of electrons
   - (n) diatomic molecules
   - (o) electronegativity
   - (p) unequal sharing of electrons in a covalent bond
   - (q) electron-dot formulas
   - (r) structural formula
   - (s) bond angle
   - (t) single bond
   - (u) double bond
   - (v) triple bond
   - (w) polyatomic ions
   - (x) ionic charge

2. Distinguish between
   - (a) valence and oxidation number
   - (b) cations and anions
   - (c) electrovalent or ionic bond and covalent bond
   - (d) covalent bond and coordinate covalent bond
   - (e) a formula unit and a molecule
   - (f) an unshared pair of electrons and an unpaired electron
   - (g) equal and unequal sharing of electrons in a covalent bond
   - (h) a single bond and a double bond
   - (i) a double bond and a triple bond
   - (j) a monatomic ion and a polyatomic ion

3. Explain the meaning of the following symbols:
   - (a) — in regard to bonding
   - (b) → in regard to bonding
   - (c) 2^+ in E^{2+}
   - (d) $\cdot\cdot$ in $-\overset{\cdot\cdot}{E}-$

PROBLEMS

Oxidation Numbers   (See Section 6-1)

4. Calculate the oxidation number for the element indicated in each of the following compounds or ions:
   - (a) Cl in HClO
   - (b) Mn in MnO_2
   - (c) N in HNO_2
   - (d) I in IO_4^{1-}
   - (e) S in HSO_3^{1-}
   - (f) Bi in BiO_3^{1-}

(g)   S in SO_4^{2-} (h)   As in AsO_4^{3-}

(i)   I in IO_2^{1-} (j)   P in $P_2O_7^{4-}$

5.   Calculate the oxidation number for the element indicated in each of the following compounds or ions:

(a)   Cl in $HClO_4$ (b)   Cl in Cl_2O

(c)   P in $H_4P_2O_7$ (d)   B in $H_2B_4O_7$

(e)   Mn in MnO_4^{1-} (f)   Sb in SbO_3^{3-}

(g)   Mn in MnO_4^{2-} (h)   Cr in CrO_4^{2-}

(i)   Cl in ClO^{1-} (j)   Cr in $Cr_2O_7^{2-}$

Electrovalent or Ionic Bond   (See Section 6-3)

6.   The radius of the Mg atom is 1.36 Å and that of Mg^{2+} is 0.65 Å. Explain this change in size.

7.   The radius of the O atom is 0.60 Å and that of O^{2-} is 1.40 Å. Explain this change in size.

8.   Diagram the ionic structure for each of the following ions, indicating the number of protons and neutrons in the nucleus and arranging the electrons in principal energy levels.

(a) $_1^1H^{1+}$ (b) $_3^7Li^{1+}$

(c) $_{12}^{24}Mg^{2+}$ (d) $_{11}^{23}Na^{1+}$

(e) $_{13}^{27}Al^{3+}$ (f) $_9^{19}F^{1-}$

(g) $_8^{16}O^{2-}$ (h) $_{16}^{32}S^{2-}$

(i) $_7^{14}N^{3-}$ (j) $_{15}^{31}P^{3-}$

9.   Write the electronic configuration in sublevels for the following ions (see 4-8):

(a) $_3^7Li^{1+}$ (b) $_4^9Be^{2+}$

(c) $_{11}^{23}Na^{1+}$ (d) $_{12}^{24}Mg^{2+}$

(e) $_{20}^{40}Ca^{2+}$ (f) $_9^{19}F^{1-}$

(g) $_{17}^{35}Cl^{1-}$ (h) $_8^{16}O^{2-}$

(i) $_{16}^{32}S^{2-}$ (j) $_{35}^{81}Br^{1-}$

Covalent Bond   (See Section 6-4)

10.   When Cl atoms unite to form a chlorine molecule (Cl_2), 3.42×10^3 J of energy is evolved in the formation of 1.00 g of gaseous chlorine. How many joules of energy would be required to break the Cl—Cl bond in 1.00 g of gaseous chlorine and to form the chlorine atoms?

11.   Place a $\delta^{(+)}$ above the atom or atoms that are relatively positive and a $\delta^{(-)}$ above the atom or atoms that are relatively negative in the following covalent bonded molecules (see Figure 6-9):

(a)   HF   (b)   HI

(c) H_2O (d)   BrCl

(e) BCl_3 (f) $SiCl_4$

(g) PCl_5 (h) NH_3

(i) OF_2 (j) Cl_2O

Electron-Dot and Structural Formulas of Molecules and Polyatomic Ions (See Section 6-6, review Sections 4-7 and 5-3)

12. Write the electron-dot and structural formulas for the following molecules or polyatomic ions:
    (a) HF   (b) H_2S
    (c) CCl_4 (d) CS_2
    (e) N_2 (f) C_2H_4
    (g) C_2H_2 (h) SH^{1-}
    (i) CN^{1-} (j) SO_3^{2-}

13. Write the electron-dot and structural formulas for the following molecules or polyatomic ions:
    (a) Cl_2 (b) PCl_3
    (c) $CHCl_3$ (d) Cl_2O
    (e) F_2 (f) H_2SO_3
    (g) H_2CO_3 (h) HNO_3
    (i) PO_4^{3-} (j) $P_2O_7^{4-}$
                               (*Hint:* $[O_3-P-O-P-O_3]^{4-}$)

Writing Formulas (See Section 6-7)

14. Write the correct formula for the compound formed by the combination of the following ions:
    (a) potassium (K^{1+}) and bromide (Br^{1-})
    (b) mercury(II) (Hg^{2+}) and iodide (I^{1-})
    (c) magnesium (Mg^{2+}) and nitride (N^{3-})
    (d) ferric (Fe^{3+}) and chloride (Cl^{1-})
    (e) cadmium (Cd^{2+}) and oxide (O^{2-})
    (f) calcium (Ca^{2+}) and phosphide (P^{3-})
    (g) lithium (Li^{1+}) and hydride (H^{1-})
    (h) barium (Ba^{2+}) and nitrate (NO_3^{1-})
    (i) aluminum (Al^{3+}) and perchlorate (ClO_4^{1-})
    (j) barium (Ba^{2+}) and permanganate (MnO_4^{1-})

15. Write the correct formula for the compound formed by the combination of the following ions:
    (a) silver (Ag^{1+}) and iodide (I^{1-})
    (b) strontium (Sr^{2+}) and oxide (O^{2-})
    (c) cuprous (Cu^{1+}) and bromide (Br^{1-})
    (d) stannous (Sn^{2+}) and hydrogen sulfite (HSO_3^{1-})
    (e) zinc (Zn^{2+}) and bicarbonate (HCO_3^{1-})
    (f) ferric (Fe^{3+}) and carbonate (CO_3^{2-})
    (g) ferrous (Fe^{2+}) and phosphate (PO_4^{3-})
    (h) aluminum (Al^{3+}) and phosphate (PO_4^{3-})
    (i) mercuric (Hg^{2+}) and cyanide (CN^{1-})
    (j) ammonium (NH_4^{1+}) and dichromate ($Cr_2O_7^{2-}$)

Using the Periodic Table, Oxidation Numbers (See Section 6-8)

16. Using the Periodic Table, indicate a maximum positive oxidation number for each of the following elements. For those elements that are nonmetals, give *both* the

maximum positive oxidation number and the maximum negative oxidation number. (If you do not know the symbol for the element, look it up inside the front cover of this text.)

(a) barium   (b) cesium
(c) sulfur   (d) chlorine
(e) aluminum   (f) selenium
(g) astatine   (h) nitrogen
(i) gallium   (j) osmium

17. Using the Periodic Table to determine the oxidation numbers, predict the formulas of the binary compounds formed from the following combinations of elements. (If you do not know the symbol for the element, look it up inside the front cover of this text.)

(a) calcium and sulfur   (b) cesium and phosphorus
(c) sodium and nitrogen   (d) strontium and selenium
(e) indium and oxygen   (f) magnesium and arsenic
(g) aluminum and sulfur   (h) barium and iodine
(i) thallium and sulfur   (j) sodium and tellurium

Using the Periodic Table, Predicting Properties   (See Section 6-8)

18. Predict the missing value for the following:

(a)
Element	Radius (Å)
K	2.02
Rb	2.16
Cs	?

(b)
Element	Density (g/mL)
Ca	1.54
Sr	2.60
Ba	?

(c)
Element	mp (°C)
Ca	845
Sr	?
Ba	$71\bar{0}$

(d)
Compound	Density (g/mL)
B_2S_3	1.55
Al_2S_3	2.37
Ga_2S_3	3.50
In_2S_3	?

Using the Periodic Table, Predicting Formulas   (See Section 6-8)

19. The following are some examples of compounds and their formulas:
sodium sulfate, Na_2SO_4
magnesium phosphate, $Mg_3(PO_4)_2$
aluminum oxide, Al_2O_3
Using the Periodic Table, write the formulas for the following compounds. (*Hint:* Note the endings of each name for the compound.)

(a) cesium sulfate (b) gallium oxide
(c) magnesium arsenate (d) aluminum sulfide
(e) sodium selenate (f) barium arsenate
(g) thallium(III) sulfide (h) indium selenide
(i) rubidium selenate (j) indium sulfide

Using the Periodic Table Predicting Bonding (See Section 6-8)

20. Using the Periodic Table, classify the following compounds as essentially ionic or covalent:

(a) MgS (b) Fe_2O_3
(c) N_2O_3 (d) BiF_3
(e) C_2H_2 (f) Na_2SO_4
(g) KI (h) P_4O_{10}
(i) CS_2 (j) $Ni(NO_3)_2$

General Problems

21. Consider the undiscovered element atomic number 114:
(a) In what group would it be placed?
(b) How many valence electrons would it have?
(c) Would it be more metallic or more nonmetallic than its predecessor in the same group?
(d) What element would it most likely resemble in properties?
(e) Suppose that element X, atomic number 114, forms the XO_3^{2-} ion. Write the electron-dot and structural formula for the XO_3^{2-} ion.

22. Consider the undiscovered element atomic number 119:
(a) In what group would it be placed?
(b) How many valence electrons would it have?
(c) What would be its oxidation number?
(d) What element would it most likely resemble in properties?
(e) Suppose that element Y, atomic number 119, reacts with bromine. What is the formula of this compound? Would the bonding in this compound be primarily ionic or covalent?

Readings

Pauling, Linus, *The Nature of the Chemical Bond*, 3rd ed. Ithaca, N.Y.: Cornell University Press, 1960, pp. 97–102. Describes the ionic character of bonds in relation to Pauling's electronegativities.

Sanderson, R. T., "The Nature of 'Ionic' Solids." *J. Chem. Educ.*, 1967, v. 44, p. 516. Discusses the fact that even the most ionic-type compound has some sort of covalent bonding.

7

Chemical Nomenclature
of Inorganic Compounds

Various common chemicals found in the home. Vinegar (dilute solution of acetic acid in water), sugar or sucrose, baking soda or sodium bicarbonate, and salt or sodium chloride.

TASKS

1. Memorize the names and formulas of ions in Table 6-1 (cations), Table 6-2 (anions), and Table 6-4 (polyatomic ions). You may use the Periodic Table unless instructed otherwise by your instructor.

2. Memorize the formulas of the common names of compounds given in Table 7-3, unless instructed otherwise by your instructor.

OBJECTIVES

1. Given the following terms, define each term and give a distinguishing characteristic of each:
    (a) acid (Section 7-6)
    (b) base (Section 7-6)
    (c) salt (Section 7-6)
    (d) acid salt (Section 7-6)
    (e) hydroxy salt (Section 7-6)
    (f) normal salt (Section 7-6)

2. Given the name, write the formula for the following compounds:
    (a) two nonmetals—binary compound (Section 7-2, Problem 6)
    (b) a metal ion with a fixed or one with variable oxidation number and a nonmetal ion—binary compound (Section 7-3, Problem 8)
    (c) a metal ion with a fixed or one with variable oxidation number and a polyatomic ion—ternary compound (Section 7-4, Problem 10)
    (d) a metal ion with a fixed or one with variable oxidation number and an oxy-halogen ion—special ternary compound (Section 7-5, Problem 12)
    (e) an acid, a base, and a salt (Section 7-6, Problem 14)

3. Given the formula, write the name for the following compounds:
    (a) two nonmetals—binary compound (Section 7-2, Problem 7)
    (b) a metal ion with a fixed or one with variable oxidation number and a nonmetal ion—binary compound (Section 7-3, Problem 9)
    (c) a metal ion with a fixed or one with variable oxidation number and a polyatomic ion—ternary compound (Section 7-4, Problem 11)
    (d) a metal ion with a fixed or one with variable oxidation number and an oxy-halogen ion—special ternary compound (Section 7-5, Problem 13)
    (e) an acid, a base, and a salt (Section 7-6, Problem 15)

4. Given the formula, identify various compounds as (1) an acid, (2) a base, (3) an acid salt, (4) a hydroxy salt, or (5) a normal salt (Section 7-6, Problems 16 and 17).

5. Given the common name for a compound, write the formula of the compound (Section 7-7, Problem 18).

One of the goals of a beginning chemistry course is to teach the student chemical nomenclature—to name compounds and, given a name, to write the formula of the compound. From this beginning, the student should start to learn the properties of compounds. We all know that table salt is sodium chloride ($NaCl$), but did you know that a Tums tablet (see 1-2) consists of calcium carbonate ($CaCO_3$), magnesium carbonate ($MgCO_3$), magnesium trisilicate ($Mg_2Si_3O_8$), as well as inert ingredients? Tums is an antacid (decreases acid in the stomach) because of these ingredients. Its action as an antacid can be illustrated by a chemical equation, as we shall see in Chapter 9.

In the present chapter, we shall apply the names and formulas of the cations (Table 6-1), anions (Table 6-2), and polyatomic ions (Table 6-4) that you learned in Chapter 6 to name compounds and to write formulas of the compounds.

There are two kinds of names in chemical nonmenclature: the **systematic chemical name** and the **common name.** The systematic chemical names are used most often, but there are a few compounds whose common names still persist, such as water (H_2O) and ammonia (NH_3). In this chapter, we first consider the systematic chemical names and then the nonsystematic common names.

7-1

Systematic Chemical Names

Systematic chemical names of inorganic compounds were developed by a group of chemists who were members of the Commission on the Nomenclature of Inorganic Chemistry of the International Union of Pure and Applied Chemistry (IUPAC), which first met in 1921. They developed rules for naming inorganic compounds and met periodically to revise and update this nomenclature.

The names of inorganic compounds are constructed so that every compound can be named from its formula and each formula has a name peculiar to that formula. The more *positive portion*, that is, the metal, the positive polyatomic ion, the hydrogen ion, or the less electronegative nonmetal, is named and written **first**. The more *negative portion*, that is, the more electronegative nonmetal or negative polyatomic ion, is named and written **last**. In this discussion, we shall divide the compounds into binary (two different elements), ternary and higher compounds (three or more different elements), special ternary compounds, and acids, bases, and salts.

7-2

Binary Compounds Containing Two Nonmetals

For all binary compounds, the ending of the second element is **-ide.** When both elements are nonmetals (see Periodic Table), the number of atoms of *each* element is indicated in the name with Greek prefixes, as shown in Table 7-1, except in the case of mono (one), which is rarely used. When no prefix appears, one atom is assumed.

Consider the following examples of naming binary compounds of nonmetals:

Formula	Name
PCl_3	Phosphorus *trichloride*
PCl_5	Phosphorus *pentachloride*
SO_2	Sulfur *dioxide*[1]
CO	Carbon *monoxide*[2,3](*mono* is used in this case)
N_2O_4	Dinitrogen *tetroxide*[2,4]
N_2O_5	Dinitrogen *pentoxide*[2]

---

[1]Sulfur dioxide is found in polluted air and is one of the air pollutants most dangerous to human beings. Small amounts of sulfur dioxide in the air appear to increase the rusting of iron products. This decrease in the life of the product requires replacement earlier than expected and, therefore, increases cost. Hence, pollution becomes an economic problem as well as a health problem.

[2]When two vowels appear next to each other, as "oo" in monooxide, or "ao" in tetraoxide, pentaoxide, and heptaoxide, the vowel from the Greek prefix is dropped for better pronunciation.

[3]Carbon monoxide, produced primarily from incomplete combustion of gasoline in automobiles, is one of the chief air pollutants.

[4]This compound was the oxidizer for the fuel in the two small rocket engines in the space shuttle *Columbia.* These engines placed *Columbia* in its final orbit and later allowed *Columbia* to leave its orbit and return to earth. The monomer (a single unit) of this compound, NO_2 (nitrogen dioxide), along with other nitrogen oxides is found in polluted air emitted in the exhausts of trucks and automobiles. These nitrogen oxides are primarily formed from the nitrogen and oxygen in the air as they are sucked through the hot cylinders of the engine. The brown tinge that sometimes appears in polluted air on hot days is probably caused by nitrogen dioxide. This compound may be more harmful to human beings than CO (carbon monoxide), since it is considerably more soluble in the blood than CO. In environmental readings the nitrogen oxide pollutants are referred to by the formula NO_x, where x equals 1 or 2.

TABLE 7-1   Greek Prefixes

Greek Prefix	Number
mono-	1
di-	2
tri-	3
tetra-	4
penta-	5
hexa-	6
hepta-	7
octa-	8
nona- (or ennea-)[a]	9
deca-	10

[a]Ennea- is preferred to the Latin nona- by the IUPAC, but nona- is still used.

Consider the following examples of writing the formulas for binary compounds of nonmetals:

Name	Formula
Boron trichloride	BCl_3
Carbon tetrachloride	CCl_4
Dichlorine monoxide	Cl_2O
Chlorine dioxide	ClO_2
Dichlorine heptoxide[2]	Cl_2O_7
Dinitrogen oxide	N_2O

Work Problems 6 and 7.

7-3

Binary Compounds Containing a Metal and a Nonmetal

Metals with Fixed Oxidation Numbers

In these compounds, we shall consider first only metals with fixed oxidation numbers (metals that show only one oxidation state, as 1^+, 2^+, 3^+, etc.). In the names of these compounds, the metal is named first, followed by the nonmetal with the ending **-ide,** as in all binary compounds. *No Greek prefixes are used.*

Consider the following examples for naming binary compounds of metals (with fixed oxidation numbers) and nonmetals:

Formula	Name (No Greek Prefixes)
KCl	Potassium chlor*ide*
Na_2S	Sodium sulf*ide*
AgBr	Silver brom*ide*
MgO	Magnesium ox*ide*
CaH_2	Calcium hydr*ide*

In writing the formulas of compounds, you must know the ionic charges of the metal cations and the nonmetal anions. Consider the following examples for writing the formulas of binary compounds of metals (with fixed oxidation numbers) and nonmetals:

Name	Formula
Lithium fluoride	LiF (Li is 1^+; F is 1^-; see Tables 6-1 and 6-2)
Strontium iodide	SrI_2 (Sr is 2^+; I is 1^-)
Cadmium phosphide	Cd_3P_2 (Cd is 2^+; P is 3^-)
Magnesium nitride	Mg_3N_2 (Mg is 2^+; N is 3^-)
Aluminum sulfide	Al_2S_3 (Al is 3^+; S is 2^-)

Metals with Variable Oxidation Numbers

In this group of binary compounds, the metal has a variable oxidation number (metals that show more than one oxidation number when combined, as 1^+ *and* 2^+, 2^+ *and* 3^+, 2^+ *and* 4^+, etc.). In the names of these compounds, the same procedure is followed as with metals having fixed numbers, except the oxidation number of the metal must be specified. There are two methods of specifying oxidation numbers: the newer Stock system[5] and the older *-ous* or *-ic* suffix system. In the **Stock system,** the oxidation number of the metal is indicated by a Roman numeral in parentheses immediately following the name of the metal. In the *-ous* or *-ic* suffix system, the Latin stem for the metal is used with *-ous* or *-ic* suffix, the *-ous* representing the lower oxidation number and the *-ic* the higher oxidation number. Both names were given in Table 6-1. For example,

Cation	Names of Cation	
	Stock System	-ous or -ic Suffix System
Cu^{1+}	Copper(I)	Cuprous
Cu^{2+}	Copper(II)	Cupric
Fe^{2+}	Iron(II)	Ferrous
Fe^{3+}	Iron(III)	Ferric

As you may have noticed from the preceding examples, the *-ous* or *-ic* suffix can become confusing, because in copper the *-ous* represents a 1^+ oxi-

---

[5]The IUPAC prefers the Stock system.

dation number, whereas in iron it is 2^+. However, the *-ous* or *ic* suffix system is still used; hence, you should become familar with both systems.

Consider the following examples for naming binary compounds consisting of metals with a variable oxidation number:

Formula	Name
$CuCl_2$	Copper(II) chlor*ide* or cu*pric* chlor*ide;* since chloride is 1^-, then copper must be 2^+
FeO	Iron(II) ox*ide* or ferr*ous* oxide; since oxide is 2^-, then iron must be 2^+
SnF_4	Tin(IV) fluor*ide* or stan*nic* fluor*ide;* since fluoride is 1^-, then tin must be 4^+
PbS	Lead(II) sulf*ide* or plumb*ous* sulf*ide;* since sulfide is 2^-, then lead must be 2^+
HgO	Mercury(II) ox*ide* or mercu*ric* ox*ide;* since oxide is 2^-, then mercury must be 2^+

Consider the following examples for writing the formulas of binary compounds of metals (variable oxidation number) and nonmetals:

Name	Formula
Cupric phosphide	Cu_3P_2 (Cu is 2^+; P is 3^-; see Tables 6-1 and 6-2)
Iron(III) oxide	Fe_2O_3 (Fe is 3^+; O is 2^-)
Lead(IV) oxide	PbO_2 (Pb is 4^+; O is 2^-)
Cuprous chloride	CuCl (Cu is 1^+; Cl is 1^-)
Stannous fluoride[6]	SnF_2 (Sn is 2^+; F is 1^-)

Work Problems 8 and 9.

7-4

Ternary and Higher Compounds

In naming and writing the formulas of ternary and higher compounds, we follow the same procedure we followed for binary compounds, except that we use the name or formula of the polyatomic ion. Hence, a knowledge of the names and formulas of all the polyatomic ions in Table 6-4 is required. Some of the negative polyatomic ions have suffixes of **-ate,** and **-ite.** The most observable difference in the formulas of the *-ate* and the *-ite* negative polyatomic

---

[6]This tooth decay-preventative ingredient is used in several toothpastes.

ions is that the **-ate** has *one more oxygen* atom than the **-ite**.[7] For example, the formula for sul*ite* is SO_3^{2-}, whereas that of sul*ate* is SO_4^{2-}. This is true for all the negative polyatomic ions listed in Table 6-4. In this table, there are three polyatomic ions that do not have an *-ate* or *-ite* ending: NH_4^{1+}, ammonium ion—the only positive polyatomic ion in this table; OH^{1-}, hydroxide—an *-ide* ending, the same as that found in binary compounds, further considered in 7-6; and CN^{1-}, cyanide—also has an *-ide* ending, the same as binary compounds.

For metals that have a variable oxidation number, either the Stock system or the *-ous* or *-ic* suffix system may be used, but the Stock system is preferred.

Consider the following examples of naming ternary and higher compounds, specifically noting the number of oxygen atoms and the name endings:

Formula	Name
$NaNO_3$	Sodium nit*ate*[8]
$NaNO_2$	Sodium nit*ite*[8]
$Cu_3(PO_4)_2$	Copper(II) phosph*ate* (since phosphate is 3^-, Cu is 2^+) Cup*ic* phosph*ate*
$CuCN$	Copper(I) cyan*ide* (since cyanide is 1^-, Cu is 1^+) Cup*ous* cyan*ide*
K_2CO_3	Potassium carbon*ate*
$Ca(HSO_4)_2$	Calcium hydrogen sul*ate* Calcium bisul*ate*
$(NH_4)_2SO_3$	Ammonium sul*ite*
$Ba(C_2H_3O_2)_2$	Barium acet*ate*
$Fe_2(CrO_4)_3$	Iron(III) chrom*ate* (since chomate is 2^-, Fe is 3^+) Fer*ic* chrom*ate*
$AgClO_3$	Silver chlor*ate*

---

[7]A more general method of describing the difference between the *-ate* and the *-ite* is by oxidation numbers of the nonmetal, other than oxygen. The higher oxidation number is the *-ate* and the lower oxidation number is the *-ite*. In the example of sulfate and sulfite, the oxidation numbers of sulfur are 6^+ and 4^+, respectively. See 6-1 for calculating oxidation numbers.

[8]Sodium nitrite and sodium nitrate ($NaNO_3$) are both used as color fixatives and food preservatives in various meat products, such as frankfurter, bologna, and poultry. Their value has been questioned. The nitrite ion (NO_2^{1-}) is believed to react with organic compounds in the body to produce new compounds that can produce cancer, but recent studies indicate that this evidence is inconclusive. The nitrate ion (NO_3^{1-}) is not that harmful, but it may be converted to the nitrite ion in the body. The nitrite ion is very effective in preventing the formation of deadly botulism. Sodium hypophosphite ($NaH_2PO_2 \cdot H_2O$) has been proposed to replace sodium nitrite as a suppressant of botulism formation in smoked meats. Sodium nitrate is used in poultry primarily for its color fixative property in that it adds a pink color to the meat and not as a preservative. Therefore, a committee of the National Research Council recommended that sodium nitrate be eliminated from all poultry and most meat products with the possible exceptions of fermented sausages and dry-cured meats.

Consider the following examples for writing the formulas of ternary or higher compounds, specifically noting the name endings and the number of oxygen atoms:

Name	Formula
Barium cyanide	$Ba(CN)_2$ (Ba is 2^+; CN is 1^-, see Tables 6-1 and 6-4)
Iron(II) phosphate	$Fe_3(PO_4)_2$ (Fe is 2^+; PO_4 is 3^-)
Ferric sulfate	$Fe_2(SO_4)_3$ (Fe is 3^+; SO_4 is 2^-)
Cupric sulfite	$CuSO_3$ (Cu is 2^+; SO_3 is 2^-)
Ammonium bicarbonate	NH_4HCO_3 (NH_4 is 1^+; HCO_3 is 1^-)
Strontium chlorite	$Sr(ClO_2)_2$ (Sr is 2^+; ClO_2 is 1^-)
Stannous sulfate	$SnSO_4$ (Sn is 2^+; SO_4 is 2^-)
Calcium permanganate	$Ca(MnO_4)_2$ (Ca is 2^+; MnO_4 is 1^-)
Cadmium nitrate	$Cd(NO_3)_2$ (Cd is 2^+; NO_3 is 1^-)
Ferrous bisulfite	$Fe(HSO_3)_2$ (Fe is 2^+; HSO_3 is 1^-)

Work Problems
10 and 11.

7-5

Special Ternary Compounds

In Table 6-4 are listed four different polyatomic ions containing chlorine: perchlorate (ClO_4^{1-}), chlorate (ClO_3^{1-}), chlorite (ClO_2^{1-}), and hypochlorite (ClO^{1-}). We have previously mentioned the relationship of chlorite (ClO_2^{1-}) to chlorate (ClO_3^{1-}). Hypochlorite (ClO^{1-}) is related to chlorite (ClO_2^{1-}) by one less oxygen atom.[9] The prefix *hypo-* is a Greek word meaning "under"; hence, *hypo*chlorite has one atom "under" the number of oxygen atoms of chlorite. You may remember the term "hypo" by remembering that a hypodermic needle goes under (hypo) the skin (dermis). Perchlorate (ClO_4^{1-}) is related to chlorate (ClO_3^{1-}) by one more oxygen atom.[10] The prefix *per-* can be used to mean "over"; therefore perchlorate has one atom "over" the number of oxygen atoms of chlorate. These prefixes can also be applied to other oxy-halogen ions, such as those of bromine and iodine. [Fluorine does not form polyatomic ions with oxygen due to the high electronegativity of both elements (see Figure 6-9)].

Consider the following examples of naming these special ternary compounds:

---

[9]The oxidation number of chlorine in chlorite is 3^+, whereas that of chlorine in hypochlorite is 1^+; hence, the *hypo-* means a lower or "under" oxidation number than that of the *-ite* polyatomic ion.

[10]The oxidation number of chlorine in perchlorate is 7^+, whereas that of chlorine in chlorate is 5^+; hence, the *per-* means an "over" or a higher oxidation number than the *-ate* polyatmic ion.

Formula	Name
NH_4ClO_4	Ammonium perchlorate[11]
$KBrO_2$	Potassium bromite
$NaClO$	Sodium hypochlorite[12]
$Ca(IO_3)_2$	Calcium iodate

Consider the following examples of writing the formulas of these special ternary compounds:

Name	Formula
Barium hypoiodite	$Ba(IO)_2$ (Ba is 2^+; IO is 1^- as is ClO; see Tables 6-1 and 6-4)
Calcium perbromate	$Ca(BrO_4)_2$ (Ca is 2^+; BrO_4 is 1^- as is ClO_4)
Potassium chlorate	$KClO_3$ (K is 1^+; ClO_3 is 1^-)
Ferric Iodate	$Fe(IO_3)$ (Fe is 3^+; IO_3 is 1^- as is ClO_3)

Work Problems 12 and 13.

7-6

Acids, Bases, and Salts

Acids

In our previous discussion, we did not consider the case where the hydrogen ion (H^{1+}) replaced the metal ion or positive polyatomic ion. This is a special case. Hydrogen compounds have completely different properties if they are in the gaseous or liquid state (pure compounds) than if they are in water solution, and hence may be named differently.

In the gaseous or liquid state, the hydrogen compounds are sometimes named as hydrogen derivatives. For example: HCl is hydrogen chloride, HCN is hydrogen cyanide, HBr is hydrogen bromide, etc.

In water solutions, these hydrogen compounds are called acids.[13] **Acids,** therefore, are hydrogen compounds that yield hydrogen ions (H^{1+}) in water solution. For *binary* compounds, the prefix **hydro-,** meaning hydrogen or in water, is added; the **-ide** of the anion name is replaced by **-ic acid.** Therefore, hydrogen chloride in water solution is *hydro*chlor*ic acid.* The same procedure

---

[11]Ammonium perchlorate is the oxidizer in the second stage of the *Columbia* space shuttle. In this stage a solid-fuel propellent is used which is 70 percent ammonium perchlorate.

[12]Many bleaching agents contain a dilute solution (about 5 percent) of sodium hypochlorite.

[13]In 15-1, we shall define an acid and a base in more exact terms.

TABLE 7-2 Summary of the Naming of Binary and Ternary Compounds of Hydrogen in the Gas or Liquid State and in Water Solution

General		Formula	Example	
Gas or Liquid	Water Solution		Name of Gas or Liquid	Name of Water Solution
Binary and -ide endings				
Hydrogen___-ide	hydro___-ic acid	HCl	Hydrogen chloride	Hydrochloric acid
Ternary and higher				
Hydrogen___-ate	___-ic acid	H_3PO_4	Hydrogen phosphate	Phosphoric acid
Hydrogen___-ite	___-ous acid	H_3PO_3	Hydrogen phosphite	Phosphorous acid

is applied to other binary compounds and also to hydrogen cyanide (HCN), which is called *hydro*cyan*ic acid* in water solution.

For *ternary* and higher compounds, the word "hydrogen" is dropped, and the name of the polyatomic ion is used; the **-ate** or **-ite** is dropped and **-ic** or **-ous acid,** respectively, is added. Therefore, "hydrogen phos*phate*" (H_3PO_4) in water solution is phosphor*ic acid* and "hydrogen phosph*ite*" (H_3PO_3) is phosphor*ous acid*.[14] Table 7-2 summarizes these changes.

Consider the naming of the following hydrogen compounds, as a pure compound and in water solution:

	Formula	Name as a Pure Compound	Name of Water Solution
Binary			
	HBr	Hydrogen brom*ide*	*Hydrobromic acid*
	HI	Hydrogen iod*ide*	*Hydriodic acid*[15]
	H_2S	Hydrogen sulf*ide*	*Hydrosulfuric acid*[16]
Ternary and higher			
	HNO_3	Hydrogen nitr*ate*	Nitr*ic acid*
	$HC_2H_3O_2$	Hydrogen acet*ate*	Acet*ic acid*
	H_2SO_4	Hydrogen sulf*ate*	Sulfur*ic acid*
	$HClO_2$	Hydrogen chlor*ite*	Chlor*ous acid*
	$HBrO_4$	Hydrogen perbrom*ate*	*Perbromic acid*[17]
	HClO	Hydrogen *hypo*chlor*ite*	*Hypochlorous acid*[17]

[14]In each of the ternary acids involving phosphorus, "or" from phosph*or*us is reinserted in the acid name. Recent IUPAC revisions suggest that H_3PO_3 be written H_2HPO_3 and named phosphonic acid. The reason for this is that only two hydrogens of the acid can be replaced when it is neutralized with a base. The name "phosphorous acid," written as H_3PO_3, is still the most used form and, hence, will be used in this text.

[15]For ease of pronunciation, the "o" in hydro- is dropped when followed by a vowel.

[16]In each of the acids involving sulfur, "ur" from sulf*ur* is reinserted in the acid name.

[17]Note that the prefix of the negative polyatomic ion is carried over to the name of the water solution.

Consider the following examples of writing the formulas of acids:

Name	Formula
Hydrofluoric acid	HF (H is 1^+; F is 1^-; see Tables 6-1 and 6-2)
Sulfurous acid	H_2SO_3 (H is 1^+; SO_3 is 2^-; see Table 6-4)
Chloric acid	$HClO_3$ (H is 1^+; ClO_3 is 1^-)
Chromic acid	H_2CrO_4 (H is 1^+; CrO_4 is 2^-)

Bases

The compound formed with a hydroxide (OH^{1-}) polyatomic ion and a metal ion can be defined as a base. A **base**, therefore, is a substance that contains a metal ion and a hydroxide ion. Even though bases are not binary compounds, they have the ending *-ide*.

Consider the naming of the following bases:

Formula	Name
LiOH	Lithium hydrox*ide*[18]
KOH	Potassium hydrox*ide*
$Ca(OH)_2$	Calcium hydrox*ide*
$Al(OH)_3$	Aluminum hydrox*ide*

Consider the following examples of writing the formulas of bases:

Name	Formula
Ferric hydroxide	$Fe(OH)_3$ (Fe is 3^+; OH is 1^-; see Tables 6-1 and 6-4)
Barium hydroxide	$Ba(OH)_2$ (Ba is 2^+; OH is 1^-)
Magnesium hydroxide[19]	$Mg(OH)_2$ (Mg is 2^+; OH is 1^-)
Sodium hydroxide	NaOH (Na is 1^+; OH is 1^-)

---

[18]This compound was used in filters to absorb carbon dioxide in the cabin atmosphere in the Apollo space missions.

[19]This compound is found in the antacid laxative, milk of magnesia.

Salts

A **salt** is a compound formed when *one* or *more* of the hydrogen ions of an acid is replaced by a cation (metal or positive polyatomic ion), **or** when one or more of the hydroxide ions of a base is replaced by an anion (nonmetal or negative polyatomic ion). The binary compounds of metal cations with nonmetal anions and the ternary compounds of metal cations or ammonium ions with negative polyatomic ions are examples of salts. Potassium bromide (KBr), sodium nitrate ($NaNO_3$), and ammonium sulfate [$(NH_4)_2SO_4$] are examples of salts.

Compounds that are salts and that contain one or more hydrogen atoms bonded to the anion are called **acid salts**. We have previously encountered these in the salts of hydrogen carbonate (bicarbonate), hydrogen sulfate (bisulfate), and hydrogen sulfite (bisulfite) polyatomic ions. These salts were formed by the replacement of one of the two hydrogens by a metal cation from their respective acids, carbonic acid (H_2CO_3), sulfuric acid (H_2SO_4), and sulfurous acid (H_2SO_3).

In the cases of phosphoric acid (H_3PO_4) and phosphorous acid (H_3PO_3), one or two of the hydrogen ions may be replaced, or all three in the first case. If just one or two hydrogen ions are replaced an acid salt is formed. To name these acid salts, we must use Greek prefixes to denote the number of a given *cation* or *hydrogen ion*, if more than one.

Consider the naming of the following acid salts:

Formula	Name
NaH_2PO_4	Sodium *di*hydrogen phosphate
Na_2HPO_3	*Di*sodium hydrogen phosphite

Consider the following examples of writing the formulas of acid salts:

Name	Formula
Potassium hydrogen carbonate	$KHCO_3$ (K is 1^+; H is 1^+ and CO_3 is 2^-; see Tables 6-1 and 6-4)
Potassium dihydrogen phosphite	KH_2PO_3 (K is 1^+; H is 1^+; and PO_3 is 3^-)

Hydroxy salts are salts that contain one or more hydroxide ions. The hydroxide ion is part of the salt and is called a *hydroxy group*. The salt is named as other binary and ternary compounds are named. If more than one *anion* or *hydroxide ion* is present, Greek prefixes must be used.

Consider the naming of the following hydroxy salts:

Formula	Name
Ca(OH)Cl	Calcium hydroxychloride
Fe(OH)(C$_2$H$_3$O$_2$)$_2$	Iron(III) hydroxy*di*acetate

Consider the following examples of writing the formulas of hydroxy salts:

Name	Formula
Lead(II) hydroxyacetate	Pb(OH)(C$_2$H$_3$O$_2$)(Pb is 2$^+$; OH is 1$^-$; and C$_2$H$_3$O$_2$ is 1$^-$; see Tables 6-1 and 6-4)
Aluminum dihydroxychloride[20]	Al(OH)$_2$Cl(Al is 3$^+$; OH is 1$^-$; and Cl is 1$^-$)

Most salts do *not* contain hydrogen atoms bonded to the anion (acid salts) or hydroxide ions (hydroxy salts); these simple salts are called **normal salts.** For example, sodium chloride (NaCl) and potassium sulfate (K$_2$SO$_4$) are normal salts.

Now, since we have developed simplified definitions of acids and bases, acid salts, hydroxy salts, and normal salts, you should be able to identify them, given their formulas. The following are generalized formulas for acids, bases, acid salts, hydroxy salts, and normal salts.

HX = acid
H^{1+} is a hydrogen ion and X^{1-} is an anion (nonmetal ion or negative polyatomic ion) in water solution.

MOH = base
M^{1+} is a metal ion and OH^{1-} is a hydroxide ion.

MHX = acid salt
M^{1+} is a cation (metal ion or positive polyatomic ion), H^{1+} is a hydrogen ion, and X^{2-} is an anion (nonmetal ion or negative polyatomic ion).

M(OH)X = hydroxy salt
M^{2+} is a cation (metal ion or positive polyatomic ion), OH^{1-} is a hydroxide ion, and X^{1-} is an anion (nonmetal ion or negative polyatomic ion).

MX = normal salt
M^{1+} is a cation (metal ion or positive polyatomic ion) and X^{1-} is an anion (nonmetal ion or negative polyatomic ion).

[20]Aluminum hydroxy *salts* are used as the active agent in various antiperspirants.

Classify the following as (1) acids, (2) bases, (3) acid salts, (4) hydroxy salts, or (5) normal salts. Assume that all soluble compounds are in water solutions.

Formula	Classification
$CaCl_2$	(5) Normal salt
$Ca(OH)Cl$	(4) Hydroxy salt
$Ca(OH)_2$	(2) Base
Na_2SO_4	(5) Normal salt
H_2SO_4	(1) Acid
$NaHSO_4$	(3) Acid salt
$Ca_3(PO_4)_2$	(5) Normal salt
$CaHPO_4$	(3) Acid salt

Work Problems 14, 15, 16, and 17.

7-7

Common Names

Common names still persist, since the systematic names appear in some cases to be too complicated to use. Thus, shorter names are used. You are aware of the fact that the systematic name for table salt is sodium chloride, but not even the most ardent chemist asks for sodium chloride at the dinner table. The same is true for hydrogen oxide—water. Table 7-3 lists some common names and a few uses with which you may be familiar. Figure 7-1 attempts to illustrate why some of these common names may be important to you some day!

Work Problem 18.

Chapter Summary

The names and formulas of binary and ternary compounds are derived from the names and formulas of cations, anions, and polyatomic ions.

All binary compounds end in **-ide.** In naming binary compounds composed of two nonmetals Greek prefixes are used to indicate the number of atoms of each element in the compound. Other binary compounds contain a metal ion with fixed or variable oxidation numbers and a nonmetal ion. The metal ion with a variable oxidation number is named by both the *-ous* or *-ic* suffix system and the Stock system. Tables 6-1 and 6-2 must be *memorized* to write the formulas of binary compounds.

Ternary compounds are composed of hydrogen ion (H^{1+}), a metal ion with fixed or variable oxidation numbers, and a polyatomic ion or oxy-halogen ion. Hydrogen compounds that yield hydrogen ions in water solution are called **acids. Bases** are substances that contain a metal ion and a hydroxide ion. **Salts** are compounds formed when one or more of the hydrogen ions of an acid is replaced by a cation or when one or more of the hydroxide ions of

TABLE 7-3   Some Common Names, Systematic Names, Their Formulas, and Some of Their Uses

Common Name	Systematic Name	Formula	Use
Ammonia	Hydrogen nitride (trihydrogen nitride)	NH_3	Cleaner, commercial refrigerant, fertilizer
Baking soda, bicarbonate	Sodium hydrogen carbonate	$NaHCO_3$	In baking powder, some fire extinguishers, antacid
Dry Ice (solid) Carbonic gas (gas)	Carbon dioxide	CO_2	Fire extinguishers, freezing substances
Epsom salts	Magnesium sulfate heptahydrate	$MgSO_4 \cdot 7H_2O$	Strong laxative, bathing infected tissue
Laughing gas	Dinitrogen oxide	N_2O	Anesthetic
Marble, chalk, limestone	Calcium carbonate	$CaCO_3$	Make cement, antacid, and prevent diarrhea
Milk of magnesia	Magnesium hydroxide	$Mg(OH)_2$	Antacid and laxative
Muriatic acid	Hydrochloric acid	HCl	Cleaning metals, such as iron before galvanizing, stomach (digestive) acid
Oil of vitriol	Sulfuric acid	H_2SO_4	Battery acid (dilute), cleaning metals
Quicklime, lime	Calcium oxide	CaO	Make slaked lime
Slaked lime	Calcium hydroxide	$Ca(OH)_2$	Mortar, plaster
Soda lye, caustic soda	Sodium hydroxide	$NaOH$	Make soap
Table salt	Sodium chloride	$NaCl$	Seasoning
Table sugar, sucrose	β-D-Fructofuranosyl α-D-glucopyranoside	$C_{12}H_{22}O_{11}$	Sweetener
Vinegar (dilute solution, about 5%)	Acetic acid	$HC_2H_3O_2$	Salad dressing, pickling of some foods
Water	Hydrogen oxide (dihydrogen oxide)	H_2O	Drinking, washing

FIGURE 7-1

Baking soda (sodium hydrogen carbonate) puts out fires.

a base is replaced by an anion. **Acid salts** are salts that contain one or more hydrogen atoms bonded to the anion, while **hydroxy salts** are salts that contain one or more hydroxide ions. A **normal salt** is a simple salt that does not contain hydrogen atoms bonded to the anion (acid salt) or hydroxide ions (hydroxy salt). Table 6-4 must be *memorized* to write the formulas of ternary compounds.

Using the general formulas for acids, bases, acid salts, hydroxy salts, and normal salts, these compounds were identified given the formula of the compound. The common names of a few compounds and their formulas as given in Table 7-3 should be memorized, unless instructed otherwise by your instructor.

EXERCISES

1. Define or explain the following terms:
   - (a) a binary compound
   - (b) a ternary compound
   - (c) IUPAC
   - (d) an acid
   - (e) Stock system
   - (f) *-ous* or *-ic* suffix system
   - (g) salt
   - (h) acid salt
   - (i) hydroxy salt
   - (j) normal salt

2. Give the meaning of the following prefixes or suffixes:
   - (a) -ide
   - (b) -ate
   - (c) -ic acid
   - (d) -ite
   - (e) -ous acid
   - (f) -ous (cation)
   - (g) -ic (cation)
   - (h) hypo-
   - (i) hydro-
   - (j) per-

3. Distinguish between
   - (a) a binary and a ternary compound
   - (b) an acid salt and a hydroxy salt
   - (c) an acid salt and a normal salt
   - (d) a hydroxy salt and a normal salt
   - (e) a polyatomic positive ion and a polyatomic negative ion
   - (f) ammonia and ammonium ion
   - (g) baking soda and soda lye
   - (h) lime and slaked lime
   - (i) -ate and -ite
   - (j) -ic acid and -ous acid
   - (k) hypo- and hydro-

Formulas of Ions

4. Write the formula for each of the following ions:
   - (a) sulfate
   - (b) nitrate
   - (c) carbonate
   - (d) chloride
   - (e) chlorite
   - (f) sulfite
   - (g) bromate
   - (h) nitrite
   - (i) perbromate
   - (j) hydrogen carbonate

(k) chromate

(l) calcium

(m) ferric

(n) cuprous

(o) barium

(p) copper(II)

(q) tin(IV)

(r) potassium

(s) aluminum

(t) stannous

Names of Ions

5. Name each of the following ions:

(a) ClO^{1-}

(b) MnO_4^{1-}

(c) S^{2-}

(d) OH^{1-}

(e) HSO_3^{1-}

(f) CN^{1-}

(g) $Cr_2O_7^{2-}$

(h) Br^{1-}

(i) HCO_3^{1-}

(j) HSO_4^{1-}

(k) Mg^{2+}

(l) NH_4^{1+}

(m) Cd^{2+}

(n) Sn^{4+}

(o) Ag^{1+}

(p) Hg_2^{2+}

(q) Pb^{2+}

(r) H^{1+}

(s) Zn^{2+}

(t) Hg^{2+}

PROBLEMS

Binary Compounds Containing Two Nonmetals   (See Section 7-2)

6. Write the correct formula for each of the following compounds:

(a) sulfur dioxide

(b) dinitrogen tetroxide

(c) carbon dioxide

(d) tetraphosphorus decoxide

(e) carbon monoxide

(f) sulfur trioxide

(g) dinitrogen oxide

(h) dinitrogen pentoxide

(i) diphosphorus pentasulfide

(j) chlorine dioxide

7. Write the correct name for each of the following compounds:

(a) ClO_2

(b) P_2S_5

(c) N_2O_5

(d) N_2O

(e) SO_3

(f) CO

(g) P_4O_{10}

(h) CO_2

(i) N_2O_4

(j) SO_2

Binary Compounds Containing a Metal and a Nonmetal   (See Section 7-3)

8. Write the correct formula for each of the following compounds:

(a) barium chloride

(b) stannic iodide

(c) lead(II) sulfide

(d) mercurous chloride

(e) lithium iodide

(f) tin(II) fluoride

(g) mercuric bromide

(h) calcium oxide

(i) cupric oxide

(j) stannic sulfide

9. Write the correct name for each of the following compounds:
   (a) SnS_2 (b) CuO
   (c) CaO (d) $HgBr_2$
   (e) SnF_2 (f) LiI
   (g) Hg_2Cl_2 (h) PbS
   (i) SnI_4 (j) $BaCl_2$

Ternary and Higher Compounds   (See Section 7-4)

10. Write the correct formula for each of the following compounds:
    (a) silver phosphate   (b) calcium nitrate
    (c) cuprous carbonate   (d) lithium sulfite
    (e) aluminum sulfate   (f) aluminum chromate
    (g) cadmium carbonate   (h) ammonium sulfate
    (i) magnesium cyanide   (j) ferrous nitrite

11. Write the correct name for each of the following compounds:
    (a) $Fe(NO_2)_2$ (b) $Mg(CN)_2$
    (c) $(NH_4)_2SO_4$ (d) $CdCO_3$
    (e) $Al_2(CrO_4)_3$ (f) $Al_2(SO_4)_3$
    (g) Li_2SO_3 (h) Cu_2CO_3
    (i) $Ca(NO_3)_2$ (j) Ag_3PO_4

Special Ternary Compounds   (See Section 7-5)

12. Write the correct formula for each of the following compounds:
    (a) sodium chlorite   (b) cadmium iodate
    (c) iron(III) perbromate   (d) potassium hypochlorite
    (e) calcium chlorate   (f) magnesium bromate
    (g) calcium hypochlorite   (h) magnesium perchlorate
    (i) cupric chlorate   (j) lithium perbromate

13. Write the correct name for each of the following compounds:
    (a) $LiBrO_4$ (b) $Cu(ClO_3)_2$
    (c) $Mg(ClO_4)_2$ (d) $Ca(ClO)_2$
    (e) $Mg(BrO_3)_2$ (f) $Ca(ClO_3)_2$
    (g) $KClO$ (h) $Fe(BrO_4)_3$
    (i) $Cd(IO_3)_2$ (j) $NaClO_2$

Acids, Bases, and Salts   (See Section 7-6)

14. Write the correct formula for each of the following compounds:
    (a) hypochlorous acid   (b) potassium hydrogen sulfite
    (c) magnesium hydroxychloride   (d) barium hydroxide
    (e) bromic acid   (f) bismuth hydroxysulfate
    (g) nitric acid   (h) potassium dihydrogen phosphate
    (i) calcium hydroxide   (j) sulfuric acid

15. Write the correct name for each of the following compounds:
    (a) H_2SO_4 in water
    (b) $Ca(OH)_2$
    (c) KH_2PO_4
    (d) HNO_3 in water
    (e) $Bi(OH)(SO_4)$
    (f) $HBrO_3$ in water
    (g) $Ba(OH)_2$
    (h) $Mg(OH)Cl$
    (i) $KHSO_3$
    (j) $HClO$ in water

16. Classify each of the following compounds as (1) an acid, (2) a base, (3) an acid salt, (4) a hydroxy salt, or (5) a normal salt. Assume that all soluble compounds are in water solution.
    (a) $H_2Cr_2O_7$
    (b) $Fe(OH)(C_2H_3O_2)_2$
    (c) NH_4HCO_3
    (d) $Sr(HCO_3)_2$
    (e) $HMnO_4$
    (f) $AlCl_3$
    (g) $HC_2H_3O_2$
    (h) $Ca(OH)_2$
    (i) $K_2Cr_2O_7$
    (j) $BiCl_3$

17. Classify each of the following compounds as (1) an acid, (2) a base, (3) an acid salt, (4) a hydroxy salt, or (5) a normal salt. Assume that all soluble compounds are in water solution.
    (a) $MgCl_2$
    (b) $Ca(HCO_3)_2$
    (c) $Pb_3(OH)_2(CO_3)_2$
    (d) $Sr(OH)_2$
    (e) $MgHPO_4$
    (f) $PbCO_3$
    (g) $Hg(C_2H_3O_2)_2$
    (h) H_3PO_4
    (i) NH_4Br
    (j) $LiOH$

Common Names   (See Section 7-7)

18. Write the correct formula for each of the following compounds:
    (a) vinegar
    (b) marble
    (c) table salt
    (d) baking soda
    (e) milk of magnesia
    (f) ammonia

General Problems

19. Write the correct formula for each of the following compounds:
    (a) stannous phosphate
    (b) silver permanganate
    (c) calcium hypoiodite
    (d) magnesium sulfite
    (e) sodium oxalate
    (f) perchloric acid
    (g) phosphorus trifluoride
    (h) cadmium nitrate
    (i) lead(II) phosphate
    (j) strontium hydrogen carbonate
    (k) calcium nitride
    (l) mercuric chloride
    (m) gold(III) oxide
    (n) lead(IV) oxide
    (o) phosphorus pentachloride
    (p) disodium hydrogen phosphate

20. Write the name for each of the following compounds:
    (a) $CaHPO_4$
    (b) $Mg(ClO_3)_2$
    (c) $PbSO_4$
    (d) $CaCr_2O_7$
    (e) $Ba(OH)_2$
    (f) $Sn(HCO_3)_2$

(g) $K_2C_2O_4$ (h) $Li_2Cr_2O_7$

(i) $HC_2H_3O_2$ in water   (j) $ZnCl_2$

(k) BiF_3 (l) $AuBr_3$

(m) Cd_3P_2 (n) $LiOH$

(o) CaC_2O_4 (p) SnS_2

21. Complete the following table by writing the correct formulas for the compounds formed by the following cations and anions.

Cations	Chloride	Carbonate	Anion Sulfate	Phosphate
Sodium	————	————	————	————
Calcium	————	————	————	————
Iron(III)	————	————	————	————
Copper(II)	————	————	————	————

Readings

Powell, Virginia, *Chemical Formulas and Names*. Englewood Cliffs, N.J.: Prentice-Hall., 1977. A programmed unit for writing formulas and naming compounds. This unit may be helpful as a supplement to your text.

Silverman, Alexander, "Nomenclature of Inorganic Chemistry." *J. Am. Chem. Soc.*, 1960, v. 82, p. 5523. The American version of the rules for nomenclature of inorganic chemistry, as developed by the International Union of Pure and Applied Chemistry, Commission on Nomenclature of Inorganic Chemistry, after the Paris, France meeting in 1957.

8

Calculations Involving Elements and Compounds

$320 g \ K_2SO_4 \times \dfrac{1 mole \ K_2SO_4}{174 g \ K_2SO_4}$

$= 1.84 \ moles \ K_2SO_4$

A calculation of the number of moles of a substance in a given quantity of the substance.

TASKS

1. Memorize Avogadro's number, **6.02 × 10²³** (Section 8-2).

2. Memorize the molar volume of a gas: 1 mol (6.02 × 10²³ molecules of any gas occupies a volume of **22.4 liters** at 0°C and 76$\overline{0}$ torr (STP conditions) (Section 8-3).

OBJECTIVES

1. Given the following terms, define each term and give a distinguishing characteristic of each:
    (a) mole (Section 8-2)
    (b) Avogadro's number (Section 8-2)
    (c) molar volume of a gas (Section 8-3)
    (d) empirical formula (Section 8-5)
    (e) molecular formula (Section 8-5)

2. Given the name or formula of a compound and the table of approximate atomic masses, calculate the formula mass or molecular mass of the compound (Problem Examples 8-1, 8-2, and 8-3, Problems 3 and 4).

3. Given one mole of formula units or molecules in a compound, determine the number of moles of ions or atoms of each element present (Section 8-2, Problem 5).

4. Given an amount of a substance (either as mass or the number of elementary units), its name or formula, and the table of approximate atomic masses, calculate the moles of elementary units in the substance (Problem Examples 8-4, 8-5, 8-6, 8-7, and 8-8, Problems 6 and 7).

5. Given the number of moles or elementary units in a given quantity of a substance, its name or formula, and the table of approximate atomic masses, calculate the mass of the substance (Problem Examples 8-9, 8-10, and 8-11, Problems 8 and 9).

6. Given the number of moles or mass of a substance, calculate the actual number of units in the given quantity (Problem Examples 8-12 and 8-13, Problems 10 and 11).

7. Given an isotope of an element and its exact atomic mass, calculate the mass in grams for one atom of the isotope (Problem Example 8-14, Problem 12).

8. Given the volume of a gas at STP, its name or formula, and the table

of approximate atomic masses, calculate the number of moles or the mass of the gas (Problem Examples 8-15 and 8-16, Problem 13).

9. Given the mass and volume of a gas at STP, or its density at STP, calculate the molecular mass of the gas (Problem Examples 8-17 and 8-18, Problem 14).

10. Given the name or formula of a gas and the table of approximate atomic masses, calculate the density of the gas at STP (Problem Example 8-19, Problem 15).

11. Given the mass of a gas, its name or formula, and the table of approximate atomic masses, calculate the volume of the gas at STP (Problem Example 8-20, Problem 16).

12. Given the name, formula, or experimental analysis of a compound, and the table of approximate atomic masses, calculate the percent composition of the compound. If the sample is impure, calculate the percent of any given element present (Problem Examples 8-21, 8-22, and 8-23, Problems 17, 18, and 19).

13. Given the mass of a sample of a compound, its name or formula, and the table of approximate atomic masses, calculate the mass of any element in the sample (Problem Example 8-24, Problem 20).

14. Given the percent composition or experimental analysis of a compound and the table of approximate atomic masses, calculate the empirical formula of the compound (Problem Examples 8-25 and 8-26, Problems 21 and 22).

15. Given the percent composition or experimental analysis of a compound, the table of approximate atomic masses, and the formula or molecular mass, or the density of the gas at STP, calculate the empirical and molecular formula of the compound (Problem Examples 8-27 and 8-28, Problems 23, 24, 25, and 26).

In the previous chapters, we have considered a general description of elements and compounds with few quantitative calculations. In this chapter, we shall again consider elements and compounds, but in regard to quantitative calculations. In our calculations we shall use the *factor-unit* method of problem solving introduced in Chapter 2 (2-9), and we suggest that you review that method.

8-1

Calculation of Formula or Molecular Masses

In 3-8, we identified the subscripts in a formula of a compound as representing the number of atoms of the respective elements in a molecule or formula unit of a compound. For example, in a molecule of ethyl alcohol (C_2H_6O), there are 2 atoms of carbon, 6 atoms of hydrogen, and 1 atom of oxygen. In 4-1, we introduced the atomic mass scale (atomic weight), based on an arbitrarily as-signed value of exactly 12 amu for carbon-12. The atomic masses of all the elements are *related* to this standard. Precise relative masses of the elements are found inside the front cover of this text, and approximate relative atomic masses of the elements are found inside the back cover. The formula masses or molecular masses of compounds can be calculated from the atomic masses of the elements.

The term "formula mass" is used for compounds that are written as *for-mula units;* that is, the compound exists as ions and has primarily electrovalent or ionic bonding (6-3). The term "molecular mass" is applied to compounds that exist as *molecules* and have primarily covalent bonding (6-4). In 6-8, we showed how to use the Periodic Table to predict the type of bonding in a compound. The methods for calculating formula masses and molecular masses are the *same*, except that formula masses are for compounds described with formula units and molecular masses are for compounds existing as molecules.

Consider the following examples of the calculation of formula or molecular masses of compounds:

Problem Example 8-1

Calculate the formula mass of sodium sulfate.[1]

SOLUTION: The formula for sodium sulfate is Na_2SO_4. In this formula unit, there are 2 atoms of sodium, 1 atom of sulfur, and 4 atoms of oxygen. Hence, the formula mass is calculated as follows:

$$2 \text{ atoms Na} \times \frac{23.0 \text{ amu}}{1 \text{ atom Na}} = 46.0 \text{ amu}$$

$$1 \text{ atom S} \times \frac{32.1 \text{ amu}}{1 \text{ atom S}} = 32.1 \text{ amu}$$

$$4 \text{ atoms O} \times \frac{16.0 \text{ amu}}{1 \text{ atom O}} = 64.0 \text{ amu}$$

$$\text{formula mass of } Na_2SO_4 = \overline{142.1 \text{ amu}} \quad \textit{Answer}$$

The answer is expressed to the least place of all the numbers that are added (2-2), which, in this example, is the ten's decimal place. The calculation can be simplified as follows:

$$2 \times 23.0 = 46.0 \text{ amu}$$
$$1 \times 32.1 = 32.1 \text{ amu}$$
$$4 \times 16.0 = \underline{64.0 \text{ amu}}$$
$$\text{formula mass of } Na_2SO_4 = 142.1 \text{ amu} \quad \textit{Answer}$$

[1] In all calculations involving atomic masses, the Table of Approximate Atomic Masses inside the back cover of this text will be used, unless otherwise stated. Refer to this table for the atomic masses of the elements in solving problems.

Problem Example 8-2

Calculate the molecular mass of ethyl alcohol (C_2H_6O).

SOLUTION

$$2 \times 12.0 = 24.0 \text{ amu}$$
$$6 \times 1.0 = 6.0 \text{ amu}$$
$$1 \times 16.0 = \underline{16.0 \text{ amu}}$$
$$\text{molecular mass of } C_2H_6O = 46.0 \text{ amu} \quad \textit{Answer}$$

Notice that the same method is used in solving for molecular mass as is used in solving for formula mass.

Since the atomic masses in amu represents a very small quantity of material (see 4-1), it is more practical to deal with a large number of unit particles. In dealing with a large number of these particles, amounts of material equal to the formula mass or molecular mass are expressed in the units grams, pounds, or tons. These respective amounts of material are called the gram, pound, or ton formula mass or molecular mass. Consider the following example:

Problem Example 8-3

Calculate the gram-molecular mass of benzene (C_6H_6).[2]

SOLUTION

$$6 \times 12.0 = 72.0 \text{ amu}$$
$$6 \times 1.0 = \underline{6.0 \text{ amu}}$$
$$\text{molecular mass of } C_6H_6 = 78.0 \text{ amu}$$
$$\text{gram-molecular mass of } C_6H_6 = 78.0 \text{ g} \quad \textit{Answer}$$

Work Problems 3
and 4.

This amount of benzene (78.0 g) is equal to the mass of 6.02×10^{23} molecules of benzene, as we will discuss in the next section.

8-2

Calculation of Moles of Units. Avogadro's Number (N)

In our discussion of atomic masses, the standard used was that of carbon-12. Carbon-12 is also used to define a new term—the mole. The **mole** (abbreviated **mol**) is the *amount* of a substance containing the *same number* of elementary units (atoms, formula units, molecules, or ions) as there are atoms in *exactly 12 g of carbon-12*. Now this poses another question: How many atoms are there in exactly 12 g of carbon-12? Experimentally, by diffraction of X rays and

---

[2]The term **molar mass** is a general term used to include gram-molecular mass, gram-formula mass, and gram-atomic mass. It is the mass in grams of *one* mole of a substance as described in 8-2.

FIGURE 8-1

Counting Avogadro's number, *N,* of peas. If all the people now alive on the earth (4.7 billion) started counting Avogadro's number, *N,* of peas at a rate of two peas per second, it would take approximately 2.0 million (2,000,000) years. That is a lot of peas!

| 0.1 million | 1.0 million | 2.0 million | 2.2 million |
| (100,000) years | (1,000,000) years | (2,000,000) years | (2,200,000) years |

other methods, the number of atoms in exactly 12 g of carbon-12 has been found to be 6.02×10^{23} atoms. This number 6.02×10^{23} or 602,000,000,000,000,000,000,000, is called **Avogadro's** (ä′vo·gä′dro) **number** (*N*) and is named in honor of the Italian physicist and chemist Amedeo Avogadro (1776–1856). Avogadro's number is also the equivalence factor between the mass units, gram and amu, 1 gram being equal to 6.02×10^{23} amu.

Figure 8-1 may be of some help in understanding the meaning of this extremely large number. In *1 mol of carbon-12 atoms there are* 6.02×10^{23} *atoms, and this number of atoms has a mass of exactly 12 g, the atomic mass for carbon-12 expressed in grams* (see Figure 8-2). For 1 mol (6.02×10^{23}) of oxygen atoms (note the same number of oxygen atoms as of carbon-12), there is a mass of 16.0 g (to three significant digits), the atomic mass of oxygen expressed in grams. You should note that for the same number of atoms, oxygen has a greater mass; hence, an oxygen atom is heavier than a carbon-12 atom. The same kind of comparison can be made for any element. *One mole of atoms*[3]

FIGURE 8-2

One mole of carbon-12 atoms (6.02×10^{23} atoms, *N*) has a mass of exactly 12 g (0.012 kg).

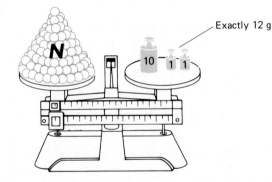

Exactly 12 g

of any element contains 6.02×10^{23} *atoms of the element and is equal to the atomic mass of the element expressed in grams:*

$$1 \text{ mol of atoms of an element} = 6.02 \times 10^{23} \text{ atoms of the element}$$
$$= \text{atomic mass of the element in grams}$$

The mole has been referred to as the "chemist's dozen." It is a quantity of matter based on a certain number (6.02×10^{23}) of elementary units per mole, just like a dozen is defined as 12 units per dozen or a gross as 144 units per gross. The total mass of a substance in a mole of the substance is related to the mass of each individual elementary unit of the substance. This mass can be expressed as *amu* or more conveniently as *grams*.

The reasoning we just applied to atoms of an element can also be applied to formula units and to molecules of a compound. Therefore, in *one mole of a compound there are 6.02×10^{23} formula units or molecules, and this number of formula units or molecules has a mass equal to the formula or molecular mass expressed in grams:*

$$1 \text{ mol of a compound} = 6.02 \times 10^{23} \text{ formula units or molecules}$$
$$\text{of the compound} = \text{formula or molecular mass in grams}$$

For 1 mol (6.02×10^{23} molecules) of water (H_2O), there is a mass of 18.0 g ($2 \times 1.0 + 1 \times 16.0 = 18.0$ amu, to the tenths place), the molecular mass of water expressed in grams (see Figure 8-3). For 1 mol (6.02×10^{23} formula units) of sodium sulfate (Na_2SO_4), there is a mass of 142.1 g ($2 \times 23.0 + 1 \times 32.1 + 4 \times 16.0 = 142.1$ amu, to the tenths place), the formula mass of sodium sulfate expressed in grams.

The same reasoning can be applied to *ions* or to *any units.* Therefore, in 1 *mol of ions there are 6.02×10^{23} ions, and this number of ions has a mass equal*

FIGURE 8-3

One mole of water molecules, (6.02×10^{23} molecules, *N*) has a mass of 18.0 g (to three significant digits).

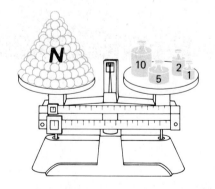

---

[3]The term "moles of atoms" of an element also has had historically another name—"gram-atoms"—but in this text we shall use "moles of atoms" to express this quantity.

to the atomic or formula mass of the ion expressed in grams. For 1 mol or 6.02×10^{23} sodium ions, there is a mass of 23.0 g (to three significant digits), the atomic mass of sodium expressed in grams. For 1 mol or 6.02×10^{23} sulfate ions, there is a mass of 96.1 g ($1 \times 32.1 + 4 \times 16.0 = 96.1$ amu, to the tenths place), the formula mass of sulfate ions expressed in grams:

$$1 \text{ mol of ions} = 6.02 \times 10^{23} \text{ ions} = \text{atomic or formula mass of the ions in grams}$$

In all the preceding cases regarding *moles,* the mass was expressed in *grams.* The mass can be expressed in any mass unit, such as *milligrams;* then the term applied is *millimole.* The millimole is frequently used in chemistry to measure small quantities of substances. A *millimole* (mmol) is equal to 0.001 (1/1000) mol, and the mass may be expressed in *milligrams* instead of grams. One millimole of adenosine triphosphate (ATP, molecular mass = 507 amu) would have a mass of 507 mg (or 0.507 g).

We stated that in the formulas of compounds, the subscripts represented the number of atoms of each element in a formula unit or molecule of the compound (3-8). These subscripts also represent the **number of moles of ions or atoms** of the elements in **1 mol of molecules or formula units** of the compound. Let us consider the case of water (H_2O), consisting of 2 atoms of hydrogen and 1 atom of oxygen. Water has a molecular mass of 18.0 amu, consisting of 2.0 amu of hydrogen atoms and 16.0 amu of oxygen atoms. One mole of water molecules has a mass of 18.0 g, consisting of 2.0 g of hydrogen atoms and 16.0 g of oxygen atoms. Hence, the moles of atoms of each element in *one* mole of water molecules is

$$2.0 \text{ g H atoms} \times \frac{1 \text{ mol H atoms}}{1.0 \text{ g H atoms}} = 2 \text{ mol H atoms}$$

$$16.0 \text{ g O atoms} \times \frac{1 \text{ mol O atoms}}{16.0 \text{ g O atoms}} = 1 \text{ mol O atoms}$$

Work Problem 5.   Therefore, in *one mole of molecules or formula units of a compound, the subscripts represent the number of moles of ions or atoms of the elements present.*

Problem Example 8-4

Calculate the number of moles of oxygen *atoms* in 24.0 g of oxygen.

SOLUTION: The atomic mass of oxygen is 16.0 amu; therefore, 1 mol of oxygen *atoms* has a mass of 16.0 g, and the number of moles of oxygen atoms is calculated as

$$24.0 \text{ g O} \times \frac{1 \text{ mol O atoms}}{16.0 \text{ g O}} = 1.50 \text{ mol oxygen atoms} \quad Answer$$

(For a review of significant digits, see 2-2, and also Problem Example 2-11 in 2-9.)

Problem Example 8-5

Calculate the number of moles of oxygen *molecules* in 24.0 g of oxygen.

SOLUTION: The formula for an oxygen molecule is O_2 (see 6-4), and it has a molecular mass of 32.0 amu (2×16.0). Therefore, 1 mol of oxygen *molecules* has a mass of 32.0 g. The number of moles of oxygen molecules in 24.0 g of oxygen is calculated as

$$24.0 \; \cancel{g \; O_2} \times \frac{1 \; \text{mol} \; O_2 \; \text{molecules}}{32.0 \; \cancel{g \; O_2}} = 0.750 \; \text{mol oxygen molecules} \quad Answer$$

Problem Example 8-6

Calculate the number of moles of water in 24.5 g of water.

SOLUTION: The molecular mass of water (H_2O) is

$$2 \times 1.0 = 2.0 \; \text{amu}$$
$$1 \times 16.0 = \underline{16.0 \; \text{amu}}$$
$$\text{molecular mass } H_2O = 18.0 \; \text{amu}$$

Therefore, 1 mol of water molecules has a mass of 18.0 g. The number of moles of water in 24.5 g of water is

$$24.5 \; \cancel{g \; H_2O} \times \frac{1 \; \text{mol} \; H_2O}{18.0 \; \cancel{g \; H_2O}} = 1.36 \; \text{mol} \; H_2O \quad Answer$$

Problem Example 8-7

Calculate the number of moles of sodium ions in 1.30 mol of sodium sulfate.

SOLUTION: The formula of sodium sulfate is Na_2SO_4 (see Chapter 7, 7-4). Since there are 2 ions of Na in one formula unit of Na_2SO_4, there will be 2 mol of Na ion in 1 mol of Na_2SO_4. The number of moles of Na ions in 1.30 mol of Na_2SO_4 is calculated as

$$1.30 \; \cancel{\text{mol} \; Na_2SO_4} \times \frac{2 \; \text{mol Na ions}}{1 \; \cancel{\text{mol} \; Na_2SO_4}} = 2.60 \; \text{mol} \; Na^{1+} \quad Answer$$

(The moles of atoms or ions in 1 mol of formula units or molecules are regarded as exact values and are not considered in computing significant digits.)

Problem Example 8-8

Calculate the number of moles of water in 9.65×10^{23} molecules of water.

SOLUTION: One mole of water molecules contain 6.02×10^{23} molecules. Therefore, the number of moles in 9.65×10^{23} molecules of water is

Work Problems 6
and 7.

$$9.65 \times 10^{23} \text{ molecules } H_2O \times \frac{1 \text{ mol } H_2O}{6.02 \times 10^{23} \text{ molecules } H_2O}$$
$$= 1.60 \text{ mol } H_2O \quad \textit{Answer}$$

Problem Example 8-9

Calculate the number of grams of sodium sulfate in 1.30 mol of sodium sulfate.

SOLUTION: The formula mass of Na_2SO_4 is 142.1 amu (see Problem Example 8-1). Therefore, 1 mol of sodium sulfate = 142.1 g, and the mass of 1.30 mol is calculated as

$$1.30 \text{ mol } Na_2SO_4 \times \frac{142.1 \text{ g } Na_2SO_4}{1 \text{ mol } Na_2SO_4} = 185 \text{ g } Na_2SO_4 \quad \textit{Answer}$$

Problem Example 8-10

Calculate the number of grams of oxygen present in 1.30 mol of sodium sulfate.

SOLUTION: Since there are 4 atoms of oxygen in one formula unit of Na_2SO_4, there will be 4 mol of oxygen atoms in 1 mol of Na_2SO_4. The atomic mass of oxygen is 16.0 amu, and 1 mol of oxygen atoms has a mass of 16.0 g. Therefore, the number of grams of oxygen present in 1.30 mol of Na_2SO_4 is calculated as

$$1.30 \text{ mol } Na_2SO_4 \times \frac{4 \text{ mol O atoms}}{1 \text{ mol } Na_2SO_4} \times \frac{16.0 \text{ g O}}{1 \text{ mol O atoms}}$$
$$= 83.2 \text{ g oxygen} \quad \textit{Answer}$$

Problem Example 8-11

Calculate the number of grams of sodium sulfate in 4.54×10^{23} formula units of sodium sulfate.

SOLUTION: In 1 mol of sodium sulfate there are 6.02×10^{23} formula units with a mass of 142.1 g (see Problem Examples 8-1 and 8-9). Therefore, in 4.54×10^{23} formula units of Na_2SO_4, there are

$$4.54 \times 10^{23} \text{ formula units } Na_2SO_4 \times \frac{1 \text{ mol } Na_2SO_4}{6.02 \times 10^{23} \text{ formula units } Na_2SO_4}$$

Work Problems 8
and 9.

$$\times \frac{142.1 \text{ g } Na_2SO_4}{1 \text{ mol } Na_2SO_4} = 107 \text{ g } Na_2SO_4 \quad \textit{Answer}$$

Problem Example 8-12

Calculate the number of formula units of sodium sulfate in 1.30 mol of sodium sulfate.

SOLUTION: There are 6.02×10^{23} formula units in 1 mol of formula units of a compound; and, in 1.30 mol of sodium sulfate, there are

$$1.30 \; \cancel{\text{mol Na}_2\text{SO}_4} \times \frac{6.02 \times 10^{23} \text{ formula units Na}_2\text{SO}_4}{1 \; \cancel{\text{mol Na}_2\text{SO}_4}}$$

$$= 7.83 \times 10^{23} \text{ formula units Na}_2\text{SO}_4 \quad \textit{Answer}$$

Problem Example 8-13

Calculate the number of water molecules in 4.50 g of water.

SOLUTION: The molecular mass of water (H_2O) is 18.0 amu (see Problem Example 8-6). Therefore, 1 mol of water molecules has a mass of 18.0 g, and 1 mol of water molecules contains 6.02×10^{23} molecules. The number of water molecules in 4.50 g of water is calculated as

Work Problems 10 and 11.

$$4.50 \; \cancel{\text{g H}_2\text{O}} \times \frac{1 \; \cancel{\text{mol H}_2\text{O molecules}}}{18.0 \; \cancel{\text{g H}_2\text{O}}} \times \frac{6.02 \times 10^{23} \text{ molecules H}_2\text{O}}{1 \; \cancel{\text{mol H}_2\text{O molecules}}}$$

$$= 1.50 \times 10^{23} \text{ molecules H}_2\text{O} \quad \textit{Answer}$$

Problem Example 8-14

Using exact atomic masses (see the inside front cover of this text), calculate the mass in grams of 1 atom of hydrogen.

SOLUTION: One mole of hydrogen atoms contains 6.02×10^{23} atoms of hydrogen and has an atomic mass of exactly 1.0080 amu; hence, the mass of 1 atom is calculated as

$$\frac{1.0080 \text{ g H}}{1 \; \cancel{\text{mol H atoms}}} \times \frac{1 \; \cancel{\text{mol H atoms}}}{6.02 \times 10^{23} \text{ atoms H}} = 0.167 \times 10^{-23} \; \frac{\text{g H}}{1 \text{ atom H}}$$

$$= 1.67 \times 10^{-24} \; \frac{\text{g H}}{1 \text{ atom H}} \quad \text{(in scientific notation; see 2-4)} \quad \textit{Answer}$$

In 4-1, we stated that the mass of an atom was calculated by indirect methods, such as we have used here to calculate the mass of the hydrogen atom. Also in 4-1 is given the mass of an oxygen atom and a carbon atom; see if you can calculate these values. (*Hint:* Use the exact atomic mass units.)

Work Problem 12.

8-3

Molar Volume of a Gas and Related Calculations

For any gas it has been experimentally determined that 6.02×10^{23} molecules (Avogadro's number, N) of a gas or 1 mol of gas *molecules* occupies a volume of **22.4 L at 0°C (273 K) and a pressure of $76\overline{0}$ torr (mm Hg).**[4] The conditions of 0°C and $76\overline{0}$ torr are defined as *standard temperature and pressure* (STP) or *standard conditions* (SC). This volume of **22.4 L** occupied by 1 mol of any gas *molecules* at 0°C and $76\overline{0}$ torr is called the **molar volume of a gas** and is approximately the volume occupied by *three* standard basketballs (see Figure 8-4). This molar volume of a gas relates the mass of a gas to its volume at STP and can be used in various types of calculations.

Moles or Mass

Problem Example 8-15

Calculate the number of moles of oxygen *molecules* in 5.60 L of oxygen at STP.

SOLUTION: In 1 mol of O_2 *molecules* at STP, there is a volume of 22.4 L. Therefore, the number of moles of O_2 molecules in 5.60 L is calculated as

$$5.60 \; \cancel{L \, O_2} \times \frac{1 \text{ mol } O_2}{22.4 \; \cancel{L \, O_2}} = 0.250 \text{ mol oxygen} \quad \textit{Answer}$$

Problem Example 8-16

Calculate the number of grams of oxygen in 5.60 L of oxygen at STP.

FIGURE 8-4

Molar volume of a gas. 6.02×10^{23} molecules of any gas (1 mol of gas molecules) occupies a volume of **22.4** liters at 0°C and $76\overline{0}$ torr (STP conditions). This volume is approximately the volume occupied by **three** standard basketballs.

---

[4]In Chapter 11 we shall consider the units of pressure and will also evaluate the effect of temperature and pressure on the volume of a gas.

SOLUTION: Oxygen is written as O_2 (a diatomic molecule, see 6-4. Calculate moles of O_2 *molecules* as in Problem Example 8-15, and then from the molecular mass of O_2 ($2 \times 16.0 = 32.0$ amu), the mass in grams of 5.60 L of oxygen molecules at STP can be calculated as

Work Problem 13.

$$5.60 \text{ L O}_2 \times \frac{1 \text{ mol O}_2}{22.4 \text{ L O}_2} \times \frac{32.0 \text{ g O}_2}{1 \text{ mol O}_2} = 8.00 \text{ g oxygen} \textit{Answer}$$

Molecular Mass (molar mass)

The molecular mass of a gas can be calculated by solving for **grams per mole** (g/mol) of the gas which is numerically equal to the molecular mass in amu.

Problem Example 8-17

Calculate the molecular mass of a gas if 5.00 L measured at STP has a mass of 9.85 g.

SOLUTION: Solving for g/mol, the molecular mass is calculated as

$$\frac{9.85 \text{ g}}{5.00 \text{ L STP}} \times \frac{22.4 \text{ L STP}}{1 \text{ mol}} = 44.1 \text{ g/mol}$$
$$\text{molecular mass} = 44.1 \text{ amu} \textit{Answer}$$

Problem Example 8-18

The density of a certain gas is 1.30 g/L at STP. Calculate the *gram*-molecular mass of the gas.

SOLUTION: Solving for g/mol, the gram-molecular mass is calculated as

Work Problem 14.

$$\frac{1.30 \text{ g}}{1 \text{ L STP}} \times \frac{22.4 \text{ L STP}}{1 \text{ mol}} = 29.1 \text{ g/mol}$$
$$\text{gram-molecular mass (molar mass)} = 29.1 \text{ g} \textit{Answer}$$

Density

Problem Example 8-19

Calculate the density of oxygen gas at STP.

SOLUTION: The units of density for a gas are g/L. Hence, from the molecular mass of O_2 (32.0 amu), the density is calculated as

$$\frac{32.0 \text{ g O}_2}{1 \text{ mol O}_2} \times \frac{1 \text{ mol O}_2}{22.4 \text{ L STP}} = 1.43 \text{ g/L at STP} \textit{Answer}$$

Problem Example 8-20

Calculate the volume (in liters) that 5.00 g of oxygen would occupy at STP.

SOLUTION: Remember that oxygen is written as O_2. Thus, by calculating the number of moles of oxygen *molecules* and then using the molar volume, you can calculate the volume occupied by 5.00 g of oxygen molecules at STP as

Work Problems 15 and 16.

$$5.00 \text{ g } O_2 \times \frac{1 \text{ mol } O_2}{32.0 \text{ g } O_2} \times \frac{22.4 \text{ L } O_2 \text{ STP}}{1 \text{ mol } O_2} = 3.50 \text{ L oxygen at STP} \quad \textit{Answer}$$

8-4

Calculation of Percent Composition of Compounds

Percent means parts per hundred. For example, if your school has a student enrollment of 1000 and there are 400 men students, the percent of men students is 40 ($\frac{400}{1000} \times 100 = 40$ percent) or 40 (men students) *per hundred* (students). In the same manner, the percent composition of each element in a compound can be calculated. The exact numbers, as 400 and 1000, may or may not be given, however. If not, then the formula will be given and from it the molecular or formula mass can be calculated; thus, the percent composition of each element in the compound can be calculated. Any units such as amu, g, lb, etc., may be assigned to the molecular or formula mass, as long as the same units are used throughout the entire calculation.

Consider the following examples:

Problem Example 8-21

Calculate the percent composition of ethyl chloride (C_2H_5Cl).

SOLUTION: The molecular mass of C_2H_5Cl is calculated as 64.5 amu.

$$2 \times 12.0 = 24.0 \text{ amu}$$
$$5 \times 1.0 = 5.0 \text{ amu}$$
$$1 \times 35.5 = \underline{35.5 \text{ amu}}$$
$$\text{molecular mass of } C_2H_5Cl = 64.5 \text{ amu}$$

The percent of each element in the compound is calculated by dividing the contribution of each element (amu) by the molecular mass (amu) and multiplying by 100.

$$\% \text{ carbon:} \quad \frac{24.0 \text{ amu}}{64.5 \text{ amu}} \times 100 = 37.2\% \text{ C}$$

$$\% \text{ hydrogen:} \quad \frac{5.0 \text{ amu}}{64.5 \text{ amu}} \times 100 = 7.8\% \text{ H}$$

$$\% \text{ chlorine:} \quad \frac{35.5 \text{ amu}}{64.5 \text{ amu}} \times 100 = 55.0\% \text{ Cl}$$

Problem Example 8-22

A student found that 1.00 g of a metal combined with 0.65 g of oxygen to form an oxide of the metal. Calculate the percent of metal in the oxide.

SOLUTION: The total mass of the oxide is 1.65 g (1.00 g + 0.65 g = 1.65 g). The percent of metal in the oxide is then calculated by dividing the mass of the metal in grams by the total mass of the oxide in grams and multiplying by 100 to get percent.

$$\text{\% metal:} \quad \frac{1.00 \text{ g metal}}{1.65 \text{ g oxide}} \times 100 = 60.6\% \text{ metal}$$

Problem Example 8-23

A crude sample of zinc sulfide has a mass of 8.00 g. It contains 5.00 g of zinc sulfide. What is the percent of zinc in the crude sample?

SOLUTION: The formula mass of zinc sulfide (ZnS) is 97.5 amu.

$$1 \times 65.4 = 65.4 \text{ amu}$$
$$1 \times 32.1 = \underline{32.1 \text{ amu}}$$
$$\text{formula mass ZnS} = 97.5 \text{ amu}$$

In 97.5 g of ZnS there are 65.4 g of Zn, and the percent of zinc in 8.00 g of crude zinc sulfide is calculated as

$$\frac{5.00 \text{ g ZnS}}{8.00 \text{ g crude ZnS}} \times \frac{65.4 \text{ g Zn}}{97.5 \text{ g ZnS}} \times 100 = 41.9\% \text{ Zn} \quad \textit{Answer}$$

Problem Example 8-24

Calculate the number of grams of carbon in 17.6 g of carbon dioxide.

SOLUTION: The molecular mass of CO_2 is calculated as 44.0 amu.

$$1 \times 12.0 = 12.0 \text{ amu}$$
$$2 \times 16.0 = \underline{32.0 \text{ amu}}$$
$$\text{molecular mass of } CO_2 = 44.0 \text{ amu}$$

In 44.0 g of CO_2 there are 12.0 g of C, and the number of grams of carbon in 17.6 g of CO_2 is calculated as

Work Problems 17, 18, 19, and 20.

$$17.6 \text{ g CO}_2 \times \frac{12.0 \text{ g C}}{44.0 \text{ g CO}_2} = 4.80 \text{ g C} \quad \textit{Answer}$$

8-5

Calculation of Empirical (Simplest) and Molecular Formulas

In the preceding section (8-4), the formulas of the compounds were given and you were asked to calculate the percent composition. In this section, the percent composition will be given and you will be asked to determine the formulas.

The **empirical formula (simplest formula)** of a compound is the formula containing the *smallest whole number ratio of the atoms* that are present in a molecule or formula unit of the compound. This empirical formula is found from the percent composition of the compound, which is determined *experimentally* from analysis of the compound in the laboratory. The empirical formula gives only the ratio of the atoms present expressed as *small whole numbers*.

The **molecular formula** of the compound is the *true* formula and contains the *actual* number of atoms of each element present in one molecule of the compound (see 3-8). The molecular formula is determined from the empirical formula *and* the molecular mass of the compound, which may be determined experimentally by various methods.

A simple analogy may help to illustrate these two types of formulas. In your school, the ratio of men to women may be $2:1$ (empirical formula), but the actual number of men to women may be $800:400$ (molecular formula). In the case of hydrogen peroxide, the empirical formula is HO (1 atom H:1 atom O), but the molecular formula is H_2O_2 (2 atoms H:2 atoms O).

In some cases, both the empirical and molecular formulas are the same, as in the case of H_2O. The true formulas of compounds existing as *molecules* (covalent compounds) are always referred to as *molecular formulas*. For those compounds written as *formula units* (ionic compounds), there are no molecular formulas, because these compounds do not exist as molecules. Hence, their formulas are called *empirical formulas*.

For new compounds prepared in the laboratory, the process of determining the empirical formula by chemical analysis, and then the molecular formula from the empirical formula and the molecular mass, is often followed before this new compound can be identified. The process at one time took days, but now, with modern instrumental equipment, it can usually be done in about 30 minutes.

Consider the following examples of calculation of empirical formulas:

Problem Example 8-25

Determine the empirical formula for the compound containing 32.4 percent sodium, 22.6 percent sulfur, and 45.1 percent oxygen.[5]

---

[5]The difference here of 0.1 percent between 100.1 percent (32.4 + 22.6 + 45.1) and exactly 100 percent emphasizes the experimental nature of these values and is due to experimental error.

SOLUTION: In exactly 100 g of the compound there would be 32.4 g of Na, 22.6 g of S, and 45.1 g of O. The first step is to calculate the moles of *atoms* of each element present, as follows:

$$32.4 \ \cancel{g \ Na} \times \frac{1 \ mol \ Na \ atoms}{23.0 \ \cancel{g \ Na}} = 1.41 \ mol \ Na \ atoms$$

$$22.6 \ \cancel{g \ S} \times \frac{1 \ mol \ S \ atoms}{32.1 \ \cancel{g \ S}} = 0.704 \ mol \ S \ atoms$$

$$45.1 \ \cancel{g \ O} \times \frac{1 \ mol \ O \ atoms}{16.0 \ \cancel{g \ O}} = 2.82 \ mol \ O \ atoms$$

The elements are combined in a ratio of 1.41 mol of Na atoms to 0.704 mol of S atoms to 2.82 mol of O atoms as

$$Na_{1.41 \ mol \ of \ atoms}, \ S_{0.704 \ mol \ of \ atoms}, \ O_{2.82 \ mol \ of \ atoms}$$

The empirical formula must express these relationships in terms of *small whole numbers*.

The second step, then, is to express these relationships in *small whole numbers* by dividing each value by the smallest one, as follows:

$$For \ Na: \quad \frac{1.41}{0.704} \approx 2$$

$$For \ S: \quad \frac{0.704}{0.704} = 1$$

$$For \ O: \quad \frac{2.82}{0.704} \approx 4$$

Hence, the elements are combined in a ratio of 2 mol of Na atoms to 1 mol of S atoms to 4 mol of O atoms, and the empirical formula is Na_2SO_4. *Answer.*

Problem Example 8-26

Calculate the empirical formula for the compound of composition 26.6 percent potassium, 35.4 percent chromium, and 38.1 percent oxygen.

SOLUTION: First calculate the moles of *atoms* of each element in exactly 100 g of the compound, as follows:

$$26.6 \ \cancel{g \ K} \times \frac{1 \ mol \ K \ atoms}{39.1 \ \cancel{g \ K}} = 0.680 \ mol \ K \ atoms$$

$$35.4 \ \cancel{g \ Cr} \times \frac{1 \ mol \ Cr \ atoms}{52.0 \ \cancel{g \ Cr}} = 0.681 \ mol \ Cr \ atoms$$

$$38.1 \ \cancel{g \ O} \times \frac{1 \ mol \ O \ atoms}{16.0 \ \cancel{g \ O}} = 2.38 \ mol \ O \ atoms$$

Second, reduce these values to simpler numbers by dividing each one by the smallest value, as follows:

$$\text{For K:} \quad \frac{0.680}{0.680} = 1$$

$$\text{For Cr:} \quad \frac{0.681}{0.680} \simeq 1$$

$$\text{For O:} \quad \frac{2.38}{0.680} = 3.5$$

These relative ratios may be converted to small whole numbers by **multiplying by 2**; the empirical formula is then $K_2Cr_2O_7$. *Answer*

If the ratio of these numbers end in 0.5, then *all* the numbers must be multiplied by **2** to obtain whole numbers. If the ratio of these numbers end in 0.33 (0.33 . . . , the fraction ⅓), then *all* the numbers must be multiplied by **3** to obtain whole numbers.

Consider the following examples of the calculation of molecular formulas:

Problem Example 8-27

An oxide of nitrogen gave the following analysis: 3.04 g of nitrogen combined with 6.95 g of oxygen. The molecular mass of this compound was found by experimentation to be 91.0 amu. Determine its molecular formula.

SOLUTION: The empirical formula is calculated from the analysis, the same as from the percent composition, first by calculating the moles of atoms of nitrogen and oxygen, as follows:

$$3.04 \text{ g N} \times \frac{1 \text{ mol N } atoms}{14.0 \text{ g N}} = 0.217 \text{ mol N } atoms$$

$$6.95 \text{ g O} \times \frac{1 \text{ mol O } atoms}{16.0 \text{ g O}} = 0.434 \text{ mol O } atoms$$

Second, these values are reduced to small whole numbers by dividing by the smallest value, as follows:

$$\text{For N:} \quad \frac{0.217}{0.217} = 1$$

$$\text{For O:} \quad \frac{0.434}{0.217} = 2$$

Therefore, the empirical formula is NO_2. The molecular formula will be equal either to the empirical formula or to some multiple (2, 3, 4, etc.) of it. The empirical formula mass of NO_2 is calculated as

$$1 \times 14.0 = 14.0 \text{ amu}$$
$$2 \times 16.0 = \underline{32.0 \text{ amu}}$$
$$\text{empirical formula mass} = 46.0 \text{ amu}$$

The molecular mass as given in the problem was 91.0 amu.[6] The multiple of the empirical formula is found to be approximately 2:

$$\frac{\text{molecular mass}}{\text{empirical formula mass}} = \frac{91.0 \ \text{amu}}{46.0 \ \text{amu}} = 1.98 \text{ or approximately 2}$$

Therefore, the molecular formula is

$$(NO_2)_2 = N_2O_4 \quad Answer^7$$

Problem Example 8-28

A hydrocarbon has the following composition: carbon = 82.7 percent and hydrogen = 17.4 percent. The density of its vapor at STP is 2.60 g/L. Calculate the molecular formula of the hydrocarbon.

SOLUTION: The empirical formula is calculated from the percent composition first by calculating the moles of atoms of carbon and hydrogen, as follows:

$$82.7 \ \text{g C} \times \frac{1 \ \text{mol C atoms}}{12.0 \ \text{g C}} = 6.89 \ \text{mol C atoms}$$

$$17.4 \ \text{g H} \times \frac{1 \ \text{mol H atoms}}{1.0 \ \text{g H}} = 17.4 \ \text{mol H atoms}$$

---

[6]The difference between an exact value of 92.0 amu and this value of 91.0 amu results from *experimental error* in the determination of the molecular mass.

[7]Alternate method to Problem Example 8-27. The total weight of the compound is the mass of nitrogen plus the mass of oxygen; that is, 3.04 g of nitrogen + 6.95 g of oxygen = 9.99 g of compound. The moles of atoms of each element in *one* mole of the compound, is calculated as follows:

$$\frac{91.0 \ \text{g compound}}{1 \ \text{mol compound}} \times \frac{3.04 \ \text{g N}}{9.99 \ \text{g compound}} = \frac{27.7 \ \text{g N}}{1 \ \text{mol compound}}$$

$$\frac{27.7 \ \text{g N}}{1 \ \text{mol compound}} \times \frac{1 \ \text{mol N atoms}}{14.0 \ \text{g N}} = \frac{1.98 \ \text{mol N atoms}}{1 \ \text{mol compound}}$$

$$\text{or approximately } \frac{2 \ \text{mol N atoms}}{1 \ \text{mol compound}}$$

$$\frac{91.0 \ \text{g compound}}{1 \ \text{mol compound}} \times \frac{6.95 \ \text{g O}}{9.99 \ \text{g compound}} = \frac{63.3 \ \text{g O}}{1 \ \text{mol compound}}$$

$$\frac{63.3 \ \text{g O}}{1 \ \text{mol compound}} \times \frac{1 \ \text{mol O atoms}}{16.0 \ \text{g O}} = \frac{3.96 \ \text{mol O atoms}}{1 \ \text{mol compound}}$$

$$\text{or approximately } \frac{4 \ \text{mol O atoms}}{1 \ \text{mol compound}}$$

Therefore, the molecular formula is N_2O_4.

Second, these values are reduced by dividing by the smallest value, as follows:

$$\text{For C: } \frac{6.89}{6.89} = 1$$

$$\text{For H: } \frac{17.4}{6.89} \approx 2.5$$

These relative ratios may be converted to small whole numbers by multiplying by 2; hence, the empirical formula is C_2H_5. You may obtain the molecular formula by knowing the molecular mass. The molecular mass can be calculated as follows (see Problem Example 8-18):

$$\frac{2.60 \text{ g}}{1 \text{ L STP}} \times \frac{22.4 \text{ L STP}}{1 \text{ mol}} = 58.2 \text{ g/mol}$$

Hence, the molecular mass is 58.2 amu. The molecular formula will be either equal to the empirical formula or to some multiple of it. The empirical formula mass of C_2H_5 is calculated as

$$2 \times 12.0 = 24.0 \text{ amu}$$
$$5 \times 1.0 = \underline{5.0 \text{ amu}}$$
$$\text{empirical formula mass} = 29.0 \text{ amu}$$

The multiple of the empirical formula is found to be approximately 2:

$$\frac{\text{molecular mass}}{\text{empirical formula mass}} = \frac{58.2 \text{ amu}}{29.0 \text{ amu}} = 2.07, \text{ or approximately 2}$$

Work Problems 21, 22, 23, 24, 25, and 26.

Hence, the molecular formula is

$$(C_2H_5)_2 = C_4H_{10} \quad \textit{Answer}[8]$$

---

[8]Alternate method to Problem Example 8-28. Calculate the molecular mass of the compound, as follows:

$$\frac{2.60 \text{ g}}{1 \text{ L STP}} \times \frac{22.4 \text{ L STP}}{1 \text{ mol}} = 58.2 \text{ g/mol}$$

Calculate the moles of atoms of each element in one mole of the compound, as follows:

$$\frac{58.2 \text{ g compound}}{1 \text{ mol compound}} \times \frac{82.7 \text{ g C}}{100 \text{ g compound}} = \frac{48.1 \text{ g C}}{1 \text{ mol compound}}$$

$$\frac{48.1 \text{ g C}}{1 \text{ mol compound}} \times \frac{1 \text{ mol C atoms}}{12.0 \text{ g C}} = \frac{4.01 \text{ mol C atoms}}{1 \text{ mol compound}}$$

$$\frac{58.2 \text{ g compound}}{1 \text{ mol compound}} \times \frac{17.4 \text{ g H}}{100 \text{ g compound}} = \frac{10.1 \text{ g H}}{1 \text{ mol compound}}$$

$$\frac{10.1 \text{ g H}}{1 \text{ mol compound}} \times \frac{1 \text{ mol H atoms}}{1.0 \text{ g H}} = \frac{10.1 \text{ mol H atoms}}{1 \text{ mol compound}}$$

Therefore, the molecular formula is C_4H_{10}.

Chapter Summary

The *mole* is defined as the amount of a substance containing the same number of elementary units (atoms, formula units, molecules, or ions) as there are in exactly 12 g of carbon-12 atoms. In exactly 12 g of carbon-12 atoms there are **6.02 × 10²³** atoms (*Avogadro's number*). One mole of atoms of an element is that amount of the element which has a mass equal to its atomic mass in grams. One mole of a compound or ion is that amount of the compound or ion which has a mass equal to the molecular or formula mass of the compound or ion expressed in grams. The number of units in one mole of atoms, formula units, molecules, or ions is equal to Avogadro's number (6.02 × 10²³). For gases, 1 mol of gas molecules occupies a volume of **22.4** L at standard temperature and pressure (*STP*, 0°C, and 76̄0 torr). This volume is referred to as the *molar volume of a gas*. Using the molar volume, the molecular mass of a gas, or the density of a gas at STP can be calculated.

From the name, formula, or experimental analysis, the percent composition of a compound can be calculated. Using the percent composition, the amount of any element in a compound can be determined.

The *empirical formula* of a compound gives the smallest whole number ratio of atoms in a molecule or formula unit of a compound. This formula can be determined from the percent composition, which is obtained by experimentation.

The *molecular formula* of a compound is the true formula and contains the actual number of atoms of each element in one molecule of the compound. Using the empirical formula and the molecular mass of the compound which is obtained by experimentation, the molecular formula can be determined.

EXERCISES

1. Define or explain the following terms:
   (a) formula mass
   (b) molecular mass
   (c) gram-formula mass
   (d) gram-molecular mass
   (e) molar mass
   (f) moles of atoms (gram-atoms)
   (g) moles of formula units
   (h) moles of molecules
   (i) moles of ions
   (j) Avogadro's number (*N*)
   (k) molar volume
   (l) STP
   (m) percent composition of a compound
   (n) empirical formula (simplest formula)
   (o) molecular formula

2. Distinguish between
   (a) formula mass and molecular mass
   (b) moles of atoms and moles of molecules
   (c) empirical formula and molecular formula

PROBLEMS

The atomic masses are found in the Table of Approximate Atomic Masses inside the back cover of this text.

Formula or Molecular Masses (See Section 8-1)

3. Calculate the molecular mass of each of the following compounds:
    (a) PCl_3
    (b) $C_6H_{12}O_6$
    (c) NH_3
    (d) CH_4
    (e) SO_3
    (f) N_2O_3

4. Calculate the formula mass of each of the following compounds:
    (a) KOH
    (b) ZnF_2
    (c) $Ca(OH)_2$
    (d) $Hg(NO_3)_2$
    (e) $Ca_3(PO_4)_2$
    (f) $Al_2(SO_4)_3$

Moles of Units and Avogadro's Number (See Section 8-2)

5. Give the number of moles of atoms of each element in one mole of the following formula units or molecules of compounds:
    (a) CO_2
    (b) N_2O_4
    (c) $Al_2(SO_4)_3$
    (d) $Ca(OH)_2$
    (e) K_2CO_3
    (f) $Ba(C_2H_3O_2)_2$

6. Calculate the number of
    (a) moles of aluminum atoms in 6.40 g of aluminum
    (b) moles of oxygen *atoms* in 54.0 g of oxygen
    (c) moles of oxygen molecules in 54.0 g of oxygen
    (d) moles of silver chloride in 51.0 g of silver chloride
    (e) moles of calcium carbonate in 2.55 g of calcium carbonate
    (f) moles of sulfuric acid in 0.150 kg of sulfuric acid
    (g) moles of iron atoms and of chlorine atoms in 1.25 mol of iron(III) chloride
    (h) moles of magnesium ions, phosphorus atoms, and oxygen atoms in 3.50 mol of magnesium phosphate
    (i) moles of sodium atoms in 1.65×10^{23} sodium atoms
    (j) moles of water molecules in 8.64×10^{24} molecules of water

7. Calculate the number of
    (a) moles of sodium atoms in 24.0 g of sodium
    (b) moles of sulfur atoms in 87.0 g of sulfur
    (c) moles of methane (CH_4) molecules in 105 g of methane
    (d) moles of sodium chloride in 4.35 g of sodium chloride
    (e) moles of calcium carbonate in 4.50 kg of calcium carbonate
    (f) moles of aluminum ions in 12.6 g of aluminum ions
    (g) moles of sulfur atoms in 0.350 mol of aluminum sulfate
    (h) moles of potassium bromide in 5.65×10^{24} formula units of potassium bromide
    (i) moles of carbon dioxide molecules in 1.50×10^{24} molecules of carbon dioxide
    (j) moles of nitrogen dioxide molecules in 6.85×10^{25} molecules of nitrogen dioxide

8. Calculate the number of
    (a) grams of carbon dioxide in 1.23 mol of carbon dioxide
    (b) grams of sodium phosphate in 1.35 mol of sodium phosphate

    (c)   grams of sodium in 1.10 mol of sodium atoms

    (d)   grams of sodium chloride in 0.135 mol of sodium chloride

    (e)   milligrams of potassium sulfate in 0.00220 mol of potassium sulfate

    (f)   grams of sugar ($C_{12}H_{22}O_{11}$) in 1.30 mol of sugar

    (g)   milligrams of ammonia in 0.0200 mol of ammonia

    (h)   grams of phosphorus in 1.40 mol of sodium phosphate

    (i)   grams of oxygen in 1.25 mol of sodium phosphate

    (j)   grams of magnesium in 3.45×10^{23} atoms of magnesium

9.   Calculate the number of

    (a)   grams of nitrogen in 3.50 mol of nitrogen molecules

    (b)   grams of copper in 2.40 mol of copper atoms

    (c)   grams of barium carbonate in 0.350 mol of barium carbonate

    (d)   milligrams of oxygen in 0.00240 mol of oxygen molecules

    (e)   grams of phosphorus in 0.305 mol of phosphorus

    (f)   milligrams of carbon in 0.00240 mol of dextrose (glucose, $C_6H_{12}O_6$)

    (g)   grams of H_2SO_4 in 2.00 mol of H_2SO_4

    (h)   kilograms of potassium chloride in 6.70 mol of potassium chloride

    (i)   grams of methane (CH_4) in 1.27×10^{21} molecules of methane

10.   Calculate the number of

    (a)   atoms in 0.400 mol of carbon atoms

    (b)   atoms in 0.0300 mol of phosphorus atoms

    (c)   molecules in 7.80 mol of methane (CH_4)

    (d)   molecules in 15.0 g of carbon dioxide

11.   Calculate the number of

    (a)   molecules in 2.30 mol of hydrogen molecules

    (b)   molecules in 16.0 g of hydrogen

    (c)   atoms in 5.00 g of hydrogen

    (d)   atoms of oxygen in 7.80 g of carbon dioxide

12.   Calculate the mass in grams to three significant digits of *one* atom of

    (a)   an isotope of helium, atomic mass = 4.00 amu

    (b)   an isotope of nickel, atomic mass = 61.9 amu

    (c)   an isotope of rubidium, atomic mass = 84.9 amu

    (d)   an isotope of mercury, atomic mass = 204 amu

Molar Volume and Related Problems   (See Section 8-3)

13.   Calculate the number of

    (a)   moles of helium molecules in 13.0 L of helium at STP

    (b)   moles of oxygen molecules in $8\overline{7}0$ mL of oxygen at STP

    (c)   moles of nitrogen molecules in 45.0 L of nitrogen at STP

    (d)   grams of carbon dioxide in 14.0 L of carbon dioxide at STP

    (e)   grams of methane (CH_4) in 6.50 L of methane at STP

    (f)   grams of carbon monoxide in 5.65 L of carbon monoxide

14.   Calculate the molecular mass of the following gases:

    (a)   3.30 L at STP has a mass of 0.572 g

    (b)   4.00 L at STP has a mass of 5.10 g

(c)   2.45 L at STP has a mass of 7.60 g
(d)   the density of a gas at STP is 0.725 g/L
(e)   the density of a gas at STP is 1.65 g/L
(f)   the density of a gas at STP is 1.80 g/L

15.   Calculate the density in g/L of the following gases at STP:
(a)   ammonia (NH_3)
(b)   ethane (C_2H_6)
(c)   acetylene (C_2H_2)
(d)   propane (C_3H_8)
(e)   hydrogen iodide (HI)
(f) X_2Y (atomic masses: X = 4.0 amu, Y = 3.2 amu)

16.   Calculate the volume in liters at STP that the following gases would occupy:
(a)   8.00 g of nitrogen
(b)   7.25 g of oxygen
(c)   2.30 g of carbon monoxide
(d)   4.60 g of chlorine
(e)   8.40 g of hydrogen chloride
(f)   6.30 g of dinitrogen oxide

Percent Composition   (See Section 8-4)

17.   Calculate the percent composition of the following compounds:
(a) CaI_2 (b) H_2S
(c) $BaCO_3$ (d) $Ca_3(PO_4)_2$
(e) C_2H_6O (f) $Fe(C_2H_3O_2)_3$

18.   Calculate the percent of the metal in the following compounds from the experimental data:
(a)   0.530 g of a metal combines with 0.400 g of oxygen
(b)   0.350 g of a metal combines with 0.380 g of oxygen
(c)   1.85 g of a metal combines with 1.30 g of sulfur
(d)   275 mg of a metal combines with 135 mg of sulfur

19.   A crude sample of lye has a mass of 12.7 g. It contains 6.85 g of sodium hydroxide. What is the percent sodium in the crude sample?

20.   Calculate the number of
(a)   grams of cadmium in 25.4 g of cadmium sulfide
(b)   grams of magnesium in 65.0 g of magnesium nitride
(c)   grams of calcium sulfide containing 5.37 g of sulfur
(d)   grams of mercury(II) oxide containing 6.40 g of mercury

Empirical and Molecular Formulas   (See Section 8-5)

21.   Determine the empirical formula for each of the following compounds:
(a)   48.0 percent zinc and 52.0 percent chlorine
(b)   19.0 percent tin and 81.0 percent iodine
(c)   25.9 percent iron and 74.1 percent bromine

(d)   62.6 percent lead, 8.5 percent nitrogen, and 29.0 percent oxygen
(e)   28.8 percent magnesium, 14.2 percent carbon, and 57.0 percent oxygen
(f)   38.8 percent calcium, 20.0 percent phosphorus, and 41.3 percent oxygen
(g)   36.5 percent sodium, 25.4 percent sulfur, and 38.1 percent oxygen
(h)   44.9 percent potassium, 18.4 percent sulfur, and 36.7 percent oxygen
(i)   7.2 percent phosphorus and 92.8 percent bromine
(j)   74.4 percent gallium and 25.6 percent oxygen

22.   Determine the empirical formula for each of the following compounds from the experimental data:
(a)   1.99 g of aluminum combines with 1.76 g of oxygen
(b)   1.07 g of carbon combines with 1.43 g of oxygen
(c)   2.95 g of sodium combines with 2.05 g of sulfur
(d)   0.500 g of sulfur combines with 0.500 g of oxygen

23.   Determine the molecular formula for each of the following compounds from the experimental data:
(a)   80.0 percent carbon, 20.0 percent hydrogen, and molecular mass of 30.0 amu
(b)   83.7 percent carbon, 16.3 percent hydrogen, and molecular mass of 86.0 amu
(c)   92.3 percent carbon, 7.7 percent hydrogen, and molecular mass of 26.0 amu
(d)   41.4 percent carbon, 3.5 percent hydrogen, 55.1 percent oxygen, and molecular mass of 116.0 amu
(e)   37.8 percent carbon, 6.3 percent hydrogen, 55.8 percent chlorine, and molecular mass of 127.0 amu

24.   Sulfadiazine, a sulfa drug used in the treatment of bacterial infections, gives, on analysis: 48.0 percent carbon, 4.0 percent hydrogen, 22.4 percent nitrogen, 12.8 percent sulfur, and 12.8 percent oxygen. The molecular mass was found to be $25\overline{0}$ amu. Calculate the molecular formula of sulfadiazine.

25.   Estrone, a female sex hormone, gives, on analysis: 80.0 percent carbon, 8.2 percent hydrogen, and 11.8 percent oxygen. The molecular mass was found to be $27\overline{0}$ amu. Calculate the molecular formula for estrone.

26.   Nicotine, a compound found from 2 to 8 percent in tobacco leaves, gives, on analysis: 74.0 percent carbon, 8.7 percent hydrogen, and 17.3 percent nitrogen. The molecular mass is found to be 162 amu. Calculate the molecular formula for nicotine.

General Problems

27.   Calculate the number of moles of H_2SO_4 in $67\overline{0}$ g of 48.0 percent sulfuric acid solution (by mass). (*Hint:* A 48.0 percent by mass sulfuric acid solution means 48.0 g of pure H_2SO_4 in $10\overline{0}$ g of solution.)

28.   How many mL of concentrated nitric acid will be needed to supply 4.20 mol of HNO_3? Concentrated nitric acid is 72.0 percent HNO_3 and has a specific gravity of 1.42. (*Hint:* A 72.0 percent HNO_3 concentrated solution means 72.0 g of pure HNO_3 in $10\overline{0}$ g of concentrated solution.)

29. When a person driving a motor vehicle has a blood alcohol level of $10\overline{0}$ mg of alcohol (C_2H_6O) per $10\overline{0}$ mL of blood, this person, in most states, is considered to be "driving while intoxicated" (DWI). How many alcohol molecules per mL of blood are required for the person to be considered (DWI)?

30. Liquid ammonia (100 percent ammonia) and pure ammonium nitrate are both used as fertilizers for their nitrogen content. Both sell for approximately $200 per ton. Based on the nitrogen content, which one would be the best buy?

31. Calculate the number of
    (a) mmol of sugar ($C_{12}H_{22}O_{11}$) in 1.26 g of sugar
    (b) grams of sugar in 11.2 mmol of sugar
    (c) molecules of sugar in 8.25 mmol of sugar

32. The blood glucose level of a normal person is about $9\overline{0}$ mg of glucose ($C_6H_{12}O_6$) per $10\overline{0}$ mL of blood. On oral ingestion of 1.00 g of glucose per kilogram of body weight, the blood glucose level rises to about $14\overline{0}$ mg of glucose per $10\overline{0}$ mL of blood.
    (a) Calculate the number of millimoles of glucose per mL of blood and the number of glucose molecules per mL of blood before and after consumption of the glucose.
    (b) The average total blood volume in a person is 5.50 L. Calculate the total number of millimoles and grams of glucose in the blood before and after consumption of the glucose.

33. In a diabetic person, the blood glucose level is about 135 mg of glucose ($C_6H_{12}O_6$) per 100 mL of blood. On oral ingestion of 1.00 g of glucose per kilogram of body weight, the blood glucose level rises to about $23\overline{0}$ mg per $10\overline{0}$ mL of blood.
    (a) Calculate the number of millimoles of glucose per mL of blood and the number of glucose molecules per mL of blood before and after consumption of the glucose.
    (b) The average total blood volume in a person is 5.50 L. Calculate the total number of millimoles and grams of glucose in the blood before and after consumption of the glucose.

34. A gaseous hydrocarbon has a density of 1.25 g/L at 0°C and 760 torr. Its composition is 85.6 percent carbon and 14.4 percent hydrogen. Calculate its molecular formula.

35. Cyclopropane, a gaseous hydrocarbon used as an anesthetic, gives, on analysis: 85.6 percent carbon and 14.4 percent hydrogen. At STP, 7.52 L of cyclopropane has a mass of 14.1 g. Calculate its molecular formula.

36. Cyanogen, a highly poisonous gas with an almondlike odor, gave, on analysis: 46.2 percent carbon and 53.8 percent nitrogen. At STP, cyanogen has a density of 2.32 g/L. Calculate its molecular formula.

Reading

Hawthorne, Robert M., Jr., "The Mole and Avogadro's Number." *J. Chem. Educ.*, 1973, v. 50, p. 282. A historical review of the definition of a mole and its association with Avogadro's number.

9

Chemical Equations

The precipitation of barium sulfate from aqueous solution. The beaker on the left contains a solution of barium nitrate in water. In the center beaker a few drops of dilute sulfuric acid has been added with the formation of a cloud of barium sulfate forming. At the right the precipitation has been completed and the solid precipitate has settled to the bottom of the beaker.

TASKS

1. Know the terms, symbols, and their meanings given in Table 9-1.
2. Be able to interpret and use the order of the electromotive or activity series given in Section 9-9, as directed by your instructor.
3. Be able to interpret and use the rules for the solubility of inorganic substances in water given in Section 9-10, as directed by your instructor.

OBJECTIVES

1. Given the following terms, define each term and give a distinguishing characteristic of each:
   (a) chemical equation (Section 9-1)
   (b) catalyst (Section 9-2)
   (c) word equation (Section 9-5)
   (d) combination reactions (Section 9-7)
   (e) basic oxide (Section 9-7)
   (f) acid oxide (Section 9-7)
   (g) decomposition reactions (Section 9-8)
   (h) single replacement reactions (Section 9-9)
   (i) double replacement reactions (Section 9-10)
   (j) neutralization reactions (Section 9-11)
2. Given the formulas of reactants and products, balance various chemical equations (Problem Examples 9-1, 9-2, 9-3, 9-4, 9-5, and 9-6, Problems 4 and 5).
3. Given the names of the reactants and products, write the formulas for the reactants and products and balance various chemical equations (word equations) (Problem Examples 9-7, 9-8, 9-9, and 9-10, Problems 6 and 7).
4. Given the formulas of the reactants and the Periodic Table, complete and balance various *combination* reaction equations (Section 9-7, Problems 8 and 9).
5. Given the formulas of the reactants and the Periodic Table, complete and balance various *decomposition* reaction equations (Section 9-8, Problems 10 and 11).
6. Given the formulas of the reactants, the Periodic Table, and the electromotive or activity series of metals, complete and balance various *single replacement* reaction equations (Section 9-9, Problems 12 and 13).

198

7. Given the formulas of the reactants, the Periodic Table, and the rules for the solubility of inorganic substances in water, complete and balance various *double replacement* reaction equations. Indicate any precipitate by (s) and any gas by (g) (Section 9-10, Problems 14 and 15).

8. Given the formulas of the reactants, the Periodic Table, and the rules for the solubility of inorganic substances in water, complete and balance various *neutralization* reaction equations. Indicate any precipitate by (s) (Section 9-11, Problems 16 and 17).

In this chapter, we shall consider the chemical properties (3-4) and the chemical changes (3-5) of elements and compounds. We shall explain the actions of some of the compounds mentioned in Chapter 7. For example, we shall discuss why Tums is used as an antacid and why sulfur dioxide as an air pollutant is dangerous to human beings. In this chapter, we shall also take up some of the chemicals you may be familiar with and study why they are used as they are. Before we can do this, we need to become proficient with a new tool— balancing chemical equations.

9-1

Definition of a Chemical Equation

A **chemical equation** is a shorthand way of expressing a chemical change (reaction) in terms of symbols and formulas. An equation for a reaction cannot be written unless the substances that are reacting and the substances that are formed are both known. For an equation to be considered correct, *it must be balanced.* We balance chemical equations because of the Law of Conservation of Mass, experimentally determined by Antoine Laurent Lavoisier (3-6). The Law of Conservation of Mass states that mass is neither created nor destroyed in ordinary chemical changes and that the total mass involved in a physical or chemical change remains unchanged. *The Law of Conservation of Mass requires the number of atoms or moles of atoms of each element to be the same on both sides of the equation.* This is the reason we balance chemical equations.

Chemical equations may be written in two general ways: as "molecular equations[1]" and as ionic equations. In this chapter, we shall consider only molecular equations; after we master the skill of balancing molecular equations, we shall consider ionic equations in Chapter 15.

9-2

Terms, Symbols, and Their Meanings

Since a chemical equation is a shorthand way of expressing a chemical change, various terms and symbols are used, just as in shorthand. In a chemical equation, the substances that combine with one another and hence are changed—

---

[1]The term "molecular equation" is used to include elements and compounds that exist not only as molecules but also those written as formula units.

the *reactants*—are written on the left. The substances that are formed and hence appear—the *products*—are written on the right. A single arrow →, or an equal sign =, or a double arrow ⇄, depending on the reaction conditions, separates the reactants from the products. A plus sign (+) separates each reactant or each product. The three physical states of substances involved in the reaction are sometimes indicated as subscripts following the formula of the substance, as follows:

1. *gas* by (g), or if a gas is a product, by an arrow pointing upward (↑): $H_{2(g)}$ or H_2 ↑

2. *liquid* by (ℓ): $H_2O_{(\ell)}$

3. *solid* by (s), or if a solid is a product precipitating or coming out of a solution, by an arrow pointing downward (↓) or by underscoring: $AgCl_{(s)}$, AgCl ↓ , or AgCl. The use of (s) is the preferred method.

Since water is often used to dissolve solids, a substance dissolved in water is indicated by the subscript (aq), meaning *aq*ueous solution, such as $NaCl_{(aq)}$. A Δ may appear above or below the arrow that separates the reactants and products, meaning that heat is necessary to make the reaction start or go, such as $\xrightarrow{\Delta}$. Also above or below the arrow may appear the symbols for an element or a compound, such as $\xrightarrow{Pt}$. These symbols denote a catalyst. A **catalyst**[2] is a substance that speeds up a chemical reaction but is recovered relatively unchanged at the end of the reaction. The various enzymes used in the digestion

TABLE 9-1 Terms, Symbols, and Their Meanings Used in Chemical Equations

Term or Symbol	Meaning
Reactants	Left side of equation
Products	Right side of equation
→, ⇄, =	Separates the products from the reactants
Subscript (g) or ↑	Gas or gas as a product
Subscript (ℓ)	Liquid
Subscript (s), ↓ , or underscoring formula	Solid or solid as a product precipitating or coming out of a solution
Subscript (aq)	Aqueous solution (dissolved in water)
Δ above or below arrow ($\xrightarrow{\Delta}$ or $\xrightarrow[\Delta]{}$)	Heat needed for reaction to start or go to completion
Symbol above or below arrow ($\xrightarrow{Pt}$ or $\xrightarrow[Pt]{}$)	Catalyst

[2]A substance that slows down a chemical reaction can be called a negative catalyst. A more appropriate term is *inhibitor*.

of foods are catalysts. One example is ptyalin in saliva, which catalyzes the breakdown of large molecules, such as starch, to smaller molecules, such as maltose. Another catalyst, chlorophyll, is used in photosynthesis to form glucose (a sugar) from carbon dioxide, water, and sunlight.

These symbols may or may not appear in the equation depending on the emphasis placed on the reactants and products in the equation. Hence, in some equations you may see many of these symbols, and in other equations you may see none. Table 9-1 summarizes these terms, symbols, and their meanings.

9-3

Guidelines for Balancing Chemical Equations

The chemical equations we consider in this chapter are balanced "by inspection." We will suggest a few guidelines for balancing chemical equations by inspection. These are guidelines and not rules, since in some equations they are not generally applicable, but for most of the simple equations you will encounter in this chapter they will be of help. Remember we are balancing the number of atoms or moles of atoms of each element, and there must be the *same number* of atoms or moles of atoms of each element on each side of the equation.

1. Write the correct formulas for the reactants and the products, with the reactants on the left and the products on the right separated by →, ⇄, or =. Each reactant and each product is separated from the others by a + sign. **Once the correct formula is written it must not be changed during the subsequent balancing operation.** (Review Chapter 7 on nomenclature to be able to write the correct formulas.)

2. Choose the compound that contains the *greatest number of atoms*, whether it is a reactant or a product. Start with the *element* in the compound that also has the greatest number of atoms. This element as a rule usually should not be hydrogen, oxygen, or a polyatomic ion. Balance the number of atoms in this compound with the corresponding atom on the other side of the equation by placing a coefficient *in front* of the formula of the element or compound. If a 2 is placed in front of H_2O, as 2 H_2O, then there are 4 atoms of H and 2 atoms of O; hence, the same number of these atoms must appear on the other side of the equation. If no number is placed in front of the formula, it is assumed to be 1. **Under no circumstances do you change the correct formula of a compound to balance the equation.**

3. Next, balance the polyatomic ions that remain the *same* on both sides of the equation. These polyatomic ions can be balanced as a single unit. In some cases you may need to go back to the coefficient you placed before the compound in number 2 and change it to balance the polyatomic ion. If this is the case, remember to adjust the coefficient on the other side of the equation accordingly.

4. Balance the H atoms and then the O atoms. If they appear in the poly-atomic ion, and you balanced them before, you need not consider them again.

5. Check all coefficients to see that they are *whole numbers* and the *lowest possible ratio*. If the coefficients are fractions, then all coefficients must be multiplied by some number so as to make them all, including the fraction, a whole number.[3] If a coefficient such as $\frac{5}{2}$ or $2\frac{1}{2}$ exists, then *all* coefficients must be multiplied by 2. The $\frac{5}{2}$ or $2\frac{1}{2}$ is then 5, a whole number. The coefficients must be reduced to the lowest possible ratios. If the coefficients are 6, 9 $\rightarrow$ 3, 12, they can all be reduced by dividing *each one* by 3 to give the lowest possible ratio of 2, 3 $\rightarrow$ 1, 4.

6. Check each atom or polyatomic ion with a $\checkmark$ above the atom or ion on both sides of the equation to ensure that the equation is balanced. As you become more proficient in balancing equations, this will not be necessary, but for the first few equations that you balance we feel that you should check each atom or ion. These $\checkmark$ are not part of the final form of the equation and are only used as a teaching device to make sure you balance each atom or ion.

9-4

Examples Involving the Balancing of Equations

Now let us apply these guidelines in balancing the following equations by inspection:

Problem Example 9-1

$$Fe_{(s)} + HCl_{(aq)} \rightarrow FeCl_{2(aq)} + H_{2(g)} \quad \text{(unbalanced)}$$

Guideline 1 need not be considered, since the formulas are given. Considering guideline 2, the compound with the greatest number of atoms besides hydrogen is $FeCl_2$, and the element we start with is Cl which has 2 atoms. To balance the Cl atoms, place a 2 in front of the HCl as 2 HCl. *The formula of HCl is not changed to balance the Cl atoms.* The equation now appears as

$$Fe_{(s)} + 2\,HCl_{(aq)} \rightarrow FeCl_{2(aq)} + H_{2(g)} \quad \text{(balanced)}$$

Guideline 3 is not applicable, because no polyatomic ions are present. For guideline 4, the H atoms are balanced and no O atoms are present. In guideline 5, all coefficients are whole numbers and the lowest possible ratio. Each atom is checked off as in guideline 6, and the final balanced equation is

---

[3]In balancing some equations it is sometimes convenient to leave the coefficients as fractions. In this text we will consider the equation balanced only if the coefficients are whole numbers in the lowest possible ratios.

$$Fe_{(s)} \; + \; 2 \; HCl_{(aq)} \rightarrow FeCl_{2(aq)} \; + \; H_{2(g)} \quad \text{(balanced)}[4]$$

Problem Example 9-2

$$Ca(OH)_{2(aq)} + H_3PO_{4(aq)} \rightarrow Ca_3(PO_4)_{2(s)} + H_2O_{(\ell)} \quad \text{(unbalanced)}$$

Considering guideline 2, the compound with the greatest number of atoms is $Ca_3(PO_4)_2$ and the element we start with is Ca which has 3 atoms. (The polyatomic PO_4^{3-} ion is balanced in guideline 3.) To balance the Ca atoms, place a 3 in front of the $Ca(OH)_2$ as $3 \; Ca(OH)_2$. The formula of $Ca(OH)_2$ is not changed to balance the Ca atoms. The equation now appears as

$$3 \; Ca(OH)_{2(aq)} + H_3PO_{4(aq)} \rightarrow Ca_3(PO_4)_{2(s)} + H_2O_{(\ell)} \quad \text{(unbalanced)}$$

Guideline 3: the polyatomic ion PO_4^{3-} appears on both sides of the equation; hence, place a 2 in front of H_3PO_4, as $2 \; H_3PO_4$ to balance the $2 \; PO_4^{3-}$ in $Ca_3(PO_4)_2$. Balance the H atoms as in guideline 4 by placing a 6 in front of the H_2O as $6 \; H_2O$, since there are 12 H atoms on the left [$3 \times 2 = 6$ from $3 \; Ca(OH)_2$ and $2 \times 3 = 6$ from $2 \; H_3PO_4$]. The O atoms are balanced by $6 \; H_2O$ because there are 6 O atoms in $3 \; Ca(OH)_2$. [The O atoms in the PO_4^{3-} are not included because they were balanced with the $Ca_3(PO_4)_2$]. The equation is now

$$3 \; Ca(OH)_{2(aq)} + 2 \; H_3PO_{4(aq)} \rightarrow Ca_3(PO_4)_{2(s)} + 6 \; H_2O_{(\ell)} \quad \text{(balanced)}$$

The coefficients are all whole numbers, and the lowest possible ratios as suggested in guideline 5. Check off each atom as in guideline 6. The final balanced equation is

$$3 \; Ca(OH)_{2(aq)} + 2 \; H_3PO_{4(aq)} \rightarrow Ca_3(PO_4)_{2(s)} + 6 \; H_2O_{(\ell)} \quad \text{(balanced)}$$

Problem Example 9-3

$$\underset{\text{butane}}{C_4H_{10(g)}} + O_{2(g)} \xrightarrow{\Delta} CO_{2(g)} + H_2O_{(g)} \quad \text{(unbalanced)}[5]$$

---

[4]If you didn't follow guideline 2, you would have changed the formula of HCl and the equation would be

$$Fe_{(s)} \; + \; H_2Cl_{2(aq)} \rightarrow FeCl_{2(aq)} \; + H_{2(g)} \quad \text{(wrong)}$$

This is wrong. Don't change the correct formulas to balance the equation.

[5]Problem Examples 9-3 and 9-4 are examples of combustion reactions. Combustion reactions are those in which organic compounds (compounds containing the element carbon) burn in oxygen forming carbon dioxide and water on complete combustion. These reactions will be discussed further in 13-5.

Considering guideline 2, the compound with the greatest number of atoms is C_4H_{10}, and we start with C which has 4 atoms. (H atoms are balanced later in guideline 4.) To balance the C atoms, place a 4 in front of the CO_2 as 4 CO_2. For guideline 4 (3 is not applicable since there are no polyatomic ions), balance the H atoms with a 5 in front of the H_2O as 5 H_2O, which gives a total of 13 O atoms in the products (8 O from 4 CO_2, and 5 O from 5 H_2O). To balance the O atoms in the reactants, you must use a fraction $\frac{13}{2}$ or $6\frac{1}{2}$ to obtain 13 O atoms in the reactants. The equation now appears as

$$C_4H_{10(g)} + \tfrac{13}{2}\,O_{2(g)} \xrightarrow{\Delta} 4\,CO_{2(g)} + 5\,H_2O_{(g)}$$

For guideline 5, the coefficients must be whole numbers. To obtain a whole number from $\frac{13}{2}$, multiply it by 2; then multiply *all* the other coefficients by 2. The coefficients are also in the lowest possible ratio, and each atom is checked off as in guideline 6. The final balanced equation is

$$2\,C_4H_{10(g)} + 13\,O_{2(g)} \xrightarrow{\Delta} 8\,CO_{2(g)} + 10\,H_2O_{(g)} \quad \text{(balanced)}$$

(Checking off each atom provides a double check and assures you that you multiplied *each* coefficient by 2.)

Problem Example 9-4

$$C_6H_{14(\ell)} + O_{2(g)} \xrightarrow{\Delta} CO_{2(g)} + H_2O_{(g)} \quad \text{(unbalanced)}$$

Referring to guideline 2, the compound with the greatest number of atoms is C_6H_{14}, and we start with C, which has 6 atoms. (H atoms are balanced later.) To balance the C atoms, place a 6 in front of CO_2, as 6 CO_2. Skip guideline 3, since there are no polyatomic ions. Next, according to guideline 4, balance the H atoms with 7 in front of the H_2O as 7 H_2O, which gives a total of 19 O atoms in the product (12 O from 6 CO_2, and 7 O from 7 H_2O). To balance the O atoms in the reactants, you must use a fraction $\frac{19}{2}$ or $9\frac{1}{2}$ to obtain 19 O atoms in the reactants. The equation now appears as

$$C_6H_{14(\ell)} + \tfrac{19}{2}\,O_{2(g)} \xrightarrow{\Delta} 6\,CO_{2(g)} + 7\,H_2O_{(g)}$$

For guideline 5, the coefficients must be whole numbers. To obtain a whole number from $\frac{19}{2}$, multiply it by 2; then multiply *all* the other coefficients by 2. The coefficients are also in the lowest possible ratio and each atom is checked off as in guideline 6. The final balanced equation is

$$2\,C_6H_{14(\ell)} + 19\,O_{2(g)} \xrightarrow{\Delta} 12\,CO_{2(g)} + 14\,H_2O_{(g)} \quad \text{(balanced)}$$

Problem Example 9-5

$$KNO_{3(s)} \xrightarrow{\Delta} KNO_{2(s)} + O_{2(g)} \quad \text{(unbalanced)}$$

Before you continue reading, try to balance this equation yourself.

The correct coefficients are $2 \rightarrow 2 + 1$. Did you get that answer? Now, let us see how we arrived at that answer. Since both K and N atoms are present as only one atom each, the next atom that has more than one atom is O; hence, by guideline 2, O atoms become our starting point. Placing a 2 in front of KNO_3 as $2\ KNO_3$, we now have 6 O atoms; likewise, placing a 2 in front of KNO_2, we balance the K and N atoms and have 6 O atoms in the products (4 O from $2\ KNO_2$, and 2 O from O_2). Since the polyatomic ion NO_3^{1-} does not appear as NO_3^{1-} on the right side of the equation, guideline 3 is not applicable. The O atoms are balanced as suggested in guideline 4, and all coefficients are whole numbers and the lowest possible ratio as suggested in guideline 5. Checking off each atom in guideline 6 gives the following balanced equation:

$$2\ \overset{\checkmark\checkmark\checkmark}{KNO_{3(s)}} \xrightarrow{\Delta} 2\ \overset{\checkmark\checkmark\checkmark}{KNO_{2(s)}} + \overset{\checkmark}{O_{2(g)}} \quad \text{(balanced)}$$

Problem Example 9-6

$$KClO_{3(s)} \xrightarrow[\Delta]{MnO_2} KCl_{(s)} + O_{2(g)} \quad \text{(unbalanced)}$$

This equation is the laboratory method for the production of oxygen, and this type of experiment is often used in a beginning chemistry course. The manganese(IV) oxide or manganese dioxide (MnO_2) acts as a catalyst to speed up the reaction. Again, try balancing this equation yourself.

The correct coefficients are $2 \rightarrow 2 + 3$. The starting point (guideline 2) is the O atoms. Placing a 2 in front of the $KClO_3$ gives 6 O atoms in the reactants. To obtain 6 O atoms in the products, place a 3 in front of the O_2 and $3\ O_2$. The K and Cl atoms in the products are balanced with a 2 in front of the KCl. The coefficients are whole numbers and the lowest possible ratio (guideline 5). Checking off each atom as in guideline 6 gives the following balanced equation:

Work Problems 4 and 5.

$$2\ \overset{\checkmark\checkmark\checkmark}{KClO_{3(s)}} \xrightarrow[\Delta]{MnO_2} 2\ \overset{\checkmark\checkmark}{KCl_{(s)}} + 3\ \overset{\checkmark}{O_{2(g)}} \quad \text{(balanced)}$$

9-5

Writing and Balancing Word Equations

Word equations are another form of chemical equation without information about coefficients. A **word equation** expresses the chemical equation in words instead of symbols and formulas. To write and balance chemical equations from word equations, we only need to apply our guidelines in 9-3 with special emphasis to 1, or the correct formulas for the elements or compounds must be written from the names. Here we apply the nomenclature you learned in Chapter 7.

Consider the following examples of changing word equations into chemical equations and balancing by inspection:

Problem Example 9-7

Word Equation

Calcium bromide + sulfuric acid → hydrogen bromide + calcium sulfate

By guideline 1, we must first write the correct formulas from each of the names of the compounds. See Tables 6-1 (cations), 6-2 (anions), and 6-4 (polyatomic ions) for the formulas of ions to refresh your memory. Also use the Periodic Table.

Chemical Equation

$$CaBr_2 + H_2SO_4 \rightarrow HBr + CaSO_4 \quad \text{(unbalanced)}$$

Once the correct formula has been written, it must not be changed to balance the equation. The starting point is the $CaBr_2$ (guideline 2). Place a 2 in front of the HBr to balance the Br atoms. The SO_4^{2-} is balanced (guideline 3) and so are the H atoms (guideline 4). The coefficients are whole numbers and in the lowest possible ratio (guideline 5). Check each atom (guideline 6), and the balanced equation appears as follows:

$$CaBr_2 + H_2SO_4 \rightarrow 2\,HBr + CaSO_4 \quad \text{(balanced)}$$

Problem Example 9-8

Word Equation

Aluminum hydroxide + phosphoric acid → aluminum phosphate + water

Write the correct formulas from the names (guideline 1).

Chemical Equation

$$Al(OH)_3 + H_3PO_4 \rightarrow AlPO_4 + H_2O \quad \text{(unbalanced)}$$

The starting point in balancing the equation is the H atoms (guideline 2), since there are a total of 6 H atoms in the reactants. To balance the H atoms in the products, place a 3 in front of the H_2O as 3 H_2O, giving 6 H atoms. The PO_4^{3-} are balanced (guideline 3), the H atoms have just been balanced (guideline 4), and the O atoms not appearing with the PO_4^{3-} are balanced with the 3 H_2O. The coefficients are whole numbers and in the lowest possible ratio (guideline 5). Check each atom (guideline 6), and the balanced equation appears as follows:

$$Al(OH)_3 + H_3PO_4 \rightarrow AlPO_4 + 3\,H_2O \quad \text{(balanced)}$$

Problem Example 9-9

When hydrochloric acid is added drop-by-drop to limestone (calcium carbonate), a gas (carbon dioxide), water, and a salt (calcium chloride) are produced. Writing the word equation from these experimental results gives

Word Equation

Hydrochloric acid + calcium carbonate →

$$\text{carbon dioxide}_{(g)} + \text{water} + \text{calcium chloride}$$

We now write the correct formulas from the names (guideline 1).

Chemical Equation

$$\text{HCl} + \text{CaCO}_3 \rightarrow \text{CO}_{2(g)} + \text{H}_2\text{O} + \text{CaCl}_2 \quad \text{(unbalanced)}$$

The starting point in the balancing of the equation is CaCl_2, since there are 2 Cl atoms (guideline 2). Place a 2 in front of the HCl to balance the Cl atoms. Next, check each of the other elements in the equation making sure they are balanced. Check each atom (guideline 6) to give the following balanced equation:

$$2\ \text{HCl} + \text{CaCO}_3 \rightarrow \text{CO}_{2(g)} + \text{H}_2\text{O} + \text{CaCl}_2 \quad \text{(balanced)}$$

Problem Example 9-10

An insufficient amount of oxygen gas on combustion of the liquid C_8H_{18} gives carbon monoxide gas and water vapor.[6]

Word Equation

$$\text{Oxygen}_{(g)} + \text{C}_8\text{H}_{18(\ell)} \rightarrow \text{carbon monoxide}_{(g)} + \text{water}_{(g)}$$

We now write the correct formulas from the names (guideline 1).

Chemical Equation

$$\text{O}_{2(g)} + \text{C}_8\text{H}_{18(\ell)} \rightarrow \text{CO}_{(g)} + \text{H}_2\text{O}_{(g)} \quad \text{(unbalanced)}$$

---

[6]This equation is also an example of a combustion reaction. These reactions will be discussed in detail in Chapter 13 (13-5). Gasoline is a mixture of carbon and hydrogen-containing compounds (hydrocarbons), one of which has the formula C_8H_{18}. In footnote 3, Chapter 7, we mentioned that carbon monoxide is a major air pollutant. Carbon monoxide is released into the air by incomplete combustion of gasoline from automobiles and is harmful to humans in that the carbon monoxide has a greater affinity for the hemoglobin in the red blood cells than does oxygen. Thus, the hemoglobin is "tied up" by the carbon monoxide and is not able to carry oxygen. Carbon monoxide hence robs the tissues of oxygen required for survival.

Balancing the carbon atoms (guideline 2) by placing an 8 in front of the CO as 8 CO, and then balancing the H atoms (guideline 4) with a 9 in front of the H_2O as 9 H_2O, require 17 O atoms in the reactants. Place $\frac{17}{2}$ or $8\frac{1}{2}$ in front of the O_2 as $\frac{17}{2}$ O_2 (guideline 4) to obtain the needed 17 O atoms. The following equation results:

$$\tfrac{17}{2}\ O_{2(g)} + C_8H_{18(\ell)} \rightarrow 8\ CO_{(g)} + 9\ H_2O_{(g)}$$

Work Problems 6 and 7.

The coefficients are not whole numbers (guideline 5), hence *all* coefficients must be multiplied by 2. Check each atom (guideline 6), and obtain the following balanced equation:

$$17\ O_2 + 2\ C_8H_{18} \rightarrow 16\ CO_{(g)} + 18\ H_2O \quad \text{(balanced)}$$

9-6

Completing Chemical Equations. The Five Simple Types of Chemical Reactions

Now, we shall not only balance the chemical equation but also complete it by writing the products. To write the products in a chemical equation, we must consider a few generalizations about ordinary chemical reactions; hence, we divide the ordinary chemical reactions that we consider in this chapter into five simple types for which we can write chemical equations:

1. combination reactions

2. decomposition reactions

3. single replacement reactions

4. double replacement reactions

5. neutralization reactions

Another type of reaction we shall consider later is the more complex oxidation-reduction reaction. Special techniques are required to write balanced equations for these oxidation-reduction reactions (see Chapter 16). In general, these equations cannot be balanced "by inspection." However, most combination, decomposition, and replacement reactions are simple cases of oxidation-reduction reactions that can be balanced "by inspection."

9-7

Combination Reactions

In **combination reactions,** two or more substances (either elements or compounds) react to produce *one* substance. Combination reactions are also called *synthesis reactions*. This reaction can be shown by a general equation,

$$A + Z \rightarrow AZ \text{ where } A \text{ and } Z \text{ are elements or compounds}$$

Consider the following equations of combination reactions:

1.   Metal + oxygen $\xrightarrow{\Delta}$ metal oxide (Card 1)[7]

$$2\ Mg_{(s)} + O_{2(s)} \xrightarrow{\Delta} 2\ MgO_{(s)}$$

The formula of the product is determined from a knowledge of the ionic charges of Mg and O in the combined state. Magnesium in the combined state has an ionic charge of 2^+ and oxygen 2^-. Hence, the correct formula of the metal oxide is MgO.[8]

2.   Nonmetal + oxygen $\xrightarrow{\Delta}$ nonmetal oxide   (Card 2)

a. $$S_{(s)} + \underset{\text{limited}}{O_{2(g)}} \xrightarrow{\Delta} SO_{2(g)}$$

and, with an *excess* of oxygen,

$$2\ S_{(s)} + 3\ \underset{\text{excess}}{O_{2(g)}} \xrightarrow{\Delta} 2\ SO_{3(g)}$$

The formula of the product can only be determined by a knowledge of the oxides of the nonmetal, that is, SO_2 and SO_3. Therefore, these two oxides should be included on the product side of your card. With limited oxygen the oxide is formed that has the smaller amount of oxygen in its formula, that is SO_2. With excess oxygen the oxide is formed that has the greater amount of oxygen in its formula, that is, SO_3.

b. $$2\ C_{(s)} + \underset{\text{limited}}{O_{2(g)}} \xrightarrow{\Delta} 2\ CO_{(g)}$$

and, with an excess of oxygen,

$$C_{(s)} + \underset{\text{excess}}{O_{2(g)}} \xrightarrow{\Delta} CO_{2(g)}$$

Again, a knowledge of the oxides of carbon is used to determine the formulas of the products, that is, CO and CO_2. Put these oxides on your card. Limited oxygen gives CO, while excess oxygen gives CO_2.

---

[7]Some students find it helpful to make flash cards of these reactions with the reactants on one side and the products on the other side. For those that plan to do this, we will mark the reaction by writing "card" and then its number next to the equation. To be able to complete and balance equations, you will need to apply the general principles you learned in Chapter 6 on the use of the Periodic Table (see 6-8). Therefore, refer to the Periodic Table as you study this material.

[8]Do not balance the equation by changing the formula of MgO, that is,

$$Mg_{(s)} + O_{2(g)} \xrightarrow{\Delta} MgO_2 \text{ (wrong)}$$

This equation is wrong because the formula of magnesium oxide is incorrect.

3.
$$\text{Metal } + \text{ nonmetal} \rightarrow \text{salt} \quad \text{(Card 3)}$$

$$2 \, Na_{(s)} + Cl_{2(g)} \rightarrow 2 \, NaCl_{(s)}$$

The formula of the product (NaCl) is determined from a knowledge of the ionic charges of Na and Cl in the combined state, that is, Na^{1+} and Cl^{1-}.

4.
$$\text{Water } + \text{ metal oxide} \rightarrow \text{base (metal hydroxide)} \quad \text{(Card 4)}$$

$$H_2O_{(\ell)} + MgO_{(s)} \rightarrow Mg(OH)_{2(s)}$$

The formula of the product is determined from a knowledge of the ionic charge of Mg in the combined state and the ionic charge on the hydroxide ion, that is Mg^{2+} and OH^{1-}—hence, $Mg(OH)_2$. Due to the formation of a base (metal hydroxide) from a metal oxide and water, the metal oxide is sometimes called a **basic oxide.**

5.
$$\text{Water } + \text{ nonmetal oxide} \rightarrow \text{oxyacid[9]} \quad \text{(Card 5)}$$

a.
$$H_2O_{(\ell)} + SO_{2(g)} \rightarrow H_2SO_{3(aq)}$$

The formula of the product is determined from the oxidation number (see 6-1) of S in SO_2, which is 4^+. This S atom forms an acid with the *same* oxidation number in the product; hence, the formula of the acid is H_2SO_3 (oxidation number of S = 4^+) and *not* H_2SO_4 (oxidation number of S = 6^+). The oxidation number is the *same* in the reactants as in the products for all such reactions of nonmetal oxides with water that we will consider in this chapter. Due to the formation of an acid from nonmetal oxides and water, such a nonmetal oxide is called an **acid oxide.** Sulfur dioxide, an air pollutant, is harmful to people partly because it combines with moisture in the eyes, throat, and lungs to form sulfurous acid (H_2SO_3), as shown by the preceding equation.

b.
$$H_2O_{(\ell)} + SO_{3(g)} \rightarrow H_2SO_{4(aq)}{}^{10}$$

---

[9]Sulfur oxides and nitrogen oxides react with water to form their respective acids. During precipitation as rain, snow, sleet, or fog these acids fall back on the earth and are referred to as "acid rain."

[10]The Monsanto Company and many other chemical companies have developed processes to control sulfur oxides that are air pollutants. The Monsanto process oxidizes the sulfur oxides to SO_3 and then converts this gas to dilute sulfuric acid, as shown in the equation. The oxygen in the air can also oxidize sulfur dioxide to sulfur trioxide. Therefore, on foggy days, with polluted air, you may be inhaling a dilute solution of sulfuric acid. High-sulfur coal can also be cleaned by a similar process developed by General Motors. When the coal is burned, sulfur dioxide gas is produced. The sulfur dioxide gas is trapped and converted by various reactions to calcium sulfate and calcium sulfite. These salts are used as landfills. The process removes about 90 percent of the sulfur dioxide from the gas.

The formula of the product is determined from the oxidation number of S in SO_3, which is 6^+, and this S atom forms an acid with the same oxidation number. Therefore, the formula of the acid is H_2SO_4. On your card you should write SO_2 forms SO_3^{2-} (oxidation number of S $= 4^+$), and SO_3 forms SO_4^{2-} (oxidation number of S $= 6^+$).

6. Metal oxide (basic oxide) + nonmetal oxide (acid oxide) $\rightarrow$ salt

(Card 6)

$$MgO_{(s)} + SO_{3(g)} \rightarrow MgSO_{4(s)}$$

The SO_3 will form the SO_4^{2-} polyatomic ion, as mentioned, and thus the correct formula for the salt based on the ionic charge of Mg in the combined state and on the ionic charge on the SO_4^{2-} ion is $MgSO_4$. If the gas were SO_2, what would be the product? (See Card 5.)

7.   Ammonia + acid $\rightarrow$ ammonium salt   (Card 7)

$$NH_{3(g)} + HCl_{(g)} \rightarrow NH_4Cl_{(s)}$$

The ammonia gas reacts with the hydrogen chloride gas from hydrochloric acid to form the ammonium salt, ammonium chloride. The correct formula for ammonium chloride can be written only from a knowledge of the ionic charges of the ammonium ion and the chloride ion, that is, NH_4^{1+} and Cl^{1-} — hence, NH_4Cl. The unshared pair of electrons on nitrogen of the NH_3 molecule forms a coordinate covalent bond (see 6-5) with the proton (H^{1+}) from the HCl molecule to give the NH_4^{1+} ion. The thin film you often see on the reagent bottles or on the windows of the laboratory is due in part to the formation of solid ammonium chloride, as the preceding equation shows.

Work Problems 8 and 9.

9-8

Decomposition Reactions

In **decomposition reactions,** one substance undergoes a reaction to form two or more substances. The substance broken down is always a compound, and the products may be elements or compounds. Heat is often necessary for this process. This reaction can be represented by a general equation,

$$AZ \rightarrow A + Z \text{ where } A \text{ and } Z \text{ are elements or compounds}$$

In general, a prediction of the products in a decomposition reaction can only be determined by a knowledge of each individual reaction; therefore, you will need to make cards for each individual reaction.

Consider the following equations of decomposition reactions:

1. Some compounds decompose to yield oxygen:

a.
$$2 \, HgO_{(s)} \xrightarrow{\Delta} 2 \, Hg_{(\ell)} + O_{2(g)}{}^{11} \quad \text{(Card 8)}$$

The red mercury(II) oxide, when heated, forms droplets of mercury along the edge of the test tube, and oxygen, which supports combustion (see 3-3 and Figure 9-1), is evolved. The production of oxygen is tested by the use of a glowing splint being put into the test tube. The glowing splint catches on fire and burns. For all equations, the order for writing the products makes no difference; that is, O_2 may be written first instead of Hg. Make sure the equation is balanced

b.
$$2 \, KNO_{3(s)} \rightarrow 2 \, KNO_{2(s)} + O_{2(g)} \quad \text{(Card 9)}$$

This reaction is used for the production of oxygen in the laboratory. Once you write the products, try balancing the equation. Do *not* memorize the coefficients.

c.
$$2 \, KClO_{3(s)} \xrightarrow[\Delta]{MnO_2} 2 \, KCl_{(s)} + 3 \, O_{2(g)} \quad \text{(Card 10)}$$

This reaction is the usual laboratory method for the production of oxygen.

FIGURE 9-1

The decomposition of red mercury(II) oxide to form mercury and oxygen.

Safety glasses

Mercury
Oxygen gas
Mercury (II) oxide

---

[11]Do not write the equation as

$$HgO_{(s)} \xrightarrow{\Delta} Hg_{(\ell)} + O_{(g)} \quad \text{(wrong)}$$

Oxygen exists as a diatomic molecule and is written O_2 (see 6-4).

d.
$$2\ H_2O_{(\ell)} \xrightarrow[\text{electric current}]{\text{direct}} 2\ H_{2(g)} + O_{2(g)} \quad \text{(Card 11)}$$

Electrolysis decomposes water into hydrogen and oxygen (see 3-3) if the water contains a trace of an ionic compound, such as sulfuric acid.

e.
$$2\ H_2O_{2(aq)} \xrightarrow[\text{light}]{\Delta\ \text{or}} 2\ H_2O_{(\ell)} + O_{2(g)} \quad \text{(Card 12)}$$

Hydrogen peroxide decomposes when heated or in the presence of light to yield water and oxygen.[12]

2.   Some carbonates, when heated, decompose to yield carbon dioxide:

$$CaCO_{3(s)} \xrightarrow{\Delta} CaO_{(s)} + CO_{2(g)} \quad \text{(Card 13)}$$

When limestone (calcium carbonate) is heated, carbon dioxide is one of the products. The production of carbon dioxide is also tested by the use of a glowing splint being put into the test tube. The glowing splint goes out, as carbon dioxide does not support combustion. If the carbonate were $SrCO_3$, what would be the products? (See group IIA in the Periodic Table.)

3.   Hydrates,[13] when heated, decompose to yield water:

$$MgSO_4 \cdot 7\ H_2O_{(s)} \xrightarrow{\Delta} MgSO_{4(s)} + 7\ H_2O_{(g)} \quad \text{(Card 14)}$$

When a hydrate such as Epsom salt crystals (magnesium sulfate heptahydrate) is heated, magnesium sulfate and water are produced, the water being released as a vapor. If the hydrate were $BaCl_2 \cdot 2\ H_2O$, or any other hydrate, what would be the products?

4.   Some compounds (not hydrates) decompose when heated to yield water:

$$C_{12}H_{22}O_{11(s)} \xrightarrow{\Delta} 12\ C_{(s)} + 11\ H_2O_{(g)} \quad \text{(Card 15)}$$

Sugar ($C_{12}H_{22}O_{11}$) decomposes when heated to form a caramel brown or black solid (carbon) and water (see 3-1 and 3-3) as a vapor. If the compound were $C_6H_{12}O_6$ (glucose, dextrose), what would be the products?

---

[12]Hydrogen peroxide is used to bleach cloth and hair. It is also used as an antiseptic. The release of oxygen can be observed when a dilute solution (3 percent) used as an antiseptic bubbles when it comes into contact with a wound or organic matter.

[13]Hydrates are crystalline salts that contain chemically bound water in definite proportions. Hydrates will be further discussed in Chapter 13.

5. Some compounds such as hydrogen carbonates, when heated, decompose to yield a carbonate salt, water, and carbon dioxide:

$$2\ NaHCO_{3(s)} \xrightarrow{\Delta} Na_2CO_{3(s)}\ +\ H_2O_{(g)}\ +\ CO_{2(g)} \quad (\text{Card 16})$$

When heated, baking soda (sodium hydrogen carbonate or sodium bicarbonate) decomposes to produce sodium carbonate, water vapor, and carbon dioxide.[14] If the hydrogen carbonate were $KHCO_3$, what would be the products? (See group IA in the Periodic Table.)

Work Problems 10 and 11.

9-9

Single Replacement Reactions. The Electromotive or Activity Series

In **single replacement reactions,** one element reacts by replacing another element in a compound. Single replacement reactions are also called *replacement, substitution,* or *displacement reactions.* This reaction can be represented by two general equations:

1. A metal replacing a metal ion in its salt or a hydrogen ion in an acid

$$\underset{\longrightarrow}{A\ +\ B}Z \rightarrow AZ\ +\ B \quad (\text{Card 17})$$

2. A nonmetal replacing a nonmetal ion in its salt or acid

$$\underset{\longrightarrow}{X\ +\ B}Z \rightarrow BX\ +\ Z \quad (\text{Card 18})$$

Li
K
Ba
Ca
Na
Mg
Al
Zn
Fe
Cd
Ni
Sn
Pb
(H)
Cu
Hg
Ag
Au

In the first case, replacement depends on the two metals involved, that is, *A* and *B*. It has been possible to arrange the metals in a series called the **electromotive or activity series,** so that each element in it will displace any of those following it from an aqueous solution of its salt or acid. The reasons for this series order will be discussed in Chapter 16 (16-6). Although hydrogen is not a metal, it is *included* in this series: **Li, K, Ba, Ca, Na, Mg, Al, Zn, Fe, Cd, Ni, Sn, Pb, (H), Cu, Hg, Ag, Au.** You should be able to interpret and use this series so that you can complete and balance chemical equations involving single replacement reactions. (This series is also listed on the inside back cover of this text.)

Consider the following replacement reactions:

1. $$Fe_{(s)}\ +\ CuSO_{4(aq)} \rightarrow FeSO_{4(aq)}\ +\ Cu_{(s)}$$

Iron is higher in the electromotive or activity series than copper. Hence, iron will replace the copper(II) from its salt. For metals existing in vari-

---

[14]Baking soda can be used to put out fires, because the carbon dioxide formed in its decomposition smothers flames (see Figure 7-1).

able oxidation numbers, the *lower* oxidation number is often formed. Thus, the new salt is $FeSO_4$ and not $Fe_2(SO_4)_3$.

2. $$Zn_{(s)} + 2\ AgNO_{3(aq)} \rightarrow Zn(NO_3)_{2(aq)} + 2\ Ag_{(s)}$$

Zinc is higher in the electromotive or activity series than silver; hence, zinc will replace the silver ion from its salt.

3. $$Zn_{(s)} + H_2SO_{4(aq)} \rightarrow ZnSO_{4(aq)} + H_{2(g)}$$

Zinc is higher in the electromotive or activity series than hydrogen, and thus zinc will replace the hydrogen ion from the *acid*. Note that hydrogen is a diatomic molecule (see 6-4).

4. $$Sn_{(s)} + 2\ HCl_{(aq)} \rightarrow SnCl_{2(aq)} + H_{2(g)}$$

Tin is higher in the electromotive or activity series than hydrogen, and hence tin will replace the hydrogen ion from the acid. The salt with the lower oxidation number for the metal is formed, and the new salt is $SnCl_2$, not $SnCl_4$.

5. $$Cu_{(s)} + MgCl_{2(aq)} \rightarrow no\ reaction\ (NR)$$

Copper is below magnesium in the electromotive or activity series; therefore, no reaction will occur.

6. $$2\ K_{(s)} + 2\ H_2O_{(\ell)} \rightarrow 2\ KOH_{(aq)} + H_{2(g)}$$

Potassium is higher in the electromotive or activity series and can replace *one* hydrogen atom from water to form the hydroxide and hydrogen. Writing water as H—OH, the replacement of *one* hydrogen atom from water by potassium simplifies the understanding of the equation. Only very active metals high in the electromotive series react with water, because water is (tightly) covalently bonded. The first five metals in the series—that is, lithium, potassium, barium, calcium, and sodium—replace *one* hydrogen atom from water to form the metal hydroxide and hydrogen gas (H_2).

F_2
Cl_2
Br_2
I_2

In the second case, when a nonmetal replaces a nonmetal ion from its salt or acid, replacement depends on the two nonmetals involved—that is, X and Z. A series similar to the electromotive or activity series exists for the halogen nonmetals: F_2, Cl_2, Br_2, I_2. Fluorine will replace the chloride ion from an aqueous solution of its salt; chlorine will replace bromide; and bromine will replace iodide. Notice that this halogen series follows the nonmetallic properties as given in the Periodic Table (see 5-3, number 4).

Consider the following replacement reactions:

1.
$$Cl_{2(g)} + 2 NaBr_{(aq)} \rightarrow 2 NaCl_{(aq)} + Br_{2(aq)}$$

Chlorine gas replaces bromide from an aqueous solution of its salt to yield the chloride salt and bromine. (A common error is to forget to write bromine as a diatomic molecule—see 6-4.)

2.
$$Br_{2(aq)} + 2 NaI_{(aq)} \rightarrow 2 NaBr_{(aq)} + I_{2(aq)}$$

Work Problems 12 and 13.

Bromine dissolved in water replaces iodide from an aqueous solution of its salt to yield the bromide salt and iodine dissolved in water.

9-10

Double Replacement Reactions. Rules for the Solubility of Inorganic Substances in Water

In **double replacement reactions,** *two* compounds are involved in a reaction, with the positive ion (cation) of one compound *exchanging* with the positive ion (cation) of another compound. In other words, the two positive ions exchange negative ions (anions) or partners. Double replacement reactions are also called *metathesis* (meaning "a change in state, substance, or form") or *double decomposition* reactions. This reaction can be represented by the general equation

$$\overset{\frown}{AX} + \overset{\frown}{BZ} \rightarrow AZ + BX^{15} \quad \text{(Card 19)}$$

In many double replacement reactions, a precipitate froms,[16] since one of the compounds formed is insoluble or only slightly soluble in water. This precipitate can be indicated in an equation by underscoring, as AgCl, or by a downward arrow, as AgCl ↓ . A better method is to indicate it by an (s), as AgCl$_{(s)}$. To recognize that a precipitate will form, you must be able to interpret and use the following rules for the solubility of inorganic substances in water (there are exceptions):

1. Nearly all *nitrates* (NO_3^{1-}) and *acetates* ($C_2H_3O_2^{1-}$) are *soluble.*

2. All *chlorides* (Cl^{1-}) are *soluble,* except AgCl, Hg_2Cl_2, and $PbCl_2$. ($PbCl_2$ is soluble in hot water.)

---

[15]Note that in double replacement reactions there are *four* separate particles, that is A, X, B, and Z, whereas in single replacement reactions there are only *three,* that is, A, B, and Z. Single replacement reactions depend on the electromotive or activity series, whereas double replacement reactions do not.

[16]A reaction is said to go to completion if as products the following substances are formed: (1) precipitate, (2) gas, or (3) a nonionized or only partially ionized substance, such as water. These "irreversible" reactions will be further discussed in 17-2.

3. All *sulfates* (SO_4^{2-}) are *soluble,* except $BaSO_4$, $SrSO_4$, and $PbSO_4$. ($CaSO_4$ and Ag_2SO_4 are only slightly soluble.)

4. Most of the *alkali metals* (Group IA, Li, Na, K, etc.) salts and *ammonium* (NH_4^{1+}) salts are *soluble.*

5. All the common *acids* are *soluble.*

6. All *oxides* (O^{2-}) and *hydroxides* (OH^{1-}) are **insoluble,** except those of the alkali metals and certain alkaline earth metals (Group IIA, Ca, Sr, Ba, Ra). [$Ca(OH)_2$ is only moderately soluble.]

7. All *sulfides* (S^{2-}) are **insoluble,** except those of the alkali metals, alkaline earth metals, and ammonium sulfide.

8. All *phosphates* (PO_4^{3-}) and *carbonates* (CO_3^{2-}) are **insoluble,** except those of the alkali metals and ammonium salts.

These rules will be quite useful in writing double replacement equations. They are also useful in understanding the actions of some chemicals used in everyday life. For example, vinegar (about 5 percent acetic acid) is often used to remove water spots on glass, caused by the presence of calcium, magnesium, and iron salts in hard water. Vinegar is used because the acetic acid in vinegar reacts with certain calcium, magnesium, and iron salts to form a new salt, an acetate salt, which is soluble in water (rule 1) and hence can be easily removed with water. (These rules are also given on the inside back cover of this text.)

Consider the following double replacement reactions:

1. A salt and an acid to form a precipitate:

$$AgNO_{3(aq)} + HCl_{(aq)} \rightarrow AgCl_{(s)} + HNO_{3(aq)}$$

The silver ion changes places with the hydrogen ion to form the insoluble silver chloride (AgCl, see rule 2), and the hydrogen ion reacts with the nitrate ion to form a new acid, nitric acid (NHO_3). The formation of an insoluble or slightly ionized compound acts as the driving force behind these reactions.

2. A salt and a base to form a precipitate:

$$Ni(NO_3)_{2(aq)} + 2\ NaOH_{(aq)} \rightarrow Ni(OH)_{2(s)} + 2\ NaNO_{3(aq)}$$

In the exchange of ions, a new salt, $NaNO_3$, and a new base, $Ni(OH)_2$, which is insoluble in water (see rule 6) are formed.

3. Two salts to form a precipitate:

$$2\ NaCl_{(aq)} + Pb(NO_3)_{2(aq)} \xrightarrow{\text{cold}} PbCl_{2(s)} + 2\ NaNO_{3(aq)}$$

In the exchange of ions, two new salts are formed: lead(II) chloride ($PbCl_2$, see rule 2), which is insoluble in cold water, and sodium nitrate ($NaNO_3$), which is soluble in water (see rules 1 and 4). These double replacement reactions can be summarized by two general statements:

$$Salt_1 + acid_1 \text{ or } base_1 \rightarrow salt_2 + acid_2 \text{ or } base_2$$

$$Salt_1 + salt_2 \rightarrow salt_3 + salt_4$$

In both cases, one of the two products is usually insoluble in water or weakly ionized.

4. Metal carbonate and an acid:

$$MgCO_{3(s)} + 2 HCl_{(aq)} \rightarrow MgCl_{2(aq)} + H_2O_{(\ell)} + CO_{2(g)}$$

The magnesium ion changes places with the hydrogen ion to form the salt, $MgCl_2$. The hydrogen ion reacts with the carbonate ion to form carbonic acid (H_2CO_3), which is *unstable* and which *decomposes* to form water and carbon dioxide. Magnesium carbonate is one of the ingredients in Tums; this reaction is the basis of the action of Tums as an antacid in neutralizing some acid (HCl) in the stomach. This type of double replacement reaction can be summarized as:

Work Problems 14
and 15.

Metal carbonate + acid → salt + water + carbon dioxide

9-11

Neutralization Reactions

A **neutralization reaction** is one in which an acid (or an acid oxide) reacts with a base (or basic oxide). In most of these reactions, water is one of the products. The formation of water acts as the driving force behind the neutralization, since water is only slightly ionized and heat is also given off in its formation. This reaction can be represented by a general equation,

$$HX + MOH \rightarrow MX + HOH \quad \text{(Card 20)}$$

where HX is an acid and MOH is a base. Water is one of the products.

The differences between neutralization reactions and ordinary double replacement reactions are: (1) an acid (or an acid oxide) reacts with a base (or a basic oxide) in a neutralization reaction; and (2) water is usually one of the products of a neutralization reaction.

Consider the following equations of neutralization reactions:

1. An acid and a base:

   a. $$HCl_{(aq)} + NaOH_{(aq)} \rightarrow NaCl_{(aq)} + H_2O_{(\ell)}$$

The hydrogen ion reacts with the hydroxide ion to form slightly ionized water molecules. This leaves the sodium ion (Na^{1+}) and the chloride ion (Cl^{1-}) to form an aqueous solution of sodium chloride (NaCl, see rule 2).

b.
$$H_2SO_{4(aq)} + Ba(OH)_{2(aq)} \rightarrow BaSO_{4(s)} + 2\ H_2O_{(\ell)}$$

Again, the hydrogen ions react with the hydroxide ions to form slightly ionized water molecules. This leaves the barium ion, Ba^{2+}, and the sulfate ion, (SO_4^{2-}), to form the water insoluble barium sulfate, $BaSO_4$ (see rule 3).

c.
$$2\ HCl_{(aq)} + Mg(OH)_{2(s)} \rightarrow MgCl_{2(aq)} + 2\ H_2O_{(\ell)}{}^{[17]}$$

Again, an acid and a base react to form water and a salt, $MgCl_2$. This type of neutralization reaction can be summarized as

$$\text{Acid} + \text{base} \rightarrow \text{salt} + \text{water}$$

2. An acid may also neutralize a basic oxide (metal oxide):

$$ZnO_{(s)} + 2\ HCl_{(aq)} \rightarrow ZnCl_{2(aq)} + H_2O_{(\ell)}$$

The zinc ion changes places with the hydrogen ion to form the salt, $ZnCl_2$; the hydrogen ion reacts with the oxide ion to form slightly ionized water. This type of neutralization reaction can be summarized as follows:

$$\text{Basic oxide (metal oxide)} + \text{acid} \rightarrow \text{salt} + \text{water}$$

3. A base may also neutralize an acid oxide (nonmetal oxide):

$$CO_{2(g)} + 2\ LiOH_{(s)} \rightarrow Li_2CO_{3(s)} + H_2O_{(\ell)}{}^{[18]}$$

The carbon dioxide reacts with the lithium hydroxide to form water and a salt, Li_2CO_3. The oxidation number of C in CO_2 is 4^+, the same as it is in the salt, Li_2CO_3. The correct formula of the salt is written from a knowledge of the ionic charge of lithium and the ionic charge on the car-

---

[17]This reaction occurs in the stomach when milk of magnesia [$Mg(OH)_2$] is used as an antacid. The milk of magnesia neutralizes some acid in the stomach to form a salt and water, which are less irritating to the inflamed tissue than is the acid.

[18]The absorption of carbon dioxide by lithium hydroxide is one of the reasons lithium hydroxide was used in filters in the cabin atmosphere of the Apollo space missions (see footnote 18, Chapter 7).

TABLE 9-2 Summary of the Five Simple Types of Chemical Reactions

Combination reactions:	$A + Z \rightarrow AZ$
Decomposition reactions:	$AZ \rightarrow A + Z$
Replacement reactions:	$A + BZ \rightarrow AZ + B$
	$X + BZ \rightarrow BX + Z$
Double replacement reactions:	$AX + BZ \rightarrow AZ + BX$
Neutralization reactions:[a]	$HX + MOH \rightarrow MX + HOH$

[a]HX is an acid, and MOH is a base. An acid oxide (nonmetal oxide) may be substituted for the acid, and a basic oxide (metal oxide) may be substituted for the base in a neutralization reaction. A nonmetal oxide plus a metal oxide forms a salt, but no water.

bonate, that is, Li^{1+} and CO_3^{2-} — hence, Li_2CO_3. On card 20 you should write "CO_2 forms CO_3^{2-}" (oxidation number of $C = 4^+$). This type of neutralization reaction can be summarized as

$$\text{Acid oxide (nonmetal oxide)} + \text{base} \rightarrow \text{salt} + \text{water}$$

4. A basic oxide (metal oxide) neutralizes an acid oxide (nonmetal oxide):

$$BaO_{(s)} + SO_{3(g)} \rightarrow BaSO_{4(s)}$$

Work Problems 16 and 17.

This type of reaction was previously discussed under combination reactions (see 9-7, number 6) because only *one* substance was formed. However, this reaction is also a neutralization in that a basic oxide reacts with an acid oxide to form a salt, but no water is formed in the reaction.

Table 9-2 summarizes the general reactions for the five simple types of chemical reactions discussed in this chapter.

Chapter Summary

A *chemical equation* is a shorthand way of expressing a chemical change (reaction) in terms of symbols and formulas. Chemical equations are balanced because of the Law of Conservation of Mass. This requires that the number of atoms or moles of atoms of each element to be the same on both sides of the equation. To write and balance word equations, the correct formulas for the elements or compounds must be written from the names. To do this the names and formulas of the cations (Table 6-1), anions (Table 6-2), and polyatomic ions (Table 6-4) must be known.

The five simple types of chemical reactions are: (1) combination reactions, (2) decomposition reactions, (3) single replacement reactions, (4) double replacement reactions, and (5) neutralization reactions. Chemical equations for these reactions were both *completed* and *balanced*. To complete these equations, given the reactants, the products must be known or determined.

EXERCISES

1. Define or explain the following terms:
   - (a) chemical equation
   - (b) catalyst
   - (c) word equation
   - (d) combination reactions
   - (e) decomposition reactions
   - (f) single replacement reactions
   - (g) double replacement reactions
   - (h) neutralization reactions
   - (i) electromotive or activity series

2. Explain the meaning of the following symbols or terms found in chemical equations:
   - (a) $\rightarrow$
   - (b) $\rightleftharpoons$
   - (c) $=$
   - (d) $\uparrow$
   - (e) $\downarrow$
   - (f) the _____ in *compound*
   - (g) (g)
   - (h) (ℓ)
   - (i) (s)
   - (j) (aq)
   - (k) the Δ in $\xrightarrow{\Delta}$
   - (l) the Fe in $\xrightarrow{Fe}$

3. Distinguish between
   - (a) word equations and chemical equations
   - (b) $\uparrow$ and $\downarrow$
   - (c) $\xrightarrow{\Delta}$ and $\xrightarrow{MnO_2}$

PROBLEMS

Balancing Equations   (See Sections 9-1, 9-2, 9-3, and 9-4)

4. Balance each of the following equations by inspection:
   - (a) $BaCl_{2(aq)} + (NH_4)_2CO_{3(aq)} \rightarrow BaCO_{3(s)} + NH_4Cl_{(aq)}$
   - (b) $KClO_{3(s)} \xrightarrow{\Delta} KCl_{(s)} + O_{2(g)}$
   - (c) $Al(OH)_{3(s)} + NaOH_{(aq)} \rightarrow NaAlO_{2(aq)} + H_2O_{(\ell)}$
   - (d) $Fe(OH)_{3(s)} + H_2SO_{4(aq)} \rightarrow Fe_2(SO_4)_{3(aq)} + H_2O_{(\ell)}$
   - (e) $Na_{(s)} + H_2O_{(\ell)} \rightarrow NaOH_{(aq)} + H_{2(g)}$
   - (f) $Mg_{(s)} + N_{2(g)} \rightarrow Mg_3N_{2(s)}$
   - (g) $Mg_{(s)} + O_{2(g)} \rightarrow MgO_{(s)}$
   - (h) $AgNO_{3(aq)} + CuCl_{2(aq)} \rightarrow AgCl_{(s)} + Cu(NO_3)_{2(aq)}$
   - (i) $C_2H_6O_{(\ell)} + O_{2(aq)} \rightarrow CO_{2(g)} + H_2O_{(\ell)}$
   - (j) $FeCl_{2(aq)} + Na_3PO_{4(aq)} \rightarrow Fe_3(PO_4)_{2(s)} + NaCl_{(aq)}$

5. Balance each of the following equations by inspection:
   - (a) $CaC_2 + H_2O \rightarrow C_2H_2 + Ca(OH)_2$
   - (b) $MnO_2 + Al \xrightarrow{\Delta} Al_2O_3 + Mn$
   - (c) $CaCO_3 + H_3PO_4 \rightarrow Ca_3(PO_4)_2 + CO_2 + H_2O$
   - (d) $Al + H_2SO_4 \rightarrow Al_2(SO_4)_3 + H_2$
   - (e) $P_4O_{10} + H_2O \xrightarrow{\Delta} H_3PO_4$
   - (f) $C_3H_8 + O_2 \rightarrow CO_2 + H_2O$
   - (g) $Na_2O + P_4O_{10} \rightarrow Na_3PO_4$
   - (h) $PCl_5 + H_2O \rightarrow H_3PO_4 + HCl$
   - (i) $Sb_2O_3 + NaOH \rightarrow NaSbO_2 + H_2O$
   - (j) $TiCl_4 + H_2O \rightarrow TiO_2 + HCl$

Word Equations   (See Section 9-5)

6. Change the following word equations into chemical equations and balance by inspection:
   (a) Sodium chloride + lead(II) nitrate → lead(II) chloride + sodium nitrate
   (b) Ferric oxide + hydrochloric acid → ferric chloride + water
   (c) Sodium hydrogen carbonate + phosphoric acid →
   $\qquad\qquad\qquad\qquad\qquad\qquad$ sodium phosphate + carbon dioxide + water
   (d) Mercury + oxygen → mercury(II) oxide
   (e) Calcium iodide + sulfuric acid → hydrogen iodide + calcium sulfate
   (f) Barium nitrate + sulfuric acid → barium sulfate + nitric acid
   (g) Magnesium cyanide + hydrochloric acid →
   $\qquad\qquad\qquad\qquad\qquad\qquad$ hydrogen cyanide + magnesium chloride
   (h) Iron(II) sulfide + hydrobromic acid→
   $\qquad\qquad\qquad\qquad\qquad\qquad$ iron(II) bromide + hydrogen sulfide
   (i) Sodium hydrogen sulfite + sulfuric acid→
   $\qquad\qquad\qquad\qquad\qquad\qquad$ sodium sulfate + sulfur dioxide + water
   (j) Aluminum sulfate + sodium hydroxide →
   $\qquad\qquad\qquad\qquad\qquad\qquad$ aluminum hydroxide + sodium sulfate

7. Change the following word equations into chemical equations and balance by inspection:
   (a) Iron + chlorine → iron(III) chloride
   (b) Potassium nitrate $\xrightarrow{\Delta}$ potassium nitrite + oxygen
   (c) Barium + water → barium hydroxide + hydrogen
   (d) Sodium hydroxide + sulfuric acid → sodium hydrogen sulfate + water
   (e) Ammonium sulfide + mercuric bromide →
   $\qquad\qquad\qquad\qquad\qquad\qquad$ ammonium bromide + mercuric sulfide
   (f) Zinc hydroxide + sulfuric acid → zinc sulfate + water
   (g) Tin(II) oxide + hydrochloric acid → tin(II) chloride + water
   (h) Strontium sulfite + acetic acid →
   $\qquad\qquad\qquad\qquad\qquad\qquad$ strontium acetate + water + sulfur dioxide
   (i) Hydrogen bromide + calcium hydroxide → calcium bromide + water
   (j) Bismuth sulfide + oxygen → bismuth oxide + sulfur dioxide

Completing Chemical Equations—Combination Reactions   (See Section 9-7 and the Periodic Table)

8. Complete and balance the following equations:
   (a) $Ca_{(s)} + O_{2(g)} \xrightarrow{\Delta}$
   (b) $B_{(s)} + O_{2(g)} \xrightarrow{\Delta}$
   (c) $S_{(s)} + \text{excess } O_{2(g)} \xrightarrow{\Delta}$
   (d) $Al_{(s)} + Cl_{2(g)} \xrightarrow{\Delta}$
   (e) $CaO_{(s)} + H_2O_{(\ell)} \rightarrow$
   (f) $Li_2O_{(s)} + H_2O_{(\ell)} \rightarrow$
   (g) $SO_{2(g)} + H_2O_{(\ell)} \rightarrow$
   (h) $SO_{3(g)} + H_2O_{(\ell)} \rightarrow$
   (i) $NH_{3(g)} + HBr_{(g)} \rightarrow$
   (j) $P_4O_{10(s)} + H_2O_{(\ell)} \rightarrow$
   (*Hint:* Determine the oxidation number of P in P_4O_{10} first and then the oxidation number of P in the phosphorus-containing polyatomic ions.)

9. Complete and balance the following equations:

(a) $Al + O_2 \xrightarrow{\Delta}$ (b) $C + \text{limited } O_2 \xrightarrow{\Delta}$

(c) $Si + O_2 \xrightarrow{\Delta}$ (d) $Mg + S \xrightarrow{\Delta}$

(e) $Al + N_2 \xrightarrow{\Delta}$ (f) $Na_2O + H_2O \rightarrow$

(g) $Al_2O_3 + H_2O \rightarrow$ (h) $BaO + SO_3 \rightarrow$

(i) $CaO + SO_3 \rightarrow$ (j) $N_2O_5 + H_2O \rightarrow$

                                                    (*Hint:* Determine the oxidation number of N in N_2O_5 first and then the oxidation numbers of N in nitrogen-containing polyatomic ions.)

Completing Chemical Equations—Decomposition Reactions   (See Section 9-8 and the Periodic Table)

10. Complete and balance the following equations:

(a) $HgO_{(s)} \xrightarrow{\Delta}$

(b) $KNO_{3(s)} \xrightarrow{\Delta}$

(c) $H_2O_{(\ell)} \xrightarrow[\text{electric current}]{\text{direct}}$

(d) $SrCO_{3(s)} \xrightarrow{\Delta}$

(e) $CdCO_{3(s)} \xrightarrow{\Delta}$

(f) $PbCO_{3(s)} \xrightarrow{\Delta}$

(g) $CaSO_4 \cdot 2\,H_2O_{(s)} \xrightarrow{\Delta}$

(h) $Al_2(SO_4)_3 \cdot 18\,H_2O_{(s)} \xrightarrow{\Delta}$

(i) $C_6H_{12}O_6 \text{ (dextrose)}_{(s)} \xrightarrow{\Delta}$

(j) $KHCO_{3(s)} \xrightarrow{\Delta}$

11. Complete and balance the following equations:

(a) $H_2O_2 \xrightarrow{\Delta}$ (b) $KClO_3 \xrightarrow{\Delta}$

(c) $MgCO_3 \xrightarrow{\Delta}$ (d) $BaCl_2 \cdot 2\,H_2O \xrightarrow{\Delta}$

(e) $Na_2CO_3 \cdot H_2O \xrightarrow{\Delta}$ (f) $CoSO_4 \cdot 7\,H_2O \xrightarrow{\Delta}$

(g) $C_{12}H_{22}O_{11} \xrightarrow{\Delta}$ (h) $NaHCO_3 \xrightarrow{\Delta}$

(i) $Ca(HCO_3)_2 \xrightarrow{\Delta}$ (j) $Ba(HCO_3)_2 \xrightarrow{\Delta}$

Completing Chemical Equations—Single Replacement Reactions
(See Section 9-9, the Periodic Table, and the Electromotive or Activity Series of Metals)

12. Complete and balance the following equations:

(a) $Cd_{(s)} + H_2SO_{4(aq)} \rightarrow$ (b) $Zn_{(s)} + NiCl_{2(aq)} \rightarrow$

(c) $Pb_{(s)} + HCl_{(aq)} \rightarrow$ (d) $Na_{(s)} + H_2O_{(\ell)} \rightarrow$

(e) $Ca_{(s)} + H_2O_{(\ell)} \rightarrow$ (f) $Cu_{(s)} + NaCl_{(aq)} \rightarrow$

(g) $Fe_{(s)} + CuCl_{2(aq)} \rightarrow$ (h) $Au_{(s)} + NiSO_{4(aq)} \rightarrow$

(i) $Cl_{2(g)} + NaBr_{(aq)} \rightarrow$ (j) $Br_{2(aq)} + NaCl_{(aq)} \rightarrow$

13. Complete and balance the following equations:

   (a) $Al + HCl \rightarrow$
   (b) $Cd + CuSO_4 \rightarrow$
   (c) $Al + HC_2H_3O_2 \rightarrow$
   (d) $Ba + H_2O \rightarrow$
   (e) $Al + SnCl_2 \rightarrow$
   (f) $Cu + FeCl_2 \rightarrow$
   (g) $Pb + HgBr_2 \rightarrow$
   (h) $Al + HgBr_2 \rightarrow$
   (i) $Hg + SnCl_2 \rightarrow$
   (j) $Br_2 + NaI \rightarrow$

Completing Chemical Equations–Double Replacement Reactions
(See Section 9-10, the Periodic Table, and the Rules for the Solubility of Inorganic Substances in Water)

14. Complete and balance the following equations. Indicate any precipitate by (s) and any gas by (g).

   (a) $Pb(NO_3)_{2(aq)} + HCl_{(aq)} \xrightarrow{\text{cold}}$
   (b) $Pb(NO_3)_{2(aq)} + H_2S_{(g)} \rightarrow$
   (c) $Bi(NO_3)_{3(aq)} + NaOH_{(aq)} \rightarrow$
   (d) $CdSO_{4(aq)} + KOH_{(aq)} \rightarrow$
   (e) $Pb(C_2H_3O_2)_{2(aq)} + K_2SO_{4(aq)} \rightarrow$
   (f) $MnSO_{4(aq)} + (NH_4)_2S_{(aq)} \rightarrow$
   (g) $CaCO_{3(s)} + HCl_{(aq)} \rightarrow$
   (h) $ZnCO_{3(s)} + H_3PO_{4(aq)} \rightarrow$
   (i) $BaCO_{3(s)} + HNO_{3(aq)} \rightarrow$
   (j) $FeSO_{4(aq)} + (NH_4)_2S_{(aq)} \rightarrow$

15. Complete and balance the following equations. Indicate any precipitate by (s) and any gas by (g).

   (a) $AgNO_3 + H_2S \rightarrow$
   (b) $FeCl_3 + NaOH \rightarrow$
   (c) $Pb(NO_3)_2 + K_2CrO_4 \rightarrow$
   (d) $SnCl_2 + H_2S \rightarrow$
   (e) $Bi(NO_3)_3 + H_2S \rightarrow$
   (f) $BaCl_2 + (NH_4)_2CO_3 \rightarrow$
   (g) $FeCO_3 + H_2SO_4 \rightarrow$
   (h) $Na_2CO_3 + HC_2H_3O_2 \rightarrow$
   (i) $Pb(NO_3)_2 + H_2S \rightarrow$
   (j) $Ca(NO_3)_2 + NaOH \rightarrow$

Completing Chemical Equations–Neutralization Reactions (See Section 9-11, the Periodic Table, and the Rules for the Solubility of Inorganic Substances in Water)

16. Complete and balance the following equations. Indicate any precipitate by (s).

   (a) $Zn(OH)_{2(s)} + H_2SO_{4(aq)} \rightarrow$
   (b) $Fe(OH)_{3(s)} + H_3PO_{4(aq)} \rightarrow$
   (c) $BaO_{(s)} + H_2SO_{4(aq)} \rightarrow$
   (d) $Na_2O_{(s)} + HNO_{2(aq)} \rightarrow$
   (e) $CO_{2(g)} + Ca(OH)_{2(aq)} \rightarrow$
   (f) $SO_{2(g)} + NaOH_{(aq)} \rightarrow$
   (g) $Al(OH)_{3(s)} + HCl_{(aq)} \rightarrow$
   (h) $Pb(OH)_{2(s)} + HNO_{3(aq)} \rightarrow$
   (i) $KOH_{(aq)} + CO_{2(g)} \rightarrow$
   (j) $MgO_{(s)} + HCl_{(aq)} \rightarrow$

17. Complete and balance the following equations. Indicate any precipitate by (s).

   (a) $Zn(OH)_2 + HNO_3 \rightarrow$
   (b) $Al(OH)_3 + H_3PO_4 \rightarrow$
   (c) $Ca(OH)_2 + HC_2H_3O_2 \rightarrow$
   (d) $Fe_2O_3 + H_3PO_4 \rightarrow$
   (e) $CO_2 + KOH \rightarrow$
   (f) $SO_3 + Fe(OH)_3 \rightarrow$
   (g) $BaO + HCl \rightarrow$
   (h) $MgO + CO_2 \rightarrow$
   (i) $SO_2 + KOH \rightarrow$
   (j) $H_2SO_4 + NaOH \rightarrow$

General Equations

18. (1) Complete and balance the following equations. Indicate any precipitate by (s) and any gas by (g).

    (2) Classify the following reactions as (a) combination, (b) decomposition, (c) single replacement, (d) double replacement, or (e) neutralization.

    (a) $Cd_{(s)} + Cl_{2(g)} \xrightarrow{\Delta}$
    (b) $BaCl_2 \cdot 2\ H_2O_{(s)} \xrightarrow{\Delta}$
    (c) $CaCO_{3(s)} + HNO_{3(aq)} \rightarrow$
    (d) $Cu_{(s)} + MgCl_{2(aq)} \rightarrow$

19. (1) Change the following words into chemical formulas, complete, and balance. Indicate any precipitate by (s) and any gas by (g).

    (2) Classify the following reactions as (a) combination, (b) decomposition, (c) single replacement, (d) double replacement, or (e) neutralization.

    (a) Aluminum + lead(II) chloride $\rightarrow$
    (b) Iron(II) sulfide + hydrochloric acid $\rightarrow$
    (c) Barium chloride + sodium carbonate $\rightarrow$
    (d) Cadmium oxide + hydrochloric acid $\rightarrow$

20. The U.S. Department of Agriculture is studying methods for keeping meat fresh longer in the display cases of supermarkets. One of their proposals is to have a packet of powder inserted inside the meat package. This powder contains citric acid ($H_3C_6H_5O_7$) and sodium hydrogen carbonate. As moisture builds up in the package, carbon dioxide gas is produced which escapes through the package pores. The carbon dioxide controls most of the microorganisms responsible for the spoilage of the meat and hence keeps the meat fresh longer. Complete and balance the equation described above.

21. The Acropolis in Athens, Greece, is slowly deteriorating. It is composed of marble ($CaCO_3$), which slowly reacts with sulfuric acid from air pollution to form a salt, which is washed away destroying this famous historical site. The sulfuric acid is formed from the air pollutant sulfur trioxide and water. Complete and balance the two equations described above.

22. In a recent train derailment, one boxcar contained 55-gallon drums of phosphorus trichloride. Some of these drums broke and reacted with water on the ground and in the moist air to give a dense "white gas" which was noticed over a 12-square-mile area. Complete and balance the equation described above.

23. Titanium diboride has recently been developed to be used as an extremely hard, wear-resistant coating for use on materials that must survive extremely erosive (eroding) environments. It is expected to be used as a coating on valves in coal liquefaction reactors. Titanium diboride is prepared by allowing titanium tetrachloride, boron trichloride, and hydrogen to react at atmospheric pressure to yield the diboride and hydrogen chloride. Write a balanced equation for this reaction.

Readings

Kolb, Doris, "The Chemical Equation. Part 1: Simple Reactions." *J. Chem. Educ.*, 1978, v. 55, p. 184. This article reviews writing, balancing, and completing chemical equations. It also discusses stoichiometry (Chapter 10) and ionic equations (Chapter 15). The first part of this article should be helpful to you now as a review.

Latimer, Wendell M., and Joel H. Hildebrand, *Reference Book of Inorganic Chemistry*, 3rd ed. New York: The Macmillan Company, 1951. A description of the reactions of the elements and their compounds. Includes many commercial preparations for elements and compounds.

Nechamkin, Howard, *The Chemistry of the Elements*. New York: McGraw-Hill Book Company, 1968. Also a description of the reactions of the elements and their compounds.

10

Calculations Involving Chemical Equations. Stoichiometry

The modern automobile uses the combustion of gasoline (a chemical reaction) to obtain its power. Gasoline is a complex mixture of volatile, low molecular mass hydrocarbons ranging in size from a C_6 molecule to C_{12}. An equation for the complete combustion of a typical gasoline component is

$$2\ C_8H_{18}\ (an\ octane) + 25\ O_2 \rightarrow 16\ CO_2\ (gas) + 18\ H_2O\ (vapor) + heat$$

OBJECTIVES

1. Given the following terms, define each term and give a distinguishing characteristic of each:
    (a) stoichiometry (Introduction to Chapter 10)
    (b) theoretical yield (Section 10-4)
    (c) actual yield (Section 10-4)
    (d) percent yield (Section 10-4)
    (e) Gay-Lussac's Law of Combining Volumes (Section 10-6)
    (f) heat of reaction (Section 10-7)
    (g) exothermic reaction (Section 10-7)
    (h) endothermic reaction (Section 10-7)

2. Given the mass of a reactant or product, a balanced equation, and the table of approximate atomic masses, calculate the mass of another reactant required or the mass that could be produced of another product (*mass-mass* stoichiometry problems—*direct* examples, Problem Examples 10-1 and 10-2, Problems 3, 4, 5, 6, 7, 8, 9, and 10).

3. Given the mass in grams or moles of a reactant or a product, a balanced equation, and the table of approximate atomic masses, calculate the number of moles or grams of another reactant required or moles or grams that could be produced of another product (*mass-mass* stoichiometry problems—*indirect* examples, Problem Examples 10-3, and 10-4, Problems 11, 12, 13, 14, 15, 16, 17, and 18).

4. (a) Given the mass of *two* or *more* reactants in a given reaction, a balanced equation, and the table of approximate atomic masses, determine the limiting reagent.
    (b) Given the mass of *two* or *more* reactants in a given reaction, a balanced equation, and the table of approximate atomic masses, calculate the mass that could be produced of a single product.
    (c) Given the actual yield and the theoretical yield, calculate the percent yield.
    (d) Given the amount of *two* or *more* reactants in a given reaction, a balanced equation, and the table of approximate atomic masses, calculate the amount of excess reagent remaining at the end of the reaction.

    (*Mass-mass* stoichiometry problems—*limiting reagent* examples, Problem Examples 10-5, 10-6, 10-7, and 10-8, Problems 19, 20, 21, and 22)

5. Given the volume (at STP) of a gaseous reactant or product, a balanced equation, and the table of approximate atomic masses, calcu-

late the mass in grams or moles of another reactant required or another product produced, and vice versa (*mass-volume* stoichiometry problems, Problem Examples 10-9, 10-10, and 10-11, Problems 23, 24, 25, 26, 27, and 28).

6. (a) Given the mass of *two* or *more* reactants in a given reaction, a balanced equation, and the table of approximate atomic masses, calculate the volume of a gas (at STP) that could be produced of a single product.

   (b) Given the amounts of two or more reactants in a given reaction, a balanced equation, and the table of approximate atomic masses, calculate the amount of excess reagent remaining at the end of the reaction.

   (*Mass-volume* stoichiometry problems, Problem Example 10-12, Problems 29 and 30)

7. Given the volume of a gaseous reactant or product and a balanced equation, calculate the volume of another gaseous product produced. All gases are measured at the same temperature and pressure (*volume-volume* stoichiometry problems, Problems Example 10-13, Problems 31, 32, 33, and 34).

8. (a) Given the volume of *two* or *more* gaseous reactants in a given reaction and a balanced equation, calculate the volume of a gas that could be produced of a single product. All gases are measured at the same temperature and pressure.

   (b) Given the volume of *two* or *more* gaseous reactants in a given reaction and a balanced equation, calculate the volume of the excess gaseous reactant remaining at the end of the reaction.

   (*Volume-volume* stoichiometry problems, Problem Example 10-14, Problems 35 and 36)

9. (a) Given a completed chemical equation with the heat of reaction, determine whether the reaction is exothermic or endothermic.

   (b) Given the amount of a reactant or product, the table of approximate atomic masses, and a completed chemical equation with the heat of reaction, calculate the amount of heat energy produced or needed, and vice versa.

   (*Heat of reaction* problems, Problem Examples 10-15 and 10-16, Problems 37 and 38)

Stoichiometry (pronounced stoi′kē·om′i·trē) is measurement based on the quantitative (how much?) laws of chemical combination. We shall use the coefficients in *balanced* chemical equations to relate quantities of reactants and products to each other in stoichiometric calculations.

FIGURE 10-1

Ms. Jones ponders the question, "How much of Z can we get?"

In Chapter 9, we learned how to complete and balance chemical equations. We are now going to use these balanced equations to calculate the amounts of material or energy produced or required in a given balanced chemical equation. We shall also apply the concepts discussed in Chapter 8. Before proceeding further, we urge you to review the sections in Chapter 8 on calculation of formula or molecular masses (8-1), moles of elementary units (8-2), and molar volume of a gas (8-3).

The concepts involved in stoichiometry must be applied in the chemical industry. For example, a chemical company may wish to purchase some inexpensive starting material and allow it to react with some other inexpensive material to form a new product, which it hopes to market at a higher rate than the cost of its original starting materials, hence making a tidy profit. Two of the many questions needing answers are: (1) starting with X and Y amounts of starting material, how much product, Z, can *theoretically* be produced, and (2) how much can *actually* be produced in the company's plant operation? This second question is answered by chemists working in an applied research laboratory and by chemical engineers working in a small-scale chemical manufacturing plant called the pilot plant. We can answer the first question, the theoretical amount, by performing suitable chemical calculations, which we shall consider in this chapter. Figure 10-1 illustrates the questions facing research and production personnel in the company.

10-1

Information Obtained from a Balanced Equation

A completed and *balanced* equation affords more information than simply which substances are reacting and which are products. It also gives the quantities involved, and it is very useful in carrying out certain calculations. Let us consider the oxidation of ethane gas to produce carbon dioxide and water:

$$2 \, \underset{\text{ethane}}{C_2H_{6(g)}} + 7 \, O_{2(g)} \xrightarrow{\Delta} 4 \, CO_{2(g)} + 6 \, H_2O_{(g)}$$

This balanced equation affords the following information:

1. **Reactants and products:** C_2H_6 (ethane) reacts with O_2 (oxygen) when sufficient heat (Δ) is applied to produce CO_2 (carbon dioxide) and H_2O (gaseous water).

2. **Molecules of reactants and products:** 2 molecules of C_2H_6 need 7 molecules of O_2 to react and produce 4 molecules of CO_2 and 6 molecules of H_2O.

3. **Moles of reactants and products:** 2 mol of C_2H_6 molecules need 7 mol of O_2 molecules to react and produce 4 mol of CO_2 molecules and 6 mol of H_2O molecules.

4. **Volumes of gases:** 2 volumes of C_2H_6 need 7 volumes of O_2 to react and produce 4 volumes of CO_2 and 6 volumes of H_2O, if all volumes are measured as *gases* at the same temperature and pressure.

5. **Relative masses of reactants and products:**[1] 60.0 g of C_2H_6

$$\left(2 \text{ mol } C_2H_6 \times \frac{30.0 \text{ g } C_2H_6}{1 \text{ mol } C_2H_6} = 60.0 \text{ g } C_2H_6 \right)$$

needs 224 g of O_2

$$\left(7 \text{ mol } O_2 \times \frac{32.0 \text{ g } O_2}{1 \text{ mol } O_2} = 224 \text{ g } O_2 \right)$$

to react and produce 176 g of CO_2

$$\left(4 \text{ mol } CO_2 \times \frac{44.0 \text{ g } CO_2}{1 \text{ mol } CO_2} = 176 \text{ g } CO_2 \right)$$

and 108 g of H_2O

$$\left(6 \text{ mol } H_2O \times \frac{18.0 \text{ g } H_2O}{1 \text{ mol } H_2O} = 108 \text{ g } H_2O \right).$$

Note that the sum of the reactants (60.0 g + 224 g = 284 g) is equal to the sum of the products (176 g + 108 g = 284 g), obeying the Law of Conservation of Mass (see 3-6).

---

[1]The coefficients used to calculate the number of atoms or masses of reactants and products are regarded as exact values and are not considered in computing significant digits. They have as many significant digits as are necessary to match those in a measurement. For the atomic masses (see Table of Approximate Atomic Masses inside the back cover of this text), the molecular masses of the substances involved in the reaction are calculated as C_2H_6 = 30.0 amu, O_2 = 32.0 amu, CO_2 = 44.0 amu, and H_2O = 18.0 amu.

10-2

The Mole Method of Solving Stoichiometry Problems. The Three Basic Steps

There are a number of methods available for solving stoichiometry problems. The method that we consider the best is the **mole method,** which is an application of our general method of problem solving—the **factor-unit method.** Three basic steps are involved in working problems by the *mole method:*[2]

Step I: Calculate **moles** of elementary units (atoms, formula units, molecules, or ions) of the element, compound, or ion from the mass or volume (if gases) of the known substance or substances in the problem.

Step II: Using the coefficients of the substances in the balanced equation, calculate **moles** of the unknown quantities in the problem.

Step III: From **moles** of the unknown quantities calculated, determine the mass or volume (for gases) of these unknowns and of the units requested by the problem.

As you can see, the key to this method is the **mole.** Think **MOLES!**

The application of these three basic steps is shown in diagram form in Figure 10-2.

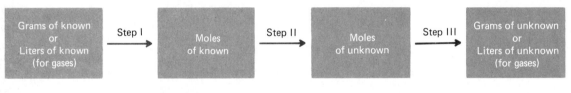

FIGURE 10-2

The three basic steps in solving stoichiometry problems. (Prior to Step I and after Step III an additional calculation may be required to convert to or from some mass measurement other than grams.)

10-3

Types of Stoichiometry Problems

There are three types of stoichiometry problems:

1. mass-mass ("weight-weight")

2. mass-volume or volume-mass ("weight"-volume or volume-"weight")

3. volume-volume

---

[2]Before you read on, turn to Problem Examples 8-6, 8-9, 8-15, and 8-20, and study these examples. You must know: (1) how to calculate moles given grams or liters of a gas at STP, and (2) how to calculate grams or liters of a gas at STP given moles. Prior to Step I and after Step III, an additional calculation may be required to convert from or to some mass measurement other than grams.

We now will apply the three basic steps to these three types of stoichiometry problems.

10-4

Mass-Mass Stoichiometry Problems

In this type of problem, the quantities of both the known and unknown are given or asked for in mass units. We shall first consider some *direct examples*, in which the known is expressed in mass units, as grams, and the unknown is asked for in mass units, as grams. These *direct examples* involve all three basic steps. We emphasize again that the equation must be **balanced** before the calculation is begun.

Direct
Examples

Problem Example 10-1

Calculate the number of grams of oxygen required to burn 72.0 g of C_2H_6 to CO_2 and H_2O. The equation for the reaction is

$$2\ C_2H_6 + 7\ O_2 \rightarrow 4\ CO_2 + 6\ H_2O$$

SOLUTION: Since the equation is balanced, we can proceed to calculate the molecular masses of the substances involved in the calculation, which are O_2 and C_2H_6.

$$O_2 = 32.0\ \text{amu}$$

$$C_2H_6 = 30.0\ \text{amu}$$

Organize the data:

Known: 72.0 g of C_2H_6

Unknown: g of O_2 required

Since the calculation will involve going from g of known to g of unknown, all three basic steps will be involved.

Step I: Calculate the moles of C_2H_6 molecules given. Since 1 mol has a mass of 30.0 g of C_2H_6,

$$72.0\ \cancel{\text{g }C_2H_6} \times \frac{1\ \text{mol }C_2H_6}{30.0\ \cancel{\text{g }C_2H_6}} = \frac{72.0}{30.0}\ \text{mol }C_2H_6\ \text{given}$$

Step II: Calculate the moles of oxygen molecules needed. From the balanced equation, the relation of C_2H_6 to O_2 is given as 2 mol of C_2H_6 to 7 mol of O_2. Therefore,

$$\frac{72.0}{30.0}\ \cancel{\text{mol }C_2H_6} \times \frac{7\ \text{mol }O_2}{2\ \cancel{\text{mol }C_2H_6}} = \frac{72.0}{30.0} \times \frac{7}{2}\ \text{mol }O_2\ \text{needed}$$

Step III: Calculate g of oxygen needed. Since 1 mol of O_2 has a mass of 32.0 g of O_2,

$$\frac{72.0}{30.0} \times \frac{7}{2}\,\text{mol}\,O_2 \times \frac{32.0\text{ g }O_2}{1\,\text{mol}\,O_2} = \frac{72.0}{30.0} \times \frac{7}{2} \times 32.0\text{ g }O_2$$

$$= 268.9\text{ g }O_2 = 269\text{ g }O_2 \quad \textit{Answer}$$

Since our given quantity (72.0 g of C_2H_6) was expressed to three significant digits, our answer is also expressed to three significant digits (269 g of O_2). [The mole relations (coefficients) are regarded as exact values and are not considered in computing significant digits.] The complete solution may be written as follows:

$$72.0\text{ g }C_2H_6 \times \underbrace{\frac{1\,\text{mol}\,C_2H_6}{30.0\text{ g }C_2H_6}}_{\text{Step I}} \times \underbrace{\frac{7\,\text{mol}\,O_2}{2\,\text{mol}\,C_2H_6}}_{\text{Step II}} \times \underbrace{\frac{32.0\text{ g }O_2}{1\,\text{mol}\,O_2}}_{\text{Step III}} = 269\text{ g }O_2 \quad \textit{Answer}$$

Problem Example 10-2

Calculate the number of grams of chlorine produced by the reaction of 22.1 g of manganese(IV) oxide with excess hydrochloric acid.

$$MnO_2 + 4\,HCl \rightarrow MnCl_2 + Cl_2 + 2\,H_2O$$

SOLUTION: The equation is balanced, so we can proceed to calculate the formula mass of MnO_2 and the molecular mass of Cl_2. The formula mass of MnO_2 is 87.0 amu and the molecular mass of Cl_2 is 71.0 amu. The known quantity is 22.1 g of MnO_2, and the unknown quantity is the grams of Cl_2 obtainable.

Work Problems 3, 4, 5, 6, 7, 8, 9, and 10.

$$22.1\text{ g }MnO_2 \times \underbrace{\frac{1\,\text{mol}\,MnO_2}{87.0\text{ g }MnO_2}}_{\text{Step I}} \times \underbrace{\frac{1\,\text{mol}\,Cl_2}{1\,\text{mol}\,MnO_2}}_{\text{Step II}} \times \underbrace{\frac{71.0\text{ g }Cl_2}{1\,\text{mol}\,Cl_2}}_{\text{Step III}} = 18.0\text{ g }Cl_2 \quad \textit{Answer}$$

Indirect Examples

Sometimes the information supplied is in moles and we do not need to calculate it, or sometimes the information to be determined must be expressed in moles and again we do not need to calculate mass in grams. Therefore, Step I or III, or even both, may be eliminated. These *indirect examples* will serve to illustrate how one, two, or both steps may be eliminated.

Problem Example 10-3

Given the following balanced equation:

$$2\,Na + 2\,H_2O \rightarrow 2\,NaOH + H_2$$

If 0.145 mol of Na atoms reacts with water, calculate (a) the number of moles and (b) the number of grams of H_2 molecules produced.

SOLUTION

(a) In this part of the problem, the known must be carried only through Step II to get the answer; hence, Steps I and III can be eliminated.

$$0.145 \ \cancel{\text{mol Na}} \times \frac{1 \ \text{mol H}_2}{2 \ \cancel{\text{mol Na}}} = 0.0725 \ \text{mol H}_2 \quad Answer$$

Step II

(b) From the answer to (a) and the molecular mass of H_2 (2.0 amu), we can calculate the answer to (b) by applying Step III.

$$0.0725 \ \cancel{\text{mol H}_2} \times \frac{2.0 \ \text{g H}_2}{1 \ \cancel{\text{mol H}_2}} = 0.145 \ \text{g H}_2 \quad Answer$$

Step III

Problem Example 10-4

Calculate the number of moles of O_2 produced by heating 1.65 g of $KClO_3$.

SOLUTION

The equation must be written and balanced before any calculation can be made. The balanced equation is as follows:

$$2 \ KClO_3 \xrightarrow{\Delta} 2 \ KCl + 3 \ O_2$$

The known quantity, 1.65 g of $KClO_3$, is given in grams. Therefore, Step I is needed to calculate moles of $KClO_3$. Step II converts the moles of $KClO_3$ to moles of O_2, and hence Step III is not needed. The formula mass of $KClO_3$ is 122.6 amu, as calculated from the atomic masses.

$$1.65 \ \cancel{\text{g KClO}_3} \times \frac{1 \ \cancel{\text{mol KClO}_3}}{122.6 \ \cancel{\text{g KClO}_3}} \times \frac{3 \ \text{mol O}_2}{2 \ \cancel{\text{mol KClO}_3}} = 0.0202 \ \text{mol O}_2 \quad Answer$$

Step I   Step II

Work Problems 11, 12, 13, 14, 15, 16, 17, and 18.

Limiting Reagent Examples

   The elimination of these same steps can also be applied to mass-volume stoichiometry problems (10-5).
   In carrying out chemical reactions, the quantities of reactants are not usually used in exact stoichiometric amounts. That is, one reactant may be used in excess of that theoretically needed for a complete reaction to take place according to the balanced equation. In such cases, the amount of product obtained is dependent upon the reactant, which is termed the **limiting reagent** or the first reactant to be entirely consumed. Many times the more expensive reactant is the limiting reagent and the cheaper reactant is in excess. The principle of limiting reagent is analogous to a party attended by both men and

women. If there are seven (7) men, but only six (6) women, the maximum number of couples we could have would be six (6). The number of women *limit* the number of couples we could obtain, so in this case the women are the limiting reagent and the men are in excess. Another analogy: You have $4.50 and wish to buy as many candy bars as you can. The store has 96 candy bars in stock at $0.25 each. Your $4.50 limits you to purchasing 18 candy bars

$$\$4.50 \times \frac{1 \text{ candy bar}}{\$0.25} = 18 \text{ candy bars}$$

and the store has 78 bars (96 bars − 18 bars = 78 bars) left in stock. If the store has only 14 bars in stock, your purchase would be limited to 14 candy bars and you would have $1.00 left

$$\$4.50 - \left(14 \text{ candy bars} \times \frac{\$0.25}{1 \text{ candy bar}} \right) = \$4.50 - \$3.50 = \$1.00$$

In the first case your $4.50 was the limiting reagent and in the second the store's stock of candy bars was the limiting reagent. The following *limiting reagent* examples will serve to illustrate our point.

Problem Example 10-5

A 50.0-g sample of calcium carbonate is allowed to react with 35.0 g of H_3PO_4. (a) How many grams of calcium phosphate could be produced? (b) Calculate the number of moles of excess reagent at the end of the reaction.

$$3 \text{ } CaCO_3 + 2 \text{ } H_3PO_4 \rightarrow Ca_3(PO_4)_2 + 3 \text{ } CO_2 + 3 \text{ } H_2O$$

SOLUTION

(a) The formula masses of the substances involved in the calculation are calculated from the atomic masses as $CaCO_3 = 100.1$ amu, $H_3PO_4 = 98.0$ amu, and $Ca_3(PO_4)_2 = 31\overline{0}$ amu. The question is: Which one of the reactants, $CaCO_3$ or H_3PO_4, is the limiting reagent? We answer as follows:
1. Calculate the moles of each used as in Step I.

$$50.0 \text{ g } CaCO_3 \times \frac{1 \text{ mol } CaCO_3}{100.1 \text{ g } CaCO_3} = 0.500 \text{ mol } CaCO_3$$

$$35.0 \text{ g } H_3PO_4 \times \frac{1 \text{ mol } H_3PO_4}{98.0 \text{ g } H_3PO_4} = 0.357 \text{ mol } H_3PO_4$$

2. Calculate the moles of product that could be produced from each reactant as in Step II.

$$0.500 \text{ mol } \cancel{CaCO_3} \times \frac{1 \text{ mol } Ca_3(PO_4)_2}{3 \text{ mol } \cancel{CaCO_3}} = 0.167 \text{ mol } Ca_3(PO_4)_2$$

$$0.357 \text{ mol } \cancel{H_3PO_4} \times \frac{1 \text{ mol } Ca_3(PO_4)_2}{2 \text{ mol } \cancel{H_3PO_4}} = 0.178 \text{ mol } Ca_3(PO_4)_2$$

3. *The reactant that gives the least number of moles of the **product** is the limiting reagent.* Notice that you carry *both* reactants through Steps I and II. You use the reactant that gives the least number of moles of product. Hence, in this example, **CaCO_3** is the **limiting reagent** [0.167 mol vs. 0.178 mol of $Ca_3(PO_4)_2$] and the H_3PO_4 is in excess.

Using $CaCO_3$, the number of grams of $Ca_3(PO_4)_2$ that could be produced would be

$$0.167 \text{ mol } \cancel{Ca_3(PO_4)_2} \times \frac{31\bar{0} \text{ g } Ca_3(PO_4)_2}{1 \text{ mol } \cancel{Ca_3(PO_4)_2}} = 51.8 \text{ g } Ca_3(PO_4)_2 \quad Answer$$

(b) The amount of excess H_3PO_4 is equal to 0.357 mol of H_3PO_4 present at the start of the reaction (see Step I) **minus** the amount which is consumed by reacting with the limiting reagent $(CaCO_3)$. The amount consumed is

$$\left[0.500 \text{ mol } \cancel{CaCO_3} \times \frac{2 \text{ mol } H_3PO_4}{3 \text{ mol } \cancel{CaCO_3}} = 0.333 \text{ mol } H_3PO_4 \right]$$

and the amount in excess is

$$0.357 \text{ mol } H_3PO_4 - 0.333 \text{ mol } H_3PO_4 \qquad = 0.024 \text{ mol } H_3PO_4 \text{ in excess} \quad Answer$$

Problem Example 10-6

The following reaction was carried out:

$$3\,A + 4\,B \rightarrow 2\,C + 3\,D + 3\,E$$

The molecular masses in amu are: $A = 100.0$, $B = 150.0$, and $C = 85.0$. A 50.0-g sample of A is allowed to react with 75.0 g of B. (a) How many grams of C could be produced? (b) Calculate the number of moles of the excess reagent remaining at the end of the reaction.

SOLUTION

(a) Determine the limiting reagent:

1. Calculate the moles of each used as in Step I.

$$50.0 \text{ g } A \times \frac{1 \text{ mol } A}{100.0 \text{ g } A} = 0.500 \text{ mol } A$$

$$75.0 \text{ g } B \times \frac{1 \text{ mol } B}{150.0 \text{ g } B} = 0.500 \text{ mol } B$$

2. Calculate the moles of product that could be produced from each reactant as in Step II.

$$0.500 \text{ mol } A \times \frac{2 \text{ mol } C}{3 \text{ mol } A} = 0.333 \text{ mol } C$$

$$0.500 \text{ mol } B \times \frac{2 \text{ mol } C}{4 \text{ mol } B} = 0.250 \text{ mol } C$$

3. The reactant that gives the *least* number of moles of *product* is the limiting reagent.

Hence, B is the limiting reagent. Using B, we find that the number of grams of C which could be produced would be

$$0.250 \text{ mol } C \times \frac{85.0 \text{ g } C}{1 \text{ mol } C} = 21.2 \text{ g } C \quad Answer$$

(b) The amount of excess A is equal to: 0.500 mol of A present at the start of the reaction (see Step I) **minus** the amount which is consumed by reacting with the limiting reagent B. The amount consumed is

$$\left[0.500 \text{ mol } B \times \frac{3 \text{ mol } A}{4 \text{ mol } B} = 0.375 \text{ mol } A \right]$$

and the amount in excess is

$$0.500 \text{ mol } A - 0.375 \text{ mol } A = 0.125 \text{ mol } A \text{ in excess} \quad Answer$$

The amounts of the various products that we have just calculated in Problem Examples 10-5 and 10-6 are called theoretical yields. The **theoretical yield** is the amount of product obtained when we assume that all the limiting reagent forms products with none of it left over and that none of the product is lost in its isolation and purification. But, such is not generally the case. In organic reactions particularly, side reactions occur, giving minor products in addition to the major one. Also, some of the product is lost in the process of its isolation and purification and in transferring it from one container to another. In the chemical industry, this loss in isolation and purification is often minimized by a continuous process in which the materials used in isolation and purification are recycled. The amount that is actually obtained is called the **actual yield**. The **percent yield** is the percent of the theoretical yield that is actually obtained, calculated as follows:

$$\% \text{ yield} = \frac{\text{actual yield}}{\text{theoretical yield}} \times 100$$

Problem Example 10-7

If 20.7 g of *C* is actually obtained in Problem Example 10-6, what is the percent yield?

SOLUTION

$$\frac{20.7 \ \text{g C} = \text{actual yield}}{21.2 \ \text{g C} = \text{theoretical yield}} \times 100 = 97.6\% \quad Answer$$

Problem Example 10-8

The industrial preparation of ethylene glycol used as an automobile antifreeze and in the preparation of a polyester fiber (Dacron; see 19-11) is as follows:

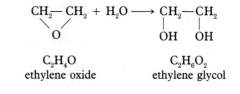

$$C_2H_4O \qquad\qquad C_2H_6O_2$$
ethylene oxide   ethylene glycol

If 165 g of ethylene oxide is allowed to react with 75.0 g of water, calculate (a) the theoretical yield of ethylene glycol in grams, (b) the number of moles of excess reagent remaining at the end of the reaction, and (c) if 215 g of ethylene glycol is actually obtained, calculate the percent yield.

SOLUTION

(a) The molecular masses of the substances involved in the calculation are calculated from the atomic masses as $C_2H_4O = 44.0$ amu, $H_2O = 18.0$ amu, and $C_2H_6O_2 = 62.0$ amu. Now, we must determine whether ethylene oxide (C_2H_4O) or water (H_2O) is the limiting reagent.

1. Calculate the moles of each used as in Step I.

$$165 \ \text{g C}_2\text{H}_4\text{O} \times \frac{1 \ \text{mol C}_2\text{H}_4\text{O}}{44.0 \ \text{g C}_2\text{H}_4\text{O}} = 3.75 \ \text{mol C}_2\text{H}_4\text{O}$$

$$75.0 \ \text{g H}_2\text{O} \times \frac{1 \ \text{mol H}_2\text{O}}{18.0 \ \text{g H}_2\text{O}} = 4.17 \ \text{mol H}_2\text{O}$$

2. Calculate the moles of product that could be produced from each reactant as in Step II.

$$3.75 \ \text{mol C}_2\text{H}_4\text{O} \times \frac{1 \ \text{mol C}_2\text{H}_6\text{O}_2}{1 \ \text{mol C}_2\text{H}_4\text{O}} = 3.75 \ \text{mol C}_2\text{H}_6\text{O}_2$$

$$4.17 \ \text{mol H}_2\text{O} \times \frac{1 \ \text{mol C}_2\text{H}_6\text{O}_2}{1 \ \text{mol H}_2\text{O}} = 4.17 \ \text{mol C}_2\text{H}_6\text{O}_2$$

3. The reactant that gives the *least* number of moles of product is the limiting reagent.

Hence, ethylene oxide (C_2H_4O) is the limiting reagent and water is in excess. Naturally, you would expect this since water is considerably cheaper than any organic substance such as ethylene oxide which is derived from oil. Water is cheaper than oil!

Using ethylene oxide, we find that the number of grams of ethylene glycol which could be produced would be

$$3.75 \text{ mol } C_2H_4O_2 \times \frac{62.0 \text{ g } C_2H_6O_2}{1 \text{ mol } C_2H_6O_2} = 232 \text{ g } C_2H_6O_2 \quad Answer$$

(b) The amount of excess water is equal to: 4.17 mol of H_2O at the start of the reaction (see Step I) **minus** the amount which is consumed by reacting with the limiting reagent ethylene oxide. The amount consumed is

$$\left[3.75 \text{ mol } C_2H_4O \times \frac{1 \text{ mol } H_2O}{1 \text{ mol } C_2H_4O} = 3.75 \text{ mol } H_2O \right]$$

and the amount in excess is

$$4.17 \text{ mol } H_2O - 3.75 \text{ mol } H_2O = 0.42 \text{ mol } H_2O \text{ in excess} \quad Answer$$

Work Problems 19, 20, 21, and 22.

(c) The percent yield is calculated from the theoretical yield and the actual yield and is as follows:

$$\frac{215 \text{ g } C_2H_6O_2}{232 \text{ g } C_2H_6O_2} \times 100 = 92.7\% \quad Answer$$

10-5

Mass-Volume Stoichiometry Problems

Next, let us consider mass-volume stoichiometry problems[3] In these types of problems, *either* the known *or* unknown is a **gas.** The known may be given in mass units and you will be asked to calculate the unknown in volume units (if a gas), or the known will be given in volume units (if a gas) and you will be asked to calculate the unknown in mass units. In either case, you need to apply the molar volume–that is, 22.4 L per mol of any gas at STP (0°C and $76\overline{0}$ torr), discussed in Chapter 8 (8-3).

Problem Example 10-9

Calculate the volume in liters of O_2 measured at 0°C and $76\overline{0}$ torr which could be obtained by heating 28.0 g of KNO_3.

SOLUTION: The equation must first be written and balanced, as follows:

$$2 \text{ KNO}_3 \xrightarrow{\Delta} 2 \text{ KNO}_2 + O_2$$

---

[3]The volume of a gas can be calculated at non-STP conditions. This is done using the gas laws as shown in Chapter 11 (11-9, stoichiometry problems).

The formula mass of KNO_3 is calculated as 101.1 amu from the atomic masses. The conditions 0°C and 76$\overline{0}$ torr are STP conditions; hence, in Step III, the relation 1 mol of O_2 *molecules* at STP occupies 22.4 L must be used.

$$28.0 \text{ g KNO}_3 \times \frac{1 \text{ mol KNO}_3}{101.1 \text{ g KNO}_3} \times \frac{1 \text{ mol O}_2}{2 \text{ mol KNO}_3} \times \frac{22.4 \text{ L O}_2 \text{ at STP}}{1 \text{ mol O}_2}$$

$$\underbrace{\hspace{4cm}}_{\text{Step I}} \quad \underbrace{\hspace{1.5cm}}_{\text{Step II}} \quad \underbrace{\hspace{1.5cm}}_{\text{Step III}}$$

$$= 3.10 \text{ L O}_2 \text{ at STP} \quad \textit{Answer}$$

Problem Example 10-10

Calculate the number of moles of Cu atoms produced if 4250 mL of H_2 measured at 0°C and 76$\overline{0}$ torr reacts with an excess of CuO.

$$CuO + H_2 \xrightarrow{\Delta} Cu + H_2O$$

SOLUTION: The conditions 0°C and 76$\overline{0}$ torr are STP conditions; hence, in Step I, the relation 1 mol of H_2 at STP occupies 22.4 L must be used.

$$4250 \text{ mL H}_2 \text{ STP} \times \frac{1 \text{ L}}{1000 \text{ mL}} \times \frac{1 \text{ mol H}_2 \text{ STP}}{22.4 \text{ L H}_2 \text{ STP}} \times \frac{1 \text{ mol Cu}}{1 \text{ mol H}_2 \text{ STP}}$$

$$\underbrace{\hspace{6cm}}_{\text{Step I}} \quad \underbrace{\hspace{2.5cm}}_{\text{Step II}}$$

$$= 0.190 \text{ mol Cu} \quad \textit{Answer}$$

Note that in this problem, Step III is not required because the question asks for moles of Cu atoms and *not* g of Cu.

Problem Example 10-11

Calculate the number of liters of O_2 (at STP) produced by heating 0.480 mol of $KClO_3$.

SOLUTION: The balanced equation is as follows:

$$2 \text{ KClO}_3 \xrightarrow{\Delta} 2 \text{ KCl} + 3 \text{ O}_2$$

The conditions given are STP; hence, the relation 1 mol of O_2 molecules at STP occupies 22.4 L must be used.

$$0.480 \text{ mol KClO}_3 \times \frac{3 \text{ mol O}_2}{2 \text{ mol KClO}_3} \quad \frac{22.4 \text{ L O}_2 \text{ STP}}{1 \text{ mol O}_2} = 16.1 \text{ L O}_2 \text{ STP}$$

$$\underbrace{\hspace{4cm}}_{\text{Step II}} \quad \underbrace{\hspace{3cm}}_{\text{Step III}} \qquad \text{Answer}$$

Work Problems 23, 24, 25, 26, 27, and 28.

Note that in this problem, Step I is not required because the number of moles of $KClO_3$ is already given.

Problem Example 10-12

Limiting Reagent
Example

A 28.0-g sample of zinc is allowed to react with 75.0 g of H_2SO_4. (a) How many liters of hydrogen measured at STP could be produced? (b) Calculate the number of moles of excess reagent remaining at the end of the reaction.

$$Zn_{(s)} + H_2SO_{4(aq)} \rightarrow ZnSO_{4(aq)} + H_{2(g)}$$

SOLUTION

(a) Determine the limiting reagent. The atomic mass of Zn is 65.4 amu and the formula mass of H_2SO_4 is 98.1 amu.

1. Calculate the moles of each used as in Step I.

$$28.0 \text{ g Zn} \times \frac{1 \text{ mol Zn}}{65.4 \text{ g Zn}} = 0.428 \text{ mol Zn}$$

$$75.0 \text{ g } H_2SO_4 \times \frac{1 \text{ mol } H_2SO_4}{98.1 \text{ g } H_2SO_4} = 0.765 \text{ mol } H_2SO_4$$

2. Calculate the moles of product that could be produced from each reactant as in Step II.

$$0.428 \text{ mol Zn} \times \frac{1 \text{ mol } H_2}{1 \text{ mol Zn}} = 0.428 \text{ mol } H_2$$

$$0.765 \text{ mol } H_2SO_4 \times \frac{1 \text{ mol } H_2}{1 \text{ mol } H_2SO_4} = 0.765 \text{ mol } H_2$$

3. The reactant that gives the least number moles of product is the limiting reagent, in this case Zn.

$$0.428 \text{ mol } H_2 \times \frac{22.4 \text{ L } H_2(STP)}{1 \text{ mol } H_2} = 9.59 \text{ L } H_2(STP) \quad \textit{Answer}$$

(b) The amount of excess H_2SO_4 is equal to 0.765 mol of H_2SO_4 present at the start of the reaction (see Step I) **minus** the amount which is consumed by reacting with the limiting reagent (Zn). The amount consumed is

$$\left[0.428 \text{ mol Zn} \times \frac{1 \text{ mol } H_2SO_4}{1 \text{ mol Zn}} = 0.428 \text{ mol } H_2SO_4 \right]$$

and the amount in excess is

Work Problems 29
and 30.

$$0.765 \text{ mol } H_2SO_4 - 0.428 \text{ mol } H_2SO_4 = 0.337 \text{ mol } H_2SO_4 \quad \textit{Answer}$$

10-6

**Volume-Volume
Stoichiometry
Problems**

Volume-volume stoichiometry problems are based on experimentation performed by the French chemist and physicist Joseph Louis Gay-Lussac (gā'lü · sak') (1778–1850). His experimental results are stated in **Gay-Lussac's Law** of **Combining Volumes,** which states that at the **same temperature and pressure** whenever **gases** react or gases are formed, they do so in the ratio of small whole numbers by volume. This ratio of small whole numbers by *volume* is directly proportional to the values of their coefficients[4] in the balanced equation.

For example, in the following reaction

$$CH_{4(g)} + 2\ O_{2(g)} \xrightarrow{\Delta} CO_{2(g)} + 2\ H_2O_{(g)}$$

all compounds are in the gaseous state and at the same temperature and pressure. One (1) volume of CH_4 gas (methane) reacts with two (2) volumes of O_2 gas to form one (1) volume of CO_2 gas and two (2) volumes of H_2O vapor. If we had measured these volumes all at STP and assumed that they *all* remain gases at STP, we could have stated that 1 mol (22.4 L) of CH_4 gas reacts with 2 mol (44.8 L) O_2 to form 1 mol (22.4 L) of CO_2 gas and 2 mol (44.8 L) of H_2O vapor. Note that in all cases, the ratio of the *volumes* is still the same— that is, 1:2:1:2 for CH_4, O_2, CO_2, and H_2O, respectively. In solving volume-volume stoichiometry problems, Steps I and II are not necessary; only Step II is required.

Problem Example 10-13

Calculate the volume of O_2 in liters required and the volume of CO_2 and H_2O in liters formed from the complete combustion of 1.50 L of C_2H_6, all volumes being measured at 400°C and 760 torr pressure.

$$2\ C_2H_{6(g)} + 7\ O_{2(g)} \xrightarrow{\Delta} 4\ CO_{2(g)} + 6\ H_2O_{(g)}$$

SOLUTION: Since all of these substances are gases measured at the same temperature and pressure, their volumes are related to their coefficients in the balanced equation.

$$1.50\ \cancel{L\ C_2H_6} \times \frac{7\ L\ O_2}{2\ \cancel{L\ C_2H_6}} = 5.25\ L\ O_2 \quad \textit{Answer}$$

Step II

$$1.50\ \cancel{L\ C_2H_6} \times \frac{4\ L\ CO_2}{2\ \cancel{L\ C_2H_6}} = 3.00\ L\ CO_2 \quad \textit{Answer}$$

Step II

$$1.50\ \cancel{L\ C_2H_6} \times \frac{6\ L\ H_2O_{(g)}}{2\ \cancel{L\ C_2H_6}} = 4.50\ L\ H_2O_{(g)} \quad \textit{Answer}$$

Step II

Work Problems 31, 32, 33, and 34.

[4]This is the same principle we applied in mass-mass problems, except that here we use volumes instead of moles and **all** *substances are gases* and are measured at the **same temperature and pressure.**

Problem Example 10-14

Limiting Reagent
Example

A commercial preparation of methyl alcohol (wood alcohol) involves the reaction of carbon monoxide with hydrogen at 350° to 400°C and 3000 lb/in.^{2} pressure in the presence of metallic oxides, such as a chromium(III) oxide-zinc oxide mixture.

$$CO_{(g)} + 2\ H_{2(g)} \xrightarrow[\Delta,p]{Cr_2O_3\text{-}ZnO} \underset{\substack{\text{methyl} \\ \text{alcohol}}}{CH_3OH_{(g)}}$$

If 60.0 L of CO is allowed to react with 80.0 L of H_2 in a sealed container, calculate (a) the number of liters of $CH_3OH_{(g)}$ that could be produced, and (b) the number of liters of both CO and H_2 that would remain, all volumes being measured at the same temperature and pressure, and assuming that the reaction can go to completion.

SOLUTION

(a) The first part involves determining the limiting reagent. Since all of these substances are gases at the same temperature and pressure, their volumes are related to their coefficients in the balanced equation; and we can calculate directly the volume in liters of $CH_3OH_{(g)}$ that could be produced. Using 60.0 L of CO, we have

$$60.0\ \cancel{L\ CO}_{(g)} \times \frac{1\ L\ CH_3OH_{(g)}}{1\ \cancel{L\ CO}_{(g)}} = 60.0\ L\ CH_3OH_{(g)}$$

Using 80.0 L of H_2, we have

$$80.0\ \cancel{L\ H}_{2(g)} \times \frac{1\ L\ CH_3OH_{(g)}}{2\ \cancel{L\ H}_{2(g)}} = 40.0\ L\ CH_3OH_{(g)}\quad \textit{Answer}$$

The least amount of product [40.0 L $CH_3OH_{(g)}$] is what could be produced in this process (see Problem Examples 10-5 and 10-6); the H_2 is the *limiting reagent* and the *CO* is in *excess*.

(b) Therefore, all the H_2 would be used, but an excess of CO would remain. The amount of excess CO is equal to 60.0 L of CO present at the start of the reaction **minus** the amount which is consumed by reacting with the limiting reagent (H_2). The amount consumed is

$$\left[80.0\ \cancel{L\ H}_{2(g)} \times \frac{1\ L\ CO_{(g)}}{2\ \cancel{L\ H}_{2(g)}} = 40.0\ L\ CO \right]$$

and

$$60.0\ L\ CO - 40.0\ L\ CO = 20.0\ L\ CO\ \text{in excess}$$

$$0\ L\ H_2\quad \textit{Answers}$$

Work Problems 35 and 36.

10-7

**Heats
in Chemical
Reactions**

Besides the mass-mass, mass-volume, and volume-volume relations just outlined, energy relationship are also important in chemical reactions. The energy involved is usually observed as heat and is expressed as the heat of reaction. The **heat of reaction** is the number of calories or joules of heat energy evolved or absorbed in a given chemical reaction per given amount of reactants and/or products. In **exothermic reactions,** heat energy is **evolved,** whereas in **endothermic reactions,** heat energy is **absorbed.**

An example of an exothermic reaction is the combination of 2 mol of hydrogen gas with 1 mol of oxygen, forming 2 mol of water (liquid), with the *evolution* of 1.37×10^5 cal (5.73×10^5 J) of heat energy at 25°C. Thus, for this exothermic reaction, the heat of reaction is 1.37×10^5 cal (5.73×10^5 J) for the formation of 2 mol of water (liquid) or 6.85×10^4 cal (2.87×10^5 J) for 1 mol.

$$2 \, H_{2(g)} + O_{2(g)} \rightarrow 2 \, H_2O_{(l)} + 1.37 \times 10^5 \text{ cal } (5.73 \times 10^5 \text{ J) at 25°C}$$

Two common examples of exothermic reactions that you may already have discovered in the laboratory are the preparation of dilute solutions of acids or bases by adding concentrated sulfuric acid to water in the former case and by adding sodium hydroxide pellets to water in the latter. In both cases you may have noticed that the flask got warm. Also, when these two substances react (sulfuric acid and sodium hydroxide) as in neutralization reactions (see 9-11) an exothermic reaction occurs. The flask gets warm—heat is evolved.

An example of an endothermic reaction is the combination of 1 mol of hydrogen gas with 1 mol of iodine gas, forming 2 mol of gaseous hydrogen iodide, with the *absorption* 1.24×10^4 cal (5.19×10^4 J) of heat energy at 25°C. The absorption of heat energy is noted by the minus sign in the following balanced equation:

$$H_{2(g)} + I_{2(g)} \rightarrow 2 \, HI_{(g)} - 1.24 \times 10^4 \text{ cal } (5.19 \times 10^4 \text{ J) at 25°C}$$

Placing the endothermic heat energy on the reactants side gives the following balanced equation:

$$+1.24 \times 10^4 \text{ cal } (5.19 \times 10^4 \text{ J)} + H_{2(g)} + I_{2(g)} \rightarrow 2 \, HI_{(g)} \text{ at 25°C}$$

Thus, for the endothermic reaction, the heat of reaction is 1.24×10^4 cal (5.19×10^4 J) absorbed for the formation of 2 mol of gaseous hydrogen iodide, or 6.20×10^3 cal (2.60×10^4 J) absorbed for 1 mol.

A common example of the endothermic formation of a solution is illustrated by dissolving potassium iodide in water and noticing how cold the flask becomes to the touch.

This heat energy[5] can be used in stoichiometric calculations. The quantity of heat energy, either exothermic or endothermic, is related to the moles of reactants or products in the balanced equation. We therefore use the heat of reaction as we did moles in Step II of our three basic steps (see 10-2).

Problem Example 10-15

Natural gas (CH_4) burns in air to produce carbon dioxide, water vapor, and heat energy. Calculate the number of kilocalories of heat energy produced by the burning of 25.0 g of natural gas, according to the following balanced equation:

$$CH_{4(g)} + 2\ O_{2(g)} \rightarrow CO_{2(g)} + 2\ H_2O_{(g)} + 213\ kcal\ at\ 25°C$$

SOLUTION: The molecular mass of CH_4 is 16.0 amu. The relationship between methane and the heat of reaction is 1 mol of CH_4 to 213 kcal. We therefore solve this problem by using Steps I and II.

$$25.0\ \text{g } CH_4 \times \underbrace{\frac{1\ \text{mol } CH_4}{16.0\ \text{g } CH_4}}_{\text{Step I}} \times \underbrace{\frac{213\ kcal}{1\ \text{mol } CH_4}}_{\text{Step II}} = 333\ kcal \quad Answer$$

Problem Example 10-16

Calculate the number of grams of hydrogen needed to produce 7.50×10^5 cal of heat energy, according to the following balanced equation:

$$2\ H_{2(g)} + O_{2(g)} \rightarrow 2\ H_2O_{(l)} + 1.37 \times 10^5\ cal\ (at\ 25°C)$$

SOLUTION: The molecular mass of H_2 is 2.0 amu. The relationship between hydrogen and the heat of reaction is 2 mol of H_2 to 1.37×10^5 cal. We therefore solve the problem by using Steps II and III:

$$7.50 \times 10^5\ \text{cal} \times \underbrace{\frac{2\ \text{mol } H_2}{1.37 \times 10^5\ \text{cal}}}_{\text{Step II}} \times \underbrace{\frac{2.0\ \text{g } H_2}{1\ \text{mol } H_2}}_{\text{Step III}} = 21.9\ \text{g } H_2 \quad Answer$$

Work Problems 37 and 38.

---

[5]The heat energy of reactions are referred to as the enthalpy of a reaction using the symbol H. The change in *enthalpy* from reactants to products is ΔH (read "delta H," with Δ meaning change). For exothermic reactions the ΔH is **negative**, such as in the formation of water, $\Delta H = -1.37 \times 10^5$ cal at 25°C. The reason for the negative sign is that in going from the reactants (H_2 and O_2) to the product(s) (H_2O) energy is lost (negative) *by* the chemicals with this energy being given off to the outside (exothermic). For endothermic reactions the ΔH is **positive**, such as in the formation of hydrogen iodide, $\Delta H = +1.24 \times 10^4$ cal at 25°C. The reason for the positive sign is that in going from the reactants (H_2 and I_2) to the product(s) (HI) energy must be gained (positive) *from* the outside (endothermic).

Chapter Summary

Stoichiometry is measurement based on the quantitative laws of chemical combination. Before stoichiometric calculations can be made, the equation must be balanced. The coefficients in the balanced equation in stoichiometric calculations are used to represent moles or volumes (for gases measured at the same temperature and pressure) of the reactants and products. To solve stoichiometry problems the mole method was used, following the three basic steps as outlined in 10-2. The types of stoichiometry problems were divided into mass-mass, mass-volume, volume-volume problems with limiting reagent problems included in each type. In addition, the terms *theroetical yield*, *actual yield*, and *percent yield* were introduced. Problems involving these terms were solved.

The *heat of a reaction* is the number of calories or joules of heat energy evolved or absorbed in a given chemical reaction per given amount of reactants and/or products. Calculations involving heats of a reaction were considered using the mole method.

EXERCISES

1. Define or explain the following terms:
   (a) stoichiometry
   (b) formula mass
   (c) molecular mass
   (d) mass-mass stoichiometry problems
   (e) limiting reagent
   (f) excess reagent
   (g) theoretical yield
   (h) actual yield
   (i) percent yield
   (j) mass-volume stoichiometry problems
   (k) molar volume
   (l) STP
   (m) volume-volume stoichiometry problems
   (n) Gay-Lussac's Law of Combining Volumes
   (o) heat of reaction
   (p) exothermic reaction
   (q) endothermic reaction

2. Distinguish between
   (a) formula and molecular masses
   (b) theoretical and actual yields
   (c) limiting and excess reagents
   (d) exothermic and endothermic reactions
   (e) mass-mass and mass-volume stoichiometry problems
   (f) mass-volume and volume-volume stoichiometry problems
   (g) calories and joules

PROBLEMS

(*Hints:* Check each equation to make sure it is balanced and, if not, balance it. For those questions in which an equation is not given, see if you can write one. See 9-7 through 9-11 in Chapter 9 for review.)

Mass-Mass—Direct Problems (See Sections 10-1, 10-2, 10-4—direct examples)

3. Calculate the number of grams of zinc chloride that can be prepared from 34.0 g of zinc.

$$Zn_{(s)} + 2\ HCl_{(aq)} \rightarrow ZnCl_{2(aq)} + H_{2(g)}$$

4. Calculate the number of grams of hydrogen that can be produced from 6.80 g of aluminum.

$$2\ Al_{(s)} + 6\ NaOH_{(aq)} \rightarrow 2\ Na_3AlO_{3(aq)} + 3\ H_{2(g)}$$

5. How many grams of silver chloride can be prepared from 78.0 g of silver nitrate?

$$AgNO_{3(aq)} + NaCl_{(aq)} \rightarrow AgCl_{(s)} + NaNO_{3(aq)}$$

6. How many kilograms of iron(III) oxide can be obtained by roasting 975 g of iron(II) sulfide?

$$4\ FeS_{(s)} + 7\ O_{2(g)} \rightarrow 2\ Fe_2O_{3(s)} + 4\ SO_{2(g)}$$

7. Sodium hydroxide (25.0 g) is neutralized with sulfuric acid. How many grams of sodium sulfate can be formed?

$$2\ NaOH_{(aq)} + H_2SO_{4(aq)} \rightarrow Na_2SO_{4(aq)} + 2\ H_2O_{(\ell)}$$

8. How many kilograms of hydrogen sulfide can be prepared by treating $\overline{400}$ g of iron(II) sulfide with an excess of hydrochloric acid?

$$FeS_{(s)} + HCl_{(aq)} \rightarrow FeCl_{2(aq)} + H_2S_{(g)} \quad \text{(unbalanced)}$$

9. Calculate the number of grams of potassium nitrate necessary to produce 6.90 g of oxygen. (See 9-8 for the equation.)

10. Calculate the number of grams of oxygen that could be produced by heating 7.60 g of potassium chlorate. (See 9-8 for the equation.)

Mass-Mass—Indirect Problems (See Section 10-4—indirect examples)

11. Calculate the number of moles of barium sulfate that can be prepared from 60.0 g of barium chloride.

$$BaCl_{2(aq)} + Na_2SO_{4(aq)} \rightarrow BaSO_{4(s)} + NaCl_{(aq)}$$

12. Calculate the number of moles of calcium chloride that would be necessary to prepare 85.0 g of calcium phosphate.

$$3\ CaCl_{2(aq)} + 2\ Na_3PO_{4(aq)} \rightarrow Ca_3(PO_4)_{2(s)} + 6\ NaCl_{(aq)}$$

13. Sodium chloride (0.400 mol) is allowed to react with an excess of sulfuric acid. How many moles of hydrogen chloride could be formed?

$$2\ NaCl_{(aq)} + H_2SO_{4(aq)} \rightarrow Na_2SO_{4(aq)} + 2\ HCl_{(g)}$$

14. If 0.380 mol of barium nitrate is allowed to react with an excess of phosphoric acid, how many moles of barium phosphate could be formed?

$$3\ Ba(NO_3)_{2(aq)} + 2\ H_3PO_{4(aq)} \rightarrow Ba_3(PO_4)_{2(s)} + 6\ HNO_{3(aq)}$$

15. Calculate the number of grams of carbon dioxide produced from the burning of 1.65 mol of propane (C_3H_8).

$$C_3H_{8(g)} + 5\ O_{2(g)} \rightarrow 3\ CO_{2(g)} + 4\ H_2O_{(g)}$$

16. How many moles of hydrogen molecules could be produced by the reaction of 3.40 mol of sodium atoms with water?

$$2\ Na_{(s)} + 2\ H_2O_{(\ell)} \rightarrow 2\ NaOH_{(aq)} + H_{2(g)}$$

17. How many moles of HI would be necessary to produce 2.20 mol of iodine, according to the following balanced equation?

$$10\ HI + 2\ KMnO_4 + 3\ H_2SO_4 \rightarrow 5\ I_2 + 2\ MnSO_4 + K_2SO_4 + 8\ H_2O$$

18. Calculate the number of grams of water that could be produced from the burning of 1.70 mol of ethane (C_2H_6).

$$C_2H_{6(g)} + O_{2(g)} \xrightarrow{\Delta} CO_{2(g)} + H_2O_{(g)} \text{ (unbalanced)}$$

Mass-Mass—Limiting Reagent Problems (See Section 10–4—limiting reagent examples)

19. A 35.0-g sample of calcium hydroxide is allowed to react with a 54.0-g sample of H_3PO_4. (a) How many grams of calcium phosphate could be produced? (b) If 45.2 g of calcium phosphate is actually obtained, what is the percent yield?

$$3\ Ca(OH)_2 + 2\ H_3PO_4 \rightarrow Ca_3(PO_4)_{2(s)} + 6\ H_2O$$

20. Cupric sulfide (0.610 mol) is treated with 1.40 mol of nitric acid. (a) How many moles of cupric nitrate could be produced? (b) If 0.500 mol of cupric nitrate is actually obtained, what is the percent yield? (c) Calculate the number of moles of excess reagent remaining at the end of the reaction.

$$3 \ CuS_{(s)} + 8 \ HNO_{3(aq)} \rightarrow 3 \ Cu(NO_3)_{2(aq)} + 3 \ S_{(s)} + 2 \ NO_{(g)} + 4 \ H_2O_{(\ell)}$$

21. A 1.4-g sample of magnesium is treated with 8.3 g of sulfuric acid. (a) How many grams of hydrogen could be produced? (b) If 0.060 g of hydrogen is actually obtained, what is the percent yield? (c) Calculate the number of moles of excess reagent remaining at the end of the reaction.

$$Mg_{(s)} + H_2SO_{4(aq)} \rightarrow MgSO_{4(aq)} + H_{2(g)}$$

22. Iron(II) hydroxide (0.320 mol) is treated with 0.270 mol of H_3PO_4. (a) How many grams of iron(II) phosphate could be produced? (b) If 34.0 g of iron(II) phosphate is actually obtained, what is the percent yield? (c) Calculate the number of moles of excess reagent remaining at the end of the reaction. (See 9-11 for the equation.)

Mass-Volume   (See Section 10-5)

23. How many liters of hydrogen sulfide measured at STP can be prepared from 40.0 g of iron(II) sulfide? (See Problem 8 for the equation.)

24. Calculate the number of liters of hydrogen gas at STP that could be produced by the reaction of 4.80 g of magnesium with excess hydrochloric acid.

$$Mg_{(s)} + 2 \ HCl_{(aq)} \rightarrow MgCl_{2(aq)} + H_{2(g)}$$

25. How many moles of potassium chlorate could be produced from 24.7 L of chlorine gas at STP?

$$3 \ Cl_{2(g)} + 6 \ KOH_{(aq)} \rightarrow 5 \ KCl_{(aq)} + KClO_{3(aq)} + 3 \ H_2O_{(\ell)}$$

26. Calculate the number of grams of magnesium nitride necessary to produce 5.75 L of ammonia gas at STP. How many moles of magnesium hydroxide could also be formed?

$$Mg_3N_{2(s)} + 6 \ H_2O_{(\ell)} \rightarrow 3 \ Mg(OH)_{2(aq)} + 2 \ NH_{3(g)}$$

27. Calculate the number of liters of hydrogen, measured at STP, that could be produced from the reaction of 0.375 mol of aluminum atoms according to the following equation:

$$Al_{(s)} + NaOH_{(aq)} + H_2O_{(\ell)} \rightarrow NaAlO_{2(aq)} + H_{2(g)} \ (unbalanced)$$

28. How many liters of oxygen measured at STP can be obtained by heating 70.0 g of potassium chlorate? (See 9-8 for the equation.)

Mass-Volume—Limiting Reagent Problems (See Section 10-5)

29. A 49.0-g sample of iron is allowed to react with 66.0 g of sulfuric acid.
    (a) How many liters of hydrogen measured at STP could be produced?
    (b) Calculate the number of moles of excess reagent remaining at the end of the reaction (See 9-9 for the equation.)

30. A 68.0-g sample of bismuth nitrate is treated with 8.50 L of hydrogen sulfide (at STP). (a) How many grams of bismuth sulfide could be produced? (b) Calculate the number of moles of excess reagent remaining at the end of the reaction. (See 9-10 for the equation.)

Volume-Volume (See Section 10-6)

31. How many liters of nitrogen would disappear in the production of 4.20 L of gaseous ammonia, according to the following balanced equation, both gases being measured at the same temperature and pressure?

$$N_{2(g)} + 3 H_{2(g)} \rightarrow 2 NH_{3(g)}$$

32. How many liters of ammonia gas measured at STP could be formed from 49.5 L of hydrogen (measured at STP)? (See Problem 31 for the balanced equation.)

33. How many liters of gaseous nitrogen dioxide measured at STP can be prepared from 76.6 L of gaseous nitrogen oxide measured at STP?

$$NO_{(g)} + O_{2(g)} \rightarrow NO_{2(g)} \text{ (unbalanced)}$$

34. How many liters of gaseous oxygen is needed to yield 5.50 L of gaseous nitrogen dioxide, according to the equation in Problem 33, both gases being measured at the same temperature and pressure?

Volume-Volume—Limiting Reagent Problems (See Section 10-6)

35. If. 4.25 L of gaseous oxygen reacts with 3.00 L of gaseous nitrogen oxide to form gaseous nitrogen dioxide, calculate (a) the number of liters of nitrogen dioxide that could be produced and (b) the number of liters of the excess reagent that would remain at the end of the reaction. All gases are measured at the same temperature and pressure. (See Problem 33 for the equation.)

36. Given the following unbalanced equation:

$$CO_{(g)} + O_{2(g)} \rightarrow CO_{2(g)}$$

(a) Calculate the number of liters of carbon dioxide gas produced if 8.25 L of carbon monoxide gas reacts with 4.30 L of oxygen gas. All gases are measured at the same temperature and pressure.

(b) Calculate the number of liters of excess reagent remaining at the end of the reaction.

Heats in Chemical Reactions   (See Section 10-7)

37. Given the following balanced equation:

$$H_{2(g)} + F_{2(g)} \rightarrow H_2F_{2(g)} + 1.284 \times 10^5 \text{ cal}$$

(a) Is the reaction exothermic or endothermic?

(b) Calculate the number of kilocalories of heat energy produced in the reaction of 37.0 g of fluorine gas with sufficient hydrogen gas.

38. Given the following balanced equation:

$$O_{2(g)} + 2 F_{2(g)} \rightarrow 2 OF_{2(g)} - 11.0 \text{ kcal}$$

(a) Is the reaction exothermic or endothermic?

(b) Calculate the number of grams of fluorine gas needed for the reaction with 2.09 kcal of heat energy and sufficient oxygen gas.

General Problems

39. Methane gas (CH_4) burns in oxygen to produce carbon dioxide gas and water vapor.

(a) Write the balanced equation for this reaction.

(b) Calculate the number of moles of hydrogen atoms in 8.00 g of methane.

(c) Calculate the number of moles of oxygen needed to completely burn 6.25 mol of methane.

(d) Calculate the number of grams of oxygen to completely burn 8.00 g of methane.

(e) Calculate the number of liters of carbon dioxide gas at STP that could be produced from 12.0 g of methane.

(f) Calculate the number of liters of oxygen gas required to produce 5.60 L of carbon dioxide gas, both gases being measured at the same temperature and pressure.

(g) Calculate the number of grams of carbon dioxide that could be produced from 17.6 g of methane.

(h) Calculate the percent yield, if 40.7 g of carbon dioxide is actually obtained [see part (g)].

40. A 30.0-g sample of iron is dissolved in concentrated hydrochloric acid (sp gr 1.18 and 35.0 percent by mass HCl). How many milliliters of the hydrochloric acid would be necessary to dissolve the iron?

$$Fe_{(s)} + 2 HCl_{(aq)} \rightarrow FeCl_{2(aq)} + H_{2(g)}$$

(*Hint:* A 35.0 percent by mass HCl means 35.0 g of pure HCl in $\overline{100}$ g of concentrated hydrochloric acid.)

41. A 47.1-g sample of copper is dissolved in concentrated nitric acid (sp gr 1.42 and 68.0 percent by mass HNO_3). How many milliliters of the nitric acid would be necessary to dissolve the copper?

$$Cu_{(s)} + 4\ HNO_{3(aq)} \rightarrow Cu(NO_3)_{2(aq)} + 2\ NO_{2(g)} + 2\ H_2O_{(\ell)}$$

42. If 2.2 g of cadmium is allowed to react with 4.9 mL of 20.0 percent hydrochloric acid (sp gr 1.10), how many grams of hydrogen could be produced? If 0.025 g of hydrogen is actually obtained, what is the percent yield? (See 9-9 for the equation.)

43. If 0.16 mol of iron atoms is allowed to react with 180 mL of 5.0 percent hydrochloric acid (sp gr 1.02), (a) how many grams of hydrogen could be formed? (b) If 0.22 g of hydrogen is actually obtained, what is the percent yield? (c) Calculate the number of moles of excess reagent remaining at the end of the reaction. (See 9-9 for the equation. *Hint:* A 5.0 percent hydrocholoric acid solution means 5.0 g of pure HCl in $\overline{100}$ g of HCl solution.)

Reading

Nyman, C. J., G. B. King, and J. A. Weyh, *Problems for General Chemistry and Qualitative Analysis,* 4th ed. New York: John Wiley & Sons, Inc., 1980. Chapter 7 is an excellent discussion of stoichiometry. This entire book of problems is an excellent supplement for a more comprehensive study of the subjects presented in basic chemistry.

Gases

Several hot air balloons flying in the Albuquerque sky. The hot air balloon achieves its buoyancy by heating the air inside the balloon envelope using a propane burner, thus lowering its density (Charles' Law).

TASKS

1. Memorize the conditions for standard temperature and pressure (STP), that is, **0°C** and **76̄0 torr** or **1.00 atm.**
2. Memorize the ideal-gas equation (**PV = nRT**) and the value of **R (0.0821 atm · L/mol · K).**

OBJECTIVES

1. Given the following terms, define each term and give a distinguishing characteristic of each:
   (a) pressure (Section 11-2)
   (b) Boyle's Law (Section 11-3)
   (c) Charles' Law (Section 11-4)
   (d) Gay-Lussac's Law (Section 11-5)
   (e) Dalton's Law of Partial Pressures (Section 11-7)
   (f) Ideal-gas equation (Section 11-8)

2. Given the kinetic theory, list the assumptions as applied to gases and describe the distinguishing characteristics of each. Differentiate between the meaning of the terms "ideal gas" and "real gas" (Section 11-1).

3. Given a fixed mass of a given gas at constant temperature and at a stated volume and pressure, calculate the volume if the pressure is changed. Calculate the pressure if the volume is changed (Boyle's Law, Problem Examples 11-1 and 11-2, Problems 8, 9, 10, and 11).

4. Given a fixed mass of a given gas at constant pressure and at a stated volume and temperature, calculate the volume if the temperature is changed. Calculate the temperature if the volume is changed (Charles' Law, Problem Examples 11-3 and 11-4, Problems 12, 13, 14, and 15).

5. Given a fixed mass of a given gas at constant volume and at a stated temperature and pressure, calculate the pressure if the temperature is changed. Calculate the temperature if the pressure is changed (Gay-Lussac's Law, Problem Example 11-5, Problems 16, 17, 18, and 19).

6. Given a fixed mass of a given gas at a stated volume, temperature, and pressure, calculate the volume if the temperature and pressure are changed. Calculate the temperature if the volume and pressure are changed. Calculate the pressure if the volume and temperature

are changed (Combined Gas Laws, Problem Examples 11-6 and 11-7, Problems 20, 21, 22, 23, 24, 25, 26, 27, and 28).

7. (a) Given the partial pressures of a mixture of gases, calculate the total pressure of the gases. Given the total pressure of a mixture of gases, calculate the partial pressure of one gas, if the pressures of the other gases in the mixture are known.

    (b) Given a fixed mass of a given gas, collected over water, at a stated volume, temperature, and pressure, and the vapor pressure of the water at the stated temperature, calculate the volume of the *dry* gas if the temperature and pressure are changed.

    (Dalton's Law of Partial Pressures, Problem Examples 11-8 and 11-9, Problems 29, 30, 31, and 32)

8. Given three of the four variables (pressure, volume, mass in moles or grams, and temperature) in the ideal-gas equation, calculate the fourth variable (Problem Examples 11-10, 11-11, and 11-12, Problems 33, 34, 35, and 36).

9. Given the mass of a gas and its volume under any stated conditions of temperature and pressure, calculate the molecular mass of the gas (Problem Examples 11-13 and 11-14, Problems 37 and 38).

10. Given the name of a gas and the table of approximate atomic masses, write the formula of the gas and calculate the density of the gas at any stated temperature and pressure (Problem Example 11-15, Problems 39 and 40).

11. Given the mass of a reactant or product in a reaction, the reaction, and the Table of Approximate Atomic Masses, write a balanced equation for the reaction and calculate the volume of the gas, reactant, or product at any stated temperature and pressure (Problem Example 11-16, Problems 41 and 42).

In Chapter 3, we stated that there are three physical states of matter: solid, liquid, and gas. In this and the next chapter, we shall consider these physical states in relation to the properties of each state and the changes each one may undergo. We shall begin our discussion with gases, since they are the simplest of the three physical states and more is known of the gaseous state than of the other two states.

Before we consider the properties and changes that occur in the gaseous state, we must first consider the general characteristics of gases. These characteristics follow:

1. *Expansion.* Gases expand indefinitely and uniformly to fill all the space in which they are placed.

2. *Indefinite shape or volume.* A given sample of gas has no definite shape or volume, but can be made to fit the vessel in which it is placed.

3. *Compressibility.* Gases can be highly compressed; for example, many pounds of oxygen gas can be placed in pressurized tanks.

4. *Low density.* The density of a gas is small, and hence it is measured in grams per liter (g/L) in the metric system and not in grams per milliliter (g/mL) as we observed for solids and liquids (see 2-13).

5. *Mixability or diffusion.* Two or more different gases will normally mix completely and uniformly when placed in contact with each other. A leaky gas jet is an example of this characteristic. An odorous gas is added to natural gas so that it can be detected readily. The natural gas thus treated mixes with the air and the gas leak can be detected by the odor of the additive.

11-1

The Kinetic Theory

The kinetic theory has been advanced to explain the characteristics and properties of matter in general. In essence, the theory states that heat and motion are related, that particles of all matter are in **motion** to some degree, and that **heat** is an indication of this motion. This theory is applied to gases, but the following assumptions must be made:

1. Gases are composed of very small particles called molecules. The **distance** between these molecules is very **great** compared to the size of the molecules themselves, and the total volume of the molecules is only a small fraction of the entire space occupied by the gas. Therefore, in considering the volume of a gas, we are considering primarily empty space. This postulate is the basis of the high compressibility and low density of gases.

2. **No attractive forces** exist between the molecules in a gas.

3. These molecules are in a state of constant, **rapid motion,** colliding with each other and with the walls of the container in a perfectly random manner. This motion could *possibly* be compared to the motion of a small "bumper car" at an amusement park. This postulate is the basis of the complete mixing of two or more different gases. The frequency of collisions with the walls of the container accounts for the pressure of gases (11-2).

4. All of these molecular collisions are perfectly **elastic;** consequently, there is no loss of kinetic energy in the system as a whole. Some energy may be transferred from one molecule to the other molecule involved in the collision. We again refer to the "bumper car" analogy; each collision with a car appears to be perfectly elastic and the cars continue to move and collide again and again.

5. The *average* **kinetic energy** per molecule of the molecules of the gas is proportional to the **temperature** in degrees kelvin (degrees absolute), and the *average* kinetic energy per molecule for all gases is the **same** at the same temperature. Since this is the average kinetic energy, some molecules have more energy ("hotter") and others less ("colder"). Theoretically, at zero kelvin, molecular motion has ceased and the kinetic energy is considered to be zero.

Gases that conform to these assumptions are called *ideal gases,* as opposed to *real gases,* such as hydrogen, oxygen, nitrogen, and others. Under moderate conditions of temperature and pressure, real gases essentially behave as ideal gases, but if the *temperature* is very *low* or the *pressure* is very *high,* then real gases deviate considerably from ideal gases. An ideal gas is considered to have the following characteristics: (1) negligible volume of the actual molecules as compared to the volume of the gas itself (assumption 1); (2) no attractive forces between molecules (assumption 2); and (3) perfectly elastic collisions (assumption 4). By avoiding extremely low temperatures and extremely high pressures, we can consider real gases to behave as ideal gases and apply the basic gas laws (11-3, 11-4, and 11-5) and the ideal-gas equation (11-8).

11-2

Pressure of Gases

Gases exert pressure, as you have probably observed in inflating automobile tires or bicycle tires to 32 pounds (pounds per square inch, abbreviated psi) or 90 pounds, respectively. **Pressure** is defined as force per unit area. The pressure of gases is produced by the impact of the gas molecules on the walls of the container.

Gases in the atmosphere (primarily nitrogen, oxygen, and a small amount of argon plus pollutants) also exert a pressure. The atmospheric pressure is measured by a mercury *barometer,* which was first devised in 1643 by Evangelista Torricelli (tŏr·rĕ·chĕl′le (1608–1647), Italian mathematician and physicist. His barometer consisted of a glass tube at least 76 cm long sealed at one end, filled with mercury, and then inverted with the open end in a dish of mercury (see Figure 11-1). At sea level, the mercury level dropped to a height of 76.0 cm in the tube, leaving no air above the mercury level in the tube. *Regardless of the diameter of the tube, the mercury level dropped to 76.0 cm.* Now, let us consider why. Suppose we use a sealed tube of 1.00 cm^2 cross-sectional area (radius of 0.564 cm). The volume of mercury in the tube at sea level would be 76.0 cm $\times$ 1.00 cm^3 = 76.0 cm^3. The density of mercury varies with temperature (see 2-13), and at 0°C the density is 13.6 g/cm^3. Hence, the weight (wt) of mercury acting as a force is calculated as

$$76.0 \text{ cm}^3 \times \frac{13.6 \text{ g}}{1 \text{ cm}^3} = 1030 \text{ g wt}$$

FIGURE 11-1

Torricelli's mercury barometer.

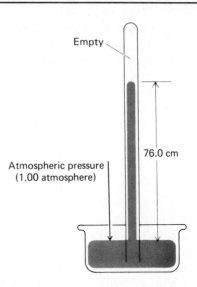

Empty

76.0 cm

Atmospheric pressure
(1.00 atmosphere)

Since pressure is defined as a force per unit area, the pressure at sea level in the barometer would be 1030 g wt/cm². If the cross-sectional area had been 2.00 cm², the force (weight) would have been twice as great but the pressure would have remained 1030 g wt/cm², since the pressure is obtained by dividing the *force by the cross-sectional area.*

This pressure, (1030 g wt/cm²) which at sea level supports a column of mercury at a height of 76.0 cm at 0°C, is called *standard pressure.* This pressure may be expressed in many other units:[1]

1. pounds per square inch (psi), 14.7 psi

2. centimeters of mercury, 76.0 cm of mercury

3. millimeters of mercury, $76\overline{0}$ mm of mercury (mm Hg)

4. torr (1 torr = 1 mm of mercury), $76\overline{0}$ torr

5. inches of mercury, 29.9 in. of mercury[2]

6. atmosphere (atm), 1.00 atm

7. pascal (Pa), 1.013×10^5 Pa[3]

[1]The millibar (mbar) is also a measure of pressure and is used primarily in meteorology. Atmospheric pressure is reported by the National Oceanic and Atmospheric Administration (formerly the Weather Bureau) in millibars. Standard pressure is 1013 millibar.

[2]Pressure is expressed in inches of mercury in weather reports, but this pressure has been corrected to sea level.

[3]The *pascal* is named in honor of Blaise Pascal (1623–1662), French scientist who formulated principles of physics related primarily to liquids. One pascal (Pa) is the pressure of a force of one newton (N) per square meter and therefore has the units N/m². See 2-12 for the definition of a newton.

Although the *pascal* is the pressure unit recommended by the International System of Units (SI) we will generally use the *torr* (named in honor of Torricelli), the *centimeter* of mercury, the *millimeter* of mercury, or the *atmosphere* as the pressure unit in this book. Conversion to atmospheres from torr, and vice versa, is done by knowing that 1 atm = $76\overline{0}$ torr. The conversion of $63\overline{0}$ torr to atmospheres is done as follows:

$$63\overline{0} \text{ torr} \times \frac{1 \text{ atm}}{76\overline{0} \text{ torr}} = 0.829 \text{ atm}$$

In the preceding discussion, we stated that Torricelli's measurement was carried out at sea level. The atmospheric pressure decreases as altitude increases (approximately 25 torr per 1000 feet). At mile-high altitude, the pressure is approximately 630 torr. You may have experienced this decrease in pressure when you drive to the mountains or go to a higher altitude where the pressure is less. When you yawn, your ears pop. Yawning equalizes the pressure on your ear drum by opening a tube from the middle portion of the ear to your mouth.

Atmospheric pressure also varies with weather conditions, as you may have noted on the TV weather reports. When there is considerable moisture in the atmosphere, the pressure may be low, due to the lower density of moist air over that of dry air.[4] This creates a low-pressure area over an entire region. Little moisture in the air creates a high-pressure area.

11-3

Boyle's Law: The Effect of Pressure Change on the Volume of a Gas at Constant Temperature

In 1660, British physicist and chemist Robert Boyle, referred to in 1-3, carried out experiments on the change in volume of a given amount of gas with the pressure of the gas at constant temperature. From his experiments he formulated the law now referred to as **Boyle's Law**. *At constant temperature, the volume of a fixed mass of a given gas is* **inversely** *proportional[5] to the pressure it exerts.* For example, if the pressure of a given volume of gas is doubled, the volume will be halved; if the pressure is halved, the volume will be doubled, as shown in Figure 11-2. Boyle's Law may be expressed mathematically as

$$V \propto \frac{1}{P} \text{ (temperature constant)}$$

---

[4]In moist air, water (molecular mass = 18.0 amu) is present in the air in place of oxygen and nitrogen (larger amu values, 32.0 amu and 28.0 amu, respectively), making moist air less dense (g/L) then dry air.

[5]Inversely proportional means that an *increase* in one variable results in a *decrease* in the other variable, or a *decrease* in one variable results in an *increase* in the other variable.

FIGURE 11-2

A demonstration of
Boyle's Law.
Temperature is
constant. (As the
volume is decreased
the frequency of
collisions is increased,
resulting in an
increase in pressure.)

2 volumes   1 volume   1/2 volume

Volume (V) is inversely $(1/P)$ proportional $(\propto)$ to the pressure (P). An equation can be written by introducing a constant of proportionality (k), the value of which depends upon the units of P and V as well as the *quantity* of gas being measured.

$$V = k \times \frac{1}{P}$$

The equation may then be expressed as

$$PV = k \qquad\qquad (11\text{-}1)$$

with the product of the pressure and volume equal to a constant at constant temperature. Since $P \times V$ is equal to a constant (k) in Equation 11-1, then different conditions of pressure and volume may be expressed for the same mass of gas at constant temperature:

$$P_{new} \times V_{new} = k = P_{old} \times V_{old} \qquad\qquad (11\text{-}2)$$

From this equation, the new pressure can be solved as

$$P_{new} = P_{old} \times V_{factor} \qquad\qquad (11\text{-}3)$$

where the new pressure is equal to the old pressure times a volume factor. The new volume can also be calculated as

$$V_{new} = V_{old} \times P_{factor} \qquad (11\text{-}4)$$

where the new volume is equal to the old volume times a pressure factor.

The evaluation of the volume factor and pressure factor can be made by considering the effect the change in volume or pressure has on the old pressure or volume, and how this change will affect the new pressure or volume. It is not necessary, then, to memorize a formula. In evaluating these factors, it is most important that **both** the numerator and denominator of the factor be expressed in the **same** units.

Consider the following problem examples:

Problem Example 11-1

A sample of gas occupies a volume of 73.5 mL at a pressure of $71\bar{0}$ torr and a temperature of $3\bar{0}°C$. What will be its volume in mL at standard pressure and $3\bar{0}°C$?

SOLUTION: In working these problems, arrange the data in an orderly form. See Figure 11-2.

$$T = \text{constant}$$

$$V_{old} = 73.5 \text{ mL} \qquad P_{old} = 71\bar{0} \text{ torr} \quad \Big| \text{ pressure increases;}$$

$$V_{new} = ? \qquad P_{new} = 76\bar{0} \text{ torr} \quad \downarrow \text{ volume decreases}$$

From Equation 11-4 (see Figure 11-2)

$$V_{new} = V_{old} \times P_{factor}$$

The pressure has increased from $71\bar{0}$ to $76\bar{0}$ torr; hence, the new volume will be decreased, and the pressure factor must be written so that the new volume will show a decrease. To reflect this decrease, we must write the pressure factor so that the ratio of pressures is less than 1 – hence, $\dfrac{71\bar{0} \text{ torr}}{76\bar{0} \text{ torr}}$.

$$V_{new} = 73.5 \text{ mL} \times \frac{71\bar{0} \text{ torr}}{76\bar{0} \text{ torr}} = 68.7 \text{ mL} \quad \textit{Answer}$$

Problem Example 11-2

The volume of a gas is 17.4 L, measured at standard pressure. Calculate the pressure of the gas in torr if the volume is changed to 20.4 L and the temperature remains constant.

SOLUTION

$$T = \text{constant}$$

$$P_{old} = 76\overline{0} \text{ torr} \qquad V_{old} = 17.4 \text{ L} \quad \Big| \quad \text{volume increases;}$$

$$P_{new} = ? \qquad V_{new} = 20.4 \text{ L} \quad \Big\downarrow \quad \text{pressure decreases}$$

$$P_{new} = P_{old} \times V_{factor}$$

The volume has increased; therefore, the new pressure must be less and the volume factor must be written so that the new pressure will be less.

Work Problems 8, 9, 10, and 11.

The ratio of volumes must be less than 1 — hence, $\dfrac{17.4 \text{ L}}{20.4 \text{ L}}$.

$$P_{new} = 76\overline{0} \text{ torr} \times \frac{17.4 \text{ L}}{20.4 \text{ L}} = 648 \text{ torr} \quad Answer$$

11-4

Charles' Law: The Effect of Temperature Change on the Volume of a Gas at Constant Pressure

Experiments carried out originally in 1787 by Jacques Charles, a French physicist (1746–1823), and refined in 1802 by Joseph Gay-Lussac (1778–1850) (see 10-6), showed that the volume of a gas is increased by $\frac{1}{273}$ of its value at 0°C for every degree rise in temperature. See Table 11-1.

Although the volume of a gas changes uniformly with changes in temperature, the volume is not directly proportional to the Celsius temperature. If a new temperature scale is devised with a zero point of -273°C (more accurately, -273.15°C) and temperatures are expressed on this scale, then the volume of a gas would be directly proportional to the temperature (refer again to Table 11-1). This scale is called the Kelvin scale and -273°C is called 0 K (see 2-11). From Table 11-1, the value 0 K is the kelvin temperature corresponding to *zero* volume of the gas, but since gases form liquids and solids on cooling, this zero value is only theoretical. These data are given in a graph in Figure 11-3 showing the theoretically 0-mL volume. To convert from °C to K,

TABLE 11-1   Relation of Temperature to Volume

t (°C)	V (mL)	T (K)
273	546	546
100	373	373
10	283	283
1	274	274
0	273	273
-1	272	272
-10	263	263
-100	173	173
-273	0 (theoretically)	0

FIGURE 11-3

Graph relating temperature to volume of a gas. Zero-milliliter volume is only theoretical since gases form liquids and solids on cooling.

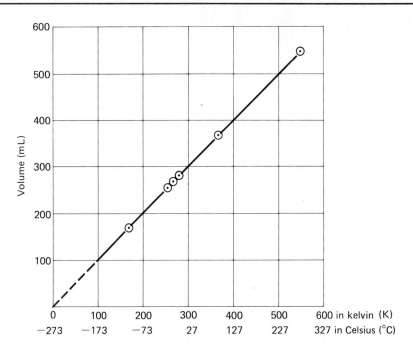

we need only add 273 to °C, as we did in Chapter 2 (see 2-11). (In this text, we shall use 273 instead of 273.15, to simplify calculations.)

$$K = °C + 273 \qquad (11\text{-}5)$$

Charles' Law states that at *constant pressure, the volume of a fixed mass of a given gas is **directly** proportional*[6] *to the kelvin temperature.* For example, if the kelvin temperature is doubled at constant pressure, the volume is doubled; and if the kelvin temperature is halved, the volume is halved (Figure 11-4).

Charles' Law may be expressed mathematically as

$$V \propto T \text{ (pressure constant)}$$

Volume (V) is directly proportional to the kelvin temperature (T). An equation can be written by introducing a constant (k) for a given sample of gas.

$$V = kT$$

---

[6]Directly proportional means that an *increase* in one variable results in an *increase* in the other variable, or a *decrease* in one variable results in a *decrease* in the other variable.

FIGURE 11-4

A demonstration of Charles' Law (temperature in kelvin). Pressure is constant. (As the temperature is increased the kinetic energy of the molecules is increased, resulting in an increase in volume.)

1/2 volume   1 volume   2 volumes

The equation may then be expressed as

$$\frac{V}{T} = \mathbf{k} \qquad (11\text{-}6)$$

with the volume divided by the kelvin temperature being equal to a constant at constant pressure. Since V/T is equal to a constant, different conditions of temperature and volume may be expressed for the same mass of a gas at constant pressure.

$$\frac{V_{new}}{T_{new}} = \mathbf{k} = \frac{V_{old}}{T_{old}} \qquad (11\text{-}7)$$

From this equation, the new temperature in kelvins can be solved as

$$T_{new} = T_{old} \times V_{factor} \qquad (11\text{-}8)$$

where the new temperature is equal to the old temperature times a volume factor. The new volume can also be calculated as

$$V_{new} = V_{old} \times T_{factor} \qquad (11\text{-}9)$$

here the new volume is equal to the old volume times a temperature factor in kelvins.

Consider the following problem examples:

Problem Example 11-3

A gas measures $15\overline{0}$ mL at 1.00 atmosphere and 27°C. Calculate its volume in mL at 0°C and 1.00 atmosphere.

SOLUTION

P = constant

$V_{old} = 15\overline{0}$ mL $\qquad t_{old} = 27°C$ $\qquad T_{old} = 27 + 273 = 3\overline{00}$ K $\Big|$ temperature decreases;

V_{new} = ? $\qquad t_{new} = 0°C$ $\qquad T_{new} = 0 + 273 = 273$ K $\Big|$ volume decreases

From Equation 11-9 (see Figure 11-4),

$$V_{new} = V_{old} \times T_{factor}$$

The temperature has decreased from $3\overline{00}$ to 273 K; hence, the new volume will be less, and the temperature factor must be written so that the new volume will be less. To reflect this decrease, we must write the temperature factor so that the ratio of temperatures is less than 1 — hence, $\dfrac{273\ \cancel{K}}{3\overline{00}\ \cancel{K}}$.

$$V_{new} = 15\overline{0}\ \text{mL} \times \frac{273\ \cancel{K}}{3\overline{00}\ \cancel{K}} = 136\ \text{mL}\quad Answer$$

Problem Example 11–4

A gas occupies a volume of 4.50 L at 27°C. At what temperature in °C would the volume be 6.00 L, the pressure remaining constant?

SOLUTION

P = constant

V_{old} = 4.50L $\Big|$ volume increases; $\qquad t_{old} = 27°C$ $\qquad T_{old} = 3\overline{00}K$

V_{new} = 6.00L $\Big|$ temperature increases $\qquad t_{new}$ = ? $\qquad T_{new}$ = ?

$$T_{new} = T_{old} \times V_{factor}$$

The volume increases; therefore, the new temperature will be greater, and the volume factor must be written so that the new temperature will be greater. The ratio of volumes must be written so that the ratio is greater than 1—hence, $\dfrac{6.00\ \cancel{L}}{4.50\ \cancel{L}}$.

$$T_{new} = 3\overline{00}\ \text{K} \times \frac{6.00\ \cancel{L}}{4.50\ \cancel{L}} = 4\overline{00}\ \text{K}$$

Work Problems 12,
13, 14, and 15.

This kelvin temperature is then converted to °C by subtracting the constant, 273.

$$400 \text{ K} = (400 - 273)°\text{C} = 127°\text{C} \quad \textit{Answer}$$

11–5

Gay-Lussac's Law: The Effect of Temperature Change on the Pressure of a Gas at Constant Volume

In 1802, Joseph Gay-Lussac published the results of his experiments, which are now known as Gay-Lussac's Law. **Gay-Lussac's Law** states that *at constant volume, the pressure of a fixed mass of a given gas is **directly** proportional to the kelvin temperature.* For example, if the kelvin temperature is doubled at constant volume, the pressure is doubled; if the kelvin temperature is halved, the pressure is halved (Figure 11–5).

This statement may be expressed mathematically as

$$P \propto T \text{ (volume constant)}$$

An equation may be written as

$$P = \mathbf{k}T$$

and also as

$$\frac{P}{T} = \mathbf{k} \tag{11-10}$$

For different conditions of pressure and temperature, an equation may be written as

$$\frac{P_{\text{new}}}{T_{\text{new}}} = \mathbf{k} = \frac{P_{\text{old}}}{T_{\text{old}}} \tag{11-11}$$

FIGURE 11-5

A demonstration of Gay-Lussac's Law (temperature in kelvin). Volume is constant. (As the temperature is increased the kinetic energy of the molecules is increased and the frequency of collisions is increased, resulting in an increase in pressure.)

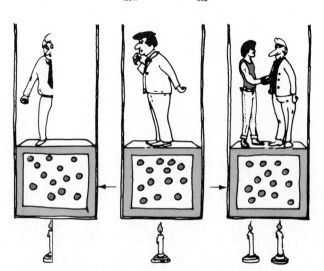

Hence, solving for P_{new} and T_{new}, the equations are

$$P_{new} = P_{old} \times T_{factor} \tag{11-12}$$

and

$$T_{new} = T_{old} \times P_{factor} \tag{11-13}$$

Consider the following problem example:

Problem Example 11-5

The temperature of 1 L of a gas originally at STP is changed to $22\overline{0}°C$ at constant volume. Calculate the final pressure of the gas in torr.

SOLUTION

$$V = \text{constant}$$

$P_{old} = 76\overline{0}$ torr	$t_{old} = 0°C$	$T_{old} = 0 + 273 = 273$ K	temperature increases;
$P_{new} = ?$	$t_{new} = 22\overline{0}°C$	$T_{new} = 22\overline{0} + 273 = 493$ K	pressure increases

From Equation 11-12 (see Figure 11-5),

$$P_{new} = P_{old} \times T_{factor}$$

Since the temperature increases, the pressure will increase, and the temperature factor must be written so that the new pressure will be greater. To reflect this increase, we must write the temperature factor so that the ratio of temperatures is greater than 1—hence, $\dfrac{493\ \cancel{K}}{273\ \cancel{K}}$.

Work Problems 16, 17, 18, and 19.

$$P_{new} = 76\overline{0}\ \text{torr} \times \frac{493\ \cancel{K}}{273\ \cancel{K}} = 1370\ \text{torr} \quad \textit{Answer}$$

11-6

The Combined Gas Laws

Boyle's and Charles' Laws can be combined into one mathematical expression:

$$\frac{P_{new}V_{new}}{T_{new}} = \frac{P_{old}V_{old}}{T_{old}} \tag{11-14}$$

Solving Equation 11-14 for V_{new}, P_{new}, and T_{new} gives

$$V_{new} = V_{old} \times P_{factor} \times T_{factor} \qquad (11\text{-}15)$$

$$P_{new} = P_{old} \times V_{factor} \times T_{factor} \qquad (11\text{-}16)$$

$$T_{new} = T_{old} \times V_{factor} \times P_{factor} \qquad (11\text{-}17)$$

Each factor in these equations (11-15 through 11-17) and its effect on the new volume, pressure, or temperature *will be considered* **separately.** In Equation 11-15, the new volume is equal to the old volume multiplied by a pressure factor and a temperature factor. If the pressure increases, the pressure ratio must be less than 1, since increasing the pressure would decrease the old volume. If the pressure decreases, the pressure ratio must be greater than 1, since a decrease in pressure would increase the old volume. If the temperature increases, the ratio of the kelvin temperatures must be greater than 1, since the temperature change would increase the old volume. Conversely, if the temperature decreases, the temperature ratio must be less than 1. By applying similar reasoning to Equations 11-16 and 11-17, we can solve for the new pressure and temperature.

These gas laws apply only when the behavior of real gases closely resembles that of an ideal gas. Under certain conditions of temperature and/or pressure, the properties of most real gases deviate markedly from those of an ideal gas. Other equations have been developed to handle such cases, but in this text we shall consider that for all practical purposes real gases generally behave like ideal gases.

The following problem examples illustrate the application of the combined gas laws to real gases behaving as ideal gases:

Problem Example 11-6

A certain gas occupies $5\overline{00}$ mL at $76\overline{0}$ torr and 0°C. What volume in mL will it occupy at 10.0 atm and $10\overline{0}$°C?

SOLUTION

$$
\begin{array}{ll}
V_{old} = 5\overline{00}\text{ mL} & P_{old} = 76\overline{0}\text{ torr} = 1.00\text{ atm} \\
V_{new} = ? & P_{new} = 10.0\text{ atm}
\end{array}
\quad
\begin{array}{l}
\text{pressure increases;} \\
\downarrow \text{volume decreases}
\end{array}
$$

$$
\begin{array}{l}
T_{old} = 0 + 273 = 273\text{ K} \\
T_{new} = 10\overline{0} + 273 = 373\text{ K}
\end{array}
\quad
\begin{array}{l}
\text{temperature increases;} \\
\downarrow \text{volume increases}
\end{array}
$$

From Equation 11-15,

$$V_{new} = V_{old} \times P_{factor} \times T_{factor}$$

Since the units of P_{old} must be the same as those of P_{new}, both pressures must be expressed in the same units. The pressure factor should make the new volume less $\left(\dfrac{1.00 \text{ atm}}{10.0 \text{ atm}}\right)$, whereas the temperature factor should make the new volume greater $\left(\dfrac{373 \text{ K}}{272 \text{ K}}\right)$. The final result is a *new volume* that is less due to the magnitude of the pressure factor.

$$V_{new} = 5\overline{0}0 \text{ mL} \times \frac{1.00 \text{ atm}}{10.0 \text{ atm}} \times \frac{373 \text{ K}}{273 \text{ K}} = 68.3 \text{ mL} \quad \textit{Answer}$$

In each case, the effect of one factor was considered *independently* of the other factor, and in each case, the effect of each factor on the old volume is considered.

Actually, what is being done when these factors are being considered independently is to consider the effect of pressure at constant (old) temperature (pressure factor) and then to consider the effect of temperature at constant (new) pressure (temperature factor). This is shown in the following diagram:

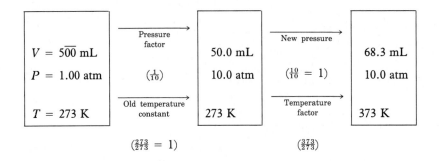

Problem Example 11-7

A certain gas occupied 20.0 L at $5\overline{0}°C$ and $78\overline{0}$ torr. Under what pressure in torr would this gas occupy 75.0 L at 0°C?

SOLUTION

$$V_{old} = 20.0 \text{ L} \ \Big| \text{ volume increases;} \qquad P_{old} = 78\overline{0} \text{ torr}$$
$$V_{new} = 75.0 \text{ L} \ \Big\downarrow \text{ pressure decreases} \qquad P_{new} = ?$$

$$T_{old} = 5\overline{0} + 273 = 323 \text{ K} \ \Big| \text{ temperature decreases;}$$
$$T_{new} = 0 + 273 = 273 \text{ K} \ \Big\downarrow \text{ pressure decreases}$$

$$P_{new} = P_{old} \times T_{factor} \times V_{factor}$$

Since the temperature decreases ($5\overline{0}$ to 0°C), the pressure will decrease, and the ratio of the kelvin temperatures will be less than 1. A decrease in pressure will also result from the volume increasing (20.0 to 75.0 L), and the ratio of volumes must be less than 1.

Work Problems 20, 21, 22, 23, 24, 25, 26, 27, and 28.

$$P_{\text{new}} = 78\overline{0} \text{ torr} \times \frac{273\,K}{323\,K} \times \frac{20.0\,L}{75.0\,L} = 176 \text{ torr}\quad Answer$$

11-7

Dalton's Law of Partial Pressures

John Dalton, whose atomic theory we referred to in 4-2, was also keenly interested in meteorology. This interest led him to study gases, and in 1801 he announced his conclusions, which are now known as **Dalton's Law of Partial Pressures.** This law states that *each gas in a mixture of gases exerts a partial pressure equal to the pressure it would exert if it were the only gas present in the same volume; the total pressure of the mixture is then the* **sum** *of the partial pressures of all the gases present.* For example, if two gases, such as oxygen and nitrogen, are present in a 1-liter flask, and the pressure of the oxygen is $25\overline{0}$ torr and that of the nitrogen is $3\overline{00}$ torr, then the total pressure is $55\overline{0}$ torr, as shown in Figure 11-6.

Dalton's Law of Partial Pressure may be expressed mathematically as

$$P_{\text{total}} = P_1 + P_2 + P_3 \ldots \tag{11-18}$$

where P_1, P_2, P_3 are the partial pressures of the individual gases in the mixture.

Consider the following problem example:

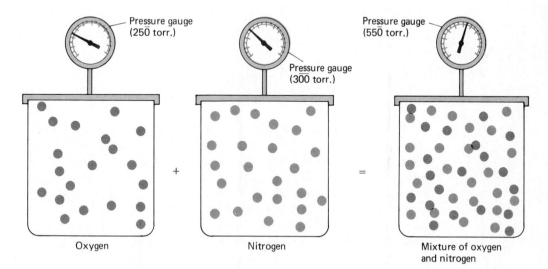

FIGURE 11-6

Dalton's Law of Partial Pressures. Partial pressure of oxygen gas (●) is $25\overline{0}$ torr, and partial pressure of nitrogen gas (●) is $3\overline{00}$ torr. When these amounts are mixed the total pressure of both gases equal $25\overline{0}$ torr + $3\overline{00}$ torr. = $55\overline{0}$ torr., assuming all temperatures are constant and equal.

Problem Example 11-8

A 1-L flask at 27°C contains a mixture of three gases, A, B, and C, at partial pressures of $30\overline{0}$ torr for A, $25\overline{0}$ torr for B, and 425 torr for C. (a) Calculate the total pressure in torr of the mixture. (b) Calculate the volume in liters at STP occupied by the gases remaining if gas A is removed selectively.

SOLUTION

(a) From Dalton's Law of Partial Pressures, the total pressure of the mixture will be equal to the sum of the individual pressures of each gas, or

$$P_{total} = P_A + P_B + P_C$$

$$P_{total} = 30\overline{0} \text{ torr} + 25\overline{0} \text{ torr} + 425 \text{ torr}$$

$$P_{total} = 975 \text{ torr}$$

(b) If gas A is removed selectively at the same temperature and volume (27°C and 1.00 L) leaving only gases B and C in the flask, the pressure will decrease by P_A or $30\overline{0}$ torr and the new pressure will be $P_{total} - P_A = 975$ torr $- 30\overline{0}$ torr $= 675$ torr. Now we are confronted with the problem of calculating a new volume at STP for a gas originally occupying 1.00 L at 27°C and 675 torr.

$$V_{old} = 1.00 \text{ L} \qquad P_{old} = 675 \text{ torr} \quad | \text{ pressure increases;}$$

$$V_{new} = ? \qquad P_{new} = 76\overline{0} \text{ torr} \quad \downarrow \text{ volume decreases}$$

$$T_{old} = 27 + 273 = 30\overline{0} \text{ K} \quad | \text{ temperature decreases;}$$

$$T_{new} = 0 + 273 = 273 \text{ K} \quad \downarrow \text{ volume decreases}$$

$$V_{new} = V_{old} \times P_{factor} \times T_{factor}$$

$$V_{new} = 1.00 \text{ L} \times \frac{675 \text{ torr}}{76\overline{0} \text{ torr}} \times \frac{273 \text{ K}}{30\overline{0} \text{ K}} = 0.808 \text{ L} \quad Answer$$

Indirectly related to Dalton's Law of Partial Pressures is an increase in red blood cell count with an increase in altitude. The percent of oxygen and nitrogen in the atmosphere is constant, but the *partial pressures* of these gases vary with altitude. At sea level the atmospheric pressure is 760 torr, and the sum of the partial pressures of the other gases excluding oxygen (primarily nitrogen with small amounts of argon and carbon dioxide) is $60\overline{0}$ torr. This gives a partial pressure for oxygen of 160 torr (760 torr $- 60\overline{0}$ torr). At mile-high altitude, the altitude of many cities in the Rocky Mountain region, the atmospheric pressure is approximately 630 torr, and the sum of the partial pressures of the other gases is $50\overline{0}$ torr. This gives a partial pressure for oxygen of only 130 torr (630 torr $- 50\overline{0}$ torr).

The body needs a given amount of oxygen for normal metabolic processes regardless of the altitude. To compensate for this reduced partial pressure of oxygen with altitude, the number of red blood cells increases and the body acclimates to this reduced partial pressure of oxygen caused by the increase in altitude. For some people this acclimation may take many days, with the person being very sleepy and groggy. For example, the number of red blood cells in a normal healthy person at sea level is approximately 4.2 million per cubic millimeter of blood, and at mile-high altitude it is approximately 5.4 million.

A more direct application of Dalton's Law of Partial Pressures is the collection of a gas over water, as shown in Figure 11-7. The gas will contain a certain amount of water vapor. The pressure exerted by the water vapor in the gas will be a constant value *at any given temperature* if sufficient time has been allowed to permit equilibrium conditions to be established. The total pressure at which the volume of the "wet" gas is measured must be equal to the sum of the gas pressure and the water vapor pressure at the temperature at which the gas is collected and measured, or, mathematically,

$$P_{total} = P_{gas} + P_{water} \text{ (Dalton's Law of Partial Pressures)} \quad (11\text{-}19)$$

The pressure of the dry gas is calculated by subtracting the known equilibrium vapor pressure of water at the temperature of the "wet" gas from the total pressure:

FIGURE 11-7

Collection of a gas over water. The water vapor is shown in the collection as ●, and the gas as ●

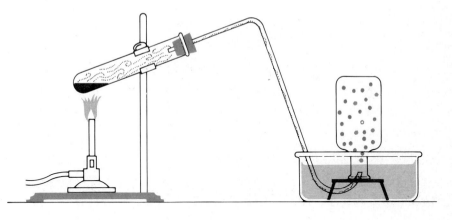

$$P_{gas} = P_{total} - P_{water} \quad (11\text{-}20)$$

The vapor pressure of water at various temperatures is found in Appendix V.

Consider the following problem example:

Problem Example 11-9

The volume of a certain gas, collected over water, is $15\overline{0}$ mL at $3\overline{0}°C$ and 720.0 torr. Calculate the volume in mL of the dry gas at STP.

SOLUTION: The first step in the calculation is to determine the pressure of the dry gas at the initial volume ($15\bar{0}$ mL) and temperature ($3\bar{0}°C$). The pressure of the wet gas (720.0 torr) is equal to the sum of the pressure of the dry gas and the vapor pressure of water at the initial temperature. From Appendix V, the vapor pressure of water at $3\bar{0}°C$ is 31.8 torr. The pressure of the dry gas is therefore equal to $P_{total} - P_{water} = 720.0$ torr $- 31.8$ torr $= 688.2$ torr. Thus, if the water vapor were removed—that is, if the gas were dry—the pressure of the gas would have measured 688.2 torr (see Figure 11-8) in a volume of $15\bar{0}$ mL at $3\bar{0}°C$. With these data, the next step is to work a combined gas law problem to calculate the volume of the dry gas at STP, as follows:

$$V_{old} = 15\bar{0} \text{ mL} \qquad P_{old} = 688.2 \text{ torr} \quad | \quad \text{pressure increases;}$$

$$V_{new} = ? \qquad\qquad P_{new} = 76\bar{0} \text{ torr} \quad \downarrow \quad \text{volume decreases}$$

$$T_{old} = 3\bar{0} + 273 = 303 \text{ K} \quad | \quad \text{temperature decreases;}$$

$$T_{new} = 0 + 273 = 273 \text{ K} \quad \downarrow \quad \text{volume decreases}$$

$$V_{new} = V_{old} \times P_{factor} \times T_{factor}$$

Work Problems 29, 30, 31, and 32.

$$V_{new} = 15\bar{0} \text{ mL} \times \frac{688.2 \text{ torr}}{76\bar{0} \text{ torr}} \times \frac{273 \text{ K}}{303 \text{ K}} = 122 \text{ mL} \quad \textit{Answer}$$

FIGURE 11-8

Pressure of a wet and dry gas.

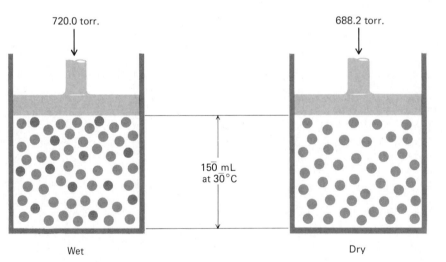

720.0 torr.		688.2 torr.
Wet	$15\bar{0}$ mL at $3\bar{0}°C$	Dry

11-8

Ideal-Gas Equation

In the gas laws—Boyle's (11-3), Charles' (11-4), and Gay-Lussac's (11-5)—the mass was fixed and one of the three variables, temperature, pressure, or volume, was also constant. Using a new equation, the **ideal-gas equation,** we can vary not only the mass, but also temperature, pressure, and volume. The **ideal-gas equation** is as follows:

$$PV = nRT \qquad\qquad (11\text{-}21)$$

where P = pressure, V = volume, n = amount of gas in *moles*, T = temperature, and R = the universal gas constant. The numerical value of R can be obtained by substituting known values of P, V, n, and T in the expression $R = PV/nT$. Since we know that at STP [0°C (273 K) and 1.00 atm] 1 mol (n = 1.00) of an ideal gas occupies 22.4 L, R is evaluated as (1.00 atm × 22.4 L/1.00 mol × 273 K) = 0.0821 atm · L/mol · K. You must know the ideal-gas equation and the numerical value of R and its *units* to work problems involving the ideal-gas equation with four variables, moles, temperature, pressure, and volume.[7]

Consider the following examples:

Problem Example 11-10

Calculate the volume in liters of 2.15 mol of oxygen gas at 27°C and 1.25 atm.

SOLUTION: Using the ideal-gas equation, $PV = nRT$, and solving for V (volume), we obtain the following equation:

$$V = \frac{nRT}{P}$$

---

[7]Using the ideal-gas equation ($PV = nRT$), holding the mass and temperature constant, and equating the constant mass and temperature along with the universal gas constant to a new constant (k), then Boyle's Law is obtained as follows:

$$(1)\ k\ \ = nRT$$
$$(2)\ PV = k \quad \text{(Equation 11-1)}$$

Holding the mass and pressure constant and equating the constant mass and pressure along with the universal gas constant to a new constant (**k**), Charles' Law is obtained:

$$(1)\ \frac{V}{T} = \frac{nR}{P}$$
$$(2)\ \mathbf{k}\ \ = \frac{nR}{P}$$
$$(3)\ \frac{V}{T} = \mathbf{k} \quad \text{(Equation 11-6)}$$

Again, applying the ideal-gas equation, holding the mass and volume constant, and equating the constant mass and volume along with the universal gas constant to a new constant (k), Gay-Lussac's Law is obtained:

$$(1)\ \frac{P}{T} = \frac{nR}{V}$$
$$(2)\ \mathbf{k}\ \ = \frac{nR}{V}$$
$$(3)\ \frac{P}{T} = \mathbf{k} \quad \text{(Equation 11-10)}$$

Substituting the values for n (2.15 mol), R (0.0821 atm $\cdot$ L/mol $\cdot$ K), T (27°C + 273 = 300 K), and P (1.25 atm), we obtain the following:

$$V = \frac{2.15 \text{ mol} \times 0.0821\dfrac{\text{atm} \cdot \text{L}}{\text{mol} \cdot \text{K}} \times 300 \text{ K}}{1.25 \text{ atm}} = 42.4 \text{ L} \quad \textit{Answer}$$

(See Appendix VII for substituting into the ideal-gas equation.)

Problem Example 11-11

Calculate the pressure in torr of 0.652 mol of oxygen gas occupying a 10.0-L cylinder at 30°C.

SOLUTION: Using the ideal-gas equation, $PV = nRT$, and solving for P (pressure), we obtain the following equation:

$$P = \frac{nRT}{V}$$

Substituting the values for n (0.652 mol), R (0.0821 atm $\cdot$ L/mol $\cdot$ K), T (30°C + 273 = 303 K), and V (10.0 L), we obtain the following:

$$P = \frac{0.652 \text{ mol} \times 0.0821 \dfrac{\text{atm} \cdot \text{L} \times 303 \text{ K}}{\text{mol} \cdot \text{K}}}{10.0 \text{ L}} = 1.62 \text{ atm}$$

From the equation, the pressure is found in *atmospheres*, but the answer is to be expressed in *torr*. Since 760 torr = 1 atm (11-2), the conversion of atmospheres to torr is as follows:

$$1.62 \text{ atm} \times \frac{760 \text{ torr}}{1 \text{ atm}} = \begin{array}{l} 1230 \text{ torr (to three significant} \\ \text{digits)} \quad \textit{Answer} \end{array}$$

Problem Example 11-12

Calculate the number of grams of oxygen gas in a 5.25-L cylinder at 27°C and 1.30 atm.

SOLUTION: Using the Ideal-gas Equation, $PV = nRT$, and solving for n (moles), we obtain the following equation:

$$n = \frac{PV}{RT}$$

Substituting the values for P (1.30 atm), V (5.25 L), R (0.0821 atm $\cdot$ L/mol $\cdot$ K), and T (27°C + 273 = 300 K), we obtain the following:

$$n = \frac{1.30 \text{ atm} \times 5.25 \text{ L}}{0.0821 \dfrac{\text{atm} \cdot \text{L}}{\text{mol} \cdot \text{K}}} = 0.277 \text{ mol } O_2 \text{ gas}$$

Converting the 0.277 mol of oxygen gas to grams using the molecular mass of 32.0 amu for oxygen, the number of grams of oxygen gas is calculated as follows:

Work Problems 33, 34, 35, and 36.

$$0.277 \text{ mol } O_2 \times \frac{32.0 \text{ g } O_2}{1 \text{ mol } O_2} = 8.86 \text{ g } O_2 \quad Answer$$

11-9

Problems Related to Gas Laws

Various types of problems can be related to the gas laws. All these problems involve calculating the *new volume when changes occur in temperature and pressure.* We shall consider three basic types of problems. Various deviations can be made in these problems, but if you follow the basic principles outlined here, you should be able to solve any similar problem.

Molecular Mass Problems[8]

In 8-3, we calculated the **molecular mass (m.m.)** of a gas by solving for grams per mole, which is numerically equal to the molecular mass in amu. This calculation involved the use of a given mass of gas, its volume at STP, and the molar volume, 22.4 L/mol of any gas at STP (0°C and 760 torr).

Generally, in the actual performance of an experiment designed to determine the molecular mass of a gas, it is difficult to measure the volume of the gas specifically at STP. Therefore, the volume is measured at some convenient

---

[8]Molecular mass problems may be solved using the ideal-gas equation (see 11-8) by substituting g/(m.m.) (g abbreviates grams; m.m. abbreviates molecular mass of the gas) for n in the ideal-gas equation.

$$PV = \frac{gRT}{\text{m.m.}} \tag{1}$$

$$\text{m.m.} = \frac{gRT}{PV} \tag{2}$$

An alternate solution to Problem Example 11-13 illustrates the application of this equation. By substituting the values of Problem Example 11-13 into Equation (2) (note that V is expressed in liters and pressure in atmospheres) we obtain the value for the molecular mass.

$$\text{m.m.} = \frac{0.600 \text{ g} \times 0.0821 \dfrac{\text{L} \cdot \text{atm}}{\text{mol} \cdot \text{K}} \times 303 \text{ K}}{630 \text{ torr} \times \dfrac{1 \text{ atm}}{760 \text{ torr}} \times 600 \text{ mL} \times \dfrac{1 \text{ L}}{1000 \text{ mL}}} = 30.0 \text{ g/mol}$$

m.m. = 30.0 amu *Answer*

temperature and pressure (wet or dry), and this volume is converted, using the gas laws, to what it would have been at STP and dry.

Consider the following examples:

Problem Example 11-13

Calculate the molecular mass of ethane gas if $6\overline{00}$ mL of the gas measured at $3\overline{0}°C$ and $63\overline{0}$ torr has a mass of 0.600 g.

SOLUTION

(1) Correct the volume ($6\overline{00}$ mL) at $3\overline{0}°C$ and $63\overline{0}$ torr to STP conditions, so we can use the molar volume of a gas in the next step:

$$V_{old} = 6\overline{00} \text{ mL} \qquad P_{old} = 63\overline{0} \text{ torr} \quad | \quad \text{pressure increases;}$$
$$V_{new} = ? \qquad P_{new} = 76\overline{0} \text{ torr} \quad \downarrow \text{ volume decreases}$$

$$T_{old} = 3\overline{0} + 273 = 303 \text{ K} \quad | \quad \text{temperature decreases;}$$
$$T_{new} = 0 + 273 = 273 \text{ K} \quad \downarrow \text{ volume decreases}$$

$$V_{new} = V_{old} \times P_{factor} \times T_{factor}$$

$$V_{new} = 6\overline{00} \text{ mL} \times \frac{63\overline{0} \text{ torr}}{76\overline{0} \text{ torr}} \times \frac{273 \text{ K}}{303 \text{ K}} = 448 \text{ mL (at STP)}$$

(2) Calculate the molecular mass:

$$\frac{0.600 \text{ g}}{448 \text{ mL (STP)}} \times \frac{1000 \text{ mL}}{1 \text{ L}} \times \frac{22.4 \text{ L (STP)}}{1 \text{ mol}} = 30.0 \text{ g/mol}$$

molecular mass = 30.0 amu   *Answer*

Problem Example 11-14

Calculate the molecular mass of a certain gas if $45\overline{0}$ mL of the gas collected over water and measured at $3\overline{0}°C$ and 720.0 torr has a mass of 0.515 g.

SOLUTION

(1) Correct the volume ($45\overline{0}$ mL) at $3\overline{0}°C$ and 720.0 torr to *dry* STP conditions, so that we can use the molar volume.

$$V_{old} = 45\overline{0} \text{ mL} \quad P_{old} = 720.0 \text{ torr} - 31.8 \text{ torr} = 688.2 \text{ torr} \quad | \quad \text{pressure}$$
(see Appendix V for vapor pressure of water) | increases;

$$V_{new} = ? \qquad\qquad\qquad P_{new} = 76\overline{0} \text{ torr} \qquad | \quad \text{volume} \downarrow \text{ decreases}$$

$$T_{old} = 3\bar{0} + 273 = 303 \text{ K} \quad | \text{ temperature decreases;}$$

$$T_{new} = 0 + 273 = 273 \text{ K} \quad \downarrow \text{ volume decreases}$$

$$V_{new} = V_{old} \times P_{factor} \times T_{factor}$$

$$V_{new} = 45\bar{0} \text{ mL} \times \frac{688.2 \text{ torr}}{76\bar{0} \text{ torr}} \times \frac{273 \text{ K}}{303 \text{ K}} = 367 \text{ mL (at STP } dry)$$

(2) Calculate the molecular mass:

$$\frac{0.515 \text{ g}}{367 \text{ mL (STP)}} \times \frac{1000 \text{ mL}}{1 \text{ L}} \times \frac{22.4 \text{ L (STP)}}{1 \text{ mol}} = 31.4 \text{ g/mol}$$

$$\text{molecular mass} = 31.4 \text{ amu} \quad Answer$$

Work Problems 37 and 38.

(We suggest that you also solve this problem using the ideal-gas equation—see footnote 8.)

Density Problems[9]

Also in 8-3 we calculated the density of a gas at STP conditions. The density of the gas need not be expressed only at STP conditions but may be calculated for any temperature and pressure. To do this we need only to calculate the volume at the new temperature and pressure or directly convert the density at STP to the new temperature and pressure. Both approaches will be given.

Consider the following problem example:

Problem Example 11-15

Calculate the density of sulfur dioxide in grams per liter at $64\bar{0}$ torr and $3\bar{0}$°C.

---

[9]**Density problems** may also be solved by using the ideal-gas equation by recalling that the density for gases is measured in grams per liter; hence, solving for g/V equal to density (D), we can obtain the following relationship:

$$PV = \frac{g}{\text{m.m.}} RT \quad \text{or} \quad \frac{P(\text{m.m.})}{RT} = \frac{g}{V}$$

and with m.m. expressed in grams per mole

$$D = \frac{g}{V} = \frac{P(\text{m.m.})}{RT} \tag{3}$$

The alternate solution to Problem Example 11-15 is obtained by substituting into Equation (3) (note that pressure is expressed in atmospheres):

$$D = \frac{64\bar{0} \text{ torr} \times \dfrac{1 \text{ atm}}{76\bar{0} \text{ torr}} \times 64.1 \dfrac{g}{\text{mol}}}{0.0821 \dfrac{\text{L} \cdot \text{atm}}{\text{mol} \cdot \text{K}} \times 303 \text{ K}} = 2.17 \text{ g/L} \quad Answer$$

SOLUTION

(1) Solve for density of the gas **at STP.** The molecular mass of sulfur dioxide (SO_2) is 64.1 amu (see Table of Approximate Atomic Masses inside the back cover of this text), and the density of the gas at STP can be calculated as

$$\frac{64.1 \text{ g } SO_2}{1 \text{ mol } SO_2} \times \frac{1 \text{ mol } SO_2}{22.4 \text{ L } SO_2 \text{ (STP)}} = 2.86 \text{ g/L (at STP)}$$

(2) Correct **1.00 L** at STP to $64\overline{0}$ torr and $3\overline{0}°$ C.

$$V_{old} = 1.00 \text{ L} \qquad P_{old} = 76\overline{0} \text{ torr} \;\bigg|\; \text{pressure decreases;}$$

$$V_{new} = ? \qquad\qquad P_{new} = 64\overline{0} \text{ torr} \;\bigg\downarrow\; \text{volume increases}$$

$$T_{old} = 0 + 273 = 273 \text{ K} \;\bigg|\; \text{temperature increases;}$$

$$T_{new} = 3\overline{0} + 273 = 303 \text{ K} \;\bigg\downarrow\; \text{volume increases}$$

$$V_{new} = V_{old} \times P_{factor} \times T_{factor}$$

$$V_{new} = 1.00 \text{ L} \times \frac{76\overline{0} \text{ torr}}{64\overline{0} \text{ torr}} \times \frac{303 \text{ K}}{273 \text{ K}} = 1.32 \text{ L}$$

(3) Calculate the density at $64\overline{0}$ torr and $3\overline{0}°$C, knowing that the 2.86 g of the gas which occupied 1 L at STP would occupy a volume of 1.32 L at $64\overline{0}$ torr and $3\overline{0}°$C.

$$\frac{2.86 \text{ g}}{1.32 \text{ L}} = 2.17 \text{ g/L} \quad \textit{Answer}$$

ALTERNATE SOLUTION TO (2) AND (3): Direct conversion of the density at STP to $64\overline{0}$ torr and $3\overline{0}°$C:

$$D_{old} = 2.86 \text{ g/L} \qquad P_{old} = 76\overline{0} \text{ torr} \;\bigg|\; \text{pressure decreases, volume increases, so}$$

$$D_{new} = ? \qquad\qquad P_{new} = 64\overline{0} \text{ torr} \;\bigg\downarrow\; \textbf{density decreases}$$

$$T_{old} \qquad\qquad\quad = 273 \text{ K} \;\bigg|\; \text{temperature increases, volume increases, so}$$

$$T_{new} \qquad\qquad\quad = 303 \text{ K} \;\bigg\downarrow\; \textbf{density decreases}$$

$$D_{new} = D_{old} \times P_{factor} \times T_{factor}$$

Since the density (mass/volume) of a given sample of gas will be *inversely* proportional to the volume, the effects of pressure and temperature changes on the density will be the *opposite* of the effects produced on the volume. If pressure increases, density increases (volume decreases); if pressure decreases, density decreases (volume increases). Increased temperature will decrease the density (volume increases), whereas decreased temperature will increase density (volume decreases).

$$D_{new} = 2.86 \text{ g/L} \times \frac{640 \text{ torr}}{760 \text{ torr}} \times \frac{273 \text{ K}}{303 \text{ K}} = 2.17 \text{ g/L} \quad \textit{Answer}$$

Work Problems 39
and 40.

Both methods are given here, so that you can choose the one you prefer.

Stoichiometry Problems[10]

In mass-volume stoichiometry problems (10-5), we expressed the volume of the gas at STP conditions. By applying the gas laws, we can express this volume at any conditions we desire.

Consider the following example:

Problem Example 11-16

Calculate the volume of oxygen in liters measured at 35°C and $63\overline{0}$ torr which could be obtained by heating 10.0 g of potassium chlorate.

SOLUTION

(1) The equation must first be written and balanced.

$$2 \text{ KClO}_3 \xrightarrow{\Delta} 2 \text{ KCl} + 3 \text{ O}_2$$

(2) Calculate the volume of oxygen at STP. From the atomic masses, the formula mass of $KClO_3$ is 122.6 amu.

$$10.0 \text{ g KClO}_3 \times \frac{1 \text{ mol KClO}_3}{122.6 \text{ g KClO}_3} \times \frac{3 \text{ mol O}_2}{2 \text{ mol KClO}_3} \times \frac{22.4 \text{ L O}_2 \text{ at STP}}{1 \text{ mol O}_2}$$

$$\underbrace{}_{\text{Step I}} \quad \underbrace{}_{\text{Step II}} \quad \underbrace{}_{\text{Step III}}$$

$$= 2.74 \text{ L O}_2 \text{ at STP}$$

---

[10]**Stoichiometry problems** related to gases may also be solved by using the ideal-gas equation, but first the number of moles of the gas must be calculated. The alternate solution to Problem Example 11-16 is as follows:
(1) Calculate the moles of oxygen.

$$10.0 \text{ g KClO}_3 \times \frac{1 \text{ mol KClO}_3}{122.6 \text{ g KClO}_3} \times \frac{3 \text{ mol O}_2}{2 \text{ mol KClO}_3} = 0.122 \text{ mol O}_2$$

(2) Use the ideal-gas equation to calculate the volume of the gas at 35°C (308 K) and $63\overline{0}$ torr.

$$V = \frac{nRT}{P} \tag{4}$$

$$V = \frac{0.122 \text{ mol} \times 0.0821 \dfrac{\text{L} \cdot \text{atm}}{\text{mol} \cdot \text{K}} \times 308 \text{ K}}{63\overline{0} \text{ torr} \times \dfrac{1 \text{ atm}}{760 \text{ torr}}} = 3.72 \text{ L} \quad \textit{Answer} \text{ (difference due to rounding off)}$$

(3) Correcting 2.74 L of O_2 to 35°C and $63\overline{0}$ torr, we have

$$V_{old} = 2.74 \text{ L} \qquad P_{old} = 76\overline{0} \text{ torr} \quad | \quad \text{pressure decreases;}$$

$$V_{new} = ? \qquad P_{new} = 63\overline{0} \text{ torr} \downarrow \text{ volume increases}$$

$$T_{old} = 0 + 273 = 273 \text{ K} \quad | \quad \text{temperature increases}$$

$$T_{new} = 35 + 273 = 308 \text{ K} \downarrow \text{ volume increases}$$

$$V_{new} = V_{old} \times P_{factor} \times T_{factor}$$

Work Problems 41
and 42.

$$V_{new} = 2.74 \text{ L} \times \frac{76\overline{0} \text{ torr}}{63\overline{0} \text{ torr}} \times \frac{308 \text{ K}}{273 \text{ K}} = 3.73 \text{ L } O_2 \text{ at 35°C and } 63\overline{0} \text{ torr} \quad \textit{Answer}$$

Chapter Summary

This chapter considered the general characteristics of gases. The characteristics and properties of gases and matter in general can be explained by the kinetic theory, which states that heat and motion are related and that particles of all matter are in motion to some degree, heat being an indication of this motion. The *kinetic theory* is applied to gases on the basis of several assumptions (11-1). Gases that obey these assumptions are called *ideal gases*, as opposed to *real gases*, such as hydrogen, oxygen, nitrogen, and others. Under moderate conditions of temperature and pressure, real gases behave essentially as ideal gases. Assuming this to be correct, the basic gas laws can be applied—*Boyle's*, *Charles'*, and *Gay-Lussac's Laws*. Problems involving these laws individually and also involving the *combined gas laws* (Boyle's and Charles' Laws) were solved.

Dalton's Law of Partial Pressures states that each gas in a mixture of gases exerts a partial pressure equal to the pressure it would exert if it were the only gas present in the same volume; therefore, the total pressure of the mixture is the sum of the partial pressures of all of the gases present. Problems involving Dalton's Law were solved including the collection of a gas over water.

The ideal-gas equation, $PV = nRT$, involving the four variables, pressure, volume, mass in moles or mass, and temperature, was considered. Problems involving the ideal-gas equation were solved.

In Chapter 8 (see 8-3) the density of a gas at STP was calculated from a knowledge of molecular mass and molecular mass from a knowledge of density of a gas at STP or the mass and volume of the gas at STP. In Chapter 10 (see 10-5, mass-volume stoichiometry) problems involving a gas at STP were solved. Now, by applying the gas laws, the molecular mass, density, and volume of a gas in a mass-volume stoichiometry problem, at *any* conditions desired, can be solved.

EXERCISES

1. Define or explain the following terms:
   (a) ideal gas
   (b) real gas
   (c) pressure
   (d) torr

(e) psi
(g) Charles' Law
(i) Dalton's Law of Partial
     Pressures
(k) STP

(f) Boyle's Law
(h) Gay-Lussac's Law
(j) molar volume

(l) ideal-gas equation

2. Distinguish between
    (a) a real gas and an ideal gas
    (b) force and pressure
    (c) Boyle's Law and Charles' Law
    (d) Gay-Lussac's Law and Gay-Lussac's Law of Combining Volumes (see 10-6)
    (e) torr and cm of mercury pressure

3. Identify the following mathematical expressions as being related to Charles' Law,
    Boyle's Law, Dalton's Law of Partial Pressures, or Gay-Lussac's Law:
    (a) $V \propto T$
    (c) $V \propto \dfrac{1}{P}$

    (b) $P_t = P_1 + P_2 + P_3$
    (d) $P \propto T$

4. List five general characteristics of gases.

5. List the five assumptions of the kinetic theory for gases in your own words as you
    understand them.

6. Under what general conditions of temperature and pressure does the behavior of
    real gases appreciably deviate from that of ideal gases?

7. Express standard pressure in cm of mercury, mm of mercury, torr, psi, atmo-
    spheres, inches of mercury, and pascals.

PROBLEMS

The atomic masses are found in the Table of Approximate Atomic Masses inside the
back cover of this text, and the vapor pressure of water at various temperatures is found
in Appendix V.

Boyle's Law (See Section 11-3)

8. A sample of gas has a volume of $39\bar{0}$ mL when measured at 25°C and $76\bar{0}$ torr.
    What volume in mL will it occupy at 25°C and 195 torr?

9. The volume of a given mass of gas is $41\bar{0}$ mL at $1\bar{0}$°C and $38\bar{0}$ torr. What will be
    its volume in mL measured at $1\bar{0}$°C and 2.00 atm?

10. What final pressure in torr must be applied to a sample of gas having a volume
     of $19\bar{0}$ mL at $2\bar{0}$°C and $75\bar{0}$ torr pressure to permit the expansion of the gas to a
     volume of $60\bar{0}$ mL at $2\bar{0}$°C?

11. The volume of a gas is 10.5 L at 10.0 atm and 273 K. Calculate the pressure in
     atm of the gas if its volume is changed to $500\bar{0}$ mL while the temperature remains
     constant.

Charles' Law (See Section 11-4)

12. A sample of gas occupies $22\bar{0}$ mL at $1\bar{0}$°C and $75\bar{0}$ torr. What volume in mL will
     the gas have at $2\bar{0}$°C and $75\bar{0}$ torr?

13. A gas occupies a volume of 90.0 mL at 27°C and 74$\overline{0}$ torr. What volume in mL will the gas have at 5°C and 74$\overline{0}$ torr?

14. A gas occupies a volume of 12$\overline{0}$ mL at 27°C and 63$\overline{0}$ torr. At what temperature in °C would the volume be 80.0 mL at 63$\overline{0}$ torr?

15. The volume of a gas is 2$\overline{00}$ mL at 3$\overline{0}$°C. At what temperature in °**F** would the volume be 26$\overline{0}$ mL, assuming the pressure remains constant?

Gay-Lussac's Law   (See Section 11-5)

16. A sample of gas occupies 10.0 L at 11$\overline{0}$ torr and 27°C. Calculate its pressure in torr if the temperature is changed to 127°C while the volume remains constant.

17. The temperature of 2$\overline{00}$ mL of a gas originally at STP is changed to −25°C at constant volume. Calculate the final pressure of the gas in torr.

18. A gas occupies a volume of 50.0 mL at 27°C and 63$\overline{0}$ torr. At what temperature in °C would the pressure be 77$\overline{0}$ torr if the volume remains constant?

19. A sample of gas occupies a volume of 5.00 L at 7$\overline{00}$ torr and 3$\overline{0}$°C. At what temperature in °C would the pressure be 62$\overline{0}$ torr if the volume remains constant?

Combined Gas Laws   (See Section 11-6)

20. A certain gas occupies a volume of 5$\overline{00}$ mL at 27°C and 74$\overline{0}$ torr. What volume in mL will it occupy at STP?

21. A certain gas has a volume of 195 mL at 2$\overline{0}$°C and 1.00 atm. Calculate its volume in mL at 6$\overline{0}$°C and 6$\overline{00}$ mm Hg.

22. A gas has a volume of 245 mL at 25°C and 6$\overline{00}$ mm Hg. Calculate its volume in mL at STP.

23. A given sample of a gas has a volume of 5.10 L at 27°C and 64$\overline{0}$ mm Hg. Its volume and temperature are changed to 2.10 L and 1$\overline{00}$°C, respectively. Calculate the pressure in mm Hg at these conditions.

24. A gas measures 31$\overline{0}$ mL at STP. Calculate its pressure in atmospheres if the volume is changed to 45$\overline{0}$ mL and the temperature to 5$\overline{0}$°C.

25. A given sample of gas has a volume of 4.40 L at 6$\overline{0}$°C and 1.00 atm pressure. Calculate its pressure in atm if the volume is changed to 5.00 L and the temperature to 27°C.

26. A gas measures 15$\overline{0}$ mL at STP. Calculate its temperature in °C if the volume is changed to 32$\overline{0}$ mL and the pressure to 95$\overline{0}$ torr.

27. A gas has a volume of 125 mL at 57°C and 64$\overline{0}$ torr. Calculate its temperature in °C if the volume is increased to 325 mL and the pressure decreased to 59$\overline{0}$ torr.

28. A gas has a volume of 2.50 L at 27°C and 1.00 atm. Calculate its temperature in °C if the volume is decreased to 2.00 L and the pressure is decreased to 0.870 atm.

Dalton's Law of Partial Pressures (See Section 11-7)

29. A mixture of gases has the following partial pressures for the component gases at 20°C in a volume of 2.00 L: oxygen, 180 torr; nitrogen, 320 torr; hydrogen, 246 torr. (a) Calculate the pressure in torr of the mixture. (b) Calculate the volume in liters at STP occupied by the gases remaining, if the hydrogen gas is selectively removed.

30. A mixture of gases has the following partial pressures for the component gases at 50°C in a volume of 450 mL: helium, 120 torr; argon, 180 torr; krypton, 60 torr; xenon, 25 torr. (a) Calculate the total pressure in torr of the mixture. (b) Calculate the volume in mL at STP occupied by the gases remaining, if the krypton is selectively removed.

31. The volume of oxygen, collected over water, is 175 mL at 25°C and 600.0 torr. Calculate the dry volume in mL of the oxygen at STP.

32. The volume of nitrogen, collected over water, is 245 mL at 25°C and 700.0 torr. Calculate the dry volume in mL of nitrogen at STP.

Ideal-Gas Equation (See Section 11-8)

33. Calculate the volume in mL of 0.0250 mol of nitrogen gas at 30°C and 1.10 atm.

34. Calculate the pressure in atm of 16.8 g of nitrogen gas occupying a 12.0-L cylinder at 35°C.

35. Calculate the temperature in °C of 0.310 mol of nitrogen gas occupying a 10.0-L cylinder at 0.925 atm.

36. Calculate the number of grams of nitrogen gas (N_2) in a 6.00-L cylinder at 27°C and 800 torr. (*Hint:* Convert the pressure in torr to atm.)

Related to Gas Laws—Molecular Mass Problems (See Section 11-9)

37. If 485 mL of a gas measured at 30°C and 600 torr has a mass of 0.384 g, what is its molecular mass?

38. A volume of 0.972 L of a gas measured at 50°C and 700 torr has a mass of 0.525 g. Calculate its molecular mass.

Related to Gas Laws—Density Problems (See Section 11-9)

39. Calculate the density of carbon dioxide in g/L at 35°C and 3.00 atm.

40. Calculate the density of methane (CH_4) in g/L at −45°C and 300 torr.

Related to Gas Laws—Stoichiometry Problems (See Section 11-9)

41. Calculate the number of mL of hydrogen gas at 25°C and 640 torr produced by the reaction of 0.520 g of magnesium with excess hydrochloric acid. (See 9-9 for the equation.)

42. Calculate the number of moles of potassium nitrate that would be required to produce 4.25 L of oxygen at $30°C$ and $71\bar{0}$ mm Hg. (See 9-8 for the equation.)

General Problems

43. Calculate the molecular mass of a gas having a density of 2.20 g/L at $3\bar{0}°C$ and $46\bar{0}$ mm Hg.

44. Calculate the number of oxygen molecules in 4.50 L of oxygen gas measured at $27°C$ and $8\bar{0}\bar{0}$ torr. (*Hint:* See 8-2 and 8-3.)

45. A hydrocarbon has a density of 2.30 g/L at $27°C$ and $5\bar{0}\bar{0}$ torr. Its composition is 83.7 percent carbon and 16.3 percent hydrogen. What is its molecular mass and molecular formula? (*Hint:* See 8-5.)

46. A hydrocarbon has a density of 0.681 g/L at $37°C$ and 438 torr. Its composition is 80.0 percent carbon and 20.0 percent hydrogen. What is its molecular mass and molecular formula? (*Hint:* See 8-5.)

47. Halothane (Fluothane) is a nonflammable, nonirritating, general anesthetic, and in many instances is superior to ethyl ether. At $57°C$ and $64\bar{0}$ torr, 0.529 g of the gas occupies a volume of 86.4 mL. Its composition is 12.2 percent carbon, 0.5 percent hydrogen, 40.5 percent bromine, 18.0 percent chlorine, and 28.9 percent fluorine. Calculate the molecular mass and molecular formula for halothane. (*Hint:* See 8-5.)

48. Given the ideal-gas equation, $PV = nRT$, calculate the value of R in the units

$$(Pa \cdot m^3) / (mol \cdot K)$$

(*Hint:* Standard pressure $= 1.013 \times 10^5$ Pa; $1m^3 = 1000$ L.)

Readings

Goldwhite, Harold, "Gay-Lussac After 200 Years." *J. Chem. Educ.*, 1978, v. 55, p. 366. This article reviews the life and contributions of Joseph Louis Gay-Lussac to the chemistry profession.

Hall, Marie Boas, "Robert Boyle." *Sci. Am.*, Aug. 1967, v. 217, p. 96. An interesting article on the life and work of Robert Boyle.

Liquids and Solids

In an oil refinery, crude oil is distilled to separate the complex mixture of hydrocarbons present into useful fractions such as gasoline, kerosene, fuel oil, and paraffin.

OBJECTIVES

1. Given the following terms, define each term and describe the distinguishing characteristic of each:
   (a) evaporation (Section 12-2)
   (b) condensation (Section 12-2)
   (c) vapor pressure (Section 12-3)
   (d) boiling point (Section 12-4)
   (e) heat of vaporization (Section 12-4)
   (f) heat of condensation (Section 12-4)
   (g) distillation (Section 12-5)
   (h) surface tension (Section 12-6)
   (i) viscosity (Section 12-6)
   (j) crystalline solid (Section 12-8)
   (k) amorphous solid (Section 12-8)
   (l) freezing point (Section 12-9)
   (m) melting point (Section 12-9)
   (n) heat of fusion (Section 12-9)
   (o) heat of solidification [(crystallization), Section 12-9]
   (p) sublimation (Section 12-10)

2. Given a graph of the vapor pressure of a liquid at various temperatures, determine the vapor pressure of the liquid from the graph at any temperature, and vice versa (Section 12-4, Problem 15).

3. Given the amount of a liquid or gas and the specific or molar heat of vaporization or condensation at its normal boiling point, calculate the amount of heat energy required or evolved in the change of state (Problem Example 12-1, Problems 16, 17, and 18).

4. Given the fact that an increase of 1 atmosphere pressure decreases the melting point of ice 0.0075°C, calculate the melting point of ice at various pressures (Section 12-9, Problem 19).

5. Given an amount of a liquid or solid and the specific or molar heat of solidification or fusion at its melting point or freezing point, calculate the amount of heat energy evolved or required in the change of state (Problem Example 12-2, Problems 20, 21, 22, and 23).

6. Given an amount of a solid, liquid, or gas, the specific or molar heat of solidification or fusion at its melting point or freezing point, the specific heat of various physical states, and the specific or molar heat of vaporization or condensation at its normal boiling point, calculate the amount of heat energy required or evolved in the various changes of state (Problem Examples 12-3 and 12-4, Problems 24, 25, and 26).

In this chapter, we shall continue our discussion of the three physical states of matter by considering liquids and solids and comparing these states with gases.

12-1

The Liquid State

Previously (see 11-1), we stated that ideal gases (according to the assumptions of the kinetic theory) were composed of molecules whose actual volumes were negligibly small compared with the distances between them and that there were **no** *attractive forces* between these molecules.

Actually, most real gases behave as ideal gases only at sufficiently high temperatures and low pressures (11-1). *Some never behave ideally.* In real gases attractive forces exist between the molecules. At large distances these forces are not significant and are largely overcome by the kinetic energy of the molecules. When the *temperature* of such a real gas is *lowered* and the *pressure* is *increased,* these attractive forces can partially overcome the kinetic energies of the molecules, and the real gas will deviate from the behavior predicted by the gas laws.

Real gases can be liquified when these attractive forces overcome the kinetic energy of the molecules sufficiently to keep the molecules confined to a relatively small volume. This can be accomplished by decreasing the temperature (lowering the kinetic energy of the molecules) and/or by applying pressure so that the molecules are brought closer together, allowing the attractive forces to become more significant (see Figure 12-1).

We can identify the general characteristics of liquids by their properties as listed. These properties are related to the *closeness* of the molecules to each other and to the *attractive forces* between them in the liquid state.

1. *Limited expansion.* Liquids do not show infinite expansibility as gases do.

2. *Shape.* Liquids have no characteristic shape and take on the shapes of the containers they occupy. This characteristic is quite evident when you spill a glass of milk!

3. *Volume.* Liquids maintain their volumes no matter what the size of the container they occupy.

4. *Compressibility.* Liquids are only slightly compressible for a given temperature or pressure change. This lack of compressibility is evident in the brake fluid of the hydraulic braking system of an automobile. If the fluid could be compressed to a considerable extent, the pressure applied to the brake pedal would compress the fluid and the brakes would not stop the

FIGURE 12-1

Molecules in the liquid state are closer together than in the gaseous state. This change of state from gas to liquid is accomplished by decreasing the temperature and/or applying pressure, allowing the attractive forces to become more significant.

Liquid   Gas

car. Instead, the pressure from your foot is transferred through the brake fluid in the system to the brake drum.[1]

5. *High density.* Liquids have considerably higher densities than gases. As we have mentioned previously (2-13), the density of a gas usually is measured in grams per liter (g/L) in the metric system, whereas liquid densities are measured in grams per milliliter (g/mL). For example, water in the liquid state at 100°C and 760 torr has a density of 0.958 g/mL, whereas water in the gaseous state under the same conditions has a density of only 0.598 g/L (0.000598 g/mL). Thus, liquid water is more dense than water vapor by a factor of 1600.

6. *Mixability or diffusion.* One liquid will mix or diffuse throughout another liquid in which it is soluble, but this mixability is much slower in liquids than in gases, as is quite evident when honey and water are mixed. Therefore, liquid molecules, like gas molecules, are also in constant motion, but liquid molecules can move only a short distance before they strike another liquid molecule, which slows their motion. Hence, the property of mixability or diffusion is slower in liquids than in gases.

12-2

Evaporation

Evaporation is the actual escape of molecules from the surface of the liquid to form a vapor in the surrounding space above the liquid. In any sample of a liquid, the kinetic energies (energies of motion) of some of the molecules are *above* average and some are *below* average, and it is these **higher** kinetic energy

---

[1]In air brakes, the air is under extremely high pressure and further compression is slight. If a leak appears in the pressure system, the braking power is lost.

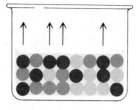

FIGURE 12-2

Evaporation of a liquid. Those molecules on the surface of the liquid which have the higher kinetic energies can escape into the space above the liquid surface. (Degree of kinetic energy is shown in shades of color.)

("hotter") molecules that can more readily overcome the attractive forces in the liquid and **escape** from its surface (see Figure 12-2).

Since the molecules of higher kinetic energy escape, the average kinetic energy of the molecules left in the liquid is *lowered*, assuming that no heat is supplied from some outside source. The lowering of the average kinetic energy results in a temperature drop in the liquid. This cooling effect is evident when you first come out of a swimming pool on a hot day and feel quite cool due to the evaporation of the water from the surface of your body. Also, the production of sweat from the sweat glands in the skin is a way in which the body cools itself and maintains a constant temperature. Evaporation can even freeze tissue. Ethyl chloride is a local anesthetic used in medicine to reduce pain near the surface of the skin. It is sprayed on the skin so that minor surface incisions can be made. Ethyl chloride evaporates so fast that it freezes the underlying tissue and results in temporary loss of feeling in that region.

The ease of escape of a molecule from the surface of a liquid is related to the strength of the attractive forces between the molecules in the liquid. For example, gasoline and alcohol (ethyl alcohol) evaporate faster than water. The attractive molecular forces in gasoline and alcohol are weaker than those in water. Alcohol is used externally as a rubbing agent on children with high fever to bring down their fever. The evaporation of the alcohol from the skin lowers the body temperature of the feverish child.

As the *temperature increases*, molecules escape more readily, because more will have sufficient energy to leave the surface of the liquid and overcome the attractive forces in the liquid. In addition, as *surface area increases*, the rate of escape *increases*, since more surface is exposed for evaporation to occur.

Condensation is the reverse of evaporation. The vapor molecules coalesce to form the liquid, and the attractive forces between molecules increase in the formation of the liquid. Heat energy is *evolved* in this process (see 12-4).

12-3

Vapor Pressure

If a quantity of liquid is placed in a *closed* container, the molecules of the liquid that are escaping from the surface are *not* removed from the immediate space above the liquid. Thus, the system will arrive at a steady state where the *rate of evaporation is equal to the rate of condensation.* When the rate of the molecules leaving from the surface (evaporation) is equal to the rate of the molecules reentering the liquid (condensation), a *dynamic equilibrium* is established (see Figure 12-3). When this dynamic equilibrium has been established, the concentration of the molecules in the space above the liquid remains constant (at constant temperature) and the vapor exerts a *definite constant pressure.* This equilibrium pressure at any fixed temperature is called the **vapor pressure** of the liquid.

FIGURE 12-3

Vapor pressure. The establishment of a dynamic equilibrium in a closed container. The rate of evaporation ↑ equals the rate of condensation ↓.

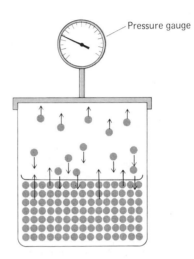

Pressure gauge

If the temperature of the liquid is increased, the vapor pressure will increase, since more molecules in the liquid will have a larger kinetic energy and will break away from the surface of the liquid. Therefore, the concentration of vapor molecules will increase before equilibrium is again established. At higher temperatures, the average kinetic energy of the vapor molecules will increase. Since the pressure is directly proportional to the concentration of the vapor molecules and their average kinetic energy, the vapor pressure will increase with an increase in temperature. This vapor pressure at various temperatures was used in calculations involving Dalton's Law of Partial Pressures (11-7). The vapor pressure of water expressed in torr, atmosphere, and pascal is given in Appendix V.

The vapor pressures of various liquids differ considerably depending upon the liquid. As we mentioned previously in 12-2, gasoline and alcohol evaporate faster than water. Hence, we would expect gasoline and alcohol to have higher vapor pressures than water at a given temperature, and that is what we observe in Figure 12-4. (Heptane is one of the many components of gasoline, and its plot of vapor pressure vs. temperature is similar to that of gasoline.) For example, from Figure 12-4, the vapor pressures of alcohol and heptane at 60°C are about 350 and 210 torr, respectively, whereas that of water at the same temperature is about 150 torr.

FIGURE 12-4

Vapor pressure changes with temperature for alcohol, heptane, and water.

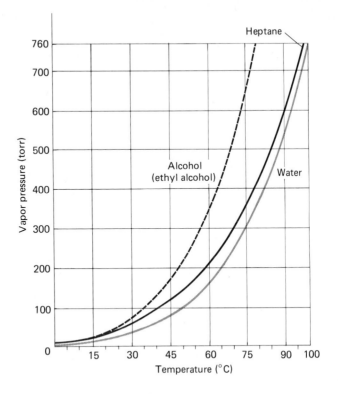

12-4

Boiling Point. Heat of Vaporization or Condensation

When the *vapor pressure* of a liquid *equals* the external pressure upon the liquid, bubbles of vapor form rapidly throughout the liquid and the liquid boils. The **boiling point** of a liquid is the temperature at which the vapor pressure of the liquid is **equal** to the external pressure acting upon the surface of the liquid. Since the atmospheric pressure varies with altitude and atmospheric conditions (11-2), standard pressure (760 torr or 1 atm) is used in reporting boiling points. When the external pressure is 760 torr, this temperature is called the normal boiling point. Therefore, the **normal boiling point** of a liquid is the temperature at which the vapor pressure of the liquid is 760 torr. From Figure 12-4, the normal boiling point of alcohol is 78°C; of heptane, 98°C; and of water, 100°C. Also, from Appendix V, the vapor pressure for water is 760.0 torr at a temperature of 100°C, which is its normal boiling point.

At higher altitudes, the atmospheric pressure is less and therefore the boiling point is lowered. At mile-high altitude, the atmospheric pressure is approximately 630 torr; hence, from Figure 12-4, the boiling points of alcohol, heptane, and water would be about 74°C, 92°C, and 95°C, respectively. Increased atmospheric pressure occurs below sea level, as in Death Valley, California, or in a pressure cooker, and thus raises the boiling point.

Work Problem 15.

Temperature is quite important in cooking foods. When cooking at a lower temperature, we must increase the length of cooking time. At sea level, where water boils at 100°C, it takes about 10 minutes to hard-boil an egg. At Pike's Peak in Colorado (14,110 ft), where water boils at about 86°C, it would take a little over 20 minutes to hard-boil the egg! Naturally, cooking time would be shortened if the boiling point were raised, as in a pressure cooker.

Evaporation and boiling involve a loss of energy by the liquid (see 12-2), and heat must be *supplied* (endothermic, see 10-7) if the temperature is to remain constant. The quantity of heat required to evaporate a unit mass of a given liquid at constant temperature and pressure is the **heat of vaporization** of the liquid. Heats of vaporization are usually measured at the normal boiling point of the liquid under 1 atmosphere pressure. If the heat of vaporization is expressed in calories per gram or joules per kilogram it is called the **specific heat of vaporization.** If it is expressed in kilocalories per mole or kilojoules (kJ) per mole, it is called the **molar heat of vaporization.** The specific heat of vaporization of water is $54\overline{0}$ calories per gram (2.26×10^6 joules per kilogram) (see Figure 12-5), and the molar heat of vaporization is 9.72 kilocalories per mole (40.7 kilojoules per mole), both at 100°C and 1 atm pressure. Since the reverse process of evaporation is called condensation, the heat *given off* (exothermic, see 10-7) in going from the vapor state to the liquid state is called the **heat of condensation** and has the *same numerical value* as the heat of vaporization. The exothermic property of the heat of condensation is evident from a steam burn. The steam condenses on the cooler skin, and in the process heat is evolved, which "cooks" the tissue and produces blisters.

FIGURE 12-5

The specific heat of vaporization ($\rightarrow$) and condensation ($\leftarrow$).

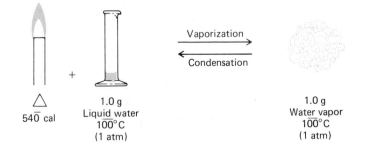

540 cal   1.0 g
            Liquid water
            $10\overline{0}$°C
            (1 atm)

1.0 g
Water vapor
$10\overline{0}$°C
(1 atm)

Vaporization
Condensation

The specific and molar heats of vaporization or condensation of various liquids are listed in Table 12-1. Consider a problem involving these heats:

Problem Example 12-1

Calculate the amount of heat in kilocalories required to vaporize 28.0 g of liquid water to steam at 100°C.

TABLE 12-1   Heats of Vaporization or Condensation of Various Liquids at Their Normal Boiling Point and 1 Atmosphere Pressure

| | Normal Boiling Point (°C) | Heat of Vaporization or Condensation | | | |
| | | Specific | | Molar | |
Liquid		cal/g	J/kg	kcal/mol	kJ/mol
Water	$10\overline{0}$	$54\overline{0}$	2.26×10^6	9.72	40.7
Alcohol	78.3	204	8.54×10^5	9.40	39.3
Heptane	98.4	76.5	3.20×10^5	7.67	32.1
Carbon tetrachloride	76.7	52.1	2.18×10^5	8.00	33.5
Benzene	80.1	94.1	3.94×10^5	7.35	30.8
Sodium chloride[a]	1465	698	2.92×10^6	40.8	171

[a]The extremely high values for the boiling point and heat of vaporization or condensation for sodium chloride over those values for the other liquids are due to the relatively strong attractive forces found in sodium chloride. These strong attractive forces are related to the ionic bonding found in sodium chloride.

SOLUTION: From Table 12-1, the specific heat of vaporization of water is $54\overline{0}$ cal/g, and the amount of heat in kilocalories is calculated as follows:

$$28.0 \text{ g} \times \frac{54\overline{0} \text{ cal}}{1 \text{ g}} \times \frac{1 \text{ kcal}}{1000 \text{ cal}} = 15.1 \text{ kcal} \quad \textit{Answer}$$

ALTERNATE SOLUTION: Using the molar heat of vaporization of water equal to 9.72 kcal/mol and the molecular mass of water equal to 18.0 amu, the calculation is as follows:

Work Problems 16, 17, and 18.

$$28.0 \text{ g} \times \frac{1 \text{ mol}}{18.0 \text{ g}} \times \frac{9.72 \text{ kcal}}{1 \text{ mol}} = 15.1 \text{ kcal} \quad \textit{Answer}$$

12-5

Distillation

Distillation is a process used in the purification of *liquids*, involving heating a liquid to its boiling point and then cooling the vapors in a condenser to form the purified liquid. A simple distillation apparatus is shown in Figure 12-6.

A common application of distillation is to separate water from dissolved salts and other nonvolatile impurities. Colorless vapors appear in the distilling flask above the impure water and are condensed to give clear colorless droplets of water called the *distillate*. The impurities remain in the distilling flask and are called the *residue*. Distilled water is prepared in this manner. The residue contains salts composed of calcium, magnesium, or iron with hydrogen carbonate (bicarbonate), carbonate, or sulfate.

Fractional distillation, which is in essence many of these distillations, is used in the petroleum industry to refine gasoline. Gasoline is a mixture of hydrocarbons (see Chapter 9, footnote 6).

FIGURE 12-6
Distillation of water.

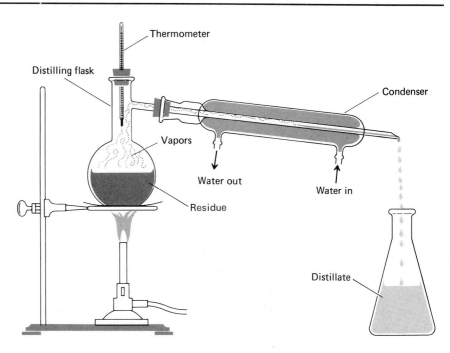

12-6

**Surface
Tension
and Viscosity**

Surface tension and viscosity are properties of liquids. **Surface tension** is the property of a liquid that tends to draw the surface molecules into the body of the liquid and hence to reduce the surface to a minimum.[2] For example, mercury, because of its high surface tension, forms droplets on a glass surface, whereas water, whose surface tension is appreciably lower than that of mercury, tends to spread out on the surface. This property of a liquid can be explained by intermolecular attractive forces (forces between molecules). Any molecule in the body of a liquid is surrounded on all sides by molecules and is attracted equally in all directions by neighboring molecules. But, a molecule at the *surface* of the liquid is attracted by other molecules *beneath* it and *not* above it. This results in an unbalanced force downward, tending to draw the surface molecules *into the body of the liquid* and to reduce the surface to a minimum (see Figure 12-7).

Some substances have greater surface tension than others because the at-

---

[2]Surface tension can be expressed in quantitative terms. It is the energy needed to increase the surface area of a liquid by a unit area. In SI the units are joules per meter squared (J/m^2). Water has a surface tension of 7.20×10^{-2} J/m^2 at 25°C. This means that 7.20×10^{-2} joule is needed to increase the surface area of a given amount of water 1 m^2 at 25°C.

FIGURE 12-7

Surface tension can be explained by intermolecular attractive forces with the • representing the intermolecular attractive forces.

tractive forces in these substances are greater (mercury vs water). Alcohol is often used to prepare an area for medical treatment because it has a low surface tension and can easily penetrate into a wound to cleanse the area. One of the reasons for the cleansing action of soap solution is that it lowers the surface tension of the water and hence allows the solution to penetrate into the creases in your hands to clean out the grease that holds the dirt there.

As the temperature increases, the average kinetic energy of the molecules increases, and this increase in energy tends to overcome the intermolecular attractive forces; hence, the surface tension decreases. As you are aware, you can wash your hands more efficiently in hot water than in cold, due in part to decreased surface tension in the hot water.

In reading the volume of a liquid in a graduated cylinder or by other means for measuring volumes, you are instructed to read at the bottom of the *meniscus,* that is, the bottom of the curved surface of the liquid. For water the surface curves upward along the walls of the cylinder (see Figure 12-8). The surface tension of the liquid is one of the factors that cause this kind of behavior in liquids.

Viscosity, a property of a liquid, is the resistance of a liquid to flow.[3] Some liquids resist flow, such as honey, whereas other liquids flow quite readily, such as water. Honey is said to have a high viscosity, while water has a low viscosity. This property can be explained by the intermolecular attractive forces. The greater these attractive forces, the more viscous is the liquid.

As the temperature increases, the average kinetic energy of the molecules increases, which breaks the attractive forces between molecules and decreases the viscosity.

Motor oil grades are based on the viscosity of the oil. The lower-viscosity oils (10W) are used in winter when it is cold, while the higher-viscosity oils (40W) are used in summer when it is hot. Multiviscosity oils (10W-40) may be used year-round depending on the engine, with the viscosity varying depending on the temperature due to the addition of various additives.

---

[3]Viscosity is expressed quantitatively in newton-second per meter squared ($N \cdot s/m^2$) units in SI. Water at 20°C has a viscosity of $1.00 \times 10^{-3} N \cdot s/m^2$.

FIGURE 12-8

Reading the volume of water in a graduated cylinder. Read at the bottom of the meniscus. The volume of water in this cylinder is 15 mL.

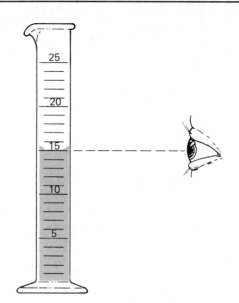

12-7

Mixability of Liquids

When two liquids are combined, they may *mix completely*, forming a solution (see 3-2); they may *partially mix*, forming two solutions; or they may *not mix* (heterogeneous mixture; see 3-2). These three cases are shown in Figure 12-9.

In the first case, where the two liquids completely mix in all proportions to form a solution, the two liquids are considered to be *miscible*. An alcohol solution in water is an example. A solution of the common antifreeze, ethylene glycol, in water is another example of two miscible liquids. In both cases, the solution is homogeneous throughout.

In the second case, one liquid partially dissolves in the other; the result is two solutions appearing as two layers. The solution with the greater density naturally is the bottom layer. These two liquids are hence considered to be *partially miscible*. Ethylene glycol and water are miscible, but if ethylene glycol is mixed with chloroform two solutions are obtained, and the two liquids are

FIGURE 12-9

(a) Two miscible liquids. (b) Two partially miscible liquids. (c) Two immiscible liquids.

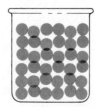

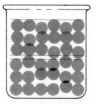

(a)   (b)   (c)

considered to be partially miscible, with the chloroform solution being the bottom layer.

In the third case, when the two liquids do not appreciably mix but form two *separate* layers of liquid, with the denser liquid being the lower layer, the two liquids are considered to be essentially *immiscible*. An example of two immiscible liquids is gasoline and water. When mixed, they form separate layers with the gasoline "floating" on the water because the water has the greater density. An oil slick at sea caused by a sinking ship or a leaking off-shore oil well is an example of oil "floating" on water.

12-8

The Solid State

In the solid state, the attractive forces between particles (molecules, ions, or atoms) are stronger than in the liquid state. These particles do not have sufficient energy to overcome the attractive forces in the solid state; hence, they are held in a relatively **fixed position** close to each other (see Figure 12-10).

We can identify the general characteristics of solids by their following properties:

1. *No expansion (at constant temperature).* As with liquids, solids do not show infinite expansion as gases do, although water, on freezing, does expand slightly.

2. *Shape.* Solids in general have definite shape. They are relatively rigid and do not flow as do gases and liquids, except under extreme pressures. Thus, they do not take the shape of the container they occupy (see Figure 12-10, again).

3. *Volume.* Solids maintain their volumes as do liquids.

4. *Compressibility.* Solids are practically incompressible, since the particles are very close to each other due to their strong attractive forces.

FIGURE 12-10

Particles in the solid state are held in a relatively fixed position due to stronger attractive forces in the solid state than in the liquid state. A solid does not take the shape of the container it occupies, while a liquid does.

Solid

Liquid

FIGURE 12-11

Crystalline solids. (a) Sodium chloride. (b) Quartz (a crystalline form of silica, silicon dioxide).

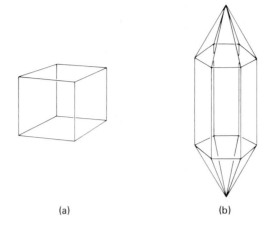

(a)   (b)

5. *High density.* Solids, like liquids, have relatively high densities.

6. *Mixability or diffusion.* Solids mix or diffuse extremely slowly, except under extreme pressures. The particles in solids have essentially permanent positions due to the strong attractive forces between the particles. Hence, the motion of solid particles is generally very slow. Of the three physical states of matter, particles in the gaseous state have the greatest motion; liquids have less and solids have the least.

The characteristic of shape should be explored further. Solids can be conveniently divided into *crystalline* and *amorphous* solids, which differ from each other in structure.

A **crystalline solid** consists of particles arranged in a definite geometric shape or form that is distinctive for a given solid. Examples of crystalline solids are sodium chloride, diamond, and quartz (a crystalline form of silica, silicon dioxide). Drawings of crystalline structures are shown in Figure 12-11.

An **amorphous solid** consists of particles arranged in an irregular manner, and thus lacks shape or form. Examples of amorphous solids are glass and many plastics, although they also are thought of as being highly viscous liquids.

Amorphous solids differ from crystalline solids in the way the solid melts. The melting point of crystalline solids are sharp with a narrow temperature range while amorphous solids have an indefinite melting point and a wide temperature range.

12-9

Melting or Freezing Point. Heat of Fusion or Solidification

As a liquid is cooled, the kinetic energy of the particles is lowered. At the *freezing point*, the particles become oriented in a definite geometric pattern characteristic of the substance, and crystals of the solid begin to separate. Removal of heat energy is necessary to allow the freer liquid particles to be "tied down" in the solid if crystallization is to continue. The *temperature* (freezing point) of the mixture of *solid* and *liquid* in dynamic equilibrium will remain **constant** until *all* the liquid has solidified.

If the solid is heated, the kinetic energy of some of the particles in the crystal will become sufficient to match the attractive forces in the crystal, and the solid will begin to melt at some temperature, the *melting point*. If melting is to continue, more heat must be applied to free the particles of the solid from each other, and the solid will change to the liquid at **constant** temperature. Therefore, the **freezing point** is equal to the **melting point** and is the temperature at which the liquid and solid forms are in dynamic equilibrium with each other. (The ⇌ is used to indicate an equilibrium.)

$$\text{solid} \underset{\text{freezing}}{\overset{\text{melting}}{\rightleftharpoons}} \text{liquid}$$

In 12-4, we observed that pressure altered the boiling point of a liquid considerably; that is, reduced pressure lowered the boiling point and increased pressure raised the boiling point. Only large pressures affect the melting or freezing point. For most substances, *increased pressure raises* the melting point; the exception is ice, the melting point of which is *lowered* because its volume in the solid state decreases in going to the liquid state. Other substances occupy a smaller volume in the *solid* state. For ice, an increase of 1 atm pressure decreases the melting point by 0.0075°C. This slight decrease in melting point with an increase in pressure is important in ice skating. The ice underneath the blades is at a temperature above its new melting point and therefore melts. The liquid layer provides the lubricant that enables the skater to skate across the ice. Another factor also important in melting the ice is the heat created by the friction between the ice and the blade. After the skate has passed over the ice, the pressure and friction decrease and the water refreezes. The skater is actually skating on liquid water. In fact, many speed skaters prefer ice that has small pools of water on it. If the ice temperature is too low, the pressure and friction are insufficient to melt the ice and skating becomes difficult.

Work Problem 19.

As with evaporation and boiling, heat must be supplied (endothermic) to convert a solid to a liquid at constant temperature. This heat is called the **heat of fusion** of a substance and is the quantity of heat necessary to convert a unit mass of a solid substance to the liquid state at the *melting point* of the substance. If the heat of fusion is expressed in calories per gram or joules per kilogram, it is called the **specific heat of fusion.** If it is expressed in kilocalories per mole or kilojoules (kJ) per mole, it is called the **molar heat of fusion.** The specific heat of fusion of water is $8\bar{0}$ calories per gram (3.3×10^5 joules per kilogram) (see Figure 12-12), and the molar heat of fusion is 1.44 kilocalories per mole (6.02 kilojoules per mole) at 0°C.

The reverse of the fusion process is solidification (crystallization). The **heat of solidification (crystallization)** is the quantity of heat evolved (exothermic) in going from the liquid state to the solid state of a unit mass of liquid at the melting point of the substance. The heat of solidification has the *same numerical value* as the heat of fusion.

FIGURE 12-12

The specific heat of fusion ($\rightarrow$) and solidification ($\leftarrow$).

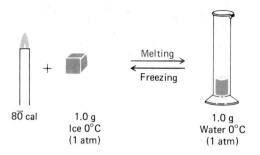

$8\overline{0}$ cal   1.0 g Ice 0°C (1 atm)   1.0 g Water 0°C (1 atm)

The specific and molar heats of fusion or solidification are listed in Table 12-2. Consider a problem involving these heats.

Problem Example 12-2

Calculate the amount of heat in kilojoules evolved when 0.510 mol of liquid benzene (C_6H_6) is converted to solid benzene at 6°C.

SOLUTION: From Table 12-2, the molar heat of solidification of benzene is 9.83 kJ/mol; hence, the amount of heat in kilojoules is calculated as follows:

$$0.510 \text{ mol} \times \frac{9.83 \text{ kJ}}{1 \text{ mol}} = 5.01 \text{ kJ} \quad \textit{Answer}$$

ALTERNATE SOLUTION: Using the specific heat of solidification of benzene equal to 1.26×10^5 J/kg and the molecular mass of benzene equal to 78.0 amu, the calculation is as follows:

Work Problems 20, 21, 22, and 23.

$$0.510 \text{ mol} \times \frac{78.0 \text{ g}}{1 \text{ mol}} \times \frac{1 \text{ kg}}{1000 \text{ g}} \times \frac{1.26 \times 10^5 \text{ J}}{1 \text{ kg}} \times \frac{1 \text{ kJ}}{1000 \text{ J}} = 5.01 \text{ kJ} \quad \textit{Answer}$$

TABLE 12-2 Heats of Fusion or Solidification of Various Solids at Their Melting Points

Solid	Melting Point (°C)	Specific		Molar	
		cal/g	*J/kg*	*kcal/mol*	*kJ/mol*
Water	0	$8\overline{0}$	3.3×10^5	1.44	6.02
Alcohol	−117	24.9	1.04×10^5	1.15	4.81
Heptane	−91	33.7	1.41×10^5	3.38	14.1
Carbon tetrachloride	−23	5.09	2.13×10^4	0.784	3.28
Benzene	6	30.1	1.26×10^5	2.35	9.83
Sodium chloride[a]	804	124	5.19×10^5	7.25	30.3

[a]The high values for sodium chloride are again due to the strong attractive forces found in the substance because of its ionic bonding. (See the footnote to Table 12-1.)

12-10

Sublimation

Substances in the solid state, like those in the liquid state, exhibit a definite vapor pressure at a given temperature and can pass directly from the solid state into the gaseous state. These particles in the solid state have obtained sufficient energy of motion to break away from the relatively fixed position in the solid state and can then pass into the gaseous state. This process is called **sublimation,** and is defined as the direct conversion of a *solid* to the *vapor* without passing through the liquid state (see Figure 12-13). Mothballs (*p*-dichlorobenzene), iodine, dry ice (solid carbon dioxide), and ice are examples of solids that can sublime. Wet clothes will dry at subzero temperatures even though they are frozen, because ice can sublime. This process occurs best with a wind blowing the water vapor away and in relatively dry climates, where the humidity is low. Freeze-dried foods are prepared by using this process.

FIGURE 12-13

Sublimation. The solid is converted directly to the vapor state without passing through the liquid state. Deposition is the reverse of sublimation.

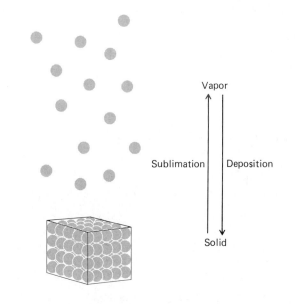

The reverse of sublimation is **deposition** (see Figure 12-13). Heat energy is evolved in this process. Ice or snow forming in clouds is an example of deposition.

12-11

Heat Energy Transformations in the Three Physical States of Matter

In this section, we summarize the various changes in state and consider the actual heat transformations accompanying these changes. Figure 12-14 summarizes these transformations and also shows the energy changes.

These changes are shown by plotting heat energy added vs. temperature, as Figure 12-15 illustrates. Heat energy is added uniformly, so time can also

FIGURE 12-14

Summary of the changes of state. Endothermic changes are shown as → while exothermic changes are shown as ←.

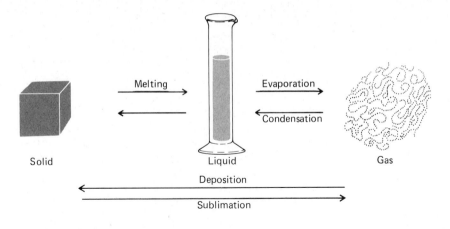

be substituted on the heat energy axis. As heat energy is added to a given solid, the temperature rises according to the **specific heat** (see 2-12) of the solid, and the kinetic energy of the particles in the solid increases, causing the particles to be set in more rapid motion. As more heat energy is added, the kinetic energy continues to increase up to the melting point, whereas at

FIGURE 12-15

Plot of heat added uniformly versus temperature to solid ⇄ liquid ⇄ gas.

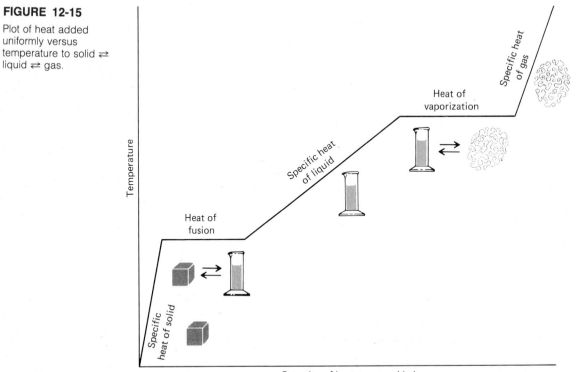

the melting point the temperature (kinetic energy) remains *constant* until all the solid is converted to liquid. At this point, the solid and liquid states are in *dynamic equilibrium,* and further addition of heat (heat of fusion) does *not* change the temperature but serves only to change the state of the substance from a solid to a liquid. The heat energy is used to break some of the attractive forces between the particles in the solid state and to convert the substance to the liquid state by increasing the freedom of movement (but not the kinetic energy) of the particles. Conversely, if heat (heat of solidification) is removed from the equilibrium mixture of solid and liquid, the temperature will again remain constant and liquid will be converted to solid at constant temperature.

After all the solid has been changed to liquid, further heating increases the kinetic energy of the particles, and hence the temperature rises according to the **specific heat** of the liquid. At the boiling point, the temperature remains *constant* until **all** the liquid is converted to vapor. At this point, the liquid and vapor states are in *dynamic equilibrium* and the heat energy applied (heat of vaporization) is used to change the state of the substance from a liquid to a vapor, with no change in temperature. The heat energy is used to break attractive forces between the particles in the liquid state and to increase the freedom of movement of the particles, and the substance is transformed to the vapor state. Removal of heat energy (heat of condensation) at this point will convert vapor to liquid at constant temperature. Further heating of the gas increases the kinetic energy, and thus the temperature rises according to the specific heat of the gas.

Now let us consider the determination of the quantity of heat energy required for a given transformation.

Problem Example 12-3

Calculate the quantity of heat energy in calories required to convert $12\overline{0}$ g of ice at 0°C to steam at $10\overline{0}$°C.

SOLUTION: Diagram the changes as follows:

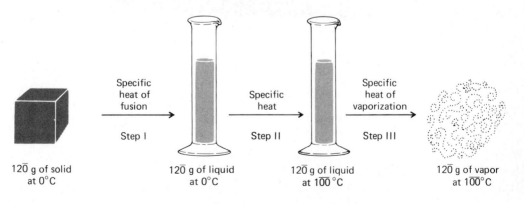

| $12\overline{0}$ g of solid at 0°C | Specific heat of fusion → Step I | $12\overline{0}$ g of liquid at 0°C | Specific heat → Step II | $12\overline{0}$ g of liquid at $10\overline{0}$°C | Specific heat of vaporization → Step III | $12\overline{0}$ g of vapor at $10\overline{0}$°C |

From Table 12-2, the specific heat of fusion of water is $8\bar{0}$ cal/g; hence, for the change in Step I,

$$120\,\bar{g} \times \frac{8\bar{0}\ \text{cal}}{1\,\bar{g}} = 9600\ \text{cal}$$

From Table 2-4, the specific heat of water is 1.00 cal/g · °C; hence, for the change in Step II,

$$\frac{1.00\ \text{cal}}{1\,\bar{g} \cdot 1\bar{\degree}\text{C}} \times 12\bar{0}\,\bar{g} \times (10\bar{0} - 0)\degree\text{C} = 12{,}\bar{0}00\ \text{cal}$$

(Review 2-12 for the calculation in Step II.)

From Table 12-1, the specific heat of vaporization is $54\bar{0}$ cal/g; hence, for the change in Step III,

$$12\bar{0}\,\bar{g} \times \frac{54\bar{0}\ \text{cal}}{1\,\bar{g}} = 64{,}800\ \text{cal}$$

The total heat energy requirement of Steps I to III is

Step I:	9,600 cal
Step II:	12,$\bar{0}$00 cal
Step III:	64,800 cal
Total heat energy required =	86,400 cal *Answer*

Note that the quantity of heat energy required in Step I (9600 cal) and Step II (12,$\bar{0}$00 cal) is nearly equal and relatively small, but that over five times this quantity of heat energy is required in Step III (64,800 cal) to convert the liquid water (10$\bar{0}$°C) to vapor (10$\bar{0}$°C).

The process can be reversed—that is, heat energy is given off on cooling. Consider the following example:

Problem Example 12-4

Calculate the quantity of heat energy in calories liberated when 45$\bar{0}$ g of steam at 100°C is changed to water at 20°C.

SOLUTION: Diagram the change as follows:

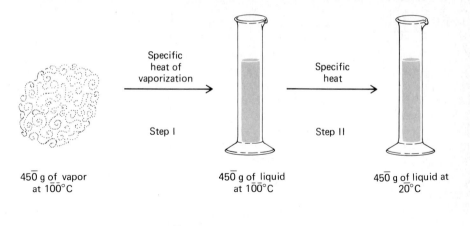

Step I: $450\ \text{g} \times \dfrac{540\ \text{cal}}{1\ \text{g}} = 243{,}000\ \text{cal}$

Step II: $\dfrac{1.00\ \text{cal}}{1\ \text{g} \cdot 1°\text{C}} \times 450\ \text{g} \times (100 - 20)°\text{C} = 36{,}000\ \text{cal evolved}$

The total of Steps I and II is

Step I: 243,000 cal

Step II: 36,000 cal

Total heat energy liberated = 279,000 cal   *Answer*

Work Problems 24, 25, and 26.

Of the two steps, the greater quantity of heat energy is liberated in Step I, which again emphasizes the scalding property of steam when it condenses on a cooler surface.

Chapter Summary

This chapter considered the general characteristics of both liquids and solids. Various terms related to liquids are *evaporation, condensation, vapor pressure, boiling point, distillation, surface tension, viscosity,* and *mixability.* There are many applications of these terms to common phenomena. Two of the applications are the reading of a graph of vapor pressure vs temperature and calculating melting-point lowering with a large increase in pressure. Terms related to solids are *melting* or *freezing point* and *solidification.*

Heat energy is required or evolved in changes of state of substances. These heat energies are *heat of vaporization* or *condensation, heat of fusion* or *solidification (crystallization),* and *specific heat.* Calculations involving changes of state using these heats were solved.

EXERCISES

1. Define or explain the following terms:
   (a) evaporation
   (b) condensation
   (c) vapor pressure
   (d) boiling point
   (e) normal boiling point
   (f) heat of vaporization or condensation
   (g) surface tension
   (h) viscosity
   (i) miscible
   (j) partially miscible
   (k) immiscible
   (l) a crystalline solid
   (m) an amorphorus solid
   (n) melting or freezing point
   (o) heat of fusion or solidification (crystallization)
   (p) sublimation
   (q) deposition

2. Distinguish between
   (a) evaporation and condensation
   (b) boiling point and normal boiling point
   (c) specific heat of vaporization and molar heat of vaporization
   (d) miscible and immiscible
   (e) miscible and partially miscible
   (f) a crystalline solid and an amorphous solid

3. List the characteristics of gases, liquids, and solids.

Evaporation   (See Sections 12-2 and 12-4)

4. Upon evaporating, rubbing alcohol (isopropyl alcohol) gives a cooling sensation to the skin. Why?

Boiling Point   (See Section 12-4)

5. Certain foods, when cooked in the mountains, never appear to be completely cooked. Explain.

Surface Tension and Viscosity   (See Section 12-6)

6. Explain why your hands get cleaner when you wash with soap and *hot* water than they do when you use just cold water.

7. The viscosity in newton-second per meter squared for castor oil is given as follows:

Temperature (°C)	Viscosity (N·s/m²)
20	9.86×10^{-1}
30	4.51×10^{-1}
40	2.31×10^{-1}

Explain this change in viscosity with temperature.

Mixability   (See Section 12-7)

8. Explain why oil in a water-oil mixture can burn.

9. Glycerine ($d^{25°} = 1.26$ g/mL) and acetone ($d^{25°} = 0.788$ g/mL) are immiscible. Which liquid will form the bottom layer when they are mixed?

Attractive Forces   (See Sections 11-1, 12-1, and 12-8)

10. Which physical state of matter exhibits the strongest attractive forces between its particles? The weakest?

Melting Point   (See Section 12-9)

11. In outdoor speed ice skating, the skate blade is very thin. Explain.

12. Mercury thermometers are of no use below $-39°C$; alcohol (ethyl alcohol) thermometers are used in extremely cold regions when the temperature dips to below $-39°C$. Why? (*Hint:* Look up in *Handbook of Chemistry and Physics,* or other suitable reference, the physical properties of the two liquids.)

Sublimation   (See Section 12-10)

13. Dry ice is usually stored in a covered box. Explain.

14. A chemistry student placed an odorous solid in an open beaker in his laboratory desk to dry until the next laboratory period. To his astonishment, when he returned to the laboratory the next week, he found only half the solid remaining. Explain.

PROBLEMS

(Refer to Tables 12-1 and 12-2 for additional data)

Boiling Point   (See Section 12-4)

15. Using the graph in Figure 12-4, determine the following:
     (a) the boiling point of heptane at 600 torr
     (b) the boiling point of water on top of Mt. Whitney, California (14,495 ft); atmospheric pressure is approximately 435 torr
     (c) the boiling point of water on top of Mt. Everest, Nepal-Tibet border (29,028 ft); atmospheric pressure is approximately 240 torr
     (d) the boiling point of heptane on top of Mt. Everest

Heat of Vaporization or Condensation   (See Section 12-4)

16. Calculate the quantity of heat in kilocalories evolved when 12.0 g of steam condenses to form water at its normal boiling point.

17. Calculate the quantity of heat in joules required to vaporize 256 g of alcohol (ethyl alcohol, C_2H_6O) at 78.3°C.

18. Calculate the quantity of heat in calories required to vaporize 0.650 mol of carbon tetrachloride (CCl_4) at its normal boiling point. Repeat the calculation for the same amount of sodium chloride at its normal boiling point. Explain the large difference in calculated values.

Melting Point   (See Section 12-9)

19. In ice skating, a person's skate blade can exert a pressure of approximately $3\overline{0}$ atm on the surface of the ice. What would be the melting point of ice at this pressure?

Heat of Fusion or Solidification (Crystallization, See Section 12-9)

20. Calculate the quantity of heat energy in joules evolved when 75 g of liquid water form ice at 0°C.

21. Calculate the quantity of heat energy in calories required to liquify 0.880 mol of solid alcohol (ethyl alcohol, C_2H_6O) at its melting point.

22. Calculate the quantity of heat energy in kilocalories required to melt, 74.8 g of sodium chloride at its melting point of 804°C.

23. Calculate the quantity of heat energy in kilocalories required to liquify 0.785 mol of solid benzene (C_6H_6) at its melting point of 6°C.

Changes of State   (See Section 12-11)

24. Calculate the quantity of heat energy in kilocalories required to convert 20.0 g of ice at 0°C to steam at exactly $10\overline{0}$°C. (Specific heat of water = 1.00 cal/g · °C.)

25. Calculate the quantity of heat energy in kilojoules required to convert 40.0 g of ice at 0°C to steam at exactly $10\overline{0}$°C. (Specific heat of water = 4.18 × 10³ J/kg · K.)

26. Calculate the quantity of heat energy to the nearest calorie required to convert 15 g of ice at −8.0°C to steam at 105°C. (Specific heats: ice = 0.500 cal/g · °C; water = 1.00 cal/g · °C; steam = 0.480 cal/g · °C.)

General Problems

27. Calculate the mass of water in grams that could be heated from 0°C to 25°C by the heat evolved on cooling 1.00 kg of water from $10\overline{0}$°C to 25°C.

28. The basis of "seeding" a hurricane involves dropping silver iodide crystals from a plane just outside the eye of the storm. The "seed" acts to condense the super-

cooled water droplets (between -20 and $0°C$) to ice, giving off heat. This heat energy then increases the temperature of the surrounding air and decreases the pressure at the edge of the eye, hopefully reducing the extremely high wind velocity in the area. In the seeding of Hurricane Debbie, silver iodide was released along a line approximately 18.0 miles long at 33,000 ft and scattered to approximately 18,000 ft. (a) Considering this "curtain of silver iodide" to be on the average approximately 5.00 miles wide due to the wind, what volume of air in cubic meters was seeded? (1 in. = 2.54 cm) (b) "On the average," this volume of air contained 2.00 g of water droplets per cubic meter. Assuming that all these droplets at $0°C$ were converted to ice at $0°C$, calculate the number of kilocalories of heat energy released in the seeding process.

Readings

Charles, R. J., "The Nature of Glasses." *Sci. Am.*, Sept., 1967, v. 217, p. 126. A discussion of the structure of glasses, including a discussion of the solidification of liquids to form crystals or glasses.

Mott, Sir Nevill, "The Solid State." *Sci. Am.*, Sept. 1967, v. 217, p. 80. A discussion of the two general forms of solids—crystalline and amorphous—and the properties of these forms.

Water

Launching of the space ship Columbia. In the initial stages thrust is obtained by the combustion of liquid hydrogen with liquid oxygen to form water and energy.

OBJECTIVES

1. Given the following terms, define each term and describe the distinguishing characteristic of each:
    (a) Hund's Rule (Section 13-1)
    (b) hybrid orbitals (Section 13-1)
    (c) sp^3 orbitals (Section 13-1)
    (d) polar bond (Section 13-2)
    (e) hydrogen bond (Section 13-3)
    (f) hydrates (Section 13-8)
    (g) hygroscopic substance (Section 13-8)
    (h) deliquescent substance (Section 13-8)
    (i) efflorescent substance (Section 13-8)

2. Given the formulas of various simple covalent molecules, determine whether or not they should show a net dipole moment (Section 13-2, Problem 10).

3. Given various physical properties of hydrogen compounds, explain these properties in terms of bonding *between* the molecules (Section 13-3, Problem 11).

4. (a) Water is formed by (1) combustion of an organic compound, (2) direct combination of hydrogen and oxygen, and (3) a neutralization reaction (an acid and a base). Write balanced equations for each of these reactions.
    (b) Water is decomposed by electrolysis, and water reacts with certain metals in the electromotive series. Write balanced equations for each of these reactions.
    (Sections 13-5 and 13-6, Problems 12 and 13)

5. Given the name or formula of a hydrate and the Table of Approximate Atomic Masses, calculate the percent water in the hydrate (Problem Example 13-1, Problem 14).

6. Given the name or formula of an anhydrous salt, the percent water (or mass of water and hydrate) in the hydrate, and the Table of Approximate Atomic Masses, calculate the formula of the hydrate (Problem Example 13-2, Problems 15 and 16).

7. Given the formulas of hydrogen-oxygen compounds, write the electron-dot and structural formulas of the compounds (Sections 13-2 and 13-9, Problem 17).

8. (a) Write balanced equations for the preparation of hydrogen peroxide from a peroxide salt and for the decomposition of hydrogen peroxide.

(b) Write balanced equations for the preparation of ozone, and for the decomposition of ozone.

(Sections 13-9 and 13-10, Problem 18)

Water is probably the most abundant and one of the most important chemical compounds familiar to you. Water covers 71 percent of the earth's surface. Approximately 60 percent of your body weight is water. Water is needed for the survival of all living organisms. In a normal diet, humans require approximately 2.7 quarts of water daily from all sources (drinks, foods containing water, and metabolic products from food), of which 1.1 quarts are obtained from beverages.

13-1

Electronic Structure of Water

A water molecule (H_2O) is composed of two hydrogen atoms and one oxygen atom, and we should first consider the electronic structure of these atoms. In sublevels, the electronic structures of hydrogen and oxygen atoms are

$$^1_1H = 1s^1$$
$$^{16}_8O = 1s^2, 2s^2\, 2p^4$$

The electronic configuration of the oxygen atom broken down in terms of orbitals (see 4-9) is

$$^{16}_8O = 1s^2, 2s^2, 2p_x^2\, 2p_y^1\, 2p_z^1$$

One electron is placed in each p orbital (p_x, p_y, p_z) before pairing up the electrons to a maximum of two electrons in each of these orbitals. This is **Hund's Rule**. Hund's Rule also applies to the filling of d and f orbitals. Bonding the electrons of hydrogen atoms ($1s$ orbitals) to the electrons in the p_y and p_z orbitals of oxygen would result in a bond angle of 90° between the hydrogen atoms, as shown in Figure 13-1. The bond angle in water is found to be 105°, which deviates considerably from the hypothetical value of 90° that would result only if the p orbitals of the oxygen were involved.

A seemingly more appropriate explanation is to assume *four new* types of orbitals, called sp^3 orbitals, which are **hybrid orbitals** (mixed orbitals). A simple analogy may help to clarify the idea of hybrid orbitals. Suppose with children's modeling clay (Pla-Doh) we make four balls having the same diameter,

FIGURE 13-1

The bond angle in water would be 90° if the molecular orbitals in the compound were formed from the overlap of the *p* orbitals from the oxygen atom and the *s* orbitals from the hydrogen atoms.

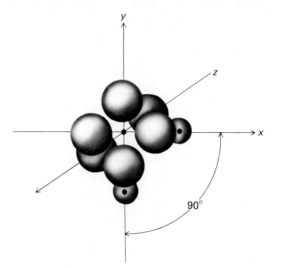

with *one* being *yellow* and *three* being *blue*. We squeeze the four balls together and mix them up by some means to form one big ball, and divide this big ball into four equal parts making four new balls, all of the same diameter. Now, none of the four new balls are yellow or blue, but *all* are a new color, *greenish blue*, consisting of *one* part *yellow* and *three* parts *blue*. This, in essence, is what we mean by sp^3 hybrid orbitals, consisting of *one part s* and *three parts p* character.

All these sp^3 orbitals have the *same* energy, which is slightly less than the energy of a *p* orbital and slightly more than the energy of an *s* orbital. Calculation of the bond angle between all four of the sp^3 orbitals gives an angle of 109.5° with the arrangement of a tetrahedron, as shown in Figure 13-2. Assuming the hybridization into four new sp^3 orbitals can occur for oxygen, we obtain the following electronic structure for the eight electrons in the oxygen atom:

FIGURE 13-2

The tetrahedral bond angle is 109.5°.

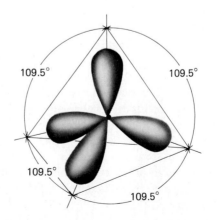

$$^{16}_{8}O \qquad 1s^2,\ 2(sp^3)^2\ 2(sp^3)^2\ 2(sp^3)^1\ 2(sp^3)^1$$

four sp^3 orbitals in principal energy level 2

One electron is placed in each of the four sp^3 orbitals before pairing up the electrons to a maximum of two electrons in each of the sp^3 orbitals (**Hund's Rule**). Each hydrogen atom ($1s$ orbital) bonds with each of the $2sp^3$ orbitals containing only one electron, as shown in Figure 13-3a. The two *unshared* pairs of electrons in the sp^3 orbitals *repel* each other (like charges repel) and repel the *shared* electron pairs of the hydrogen-oxygen bonds more than the hydrogen–oxygen bonds can themselves. This, in turn, pushes the hydrogen atoms closer together in the water molecule, accounting for the deviation of the predicted bond angle of 109.5° to the *actual* value of **105°**.

FIGURE 13-3

Electronic structure of water. (a) The structure of a water molecule using sp^3 orbitals of oxygen and s orbitals of hydrogen. Due to the repulsion of the unshared pairs of electrons, the bond angle is 105°, instead of the predicted tetrahedral bond angle of 109.5°. (b) Electron-dot formula of a water molecule, showing the unshared pairs of electrons.

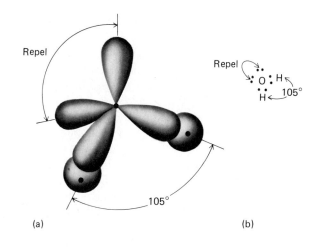

(a)   (b)

13-2

Polarity of Water

We mentioned previously (6-4) that oxygen was more electronegative than hydrogen. This difference in electronegativities of oxygen and hydrogen results in an unequal sharing of electrons in the formation of the covalent bonds between oxygen and hydrogen, with the hydrogen being partially positive, $\delta^{(+)}$, and the oxygen partially negative, $\delta^{(-)}$, as shown in Figure 13-4a. This produces a **polar bond (polar covalent bond)**. The resultant of these partially polar bonds is the formation of a dipole moment for the molecule, as shown in Figure 13-4b. A molecule having a **net dipole moment** is one in which the *centers* of positive and negative charges do not coincide, but are separated by a finite distance. An arrow symbol ($\leftrightarrow$) is used to indicate this moment, with the head of the arrow pointing toward the negative center. Compounds having molecules possessing *net dipole moments* are said to be *polar;* water is such a substance.

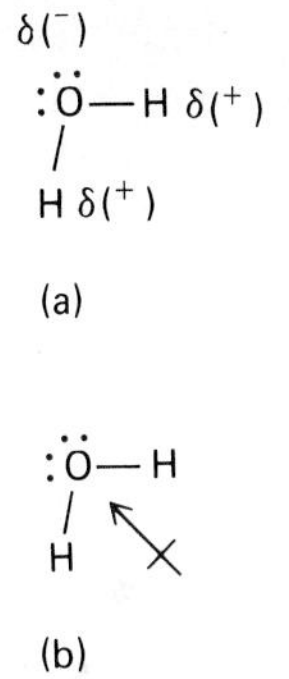

FIGURE 13-4

Polarity of water. (a) Unequal sharing of electrons in the water molecule with the hydrogen partially positive, $\delta(+)$, and the oxygen partially negative, $\delta(-)$. (b) the *net dipole moment* in a water molecule.

$$\delta(^-)$$
$$:\ddot{O}\text{---H } \delta(^+)$$
$$|$$
$$H \ \delta(^+)$$

(a)

$$:\ddot{O}\text{---H}$$
$$|\ \nwarrow$$
$$H \quad \times$$

(b)

If the hydrogen-oxygen bond angle were 180° (linear), then *no* net dipole moment would exist for the compound because the center of the positive charges would coincide with the center of negative charge, on the oxygen atom.

$$\text{H---O---H} \quad \text{dipole moment} = 0$$
$$\text{(linear)}$$

Therefore, the *shape* (geometry) of a molecule is one of the important factors in determining its dipole moment. The presence of a dipole moment in water indicates that the angle between the hydrogens is not 180°.

Work Problem 10.

13-3

Hydrogen Bonding in Water

Water has some extraordinary properties when compared with compounds formed from hydrogen and other elements in the same group as oxygen, group VIA, as shown in Table 13-1. Considering the melting and boiling points in Table 13-1, you will note that as you move from H_2Te to H_2Se to H_2S, both the melting and boiling points decrease. If we were to predict from these trends the melting and boiling points of water, we would predict a value lower

TABLE 13-1 Some Physical Properties of Hydrogen Compounds of Group VIA

Property	H_2O	H_2S	H_2Se	H_2Te
Molecular mass (amu)	18.0	34.1	81.0	129.6
Melting point (°C)	0	−86	−66	−49
Normal boiling point (°C)	100	−61	−41	−2
Molar heat of fusion				
(kcal/mol)	1.44	0.57	0.60	1.0
(kJ/mol)	6.02	2.38	2.51	4.18
Molar heat of vaporization				
(kcal/mol)	9.72	4.46	4.62	5.55
(kJ/mol)	40.7	18.7	19.3	23.2

than $-86°C$ (melting point) and lower than $-61°C$ (boiling point), probably in the neighborhood of -106 and $-81°C$, respectively (see 6-8). Certainly, based on these trends, we would not have predicted values of 0 and 100°C, respectively. Also from the table, note that the molar heats of fusion and vaporization are quite high for water.

Why, then, does water have such comparatively high melting and boiling points and molar heats of fusion and vaporization? This is explained by assuming that hydrogen bonding (a type of attractive force, see 12-1) occurs *between* water molecules to a much greater extent than between the molecules of the other hydrogen compounds (H_2S, H_2Se, H_2Te). As we noted in 13-2, water is polar. The partial negative charge, $\delta^{(-)}$, on the oxygen atom from the unshared pairs of electrons in *one* molecule of water attracts a partial positive charge, $\delta^{(+)}$, on the hydrogen atom of another molecule of water, as shown in Figure 13-5a, b, and c. This attraction results in the formation of a weak linkage

FIGURE 13-5

Hydrogen bonding in water. (a) A hydrogen bond shown with electron-dot formula. (b) A hydrogen bond shown with the Prentice-Hall model. (c) A hydrogen bond shown with the Stuart-Briegleb model.

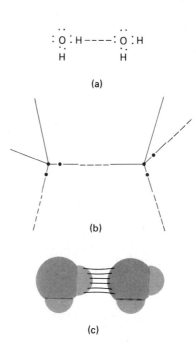

called a *hydrogen bond*. A **hydrogen bond** results when a hydrogen atom bonded to a highly electronegative atom (X) becomes bonded additionally to another electronegative atom. *Electronegative* atoms, such as F, O, and N (compare order of electronegativities in Figure 6-9), are generally involved in such bonding. In the case of water, these hydrogen bonds hold *two* or *more* molecules together in chains or clusters. This linkage is depicted with a dashed line (-----), as shown below and in Figures 13-5 and 13-6.

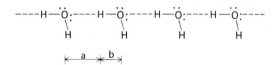

$$\underset{a}{\longleftarrow}\quad\underset{b}{\longleftarrow}$$

FIGURE 13-6

Water in the solid and liquid states contains many hydrogen bonds. (Notice that the "hydrogen bond" distance, a, is greater than they hydrogen-oxygen covalent bond distance, b.)

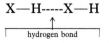

A hydrogen bond is much weaker than a covalent bond. For example, the hydrogen bond energy is 6.4 kilocalories (27 kilojoules) per mole in ice; however, that of an actual hydrogen-oxygen covalent bond is 119 kilocalories (498 kilojoules) per mole. The average hydrogen bond energy is about 5 kilocalories (20 kilojoules) per mole.

The "hydrogen bond" hydrogen atom is not in the exact center between the two electronegative oxygen atoms in the two water molecules. The hydrogen bond distance is larger than the hydrogen-oxygen covalent bond distance, as shown in Figures 13-5 and 13-6.

In all cases, in going from solid to liquid to gas, heat energy must be supplied to overcome the attractive forces in the solid or liquid. With water, the attractive forces include hydrogen bonding, which, although weaker than covalent bonding, is stronger than most attractive forces found between molecules of relatively nonhydrogen-bonded compounds. Therefore, in going from ice to liquid water at 0°C, some, but not all, hydrogen bonds are broken, and thus more energy must be supplied than for relatively nonhydrogen-bonded compounds, where the attractive forces are weaker. This, therefore, accounts for the high melting point and high molar heat of fusion for water (see Table 13-1). In going from liquid water at 0 to 100°C, there are fewer hydrogen bonds at the higher temperature than at the lower temperature. In changing from liquid water to vapor (steam) at 100°C (1 atm), nearly all the hydrogen bonds are broken, and again more energy must be supplied than for the corresponding relatively nonhydrogen-bonded compounds, accounting for the high boiling point and high molar heat of vaporization of water (see Table 13-1). Figure 13-7 depicts these transformations.

The hydrogen compounds of the other elements in group VIA (S, Se, and Te) do **not** appear to *hydrogen bond* appreciably. These elements are **not** as *electronegative* (weaker attractive forces) as oxygen, and hence the hydrogen derivatives have considerably lower melting points, boiling points, molar heats of fusion, and molar heats of vaporization (see Table 13-1).

A few years ago, several Soviet scientists claimed to have prepared water that had a boiling point of about 540°C and did not freeze, although cooled to

FIGURE 13-7

Transformation of ice to liquid water to gaseous steam, showing hydrogen bonding. Notice the fewer hydrogen bonds in the series of transformations. (Shown only in two dimensions and with a 90° bond angle in water for convenience.)

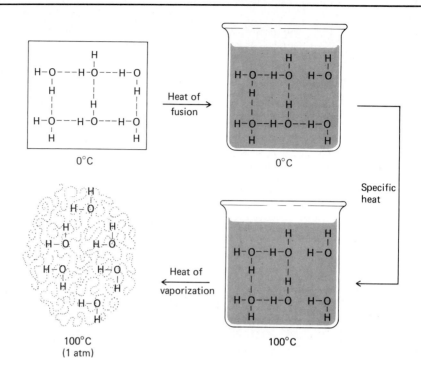

Work Problem 11.

−40°C, but formed a glasslike substance. This form of water was tabbed "polywater." Some United States scientists at that time believed that polywater was nothing more than a concentrated solution of many salts with less than 5 percent water. This latter belief has been proven correct.

13-4

Physical Properties of Water

Appearance

Pure water is colorless, odorless, and tasteless. Any change in these properties of water is due to impurities dissolved in it. An example of such an impurity is hydrogen sulfide, which is often dissolved in sulfur spring water, giving it the odor of rotten eggs.

Density

At 4°C, the density of liquid water is 1.00000 g/mL, which is its maximum density. This value is used as the standard in calculations involving specific gravity (see 2-14). At 0°C, the density of liquid water is 0.99987 g/mL, whereas ice at 0° has a density of 0.917 g/mL. Since ice is *less dense* than liquid water at the same temperature, it floats on water (see Figure 13-8). The volume of floating ice exposed to the air is only approximately 8 percent, with about 92 percent below the surface.

FIGURE 13-8

Ice floats on water.

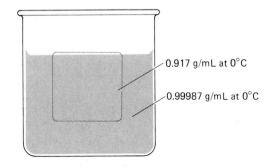

0.917 g/mL at 0°C

0.99987 g/mL at 0°C

In the ice structure, a good deal of empty space lies between the water molecules; hence, the volume increases in the change from liquid to solid (ice) and the density decreases (see Figure 13-9).

The *lower* density of ice and water at 0°C than water at temperatures above 4°C is responsible for lakes and ponds freezing from the *top down*. If ice were denser than water, as most solids are denser than their corresponding liquids, lakes and ponds would freeze from the bottom up. They would be completely frozen in the winter and fish could not live in them.

FIGURE 13-9

Structure of ice, using Prentice-Hall model. On freezing, water expands and leaves a good deal of empty space between the molecules. (Note the hexagonal "chair-like" structure of ice and also the hydrogen bonds,----.)

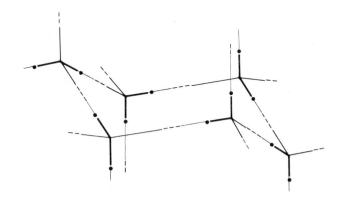

The expansion of water on freezing also accounts for the weathering of rocks. Water cleaves the rocks and thus assists in their erosion.

Table 13-2 summarizes some of the physical properties of water.

TABLE 13-2 Some Physical Properties of Water

Property	Value
Appearance	Colorless, odorless, tasteless
Normal boiling point	$100°C$
Melting point	$0°C$
Density at $4°C$	1.00000 g/mL
Specific heat of liquid	1.00 cal/g · °C (4.18×10^3 J/kg · K)
Heat of fusion:	
Specific	$8\overline{0}$ cal/g (3.3×10^5 J/kg)
Molar	1.44 kcal/mol (6.02 kJ/mol)
Heat of vaporization:	
Specific	$54\overline{0}$ cal/g (2.26×10^6 J/kg)
Molar	9.72 kcal/mol (40.7 kJ/mol)

13-5

Important Reactions Which Produce Water

Combustion

Many organic compounds burn in oxygen (air contains approximately 20 percent oxygen), forming water and carbon dioxide on complete combustion.[1] An example of this is methane (natural gas) burning in air to produce water and carbon dioxide, with the evolution of energy (exothermic; see 10-7) the most important product. This energy in the form of heat is used to heat many homes.

$$CH_{4(g)} + 2\,O_{2(g)} \rightarrow$$
$$2\,H_2O_{(g)} + CO_{2(g)} + \text{energy} \ (= 213 \text{ kcal at } 25°C) \quad (13\text{-}1)$$

The complete combustion of kerosene to produce carbon dioxide, water, and energy is another example:

$$2\,C_{12}H_{26(\ell)} + 37\,O_{2(g)\ \text{or}\ (\ell)} \rightarrow 26\,H_2O_{(g)} + 24\,CO_{2(g)} + \text{energy} \quad (13\text{-}2)$$

This reaction was used in the first stage of the Saturn V rocket to launch the astronauts to the moon in the old Apollo space missions. The oxidizer was pure liquid oxygen instead of air.

---

[1] Although there has been a greater demand for oxygen due to an increase in population and an increased per capita use of fossil fuels for energy (by combustion), the oxygen content in the air has remained constant to 0.1 percent over the past 50 years, although the carbon dioxide concentration has continued to increase. This increase in the carbon dioxide could cause an increase in the mean temperature of the earth.

Another example is the metabolism of foods. The complete combustion of sucrose (sugar) yields carbon dioxide, water, and energy:

$$C_{12}H_{22}O_{11(s)} + 12\ O_{2(g)} \rightarrow 11\ H_2O_{(g)} + 12\ CO_{2(g)} + \text{energy} \quad (13\text{-}3)$$

This energy keeps the body warm and maintains body temperature. In the preceding reaction, one tablespoonful of sugar (12 g) yields 45 kilocalories (Calories; see 2-12).

The approximate heat production for an adult male is 1600 to 1800 Cal per day. Depending on the type of activity more calories are required. For an activity such as going to college the requirement is about 3000 Cal per day for the male and about 2200 Cal per day for the female.

Combination of Hydrogen and Oxygen

Hydrogen gas combines with oxygen gas under appropriate conditions to form water:

$$2\ H_{2(g)} + O_{2(g)} \rightarrow 2\ H_2O_{(g)} + \text{energy} \quad (13\text{-}4)$$

If pure hydrogen gas is used, the hydrogen will burn smoothly *at the surface* in either oxygen gas or air to form a very hot, colorless flame and to produce water. But a mixture of hydrogen and oxygen gas (2:1) burns with explosive violence, because it contains an intimate mixture of the two and totally reacts almost instantaneously upon ignition.

This reaction of hydrogen and oxygen occurs in the initial stage of the space shuttle *Columbia*, with liquid hydrogen and liquid oxygen being used. A few seconds later solid-fuel rockets are used. This propellant consists of the following: 16 percent powdered aluminum (fuel), about 70 percent ammonium perchlorate (NH_4ClO_4, oxidizer), about 0.17 percent iron oxide powder (catalyst), 12 percent organic chemical polymer (binder), and 2 percent liquid epoxy resin (curing agent). The white "cloud" at takeoff consists mostly of aluminum oxide (Al_2O_3). These solid-fuel rockets provide more than 75 percent of the thrust necessary to lift the shuttle. To get the *Columbia* into its final orbit and days later to leave its orbit and return to earth, the *Columbia* uses monomethylhydrazine (fuel)[2] and dinitrogen tetroxide (N_2O_4, oxidizer, see footnote 4, Chapter 7).

---

[2]The structure of monomethylhydrazine is

$$\begin{array}{ccc} CH_3 & & H \\ \diagdown & & \diagup \\ & N{-}N & \\ \diagup & & \diagdown \\ H & & H \end{array}$$

Product of Neutralization Reactions

In nearly all neutralization reactions (see 9-11), water is one of the products. An example is the reaction of an acid with a base:

$$KOH_{(aq)} + HCl_{(aq)} \rightarrow KCl_{(aq)} + H_2O_{(\ell)} \qquad (13\text{-}5)$$

13-6

Reactions of Water

Li
K
Ba
Ca
Na
—
Mg
Al
Zn
Fe
Cd
—
Ni
Sn
Pb
(H)
Cu
Hg
Ag
Au

Electrolysis

When a direct electric current is passed through water containing a trace of sulfuric acid or some other ionic substance (necessary to give a conducting solution), hydrogen and oxygen gases are liberated, as was shown in Figure 3-3 (Chapter 3).

$$2\,H_2O_{(\ell)} \xrightarrow[\text{electric current}]{\text{direct}} 2\,H_{2(g)} + O_{2(g)} \qquad (13\text{-}6)$$

This reaction is just the reverse of Equation 13-4. Is this reaction exothermic or endothermic?

Reaction with Certain Metals

In 9-9, we introduced the electromotive or activity series and stated that certain metals in the series will displace any of those following it from an aqueous solution of its salt.

In the case of water, the first five metals in the series, **lithium, potassium, barium, calcium,** and **sodium,** are very active and can replace *one* hydrogen atom from water to form the *metal hydroxide* and hydrogen gas:

$$Ba_{(s)} + 2\,HOH \rightarrow Ba(OH)_{2(aq)} + H_{2(g)} \qquad (13\text{-}7)$$
$$\text{(excess)}$$

The metals from **magnesium** to **cadmium**—that is, magnesium, aluminum, zinc, iron, and cadmium—liberate hydrogen gas from *steam* under proper conditions and form the *metal oxide:*

$$Mg_{(s)} + H_2O_{(g)} \xrightarrow{\Delta} MgO_{(s)} + H_{2(g)} \qquad (13\text{-}8)$$

Work Problems 12 and 13.

The other metals in the electromotive or activity series (under normal conditions) are not active enough to react with water, neither cold nor hot.

13-7

Purification of Water

Water may contain many contaminants, depending on its source. If it is obtained from a river into which sewage has been dumped, it contains bacteria and other contaminants depending on the type and amount of sewage dumped into the river. Naturally, this water is unsafe for most purposes and must be purified considerably before it can be used for drinking purposes. Water obtained from springs and wells is usually safe for drinking purposes, but does contain salts of calcium, magnesium, or iron, with hydrogen carbonate (bicarbonate), carbonate, or sulfate, because water dissolves these salts as it seeps through the ground or rocks. Water containing little or none of these salts is called **soft water,** whereas water containing an appreciable amount of these salts is called **hard water.** Hard-water salts use up some amount of soap before a soapy lather appears. The hard-water cations react with the anion from the soap to form an insoluble "soap." This insoluble soap forms the "ring around the bathtub" and leaves a deposit on your skin! To prevent waste of soap through this reaction, hard water is frequently "softened" by the use of water softeners or agents that remove these undesirable cations.

Undesirable impurities in water may be removed in several ways. First, we shall consider purification of the highly contaminated water containing bacteria and other pollutants to produce potable or drinkable water.

1. *Settling out and filtering.* The water is allowed to stand in a reservoir to permit suspended particles, such as mud and silt, to settle. After settling, the water is filtered through sand and gravel beds to remove suspended material, such as silt, and also some bacteria. Prior to the filtering process, chemicals, such as lime and aluminum sulfate, may be added to aid in the filtering. Lime (calcium oxide) reacts with the water to form calcium hydroxide, which in turn reacts with the aluminum sulfate to form aluminum hydroxide. Aluminum hydroxide is an insoluble gelatinous precipitate that aids in removing some bacteria from the water by retaining them in the gelatinous precipitate. The following equations illustrate the reactions involved in the addition of chemicals prior to the filtering process:

$$CaO_{(s)} + H_2O_{(\ell)} \rightarrow Ca(OH)_{2(aq)} \tag{13-9}$$

$$3\ Ca(OH)_{2(aq)} + Al_2(SO_4)_{3(s)} \rightarrow 3\ CaSO_{4(aq)} + 2\ Al(OH)_{3(s)} \tag{13-10}$$

2. *Chlorination.* Chlorine is added to the water to kill any harmful bacteria that may have remained through the filtering process. Bleaching powder (a mixture of calcium hypochlorite, calcium chloride, and calcium hydroxide) is often used in place of gaseous chlorine.

Other substances may also be added to the water, depending on the amount of these substances present from natural sources. For example, small amounts (0.1 to 1 part per million; see 14-7) of *sodium fluoride* may be added to help prevent tooth decay. In the United States natural fluoride exists in water in the range 0.05 to 8 parts per million, with the least in the Northeast and the most in the Southwest.

Various other processes may be used depending on the contaminants present. Naturally, the more contaminants, the more special are the processes used before the water is safe for drinking purposes. Such processes generally increase the cost of the purification.[3]

Next, we shall consider the purification of hard water containing the salts of calcium, magnesium, iron, and hydrogen carbonate (bicarbonate), carbonate, and sulfate to produce soft water. The two general processes used are (1) distillation and (2) ion exchange.

We have mentioned distillation previously (12-5) as a process of water purification. In this process, the salts remain in the distilling flask as the *residue* and the pure water distills over as the *distillate* (see Figure 12-6). To obtain very pure water, we may have to repeat the distillation many times, depending on the amount of salts and other impurities present.

In the ion exchange process, ions from the salts are exchanged for less damaging ions from the exchanger. One type of ion exchanger uses zeolite, a hydrated sodium-aluminum silicate, which *exchanges* calcium, magnesium, or iron ions from the salts in hard water for sodium ions. The hard water containing the salt seeps through the zeolite layers in the exchanger tank, and the cations of the salts are exchanged for sodium ions.

$$Na_2Z_{(s)} \; + \; CaSO_{4(aq)} \rightarrow Na_2SO_{4(aq)} + CaZ_{(s)} \qquad (13\text{-}11)$$

(in tank, $Z =$ aluminum silicate portion of the zeolite)   (from hard water)   (in water)   (in tank)

The sodium salts are soluble (see 9-10) and do *not* precipitate the soap, nor do they interfere with lather formation. The used tank of zeolite can be regenerated by the treatment with a concentrated solution of sodium chloride, which regenerates the Na_2Z.

$$CaZ_{(s)} \; + \; 2 \, NaCl_{(aq)} \rightarrow CaCl_{2(aq)} \; + \; Na_2Z_{(s)} \qquad (13\text{-}12)$$

(in tank)   (washed out)   (in tank)

This is the basis of many commercial water softeners.

Another type of ion exchanger, removes both the cations—such as calcium, magnesium, and iron—and the anions—such as hydrogen carbonate,

---

[3]To conserve water, some communities are reusing waste water for watering golf courses and large yards. This water is sometimes treated with gaseous oxygen to purify it. The growing of grass or green plants increases the oxygen supply and decreases the carbon dioxide supply in the air by the process of photosynthesis, thereby helping in the fight against carbon dioxide pollution. A chemical equation for photosynthesis is

$$6 \, CO_{2(g)} \; + \; 6 \, H_2O \xrightarrow[\text{sunlight}]{\text{chlorophyll}} C_6H_{12}O_{6(s)} \; + \; 6 \, O_{2(g)}$$

(glucose, a sugar)

As carbon dioxide is removed, the equilibrium, $2 \, CO_{(g)} + O_{2(g)} \rightleftarrows 2 \, CO_{2(g)}$, is shifted to the right (see 17-3), decreasing the concentration of the dangerous carbon monoxide pollutant in the air.

carbonate, and sulfate—for hydrogen ions (H^{1+}) and hydroxide ions (OH^{1-}), respectively. The exchanger consists of two types of resins: One of the resins exchanges the cations for hydrogen ions (H^{1+}); the other exchanges the anions for hydroxide ions (OH^{1-}). This exchange results in the formation of water according to the following equation:

$$H^{1+} + OH^{1-} \rightarrow H_2O \qquad (13\text{-}13)$$

Water purified by this method is called *demineralized water*, since all the minerals salts have been removed.

13-8

Hydrates

In Chapter 9 (9-8), we mentioned that certain hydrates, when heated, decompose to yield water. **Hydrates** are crystalline salts that contain chemically bound water in *definite* proportions. An example of a hydrate that was given in Table 7-3 is Epsom salts, $MgSO_4 \cdot 7\ H_2O$, read as magnesium sulfate **heptahydrate**. The Greek prefixes (Table 7-1) are used to indicate the number of water molecules; for example, **2** H_2O is **dihydrate** and **3** H_2O is **trihydrate**.

The water may be held to the salt by any one of three possible mechanisms:

1. coordinate covalent (see 6-5) bonding with the metal cation

2. hydrogen bonding (see 13-3) with the anion

3. entrapment in the crystalline structure of the hydrate, like a stone or rock in a cement sidewalk

A given hydrate may have various combinations of these types of attachments to the water molecules. For example, the hydrate cobalt(II) sulfate heptahydrate, $CoSO_4 \cdot 7\ H_2O$, has six water molecules attached by coordinate covalent bonding to the metal cation (mechanism 1) and one water molecule attached by hydrogen bonding to the anion (mechanism 2), as shown in Figure 13-10.

FIGURE 13-10
Cobalt sulfate heptahydrate. Six water molecules are attached by coordinate covalent bonding with the cobalt ion, and **one** is attached by hydrogen bonding to the sulfate ion. (Notice the octahedron formed by the six water molecules about the cobalt ion.)

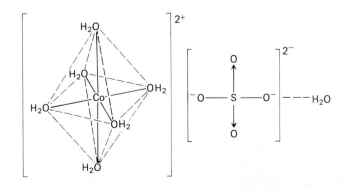

The percent of water in a hydrate can be calculated from the formula of the hydrate by a method similar to that which we used in calculating the percent composition of a compound given the formula (see 8-4).

Consider the following problem example:

Problem Example 13-1

Calculate the percent of water in calcium chloride hexahydrate.

SOLUTION: The formula mass of $CaCl_2 \cdot 6 H_2O$ is calculated as 219.1 amu.

$$1 \times 40.1 = \quad 40.1 \text{ amu}$$

$$2 \times 35.5 = \quad 71.0 \text{ amu}$$

$$6 \times 18.0 = \underline{108.0 \text{ amu}} \text{ (molecular mass } H_2O = 18.0 \text{ amu)}$$

$$\text{formula mass } CaCl_2 \cdot 6 H_2O = 219.1 \text{ amu}$$

The percent of water in the hydrate is calculated by dividing the *total* contribution of the water in amu by the formula mass in amu of the hydrate and multiplying by 100.

Work Problem 14.

$$\frac{108.0 \text{ amu}}{219.1 \text{ amu}} \times 100 = 49.3\% \text{ water} \quad \textit{Answer}$$

Knowing the percent of the salt and percent of the water in a given hydrate, we can calculate the formula of the hydrate by a method similar to that which we used in calculating the formula of a compound given the percent composition (see 8-5).

Consider the following problem example:

Problem Example 13-2

Calculate the formula of the hydrate of barium chloride containing 14.7 percent water.

SOLUTION: If 14.7 percent of the hydrate is water, the 85.3 percent $(100.0 - 14.7 = 85.3)$ must be barium chloride. Hence, in $1\overline{00}$ g of the hydrate there would be 14.7 g of water and 85.3 g of barium chloride. Therefore, the first step is to calculate the number of moles of water and of barium chloride in $1\overline{00}$ g of the hydrate, as follows:

$$14.7 \text{ g } H_2O \times \frac{1 \text{ mol } H_2O}{18.0 \text{ g } H_2O} = 0.817 \text{ mol } H_2O$$

$$85.3 \text{ g } BaCl_2 \times \frac{1 \text{ mol } BaCl_2}{208.3 \text{ g } BaCl_2} = 0.410 \text{ mol } BaCl_2$$

The second step is to express these relationships in small whole numbers by dividing by the smallest value, as follows:

$$\text{For BaCl}_2: \quad \frac{0.410}{0.410} = 1.00$$

$$\text{For H}_2\text{O}: \quad \frac{0.817}{0.410} = 1.99, \quad \text{approximately 2}$$

Work Problems 15
and 16.

Hence, the formula for the hydrate is $\text{BaCl}_2 \cdot 2 \text{ H}_2\text{O}$.   *Answer*

Associated with hydrates are a number of related substances, such as hygroscopic, deliquescent, and efflorescent substances.

Hygroscopic Substances

A **hygroscopic substance** is one that readily absorbs moisture from the air. Many finely divided substances are slightly hygroscopic. For example, sugar absorbs moisture from the air on humid days and forms a cake. Other examples are flour and salt. A small packet containing a hygroscopic substance is placed in many medicine bottles containing tablets or capsules to prevent absorption of water by the drug. On the packet is printed "Do not take."

Deliquescent Substances

A hygroscopic substance that continues to absorb water and eventually forms a *solution* is a deliquescent substance. Or, a **deliquescent substance** is one that absorbs enough moisture from the air to form a *solution*. This process is called *deliquescence*. Hence, all deliquescent substances are hygroscopic, but not all hygroscopic substances are deliquescent. An example of a deliquescent substance is anhydrous calcium chloride, which absorbs water to form first the solid hexahydrate and then absorbs more water to form a *solution*. Another example is sodium hydroxide, which readily absorbs water to form a solution. You may have already noticed this in the laboratory. If a person spilled some sodium hydroxide pellets near the balance and you followed that person, you may have placed your laboratory book or your hand on the *now*-formed sodium hydroxide solution!

Efflorescent Substances

We previously mentioned that hydrates decompose when *heated* to produce water. An **efflorescent substance** is a hydrate that loses its water of hydration when *simply* exposed to the atmosphere. This process is called *efflorescence*. Examples of efflorescent substances are washing soda, $\text{Na}_2\text{CO}_3 \cdot 10 \text{ H}_2\text{O}$, which forms $\text{Na}_2\text{CO}_3 \cdot \text{H}_2\text{O}$ upon standing in the open air, and sodium acetate trihydrate, $\text{NaC}_2\text{H}_3\text{O}_2 \cdot 3 \text{ H}_2\text{O}$, which forms the anhydrous salt, $\text{NaC}_2\text{H}_3\text{O}_2$ (see Figure 13-11).

For efflorescence to occur, the vapor pressure of water in the atmosphere must be *less* than the vapor pressure of the hydrate. If the vapor pressure in

(a)   (b)

FIGURE 13-11

Efflorescence. (a) Crystal on the left is a fresh sodium acetate trihydrate crystal grown from a saturated solution of the salt. Crystal on the right is a similar crystal which has lost water of hydration after exposure for several hours to the arid New Mexico atmosphere. (b) The same crystal after further exposure for a few minutes. The fresh crystal is rapidly becoming opaque due to the lost water of hydration (Courtesy of the University of New Mexico and Los Alamos Scientific Laboratory.)

the atmosphere is *greater* than the vapor pressure of the hydrate, efflorescence does not occur. The term efflorescence should not be confused with effervescence, which means "to bubble," as Alka-Seltzer or "baking *powder*" does in water.

13-9

Hydrogen Peroxide

A second oxide of hydrogen is hydrogen peroxide (H_2O_2). If we assign an oxidation number of 1^+ to hydrogen, then each oxygen atom has an oxidation number of 1^- in hydrogen peroxide (see 6-1).

Structure

In hydrogen peroxide, the two oxygen atoms each share a covalent bond with a hydrogen atom, and, in addition, they share one covalent bond with each other, as shown by the following electron-dot and structural formulas:

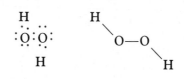

The hydrogen peroxide molecule is not planar, but has an H—O—O bond angle of approximately 102°. The other hydrogen atom extends from the plane of these three atoms at approximately 90°, as shown in Figure 13-12. Hydrogen peroxide also forms strong hydrogen bonds, as does water.

Preparation

Hydrogen peroxide can be prepared in the laboratory by treating barium peroxide (BaO_2) with an aqueous solution of sulfuric acid:

$$BaO_{2(s)} + H_2SO_{4(aq)} \rightarrow BaSO_{4(s)} + H_2O_{2(aq)} \qquad (13\text{-}14)$$

The insoluble barium sulfate is removed by filtration, and the resulting solution of hydrogen peroxide may be concentrated by careful distillation at atmospheric pressure to 30 percent. Distillation at reduced pressure gives 98 percent hydrogen peroxide.

Properties

Pure hydrogen peroxide is a pale blue, odorless, oily liquid. One of its chemical properties is its ready decomposition to water and oxygen, according to the following equation:

$$2\,H_2O_{2(aq)} \xrightarrow[\text{or catalyst}]{\text{heat, light}} 2\,H_2O_{(\ell)} + O_{2(g)} \qquad (13\text{-}15)$$

This decomposition may be accelerated by heat or light. It is for this reason that hydrogen peroxide sold in the drugstore is stored in brown bottles (which helps cut down on light absorption) and kept in a relatively cool place. A stabilizer is often added to a solution of hydrogen peroxide to prevent decom-

FIGURE 13-12

Structure of hydrogen peroxide, H_2O_2. (a) Prentice-Hall model ($\rightarrow$ points to the unshared pair of electrons). (b) Stuart-Briegleb model.

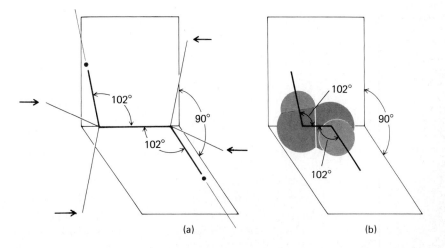

(a)   (b)

position. Other substances, such as silver, carbon, manganese(IV) oxide, saliva, dirt, and an enzyme in the blood, readily catalyze its decomposition. The release of oxygen can be observed as bubbles when hydrogen peroxide is used as an antiseptic and comes into contact with a wound.

Uses

Work Problem 17.

Relatively dilute solutions of hydrogen peroxide are used principally as antiseptics and bleaching agents. A 3 percent solution is used as an antiseptic, whereas slightly higher concentrations (6 to 30 percent) are used as bleaching agents for cloth, silk, straw, flour, and hair. A 90 percent solution has been used as a fuel in the propulsion of rockets.

13-10

Ozone

Oxygen exists in two different forms: diatomic oxygen, O_2; and triatomic oxygen or ozone, O_3. Both are gases at room temperature.

Structure

The ozone molecule consists of three oxygen atoms, the bond angle between the three oxygen atoms being 127°.

Preparation

Ozone can be prepared from oxygen by using an electrical discharge or sunlight (ultraviolet light). It can be detected by its pungent odor noticeable around certain electrical equipment.

$$3 \; O_{2(g)} \xrightarrow[\text{sunlight}]{\text{electricity or}} 2 \; O_{3(g)} \tag{13-16}$$

Ozone is also found in *smog*, and its formation is promoted by sunlight.

Properties

Ozone decomposes slowly at low temperatures, but rapidly at higher temperatures, to form diatomic oxygen. This decomposition is catalyzed by various substances, such as manganese(IV) oxide and silver.

$$2 \; O_{3(g)} \xrightarrow{\Delta, \; \text{catalyst}} 3 \; O_{2(g)} \tag{13-17}$$

Ozone is both beneficial and harmful to mankind. It is beneficial in that it protects all life on the earth from an overdose of ultraviolet light given off by the sun. There is a blanket of ozone in the upper atmosphere (10 to 30 miles up from the earth, stratosphere) which filters out this ultraviolet light. If this ozone layer is destroyed or reduced, there possibly would be a higher incidence of blindness and skin cancer.

The environmental problem raised in the use of a worldwide fleet of SSTs (supersonic transports), which operate very efficiently in this stratosphere ozone layer is that large amounts of nitrogen oxides would be given off by the aircraft. These nitrogen oxides would react with the ozone to produce oxygen which does *not* filter out the ultraviolet light. Investigation at MIT (Massachusetts Institute of Technology) confirms this situation and predicts that the ozone would be considerably reduced in this layer in the Northern Hemisphere due to the operation of the SSTs.

Another problem is in the use of chlorofluorocarbons (CFCs) in aerosol spray containers. These CFCs destroy the ozone layer. If their use continues at the present rate, the ozone layer may be reduced by 5 to 9 percent in the next 100 years. For each 1 percent decrease in ozone in the stratosphere, scientists have predicted that there will be up to a 10 percent increase in skin cancer in light-skinned people. Ways must be developed to save our stratospheric ozone layer.

Ozone is also beneficial in that it has been used with ultrasonic (noiseless) sound to purify sewage. The sound breaks up the large sewage particles into smaller particles and the ozone destroys the bacteria and viruses. Ozone by itself has also been used as a replacement for chlorine in the purification of water (see 13-7). Water treated in this manner lacks the disagreeable taste of water treated with chlorine.

Ozone is found in *smog* produced from automobile exhausts. It irritates the mucous membranes and may be quite harmful if inhaled in quantity. On the days when the smog in Los Angeles is quite heavy and the ozone concentration is high (0.35 part per million in the atmosphere), school children are excused from strenuous physical activity. This occurs on the average of about 21 days per year, mainly from July through October. On these days even jogging may be harmful to your health!

Work Problem 18.

Ozone also has a very harmful effect on many plants and trees. It produces diseases in vegetables and is known to be destructive to rubber products.

Chapter Summary

The unusual physical properties of water can be explained by its ability to *hydrogen bond*. In this chapter, the electronic structure and polarity of this abundant and necessary compound were considered. Also, the preparations and reactions of water were considered. Water may contain many contaminants, depending on its source. The purification of it to produce (1) potable or drinkable water and (2) distilled or demineralized water was considered.

Hydrates are crystalline salts that contain chemically bound water in definite proportions, which when heated decompose to yield water. Both the nomenclature of hydrates and the mechanism for holding the water molecules in the hydrate crystal were examined. Problems calculating the percent water in a hydrate along with the calculation of the formulas of hydrates were studied. Terms related to hydrates, such as *hygroscopic, deliquescent,* and *efflorescent* substances were defined.

Finally, the structure, preparation, and properties of hydrogen peroxide and ozone along with some of their applications to our technical world were discussed.

EXERCISES

1. Define or explain
   (a) sp^3 orbitals   (b) tetrahedron
   (c) Hund's Rule   (d) dipole
   (e) polar compounds   (f) hydrogen bond
   (g) hard water   (h) distillation
   (i) ion exchange   (j) demineralized water
   (k) hydrate   (l) hygroscopic substance
   (m) deliquescent substance   (n) deliquescence
   (o) efflorescent substance   (p) efflorescence

2. Distinguish between
   (a) *p* orbitals and sp^3 orbitals
   (b) soft water and hard water
   (c) H and O found in $Na_2CO_3 \cdot H_2O$ and H and O found in $C_{12}H_{22}O_{11}$
   (d) hydrate and hygroscopic substance
   (e) deliquescent substance and deliquescence
   (f) efflorescent substance and efflorescence
   (g) efflorescence and effervescence

Structure of Water   (See Sections 13-1 and 13-2)

3. At one time, water was considered to be formed by the overlap of *p* orbitals of oxygen with the *s* orbitals of hydrogen.
   (a) Based on this approach, what would be the predicted bond angle in water?
   (b) What orbitals would the unshared pairs of electrons occupy?
   (c) Would these unshared pairs of electrons then be geometrically equivalent? Explain.

4. (a) Using the sp^3 orbital approach to the structure of water, what is the predicted bond angle? (b) What is the actual bond angle? (c) Explain the difference between the two values.

Physical Properties of Water   (See Section 13-4)

5. Would ice float when placed in the following liquids? Explain.
   (a) ethyl alcohol ($d^{0°} = 0.806$ g/mL)
   (b) benzene ($d^{0°} = 0.899$ g/mL)
   (c) glycerine ($d^{0°} = 1.26$ g/mL)

6. Discuss the consequences to the earth if, on going from liquid water to solid ice, the volume of the solid did not expand.

Purification of Water   (See Section 13-7)

7. Explain how a mixture of lime and aluminum sulfate aids in the filtering process of water purification.

8. Explain briefly the operation of a commercial water softener.

Hydrates   (See Section 13-8)

9. List the three possible mechanisms by which water combines with a salt to form a hydrate.

PROBLEMS

Polarity of Molecules   (See Section 13-2)

10. Represent the *net dipole moment*, if any, for the following molecules. (*Hint:* Draw the structural formula, and then refer to the order of electronegativities of the elements.)
    (a) H_2O
    (b) CO_2
    (c) HCl
    (d) BrCl

Hydrogen Bonding in Water   (See Section 13-3)

11. Explain the apparent irregularity in the series of boiling points shown for the compounds HF, HCl, HBr. The electronegativities are as follows: F = 4.0, Cl = 3.0, and Br = 2.8.

Compound	Boiling point (°C at 1 atm)
HF	19.7
HCl	−85.0
HBr	−66.8

Important Reactions Which Produce Water and Reactions of Water
(See Sections 13-5 and 13-6)

12. Complete and balance the following equations; indicate any precipitate by (s) and any gas by (g):
    (a) $C_6H_{12}O_{6(s)} + O_{2(g)} \xrightarrow{\Delta}$
        (glucose, dextrose)   (excess)
    (b) $H_{2(g)} + O_{2(g)} \xrightarrow{\text{"spark"}}$
    (c) $Ca(OH)_{2(aq)} + H_2SO_{4(aq)} \rightarrow$
    (d) $BaO_{(s)} + HCl_{(aq)} \rightarrow$
    (e) $CO_{2(g)} + KOH_{(aq)} \rightarrow$
    (f) $H_2O_{(\ell)} \xrightarrow[\text{electric current}]{\text{direct}}$
    (g) $K_{(s)} + HOH_{(\ell)} \rightarrow$
        (excess)

(h) $Na_{(s)} + HOH_{(\ell)} \rightarrow$
    (excess)

(i) $Mg_{(s)} + H_2O_{(g)} \xrightarrow{\Delta}$

(j) $Al_{(s)} + H_2O_{(g)} \xrightarrow{\Delta}$

13. Complete and balance the following equations; indicate any precipitate by (s) and any gas by (g):

   (a) $C_2H_{6(g)} + O_{2(g)} \xrightarrow{\Delta}$
       (ethane)   (excess)

   (b) $C_2H_6O_{(g)} + O_{2(g)} \xrightarrow{\Delta}$
       (ethyl alcohol)   (excess)

   (c) $Ca(OH)_{2(aq)} + CO_{2(g)} \rightarrow$

   (d) $KOH_{(aq)} + SO_{2(g)} \rightarrow$

   (e) $KOH_{(aq)} + SO_{3(g)} \rightarrow$

   (f) $Zn(OH)_{2(s)} + HCl_{(aq)} \rightarrow$

   (g) $ZnO_{(s)} + HCl_{(aq)} \rightarrow$

   (h) $Ca_{(s)} + HOH_{(\ell)}$ (excess) $\rightarrow$

   (i) $Li_{(s)} + HOH_{(\ell)}$ (excess) $\rightarrow$

   (j) $Zn_{(s)} + H_2O_{(g)} \xrightarrow{\Delta}$

Hydrates (See Section 13-8)

14. Calculate the percent of water in the following hydrates:
    (a) magnesium sulfate heptahydrate
    (b) calcium sulfate dihydrate
    (c) zinc sulfate heptahydrate
    (d) cobalt(II) sulfate heptahydrate
    (e) calcium chloride dihydrate

15. Calculate the formula for the following hydrates:
    (a) a hydrate of copper(II) sulfate, containing 10.1 percent water
    (b) a hydrate of copper(II) sulfate, containing 25.3 percent water
    (c) a hydrate of copper(II) sulfate, containing 36.1 percent water
    (d) a hydrate of sodium carbonate, containing 14.5 percent water
    (e) a hydrate of sodium carbonate, containing 63.0 percent water

16. A student heated 1.75 g of a hydrate of calcium chloride and then found it to have a mass of 0.89 g. Calculate the formula of the hydrate.

Electron-Dot and Structural Formulas (See Sections 13-2 and 13-9)

17. Write the electron-dot and structural formulas for
    (a) H_2O (^1_1H, $^{16}_8O$)
    (b) H_2O_2

Hydrogen Peroxide and Ozone (See Sections 13-9 and 13-10)

18. Complete and balance the following equations; indicate any precipitate by (s) and any gas by (g).

(a) $H_2O_{2(aq)} \xrightarrow{\text{light}}$

(b) $BaO_{2(s)} + H_2SO_{4(aq)} \rightarrow$

(c) $O_{2(g)} \xrightarrow{\text{sunlight}}$

(d) $O_{3(g)} \xrightarrow{\text{MnO}_2}$

General Problems

19. Calculate the number of grams of water required to produce 6.25 g of oxygen by electrolysis.

20. Calculate the volume of oxygen in liters produced at STP from the electrolytic decomposition of 6.20 g of water.

21. Calculate the volume in liters of the hydrogen produced from the electrolysis of 3.20 g of water if it is collected over water at 627 torr and 27°C. (See Appendix V for additional data.)

22. In the decomposition of hydrogen peroxide, how many milliliters of oxygen can be produced at STP from 2.00 g of 6.00 percent hydrogen peroxide solution? (*Hint:* A 6.00 percent hydrogen peroxide solution contains 6.00 g of pure hydrogen peroxide in $\overline{1}00$ g of solution.)

Readings

Cook, Gerhard A., "Industrial Uses of Ozone." *J. Chem. Educ.*, 1982, v. 59, p. 392. Discusses the uses and industrial preparation of ozone.

Franks, Felix, *Polywater*. Cambridge, Mass.: The MIT Press, 1981. Reviews the history of the "polywater" controversy.

Sanders, Howard J., "Tooth Decay." *Chem. Eng. News*, February 25, 1980, p. 80. A comprehensive article on preventing tooth decay. It includes a discussion of the use of fluoride ion in preventing tooth decay and its method of action.

Solutions and Colloids

Colloids found in the home: gelatin dessert (liquid in solid); milk, butter, and mayonnaise (liquid in liquid); milk of magnesia (solid in liquid); and whipped cream (gas in liquid).

TASKS

1. Memorize the five methods for expressing concentrations of solutions given in Table 14-2.

2. Memorize the formulas for calculating the boiling-point elevation, that is, ΔT_b = molality $(m) \cdot K_b$, and the freezing-point depression, that is, ΔT_f = molality $(m) \cdot K_f$.

OBJECTIVES

1. Given the following terms, define each term and describe the distinguishing characteristic of each:
   (a) solution (Section 14-1)
   (b) solute (Section 14-1)
   (c) solvent (Section 14-1)
   (d) Henry's Law (Section 14-3)
   (e) saturated solution (Section 14-4)
   (f) unsaturated solution (Section 14-4)
   (g) supersaturated solution (Section 14-4)
   (h) percent by mass (Section 14-6)
   (i) parts per million (Section 14-7)
   (j) molarity (Section 14-8)
   (k) normality (Section 14-9)
   (l) molality (Section 14-10)
   (m) colloid (Section 14-13)
   (n) dispersed particles (Section 14-13)
   (o) dispersing medium (Section 14-13)

2. Given a graph of the solubility of a salt in a liquid at various temperatures, determine the solubility from the graph at any temperature, and vice versa (Section 14-3, Problem 10).

3. Given the solubility of a gas in a liquid at a given partial pressure, calculate the solubility at any *total* pressure (Henry's Law, Problem Example 14-1, Problems 11 and 12).

4. Given the mass of solute in a given mass of solution or mass of solvent, calculate the percent concentration of the solution (Problem Example 14-2, Problems 13 and 14).

5. Given the percent concentration of a solution by mass and the mass of the solvent, calculate the mass of the solute (Problem Example 14-3, Problem 15).

6. Given the percent concentration of a solution by mass and the mass of solute, calculate the mass of solvent or solution (Problem Example 14-4, Problems 16 and 17).

7. Given the mass of solute in a given volume of solution (density assumed = 1.00 g/mL), calculate the parts per million of the solute in the solution (Problem Example 14-5, Problems 18 and 19).

8. Given the volume of solution (density assumed = 1.00 g/mL) and the parts of solute per million, calculate the mass of solute (Problem Example 14-6, Problem 20).

9. Given the name or formula of a solute, the Table of Approximate Atomic Masses, and the mass of solute in a given volume of solution, calculate the molarity of the solution (Problem Example 14-7, Problems 21 and 22).

10. Given the name or formula of a solute, the Table of Approximate Atomic Masses, the molar concentration, and the volume of solution, calculate the mass of the solute (Problem Example 14-8, Problem 23).

11. Given the name or formula of a solute, the Table of Approximate Atomic Masses, the molar concentration, and the mass of solute needed, calculate the volume of solution required to provide this mass of solute (Problem Example 14-9, Problem 24).

12. Given the name or formula of a solute, the Table of Approximate Atomic Masses, the mass of the solute, the volume of the solution, and the use of the solute in a reaction, calculate the normality of the solution (Problem Example 14-10, Problems 25 and 26).

13. Given the name or formula of a solute, the Table of Approximate Atomic Masses, the normal concentration, the use of the solute in a reaction, and the volume of the solution, calculate the mass of the solute (Problem Example 14-11, Problem 27).

14. Given the name or formula of a solute, the Table of Approximate Atomic Masses, the normal concentration, the use of the solute in a reaction, and the mass of solute needed, calculate the volume of solution required to provide this mass of the solute (Problem Example 14-12, Problem 28).

15. Given the name or formula of a solute, the Table of Approximate Atomic Masses, and the mass of solute in a given mass of solvent, calculate the molality of the solution (Problem Example 14-13, Problems 29 and 30).

16. Given the name or formula of a solute, the Table of Approximate Atomic Masses, the molal concentration, and the mass of the *solution,* calculate the mass of the solute required (Problem Example 14-14, Problem 31).

17. Given the name or formula of a solute, the Table of Approximate Atomic Masses, the molal concentration, and the mass of the solute, calculate the mass of the *solvent* (Problem Example 14-15, Problem 32).

18. Given the molar, normal, and molal concentration of a solution, the name or formula of the solute, the Table of Approximate Atomic Masses, the solution's density, and the use of the solute in a reaction, calculate the percent by mass of the solution (Problem Example 14-16, Problem 33).

19. Given the percent by mass, normal, and molal concentration of a solution, the name or formula of the solute, the solution's density, and the use of the solute in a reaction, calculate the molarity of the solution (Problem Example 14-17, Problem 34).

20. Given the percent by mass, molar, and molal concentration of a solution, the name or formula of the solute, the solution's density, and the use of the solute in a reaction, calculate the normality of the solution (Problem Example 14-18, Problem 35).

21. Given the percent by mass, molar, and normal concentration of a solution, the name or formula of the solution, the Table of Approximate Atomic Masses, the solution's density, and the use of the solute in a reaction, calculate the molality of the solution (Problem Example 14-19, Problem 36).

22. Given the molal boiling-point-elevation constant and the boiling point of the solvent, or the molal freezing-point-depression constant and the freezing point of the solvent, and the molality of the solution or sufficient information to calculate the molality, calculate the boiling point, or the freezing point of the solution (Problem Example 14-20, Problems 37 and 38).

23. Given the boiling point or freezing point of a solution, the boiling point or freezing point of the solvent, the molal-boiling- or molal-freezing-point elevation or depression constant, and the mass of the solute dissolved in the mass of the solvent, calculate the molecular mass of the solute (Problem Example 14-21, Problems 39 and 40).

In Chapter 3, we discussed homogeneous matter as consisting of pure substances, homogeneous mixtures, and solutions, and in Chapters 11 and 12 we considered the three physical states of matter: gases, liquids, and solids. Now we are going to consider solutions. Solutions are homogeneous, but a solution can have *variable* composition usually **within certain limits.**

Colloids bridge the gap between matter in a solution and matter dispersed in a suspension. In a solution, the particles are homogeneously dispersed and *do not* settle out on standing, because they may be partially bound to solvent molecules. In a **suspension,** the particles are *not* bound to solvent molecules and *do* settle out on standing. In **colloids,** the dispersed particles are *not* appreciably bound to solvent molecules but *do not* settle out on standing. We shall discuss properties and examples of colloids at the end of this chapter, after we consider solutions in detail.

14-1

Solutions

A **solution** is homogeneous throughout. It is composed of two or more pure substances, and its composition can be *varied* usually **within certain limits.** Solutions are considered weakly bound mixtures of a solute and a solvent. The **solute** is usually the component in lesser quantity, and the **solvent** is the component in greater quantity. The solute dissolves in the solvent whatever is its physical state. The solute is then considered *soluble* in the solvent. For example, in a 5.00 percent sugar solution in water, the sugar is the solute and the water the solvent. This solution is referred to as an aqueous solution since water is the solvent. The components of a solution (solute and solvent) are dispersed as either **molecules** or **ions** in many cases *bound to molecules of the solvent.*

In an aqueous sugar solution, a given sugar molecule is surrounded by water molecules and the sugar molecule is considered to be hydrated (hydrogen-bonded by water molecules). The process of dissolving solid sugar in water involves hydrating the solute molecules until all the sugar has dissolved or reached a certain maximum concentration, as shown in Figure 14-1a.

In an aqueous sodium chloride solution, both the sodium ions (Na^{1+}) and chloride ions (Cl^{1-}) are surrounded by water molecules, and hence become hydrated. The strong ionic forces of attraction of the sodium ion (positive) for the chloride ion (negative) in the solid crystal are diminished by the energy released in the formation of the hydrated ions. The relatively negative oxygen atom (13-2) of the water molecule is attracted to the sodium ion, and the relatively positive portion of the water molecule hydrates the chloride ion, as shown in Figure 14-1b.

14-2

Types of Solutions

Solutions may exist in any one of the three physical states of matter; the more common types are a gas in a liquid, a liquid in a liquid, and a solid in a liquid. Examples of these are given in Table 14-1.

14-3

Factors Affecting Solubility and Rate of Solution

We shall consider two sets of factors in this section:

1. those factors that affect the actual **solubility** of a given solute in a solvent

2. those factors that affect the **rate** (how fast?) at which a given solute dissolves in a given solvent

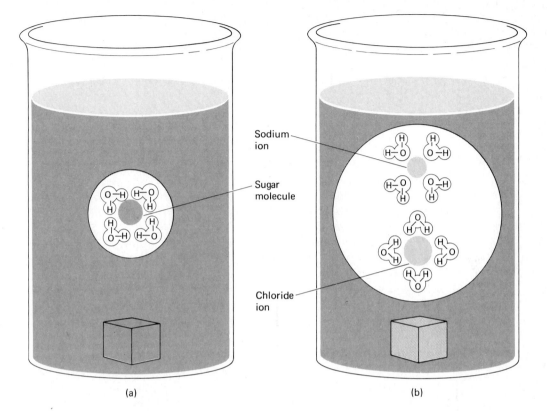

(a)　　　　　　　　　　　　(b)

FIGURE 14-1

The process of dissolving. (a) Sugar dissolving in water to form hydrated sugar molecules, $C_{12}H_{22}O_{11}(H_2O)_x$. (b) Sodium chloride dissolving in water to form hydrated sodium ions, $Na^{1+}(H_2O)_x$. (relatively negative oxygen atom of water attracted to positive sodium ions); and hydrated chloride ions, $Cl^{1-}(H_2O)_y$ (relatively positive hydrogen atom of water attracted to negative chloride ions).

TABLE 14-1　Some Types of Solutions

Solute	Solvent	Example
Gas	Liquid	Carbonated beverages (carbon dioxide in aqueous solution)
Liquid	Liquid	Antifreeze in car radiator (ethylene glycol in water)
Liquid	Solid	Dental fillings (mercury in silver)
Solid	Liquid	Sugar in water
Solid	Solid	Solder (tin in lead)

343

The actual **solubility** of a solute in a given solvent is dependent on three factors: (1) the properties of the solute and the solvent, (2) temperature, and (3) pressure.

Properties of Solute and Solvent

In the preceding chapter (13-2), we stated that water was considered a *polar compound* since its molecules exist as dipoles. Many ionic compounds such as sodium chloride are, in general, soluble in water, because these ions are attracted to one of the poles of the dipole in water, or because they are held by hydrogen bonds, or both. Carbon tetrachloride (CCl_4) is a *nonpolar compound* because the center of negative charge coincides with that of the positive charge due to the tetrahedral arrangement of the molecule, as shown in Figure 14-2. The ionic compound sodium chloride is insoluble in carbon tetrachloride since the ions find more attraction for their opposite charged ions than for the nonpolar carbon tetrachloride molecules. Covalent (nonionic) compounds are, in general, soluble in carbon tetrachloride and insoluble in water, since the water molecules are bound by hydrogen bonds more firmly to each other than they could be to nonionic, nonpolar molecules. Generally, you will find that ionic compounds are soluble in polar solvents and covalent compounds are soluble in nonpolar solvents.

Household "spot" removers generally are composed of chlorinated hydrocarbons (for example, carbon tetrachloride) and are effective solvents for relatively nonpolar organic materials, such as fats. In "dry cleaning," soiled clothes are cleaned on a large scale by using such solvents, since the fatty grease on clothes binds the dirt to the fabric.

A phrase frequently used in *like dissolves like.* Therefore, *ionic compounds (polar) are generally soluble in polar solvents, and covalent compounds (nonpolar or weakly polar) are generally soluble in nonpolar solvents.*

FIGURE 14-2

Carbon tetrachloride is a nonpolar molecule. (a) Structure of carbon tetrachloride, showing zero dipole moment since the centers of positive and negative charge coincide on the carbon atom due to the tetrahedral structure. (b) Prentice-Hall model of carbon tetrachloride. (c) Stuart-Briegleb model of carbon tetrachloride.

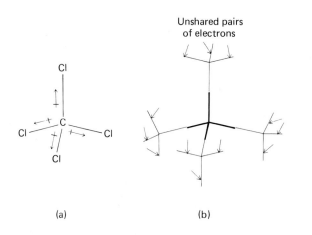

Unshared pairs of electrons

(a)   (b)   (c)

Temperature

The solubility of a gas in a liquid decreases with an increase in temperature. For example, oxygen is soluble in water to the extent of 4.89 mL of oxygen in 100 mL of water at 0°C, but at 50°C this solubility is reduced to 2.46 mL of oxygen in 100 mL of water. The lake fisherman knows that fish move away from shore as the day grows warmer. He may not be aware, however, that the sun heats the water near the shore, decreasing the solubility of oxygen in the water. The fish may move into deeper, cooler water for a better oxygen supply. You may also have observed the decrease in solubility of a gas with an increase in temperature when you heat a glass container and bubbles of gas form along the side of the container before the water boils. The water contains dissolved gases, such as oxygen, nitrogen, and carbon dioxide, which escape from the water when it is heated.

The solubility of a solid in a liquid *usually increases* with an *increase* in *temperature;* however, there are a number of exceptions. The variable effect of increasing temperature on solubility is illustrated in Figure 14-3. You will note that as the temperature increases, the solubility of KCl and KNO_3 increases, whereas that of NaCl remains relatively constant and that of $CaCrO_4$ decreases. Consider the solubility of $CaCrO_4$ in water from the graph in Figure 14-3. At

FIGURE 14-3.

Solubility of various salts with temperature.

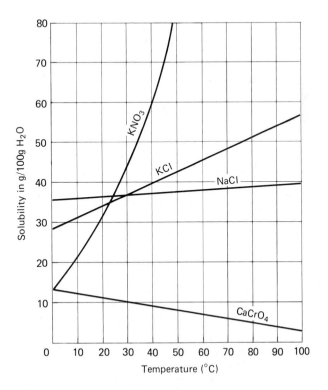

Work Problem 10.

30°C the solubility is $1\overline{0}$ g per $10\overline{0}$ g of water, while at 80°C it is about 5 g per $10\overline{0}$ g of water.

Pressure

Solutions involving only liquids and solids are not influenced to any great extent by pressure. However, solutions of gases in liquids are appreciably affected by pressure. William Henry (1774–1836), English chemist and physician, and friend of John Dalton (4-2 and 11-7), carried out experiments in 1803 on the solubility of gases in a liquid with pressure. From his experiments, he formulated the law now referred to as **Henry's Law.** This law states that the solubility of a gas in a liquid is directly *proportional* to the *partial* pressure of the gas above the liquid. For example, if the partial pressure of the gas above the liquid is *doubled*, then the *solubility* of the gas in the liquid is also *doubled*. Conversely, if the partial *pressure is reduced by one-half*, then the *solubility* of the gas in the liquid is *reduced by one-half*, and the gas escapes from the liquid. When you open a carbonated beverage, the release of the pressure allows the excess carbon dioxide gas to escape and effervescence occurs.

Consider the following problem example involving Henry's Law:

Problem Example 14-1

Calculate the solubility in grams per liter of a certain gas in water at a partial pressure of 3.50 atm and 0°C. The solubility is 0.530 g/L at a *total* pressure of 1.000 atm and 0°C.

SOLUTION: The total pressure is the sum of the partial pressures of water and the gas, according to Dalton's Law of Partial Pressures (see 11-7).

$$P_{\text{total}} = P_{\text{gas}} + P_{\text{water}}$$

P_{water} at 0°C is 0.0061 atm (see Appendix V)

$$P_{\text{gas}} = 1.000 \text{ atm} - 0.0061 \text{ atm}$$

$$P_{\text{gas}} = 0.994 \text{ atm}$$

Now, we solve the problem using the same general method we used for solving gas law problems (see Chapter 11):

Solubility$_{\text{old}}$ = 0.530 g/L P_{old} = 0.994 atm | pressure increases;

Solubility$_{\text{new}}$ = ? P_{new} = 3.50 atm ↓ solubility increases

Solubility$_{\text{new}}$ = solubility$_{\text{old}}$ × pressure factor

Work Problems 11 and 12.

$$\text{Solubility}_{\text{new}} = 0.530 \text{ g/L} \times \frac{3.50 \text{ atm}}{0.994 \text{ atm}} = 1.87 \text{ g/L} \quad \textit{Answer}$$

The **rate** at which a given solute dissolves in a given solvent is dependent on three factors: (1) particle size of solute, (2) rate of stirring, and (3) temperature.

Particle Size

Smaller solute particles have a greater surface exposed to the solvent, and hence the rate of dissolution (dissolving) is faster. A solid lump of sugar will dissolve much more slowly than the same amount of sugar in small granular form, since the small granules have a greater *total* surface area exposed to the solvent molecules.

Rate of Stirring

The rate of dissolution can also be increased by stirring. This increases the rate of more direct contact of the solvent molecules that are not yet bound to solute particles with each of the solute particles. This increased rate of solution by stirring is the reason you stir your coffee or ice tea after you add sugar.

Temperature

Regardless of the effect of temperature on the solubility of a solute in a given solvent (see Figure 14-3), an increase in temperature results in an increase in the *rate* at which a given solute dissolves in a given solvent. This increase in rate is related to the increase in kinetic energy of the solute, solvent, and solution. With increased kinetic energy of the solute, the particles more readily break away from the other solute particles. With increased kinetic energy of the solvent, the solvent molecules can interact more often with the solute particles. Hence, the rate of dissolution is increased. As the solution is being prepared, the higher temperature increases the kinetic energy of the entire solution resulting in a faster motion of all particles, both solute and solvent. The effect of temperature on rate is quite evident in the sweetening of iced tea prepared from tea leaves and hot water. If sugar is added to the hot tea, dissolution is effected more readily than if the tea is allowed to cool and *then* the same amount of sugar is added.

14-4

Saturated, Unsaturated, and Supersaturated Solutions

A **saturated solution** is one which is in dynamic equilibrium ($\rightleftarrows$) with undissolved solute. The rate of dissolution of undissolved solute is equal to the rate of crystallization of dissolved solute, as shown:

$$\text{undissolved solute} \xrightleftharpoons[\text{rate of crystallization}]{\text{rate of dissolution}} \text{dissolved solute}$$

The concentration of solute in a saturated solution may vary, depending upon temperature and other factors, and is determined experimentally. A saturated solution can be prepared by adding an *excess* of solute to a given amount of solvent and allowing sufficient time for a maximum amount of solute to dissolve with excess solute present. Therefore, the solution formed *above* the undissolved solute is a saturated solution (see Figure 14-4a). To prepare a saturated solution of potassium chloride (solubility at $2\overline{0}°C$, 34 g of solute/$10\overline{0}$ g of water; Figure 14-3) in $10\overline{0}$ g of water at $2\overline{0}°C$, we must use an *excess* of solute, and therefore we would add more than 34 g of potassium chloride to $10\overline{0}$ g of water to assure the presence of excess solute. The water and potassium chloride should be *warmed* with stirring until saturation or near saturation is achieved at the elevated temperature. The solution (in contact with undissolved solute) is then cooled at 20°C to allow excess solute to crystallize. After stirring this solution is now saturated at 20°C.

An **unsaturated solution** is one in which the concentration of solute is *less* than that of the saturated (equilibrium) solution under the *same conditions*. In an unsaturated solution, the rate of dissolution of undissolved solute present is *greater* than the rate of crystallization of dissolved solute, so that, in time, no undissolved solute remains. An unsaturated solution may be prepared by dilution of a saturated solution or by using less solute than was dissolved in forming a given saturated solution. In the case of an unsaturated solution of potassium chloride, *less* than 34 g of potassium chloride per $10\overline{0}$ g of water at $2\overline{0}°C$ would be present. A solution consisting of $2\overline{0}$ g of potassium chloride in $10\overline{0}$ g of water is an unsaturated solution at $2\overline{0}°C$.

A **supersaturated solution** is one in which the concentration of solute is *greater* than that possible in a saturated (equilibrium) solution under the *same conditions*. Such a solution is unstable and will revert to a saturated solution if a "seed" crystal of the solute is added, the excess solute crystallizing out of solution. A supersaturated solution may be prepared by filtering the undissolved solute from a saturated solution at an *elevated* temperature and then

FIGURE 14-4

Saturated and supersaturated solutions. (a) A *saturated* solution. (b) Preparing a *supersaturated* solution: (1) filtering excess solute from a hot saturated solution; (2) cooling the hot saturated filtrate to form a supersaturated solution, if no crystallization occurs.

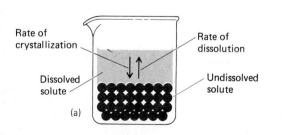

Rate of crystallization

Rate of dissolution

Dissolved solute

Undissolved solute

(a)

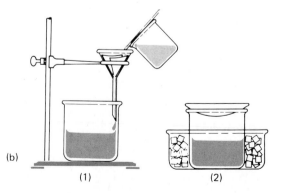

(b)

(1)   (2)

allowing this solution to cool. If crystallization does *not* occur during the cooling process, a supersaturated solution is obtained at the *lower* temperature (see Figure 14-4b). The addition of one "seed" crystal of solute to this solution will usually start crystallization of the excess solute, which will continue until a saturated solution is obtained at the lower temperature.

In the case of most solids dissolved in a liquid, crystallization of the excess solute occurs on cooling. In the case of a solution of potassium chloride saturated at $20°C$ (34 g/100 g of water), cooling to $0°C$ will result in the crystallization of 6 g of potassium chloride, since the solubility at $0°C$ is only 28 g in 100 g of water (see Figure 14-3). A solid that readily forms a supersaturated solution is sodium acetate.

The preparation of saturated solutions, and in some cases supersaturated solutions, is used to purify a particular solid by *crystallization*. Suppose that a given solid contains both water-soluble impurities and water-insoluble impurities, and that this solid is soluble in hot water but relatively insoluble in cold water. The solid may be purified by dissolving it in sufficient hot water to form a saturated (or nearly saturated) solution, filtering to remove the water insoluble impurities and finally cooling the filtrate. In the cooling process, the solid crystallizes out of solution and the water soluble impurities remain in solution. Since not all the impurities may be removed in one such operation, several *re*crystallizations may be needed to achieve a desired state of purity. The crystals of the purified solid are filtered (on a Büchner funnel), dried, and finally packaged. Sometimes, upon cooling a supersaturated solution results, and a "seed" crystal should be added to induce crystallization. This process of purification by crystallization of a solid is outlined in Figure 14-5. Industrially,

FIGURE 14-5

Purification of a solid by crystallization((1) dissolving the solid in water to form a solution; (2) filtering the solution to remove water-insoluble impurities; (3) cooling the filtrate to form crystals of purified solid with the water-soluble impurities staying in the solution; (4) filtering the crystals on a Büchner funnel; (5) allowing the crystals to dry.

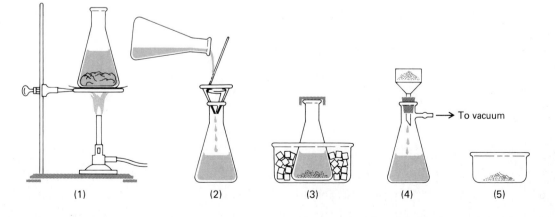

(1)   (2)   (3)   (4)   (5)

this process is carried out in the preparation of medicinal agents, such as aspirin (acetylsalicylic acid), on a large scale using automated equipment.

14-5

Concentration of Solutions

The terms "concentrated" (conc or concd) and "dilute" (dil) are sometimes used to express concentration, but these are at best very qualitative. Concentrated hydrochloric acid contains approximately 37 g of hydrogen chloride per $10\overline{0}$ g of solution, whereas concentrated nitric acid has approximately 72 g of hydrogen nitrate per $10\overline{0}$ g of solution. Dilute solutions are less concentrated, but beyond this little more can be said concerning them. A dilute solution of hydrochloric acid, for example, could be 1.00 g, 5.00 g, or 10.0 g of hydrogen chloride per $10\overline{0}$ g of solution, depending on the particular purpose intended for the acid. Obviously, more quantitative terms for expressing concentration must be used.

In the next five sections, we shall discuss the more common quantitative methods used to express the concentration of solutions. The particular method for expressing the concentration of a solution will generally be determined by the *eventual use* of the solution. These methods are as follows:

1. percent by mass

2. parts per million

3. molarity

4. normality

5. molality

Each method has an advantage over the other methods depending on the eventual use. For example, if you want to know the mass of salt in a given mass of ocean water, it is more convenient to express the concentration in percent by mass.

14-6

Percent by Mass

The **percent by mass** of a solute in a solution is the parts by mass of solute per $10\overline{0}$ parts by mass of *solution:*

$$\% \text{ by mass} = \frac{\text{mass of solute}}{\text{mass of } solution} \times 100$$

The mass of *solution* is equal to the mass of the *solute* plus the mass of the *solvent.* For example, a 20.0 percent solution of sodium sulfate would contain 20.0 g of sodium sulfate in $10\overline{0}$ g of solution (80.0 g of water), as shown in Figure 14-6.

Consider the following problem examples involving percent by mass:

FIGURE 14-6

A 20.0 percent by mass aqueous solution of sodium sulfate (Na_2SO_4).

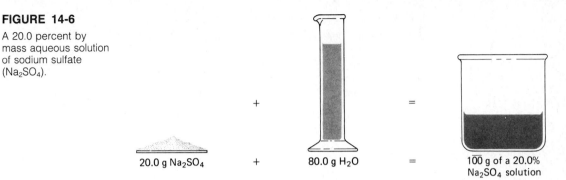

20.0 g Na_2SO_4 +   80.0 g H_2O = $10\overline{0}$ g of a 20.0% Na_2SO_4 solution

Problem Example 14-2

Calculate the percent sodium chloride if 19.0 g of sodium chloride is dissolved in enough water to make 175 g of solution.

SOLUTION: Since the total mass of the solution is 175 g, the percent of sodium chloride is readily obtained as

Work Problem 13 and 14.

$$\frac{19.0 \text{ g NaCl}}{175 \text{ g solution}} \times 100 = 10.9 \text{ parts NaCl per } 1\overline{0}\overline{0} \text{ parts of solution}$$

$$= 10.9\% \text{ NaCl} \quad Answer$$

Problem Example 14-3

Calculate the number of grams of sugar ($C_{12}H_{22}O_{11}$) that must be dissolved in 825 g of water to prepare a 20.0% sugar solution.

SOLUTION: In this solution, there would be 20.0 g of sugar for every 80.0 g of water (100.0 g of solution − 20.0 g of sugar = 80.0 g of water), and the number of grams of sugar needed for 825 g of water is calculated as

Work Problem 15.

$$825 \text{ g } H_2O \times \frac{20.0 \text{ g sugar}}{80.0 \text{ g } H_2O} = 206 \text{ g sugar needed for 825 g water} \quad Answer$$

Problem Example 14-4

Calculate the number of grams of water that must be added to 10.0 g of phenol to prepare a 2.00 percent aqueous phenol solution.

SOLUTION: In a 2.00 percent phenol solution, there are 2.00 g of phenol for every 98.0 g of water (100.0 g of solution − 2.00 g of phenol = 98.0 g of water), and the number of grams of water needed for 10.0 g of phenol is calculated as

Work Problems 16 and 17.

$$10.0 \text{ g phenol} \times \frac{98.0 \text{ g water}}{2.00 \text{ g phenol}} = 49\overline{0} \text{ g water needed for 10.0 g phenol} \quad Answer$$

In addition to percent by mass, it is occasionally convenient to express concentration as *percent by volume*. Percent by volume expresses concentration as parts by volume of the solute per $\overline{100}$ parts by volume of solution. This method is generally used to express the concentration of alcohol in alcoholic beverages. A wine may be 12.5 percent by volume, or there would be 12.5 mL of alcohol in $\overline{100}$ mL of the wine. This 12.5 percent by volume concentration of alcohol in wine is equivalent to only 10.0 percent by mass of alcohol. In chemical usage, concentrations expressed as percent are understood to mean *percent by mass*.

14-7

Parts per Million

In percent by mass the concentration was expressed as parts by mass of solute per $\overline{100}$ parts by mass of solution. In **parts per million (ppm)** the concentration is expressed as parts by mass of solute per 1,000,000 parts by mass of solution. The units of mass for the solute and solvent must be the same. This concentration is used for very dilute solutions such as water analysis or in biological preparations. In these very dilute solutions, the density of the solution is very near to that of water, with a density of the solution assumed to be 1.00 g/mL.

$$\text{parts per million (ppm)} = \frac{\text{mass of solute}}{\text{mass of solution}} \times 1,000,000$$

Problem Example 14-5

A water sample contains 3.5 mg of fluoride (F^{1-}) ions in 825 mL of the water sample. Calculate the parts per million (ppm) of the fluoride ion in the sample. (Assume that the density of the very dilute water sample is 1.00 g/mL.)

SOLUTION

Work Problems 18 and 19.

$$\frac{3.5 \text{ mg } F^{1-}}{825 \text{ mL sample}} \times \frac{1 \text{ mL sample}}{1.00 \text{ g sample}} \times \frac{1 \text{ g sample}}{1000 \text{ mg sample}} \times 1,000,000$$

$$= 4.2 \text{ ppm} \quad Answer$$

Problem Example 14-6

Calculate the number of mg of fluoride (F^{1-}) in 1.25 L of a water sample having 4.0 parts per million (ppm) fluoride ion. (Assume that the density of the very dilute water sample is 1.00 g/mL.)

SOLUTION

Work Problem 20.

$$1.25 \text{ L sample} \times \frac{1000 \text{ mL sample}}{1 \text{ L sample}} \times \frac{1.00 \text{ g sample}}{1 \text{ mL sample}} \times \frac{4.0 \text{ g F}^{1-}}{1{,}000{,}000 \text{ g sample}}$$

$$\times \frac{1000 \text{ mg F}^{1-}}{1 \text{ g F}^{1-}} = 5.0 \text{ mg F}^{1-} \quad Answer$$

14-8

Molarity

Molarity or molar concentration (abbreviated as M) is defined as the number of moles of solute per *liter* of *solution:*[1]

$$M = \text{molarity} = \frac{\text{moles of solute}}{\text{liter of } solution}$$

This method of expressing concentration is very useful when volumetric equipment (graduated cylinders, burets, etc.) is used to measure a quantity of the solution. From the volume measured, a simple calculation gives the mass of solute used.

To prepare one liter of a one-molar aqueous solution of sodium sulfate, one mole of sodium sulfate (142.1 g) is dissolved in water. *Enough* water is then added to bring the volume of the solution **to** *one liter* in a volumetric flask, as shown in Figure 14-7. An important point to note here is that no

FIGURE 14-7

A one molar (1.00 M) aqueous solution of sodium sulfate (Na_2SO_4).

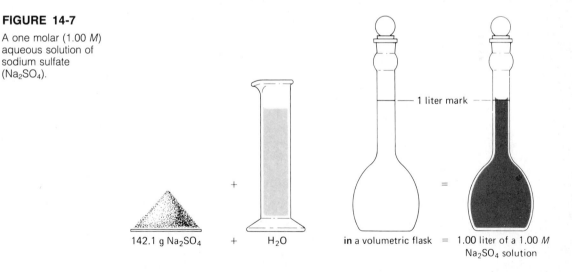

142.1 g Na_2SO_4 + H_2O in a volumetric flask   =   1.00 liter of a 1.00 M Na_2SO_4 solution

---

[1]Another term similar to molarity is "formality" (F). This term is used for solutions in which the solute exists as ions. In this text, we shall use the term "molarity" despite the type of bonding found in the solute.

information is stated as to the amount of solvent added, only that the solution is made to bring the total volume **to** one liter. The amount of water used can be calculated if the density of the solution is known.

Consider the following problem examples involving molarity:

Problem Example 14-7

(a) Calculate the molarity of an aqueous sodium chloride solution containing 284 g of NaCl in 2.20 L of solution. (b) Calculate the molarity of chloride (Cl^{1-}) in the solution.

SOLUTION

(a) The formula mass of NaCl is 58.5 amu; therefore, the molarity is calculated as

$$\frac{284 \text{ g NaCl}}{2.20 \text{ L solution}} \times \frac{1 \text{ mol NaCl}}{58.5 \text{ g NaCl}} = \frac{2.21 \text{ mol NaCl}}{1.00 \text{ L solution}} = 2.21 \text{ M} \quad Answer$$

(b) One mole of sodium chloride will form 1 mol of sodium ions and 1 mol of chloride ions, according to the following balanced equation:

$$NaCl_{(aq)} \rightarrow Na^{1+}_{(aq)} + Cl^{1-}_{(aq)}$$

In a 2.21 M sodium chloride solution there is 2.21 mol of sodium chloride per liter of solution. Hence, 2.21 mol of NaCl will form 2.21 mol of Na^{1+} and 2.21 mol of Cl^{1-}.

$$2.21 \text{ mol NaCl} \times \frac{1 \text{ mol } Cl^{1-}}{1 \text{ mol NaCl}} = 2.21 \text{ mol } Cl^{1-}$$

Work Problems 21 and 22.

The molarity of the chloride ion (Cl^{1-}) will be 2.21 M. *Answer*

Problem Example 14-8

(a) Calculate the number of grams of sodium chloride (NaCl) necessary to prepare $23\overline{0}$ mL of a 2.00 M aqueous sodium chloride solution. (b) Explain how this solution would be prepared.

SOLUTION

(a) The formula mass of NaCl is 58.5 amu. In a 2.00 M NaCl solution, there is 2.00 mol of NaCl per 1.00 L of solution. The number of grams of NaCl necessary for preparing $23\overline{0}$ mL of a 2.00 M solution is calculated as

$$23\overline{0} \text{ mL solution} \times \frac{1 \text{ L solution}}{1000 \text{ mL solution}} \times \frac{2.00 \text{ mol NaCl}}{1 \text{ L solution}} \times \frac{58.5 \text{ g NaCl}}{1 \text{ mol NaCl}}$$
$$= 26.9 \text{ g NaCl needed} \quad Answer$$

Work Problem 23.

(b) The sodium chloride (26.9 g) is dissolved in sufficient water to make the total volume of the solution equal to $23\overline{0}$ mL. *Answer*

Problem Example 14-9

Calculate the number of liters of a 6.00 M sodium hydroxide solution required to provide $41\overline{0}$ g of sodium hydroxide (NaOH).

SOLUTION: The formula mass of NaOH is 40.0 amu. In a 6.00 M NaOH solution, there is 6.00 mol of NaOH per 1.00 L of solution. The number of liters of 6.00 M solution necessary to provide $41\overline{0}$ g of NaOH is calculated.

Work Problem 24.

$$41\overline{0} \text{ g NaOH} \times \frac{1 \text{ mol NaOH}}{40.0 \text{ g NaOH}} \times \frac{1.00 \text{ L solution}}{6.00 \text{ mol NaOH}} = 1.71 \text{ L solution} \textit{Answer}$$

14-9

Normality

Normality (abbreviated as N) is defined as the number of equivalents (eq) of solute per liter of *solution:*

$$N = \text{normality} = \frac{\text{equivalents of solute}}{\text{liter of } \textit{solution}}$$

The equivalent mass in grams (one equivalent) of the solute is based on the reaction involved and is defined by either the acid-base concept or the oxidation-reduction concept, depending upon the ultimate use of the solution. However, in this text we shall limit our discussion of equivalents and normality to applications using the acid-base concept of equivalence.

   One equivalent of any acid is equal to the mass in grams of that acid capable of supplying 6.02×10^{23} (Avogadro's number; see 8-2) of hydrogen ions (1 mol). One equivalent of any base is equal to the mass in grams of that base that will combine with 6.02×10^{23} hydrogen ions (1 mol) or supply 6.02×10^{23} hydroxide ions (1 mol). Thus, *one equivalent of any acid will exactly combine with one equivalent of any base.* One equivalent of any salt is defined by the reaction the salt undergoes and is equal to the mass in grams of the salt capable of supplying 6.02×10^{23} positive charges or 6.02×10^{23} negative charges.

   The equivalent mass in grams (one equivalent) of an acid is determined by dividing the gram-formula mass of the acid by the number of moles of hydrogen ion per mole of acid *used in the reaction.* The equivalent mass in grams (one equivalent) of a base is determined by dividing the gram-formula mass of the base by the number of moles of hydrogen ions combining with one mole of the base *used in the reaction.* The equivalent mass in grams (one equivalent) of a salt is determined by dividing the gram-formula mass of the salt by the number of moles of positive or negative charges per mole of the salt *used in the reaction*. In all cases, the reaction must be considered.

Consider the following examples:

$$\left.\begin{array}{c}\text{One equivalent of } H_2SO_4 \text{ if}\\ \textbf{2 } \textbf{H}^{1+} \text{ are replaced}\end{array}\right\} = \frac{\text{gram-formula mass of } H_2SO_4}{2}$$

$$= \frac{98.1 \text{ g}}{2} = 49.0 \text{ g (equivalent mass)}^2$$

$$\left.\begin{array}{c}\text{One equivalent of } H_2SO_4 \text{ if}\\ \textbf{1 } \textbf{H}^{1+} \text{ is replaced}\end{array}\right\} = \frac{\text{gram-formula mass of } H_2SO_4}{1}$$

$$= \frac{98.1 \text{ g}}{1} = 98.1 \text{ g (equivalent mass)}$$

$$\left.\begin{array}{c}\text{One equivalent of } Ca(OH)_2 \text{ if}\\ \textbf{2 } \textbf{OH}^{1-} \text{ are replaced}\end{array}\right\} = \frac{\text{gram-formula mass of } Ca(OH)_2}{2}$$

$$= \frac{74.1 \text{ g}}{2} = 37.0 \text{ g (equivalent mass)}$$

$$\left.\begin{array}{c}\text{One equivalent of } Ca(OH)_2 \text{ if}\\ \textbf{1 } \textbf{OH}^{1-} \text{ is replaced}\end{array}\right\} = \frac{\text{gram-formula mass of } Ca(OH)_2}{1}$$

$$= \frac{74.1 \text{ g}}{1} = 74.1 \text{ g (equivalent mass)}$$

$$\left.\begin{array}{c}\text{One equivalent of } Na_2SO_4 \text{ if}\\ \textbf{2 } \textbf{Na}^{1+} \text{ are replaced}\end{array}\right\} = \frac{\text{gram-formula mass of } Na_2SO_4}{2}$$

$$= \frac{142.1 \text{ g}}{2} = 71.0 \text{ g (equivalent mass)}$$

$$\left.\begin{array}{c}\text{One equivalent of } Na_2SO_4 \text{ if}\\ \textbf{1 } \textbf{Na}^{1+} \text{ is replaced}\end{array}\right\} = \frac{\text{gram-formula mass of } Na_2SO_4}{1}$$

$$= \frac{142.1 \text{ g}}{1} = 142.1 \text{ g (equivalent mass)}$$

Since we are dividing the formula mass by whole numbers, a one normal solution (1.00 *N*) of a compound will then bear a certain whole number ratio to a one molar solution (1.00 *M*) of the same compound. One normal sodium

---

[2]Examples of reactions are as follows:

$$2 \text{ NaOH} + H_2SO_4 \rightarrow Na_2SO_4 + 2 H_2O \text{ (*both* } H^{1+} \text{ replaced)}$$

$$\text{NaOH} + H_2SO_4 \rightarrow NaHSO_4 + H_2O \text{ (*one* } H^{1+} \text{ replaced)}$$

FIGURE 14-8

A one normal (1.00 N) aqueous solution of sodium sulfate (Na_2SO_4).

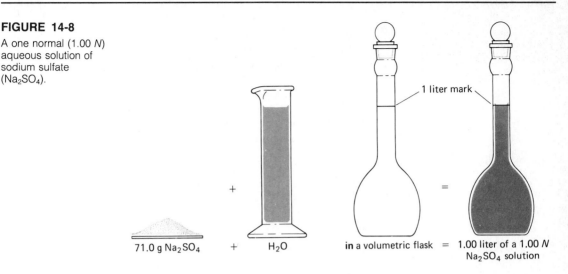

71.0 g Na_2SO_4 + H_2O in a volumetric flask   =   1.00 liter of a 1.00 N Na_2SO_4 solution

chloride (NaCl) solution converted to molarity would be one molar, since there is only one equivalent in one mole of sodium chloride. A one normal sodium sulfate (Na_2SO_4) solution replacing *both* sodium ions converted to molarity would be 0.500 *M*, because there are *two equivalents* of sodium sulfate in *one mole* of sodium sulfate.

$$\frac{1.00 \text{ eq } Na_2SO_4}{1 \text{ L solution}} \times \frac{1 \text{ mol } Na_2SO_4}{2 \text{ eq } Na_2SO_4} = \frac{0.500 \text{ mol } Na_2SO_4}{1 \text{ L solution}} = 0.500 \text{ M}$$

Notice that there will always be **ONE** mole of the substance with a variable whole number of equivalents to this one mole.

To prepare a one normal aqueous solution of sodium sulfate where *both* sodium ions are replaced, dissolve one equivalent (142.1 g/2 = 71.0 g) in water. Add *enough* water to bring the volume of the solution **to** *one liter* in a volumetric flask, as shown in Figure 14-8.

Consider the following problem examples involving normality:

Problem Example 14-10

Calculate the normality of an aqueous phosphoric acid solution containing 275 g of H_3PO_4 in 1.20 L of solution in reactions that replace all three hydrogen ions.

SOLUTION: The gram-formula mass of H_3PO_4 is 98.0 g, and since 3 mol of hydrogen ions is used per mole of the acid, 1 eq of H_3PO_4 is (98.0 g/3) = 32.7 g.

Work Problems 25 and 26.

$$\frac{275 \text{ g } H_3PO_4}{1.20 \text{ L solution}} \times \frac{1 \text{ eq } H_3PO_4}{32.7 \text{ g } H_3PO_4} = \frac{7.01 \text{ eq } H_3PO_4}{1.00 \text{ L solution}}$$

$$= 7.01 \text{ N} \quad \textit{Answer}$$

Problem Example 14-11

Calculate the number of grams of H_2SO_4 necessary to prepare $52\overline{0}$ mL of 0.100 N aqueous sulfuric acid solution in reactions that replace both hydrogen ions.

SOLUTION: The gram-formula mass of H_2SO_4 is 98.1 g, and since 2 mol of hydrogen ions is used per mole of the acid, 1 eq of H_2SO_4 is $(98.1 \text{ g}/2) = 49.0$ g. In a 0.100 N H_2SO_4 solution, there would be 0.100 of an equivalent of H_2SO_4 in 1.00 L of solution. Therefore, the number of grams of H_2SO_4 necessary for preparing $52\overline{0}$ mL of 0.100 N H_2SO_4 solution would be

Work Problem 27.

$$52\overline{0} \text{ mL solution} \times \frac{1 \text{ L solution}}{1000 \text{ mL solution}} \times \frac{0.100 \text{ eq } H_2SO_4}{1 \text{ L solution}}$$

$$\times \frac{49.0 \text{ g } H_2SO_4}{1 \text{ eq } H_2SO_4} = 2.55 \text{ g } H_2SO_4 \quad \textit{Answer}$$

Problem Example 14-12

Calculate the number of milliliters of sulfuric acid solution required to provide $12\overline{0}$ g of H_2SO_4 from a 2.00 N aqueous sulfuric acid solution in reactions that replace both hydrogen ions.

SOLUTION: A 2.00 N sulfuric acid solution would contain 2.00 eq/L of solution. Since 2 mol of hydrogen ions is used per mole of the acid, 1 eq of H_2SO_4 is equal $(98.1$ g/2$)$ to 49.0 g. Therefore, the number of milliliters is calculated as follows:

Work Problem 28.

$$12\overline{0} \text{ g } H_2SO_4 \times \frac{1 \text{ eq } H_2SO_4}{49.0 \text{ g } H_2SO_4} \times \frac{1 \text{ L solution}}{2.00 \text{ eq } H_2SO_4}$$

$$\times \frac{1000 \text{ mL solution}}{1 \text{ L solution}} = 1220 \text{ mL solution} \quad \textit{Answer}$$

14-10

Molality

Molality (abbreviated as m) is defined as the number of moles of solute per *kilogram* of **solvent**. This method of expressing concentration is based on the mass of solute (expressed as moles) per kilogram of **solvent:**

$$m = \text{molality} = \frac{\text{moles of solute}}{\text{kilogram of \textbf{solvent}}}$$

In the preparation of a one molal aqueous solution of sodium sulfate, 1 mol of sodium sulfate (142.1 g) would be dissolved in 1.000 kg ($100\overline{0}$ g) of water, as shown in Figure 14-9. Note that the total volume of the solution is not known. However, the mass of the solution is obtainable by adding the mass of the solute and the mass of the solvent. From a knowledge of the density of the solution we can calculate the total volume. In the expression of

FIGURE 14-9

A one molal (1.00 *m*) aqueous solution of sodium sulfate (Na_2SO_4).

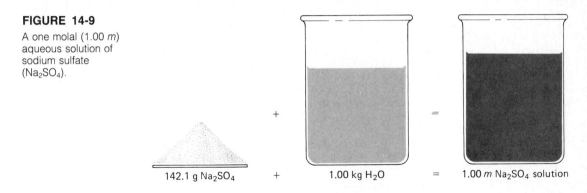

142.1 g Na_2SO_4 + 1.00 kg H_2O = 1.00 *m* Na_2SO_4 solution

concentration in terms of molality, the masses of solute and solvent must be known and their volumes are not involved.

In working problems involving molality, the use of factors relating mass of solute to mass of *solvent* and mass of solute to mass of *solution* will frequently be useful. The mass units for the solute in these factors may be expressed in grams or moles, and the mass units for the solvent or solution may be expressed in grams or kilograms. The factor and the units will be dependent on the application.

Consider the following problem examples involving molality:

Problem Example 14-13

Calculate the molality of a phosphoric acid solution containing 32.7 g of H_3PO_4 in $10\overline{0}$ g of water.

SOLUTION: The molality of the solution must express the concentration of H_3PO_4 as moles per kilogram of water. The formula mass of H_3PO_4 is 98.0 amu; hence, the molality is calculated

Work Problems 29 and 30.

$$\frac{32.7 \text{ g } H_3PO_4}{10\overline{0} \text{ g } H_2O} \times \frac{1 \text{ mol } H_3PO_4}{98.0 \text{ g } H_3PO_4} \times \frac{1000 \text{ g } H_2O}{1 \text{ kg } H_2O} = \frac{3.34 \text{ mol } H_3PO_4}{1 \text{ kg } H_2O}$$

$$= 3.34 \ m \quad Answer$$

Problem Example 14-14

Calculate the number of grams of glycerol ($C_3H_8O_3$) necessary to prepare $52\overline{0}$ g of a 2.00 *m* solution of glycerol in water.

SOLUTION: The molecular mass of glycerol is 92.0 amu. A 2.00 *m* glycerol solution would contain 2.00 mol (184 g) of glycerol in 1.000 kg ($100\overline{0}$ g) of water. The *total* mass of this solution would be 1184 g (2.00 mol $\times \dfrac{92.0 \text{ g}}{1 \text{ mol}} = 184$ g of glycerol $+ \overline{1000}$

g of water), and the mass of the glycerol necessary for $52\overline{0}$ g of a 2.00 m solution is calculated as

Work Problem 31.

$$52\overline{0} \text{ g solution} \times \frac{184 \text{ g glycerol}}{1184 \text{ g solution}} = 80.8 \text{ g glycerol} \quad \textit{Answer}$$

Problem Example 14-15

Calculate the number of grams of water that must be added to 5.80 g of H_3PO_4 in the preparation of a 0.100 m solution.

SOLUTION: The number of grams of water is asked for; therefore, the inverse factor for molality must be used. The formula mass of H_3PO_4 is 98.0 amu; therefore, the number of grams of water is calculated as

Work Problem 32.

$$5.80 \text{ g } H_3PO_4 \times \frac{1 \text{ mol } H_3PO_4}{98.0 \text{ g } H_3PO_4} \times \frac{1.00 \text{ kg } H_2O}{0.100 \text{ mol } H_3PO_4} \times \frac{1000 \text{ g } H_2O}{1 \text{ kg } H_2O}$$

$$= 592 \text{ g } H_2O \quad \textit{Answer}$$

Table 14-2 reviews the five different types of solutions discussed previously.

TABLE 14-2 Expressing Concentrations of Solutions[a]

$$\text{percent by mass} = \frac{\text{mass of solute}}{\text{mass of \textit{solution}}} \times 100$$

$$\text{parts per million (ppm)} = \frac{\text{mass of solute}}{\text{mass of \textit{solution}}} \times 1{,}000{,}000$$

$$M = \text{molarity} = \frac{\text{moles of solute}}{\text{liter of \textit{solution}}}$$

$$N = \text{normality} = \frac{\text{equivalents of solute}}{\text{liter of \textit{solution}}}$$

$$m = \text{molality} = \frac{\text{moles of solute}}{\text{kilogram of \textbf{solvent}}}$$

[a]You may find it helpful to make "flash cards" for these various methods of expressing concentrations of solutions.

14-11

Conversion of Concentration of Solutions

In some cases it is necessary to interconvert percent by mass, molarity, normality, and molality. Conversion of parts per million to these units[3] is usually not done, since the solution expressed in parts per million is extremely dilute

---

[3]Parts per million is sometimes converted to percent by mass. To convert parts per million to percent by mass, the decimal point is moved four places to the left, that is, 1,000,000/100 = 10,000. A concentration of 875 ppm converted to percent by mass is 0.0875 percent.

and concentrations in these other units would have little meaning. The following problem examples will illustrate the interconversion of these concentrations:

Problem Example 14-16

Calculate the percent by mass of the following aqueous solutions:

(a) 6.00 M sulfuric acid (H_2SO_4) solution (density of solution = 1.34 g/mL)

(b) 5.20 N sulfuric acid solution in reactions that replace both hydrogen ions (density of solution = 1.16 g/mL)

(c) 6.80 m sulfuric acid solution

SOLUTION

(a) In a 6.00 M sulfuric acid solution, there is 6.00 mol of H_2SO_4 in 1.00 L of solution. The formula mass of H_2SO_4 is 98.1 amu. In working problems of this type, there is a step requiring the conversion of volume of solution to mass of solution or vice versa. In this step, the density of the solution is used as a conversion factor and must be applied to the entire solution and not to some component (solute or solvent) of it. The density of the solution should have the units g of solution/mL of solution. The percent by mass is, therefore, calculated as follows using the density factor:

$$\frac{6.00 \text{ mol } H_2SO_4}{1 \text{ L solution}} \times \frac{1 \text{ L solution}}{1000 \text{ mL solution}} \times \frac{1 \text{ mL solution}}{1.34 \text{ g solution}} \times \frac{98.1 \text{ g } H_2SO_4}{1 \text{ mol } H_2SO_4}$$

$$\times \ 100 = 43.9\% \text{ by mass} \quad \textit{Answer}$$

(b) In a 5.20 N sulfuric acid solution, there is 5.20 eq of H_2SO_4 in 1.00 L of solution. In 1 eq of H_2SO_4 in which both hydrogen ions are replaced, there is 49.0 g. The percent by mass is therefore calculated as follows, using the density factor (1.16 g of solution/mL of solution):

$$\frac{5.20 \text{ eq } H_2SO_4}{1 \text{ L solution}} \times \frac{1 \text{ L solution}}{1000 \text{ mL solution}} \times \frac{1 \text{ mL solution}}{1.16 \text{ g solution}}$$

$$\times \ \frac{49.0 \text{ g } H_2SO_4}{1 \text{ eq } H_2SO_4} \times 100 = 22.0\% \text{ by mass} \quad \textit{Answer}$$

(c) In a 6.80 m sulfuric acid solution there is 6.80 mol of H_2SO_4 in 1.00 kg of water. The formula mass of H_2SO_4 is 98.1 amu. The *total* mass of this solution is calculated as follows:

$$\left(6.80 \text{ mol } H_2SO_4 \times \frac{98.1 \text{ g } H_2SO_4}{1 \text{ mol } H_2SO_4} \right) + 1\overline{000} \text{ g } H_2O$$

$$= 667 \text{ g } H_2SO_4 + 1\overline{000} \text{ g } H_2O = 1667 \text{ g solution}$$

Work Problem 33.   The percent by mass is calculated as follows:

$$\frac{667 \text{ g H}_2\text{SO}_4}{1667 \text{ g solution}} \times 100 = 40.0\% \text{ by mass} \quad Answer$$

Problem Example 14-17

Calculate the molarity of each of the following aqueous solutions:

(a) 9.00 percent nitric acid (HNO_3) solution (density of solution = 1.05 g/mL)

(b) 0.730 N nitric acid solution

(c) 1.50 m nitric acid solution (density of solution = 1.05 g/mL)

SOLUTION

(a) A 9.00 percent nitric acid solution contains 9.00 g of HNO_3 in $10\overline{0}$ g of solution. The formula mass of HNO_3 is 63.0 amu. The molarity is calculated using the density factor as follows:

$$\frac{9.00 \text{ g HNO}_3}{10\overline{0} \text{ g solution}} \times \frac{1 \text{ mol HNO}_3}{63.0 \text{ g HNO}_3} \times \frac{1.05 \text{ g solution}}{1 \text{ mL solution}} \times \frac{1000 \text{ mL solution}}{1 \text{ L solution}}$$

$$= 1.50 \ M \quad Answer$$

(b) In a 0.730 N nitric acid solution, there is 0.730 eq of HNO_3 in 1 L of solution. In 1 mol of HNO_3 there is just *one* equivalent of HNO_3; hence, the molarity is calculated as

$$\frac{0.730 \text{ eq HNO}_3}{1 \text{ L solution}} \times \frac{1 \text{ mol HNO}_3}{1 \text{ eq HNO}_3} = 0.730 \ M \quad Answer$$

(c) In a 1.50 m nitric acid solution, there is 1.50 mol of HNO_3 in 1.00 kg of H_2O. The *total* mass of the solution is

$$\left(1.50 \text{ mol HNO}_3 \times \frac{63.0 \text{ g HNO}_3}{1 \text{ mol HNO}_3} \right) + 100\overline{0} \text{ g H}_2\text{O}$$

$$= 94.5 \text{ g HNO}_3 + 100\overline{0} \text{ g H}_2\text{O} = 1094 \text{ g solution}$$

The molarity is calculated as follows:

Work Problem 34.
$$\frac{1.50 \text{ mol HNO}_3}{1094 \text{ g solution}} \times \frac{1.05 \text{ g solution}}{1 \text{ mL solution}} \times \frac{1000 \text{ mL solution}}{1 \text{ L solution}} = 1.44 \ M \quad Answer$$

Problem Example 14-18

Calculate the normality of each of the following aqueous solutions:

(a) 13.0 percent sulfuric acid solution in reactions where both hydrogen ions are replaced (density of solution = 1.09 g/mL)

(b) 2.75 M sulfuric acid solution in reactions where both hydrogen ions are replaced

(c) 4.10 m sulfuric acid solution in reactions where both hydrogen ions are replaced (density of solution = 1.21 g/mL)

SOLUTION

(a) A 13.0 percent sulfuric acid solution contains 13.0 g of H_2SO_4 in $1\overline{0}0$ g of solution. The formula mass of H_2SO_4 is 98.1 amu, and in 1 eq there is 49.0 g for reactions where both hydrogen ions are replaced. The normality is calculated as follows:

$$\frac{13.0 \text{ g } H_2SO_4}{1\overline{0}0 \text{ g solution}} \times \frac{1 \text{ eq } H_2SO_4}{49.0 \text{ g } H_2SO_4} \times \frac{1.09 \text{ g solution}}{1 \text{ mL solution}} \times \frac{1000 \text{ mL solution}}{1 \text{ L solution}}$$

$$= 2.89 \; N \quad Answer$$

(b) In a 0.730 N nitric acid solution there is 0.730 eq of HNO_3 in 1 L of solution. In 1 mol of HNO_3 there is just *one* equivalent of HNO_3; hence, the molarity is calculated as

$$\frac{0.730 \text{ eq } HNO_3}{1 \text{ L solution}} \times \frac{1 \text{ mol } HNO_3}{1 \text{ eq } HNO_3} = 0.730 \; M \quad Answer$$

(c) In a 4.10 m sulfuric acid solution, there is 4.10 mol of H_2SO_4 in 1.00 kg of H_2O. The *total* mass of the solution is calculated as follows:

$$\left(4.10 \text{ mol } H_2SO_4 \times \frac{98.1 \text{ g } H_2SO_4}{1 \text{ mol } H_2SO_4} \right) + 1\overline{0}00 \text{ g } H_2O$$

$$= 402 \text{ g } H_2SO_4 + 1\overline{0}00 \text{ g } H_2O = 1402 \text{ g solution}$$

The normality is calculated in reactions where both hydrogen ions are replaced as follows:

Work Problem 35.

$$\frac{4.10 \text{ mol } H_2SO_4}{1402 \text{ g solution}} \times \frac{2 \text{ eq } H_2SO_4}{1 \text{ mol } H_2SO_4} \times \frac{1.21 \text{ g solution}}{1 \text{ mL solution}}$$

$$\times \frac{1000 \text{ mL solution}}{1 \text{ L solution}} = 7.08 \; N \quad Answer$$

Problem Example 14-19

Calculate the molality of each of the following aqueous solutions:

(a) 12.0 percent phosphoric (H_3PO_4) solution

(b) 2.00 M phosphoric acid solution (density of solution = 1.10 g/mL)

(c) 3.60 N phosphoric acid solution in reactions that replace all three hydrogen ions (density of solution = 1.21 g/mL)

SOLUTION

(a) A 12.0 percent phosphoric acid solution contains 12.0 g of H_3PO_4 and 88.0 g of H_2O (100.0 g of solution − 12.0 g of H_3PO_4 = 88.0 g of H_2O). The formula mass of H_3PO_4 is 98.0 amu. The molality is calculated as follows:

$$\frac{12.0 \text{ g } H_3PO_4}{88.0 \text{ g } H_2O} \times \frac{1 \text{ mol } H_3PO_4}{98.0 \text{ g } H_3PO_4} \times \frac{1000 \text{ g } H_2O}{1 \text{ kg } H_2O} = 1.39 \ m \quad Answer$$

(b) In a 2.00 M phosphoric acid solution, there is 2.00 mol of H_3PO_4 in 1L of solution. The *total* mass of the solution is calculated as follows:

$$1.00 \text{ L solution} \times \frac{1000 \text{ mL solution}}{1 \text{ L solution}} \times \frac{1.10 \text{ g solution}}{1 \text{ mL solution}} = 11\overline{0}0 \text{ g solution}$$

The amount of water present in the solution is calculated as

$$11\overline{0}0 \text{ g solution} - \left(2.00 \text{ mol } H_3PO_4 \times \frac{98.0 \text{ g } H_3PO_4}{1 \text{ mol } H_3PO_4}\right)$$
$$= 11\overline{0}0 \text{ g solution} - 196 \text{ g } H_3PO_4 = 904 \text{ g } H_2O$$

Using the mass of water in the solution, the molality is calculated as follows:

$$\frac{2.00 \text{ mol } H_3PO_4}{904 \text{ g } H_2O} \times \frac{1000 \text{ g } H_2O}{1 \text{ kg } H_2O} = 2.21 \ m \quad Answer$$

(c) In a 3.60 N phosphoric acid solution, there is 3.60 eq of H_3PO_4 in 1 L of solution. The *total* mass of the solution is calculated as follows:

$$1.00 \text{ L solution} \times \frac{1000 \text{ mL solution}}{1 \text{ L solution}} \times \frac{1.21 \text{ g solution}}{1 \text{ mL solution}} = 1210 \text{ g solution}$$

In 1 eq of H_3PO_4, where all three hydrogen ions are replaced, there is 32.7 g of H_3PO_4 (98.0 g/3). The amount of water present in the solution is calculated as

$$1210 \text{ g solution} - \left(3.60 \text{ eq } H_3PO_4 \times \frac{32.7 \text{ g } H_3PO_4}{1 \text{ eq } H_3PO_4}\right)$$
$$= 1210 \text{ g solution} - 118 \text{ g } H_3PO_4 = 1092 \text{ g } H_2O$$

The molality is calculated as follows:

Work Problem 36.

$$\frac{3.60 \text{ eq } H_3PO_4}{1092 \text{ g } H_2O} \times \frac{1 \text{ mol } H_3PO_4}{3 \text{ eq } H_3PO_4} \times \frac{1000 \text{ g } H_2O}{1 \text{ kg } H_2O}$$

$$= 1.10 \ m \quad Answer$$

14-12

Colligative Properties of Solutions

Colligative properties of solutions are properties of solutions that depend only on the *number* of particles of solute present in the solution and not on the actual identity of these solute particles. Both sucrose $(C_{12}H_{22}O_{11})$ and urea (CH_4N_2O) act the same in a given solvent in regard to changing the boiling and freezing points of the solvent, but have completely different properties, such as solubility and density. These colligative properties are (1) vapor-pressure lowering, (2) boiling-point elevation, and (3) freezing-point depression. In this discussion we will consider only the effect of nonvolatile and nonionized solute particles on the solvent since volatile and ionized solute particles have a more complicated effect on the solvent.

Vapor-Pressure Lowering

If a nonvolatile solute is placed in a volatile solvent, the vapor pressure (see 12-3) of the volatile solvent is lowered from that of the pure solvent. The nonvolatile solute blocks the actual escape of the volatile solvent molecules from the surface of the liquid. By blocking the escape of the solvent molecules, the vapor pressure of the solvent is lowered. The same effect would occur with sucrose as with urea. For example, 1.00 mol of sucrose (342 g) placed in $10\overline{00}$ g of water will lower the vapor pressure of pure water at 25°C from 23.8 torr to 23.4 torr (lowered 0.4 torr); 1.00 mol of urea (60.0 g) will also lower the vapor pressure of $10\overline{00}$ g of pure water at 25°C from 23.8 torr to 23.4 torr.

Boiling-Point Elevation

Previously (see 12-4), we defined the boiling point of a liquid as the temperature at which the vapor pressure of the liquid is equal to the external pressure acting upon the liquid. In a solution of a nonvolatile solute in a volatile solvent, the vapor pressure is lowered; therefore, for the vapor pressure of the volatile solvent to reach the external pressure, more heat must be applied. This *raises* the boiling point of the solution over that of the pure volatile solvent. For example, 1.00 mol of sucrose in $10\overline{00}$ g of water (1.00 *m*) will raise the normal boiling point (at 1 atm) of water from 100.00°C to 100.52°C.

To calculate the boiling-point elevation over that of the pure solvent, we need to consider the molality of the solution (see 14-10) and the **molal boiling-point-elevation constant** (K_b). This constant for various solvents is given in Table 14-3. The units are °C per molal solution. For example, water has a K_b of 0.52 (0.52°C/*m*). This means that a 1.00 *m* solution of any nonvolatile solute in water will result in the elevation of the boiling point (at 1 atm) of pure water from 100.00°C to 100.52°C. The equation for calculating the boiling-point elevation is

$$\Delta T_b = \text{molality } (m) \cdot K_b \qquad (14-1)$$

with ΔT_b being the boiling-point elevation and K_b, the boiling-point-elevation constant for the volatile solvent.

TABLE 14-3 Molal Boiling-Point-Elevation and Freezing-Point-Depression Constants for Various Solvents

Solvent	Normal Boiling Point (°C)	K_b (°C/m)	Freezing Point (°C)	K_f (°C/m)
Water	100.00	0.52	0.00	1.86
Benzene	80.10	2.53	5.50	5.12
Ethyl alcohol (grain alcohol)	78.50	1.22	−117.30	1.99

Freezing-Point Depression

For a liquid to freeze the liquid particles must be "tied down" in the solid and hence the attractive forces increase in the solid state (see 12-9). In a solution of a nonvolatile solute in a solvent, the nonvolatile solute prevents the solvent molecules from developing strong attractive forces and hence prevents them from being "tied down." This means that to finally form a solid, heat must be removed and a lower temperature is needed. This *lowers* the freezing point of the solution over that of the pure solvent. For example, 1.00 mol of sucrose in 1000 g of water (1.00 molal) will lower the freezing point of water from 0.00°C to −1.86°C (see Table 14-3).

To calculate the freezing-point depression over that of the pure solvent, we need to consider the molality of the solution (14-8) and the **molal freezing-point-depression constant** (K_f). This constant for various solvents is given in Table 14-3. The units are °C per molal solution. For example, water has a K_f of 1.86 (1.86°C/m). This means that a 1.00 m solution of any nonvolatile solute will *lower* the freezing point of pure water from 0.00°C to −1.86°C. The equation for calculating the freezing-point lowering is

$$\Delta T_f = \text{molality } (m) \cdot K_f \tag{14-2}$$

with ΔT_f being the freezing-point lowering and K_f the freezing-point-depression constant for the solvent.

The principle of freezing-point depression is applied to melt ice on the highways or sidewalks in the winter. A salt is used, such as sodium chloride or calcium chloride. The freezing point of the mixture (ice-salt) is lower than that of pure ice and the ice melts if the outside temperature is not lower than the freezing point of the mixture. The principle of freezing-point depression and boiling-point elevation is applied in raising the boiling point and lowering the freezing point of the solution in the radiator of your car. The solute (ethylene glycol) lowers the freezing point of the solution in the winter and raises the boiling point of the solution in the summer. Since some of the additives in the antifreeze do break down, the radiator should be flushed and the antifreeze replaced from time to time.

Using Equations 14-1 and 14-2, we will now calculate the boiling-point elevation and the freezing-point depression of various solutions. Consider the following problem example:

Problem Example 14-20

Calculate the boiling point (1 atm) and the freezing point of the following solutions:

(a) a 2.50 m aqueous sugar $(C_{12}H_{22}O_{11})$ solution

(b) a sugar solution containing 4.27 g of sugar $(C_{12}H_{22}O_{11})$ in 50.0 g of water

(c) a nitrobenzene solution containing 9.75 g of nitrobenzene $(C_6H_5NO_2)$ in 175 g of benzene

(d) a urea solution containing 9.75 g of urea (CH_4N_2O) in $25\overline{0}$ g of ethyl alcohol

SOLUTION

(a) The boiling-point elevation is calculated as follows from Equation 14-1:

$$\Delta T_b = \text{molality } (m) \cdot K_b = 2.50 \, m \times \frac{0.52°C}{1 \, m} = 1.3°C$$

(The value of K_b for water is found in Table 14-3.) Therefore, the boiling point of the solution (1 atm) is 101.3°C (100.00°C + 1.3°C = 101.3°C). *Answer*
The freezing-point depression is calculated as follows from Equation 14-2:

$$\Delta T_f = \text{molality } (m) \cdot K_f = 2.50 \, m \times \frac{1.86°C}{1 \, m} = 4.65°C$$

(The value of K_f for water is found in Table 14-3.) Therefore, the freezing point of the solution is $-4.65°C$ (0.00°C $-$ 4.65°C = $-4.65°C$). *Answer*

(b) Before the boiling-point elevation and freezing-point depression can be calculated, the molality of the sugar solution must be determined. The molecular mass of sugar $(C_{12}H_{22}O_{11})$ is 342 amu, and the molality of this solution is calculated as follows:

$$\frac{4.27 \, g \, C_{12}H_{22}O_{11}}{50.0 \, g \, H_2O} \times \frac{1 \, \text{mol} \, C_{12}H_{22}O_{11}}{342 \, g \, C_{12}H_{22}O_{11}} \times \frac{1000 \, g \, H_2O}{1 \, \text{kg} \, H_2O}$$

$$= \frac{0.250 \, \text{mol} \, C_{12}H_{22}O_{11}}{1 \, \text{kg} \, H_2O} = 0.250 \, m$$

The boiling-point elevation is calculate as follows from Equation 14-1:

$$\Delta T_b = \text{molality } (m) \cdot K_b = 0.250 \, m \times \frac{0.52°C}{1 \, m} = 0.13°C$$

(The value of K_b for water is found in Table 14-3.) Therefore, the boiling point of the solution (1 atm) is 100.13°C (100.00°C + 0.13°C = 100.13°C). *Answer*

The freezing-point depression is calculated as follows from Equation 14-2:

$$\Delta T_f = \text{molality } (m) \cdot K_f = 0.250 \, m \times \frac{1.86°C}{1 \, m} = 0.465°C$$

(The value of K_f for water is found in Table 14-3). Therefore, the freezing point of the solution is $-0.46°C$ (0.00°C $-$ 0.465°C $=$ $-0.46°C$, expressed to the hundredths place). *Answer*

(c) Calculating the molality of the nitrobenzene ($C_6H_5NO_2$, molecular mass = 123.0 amu) solution as follows:

$$\frac{9.75 \text{ g } C_6H_5NO_2}{175 \text{ g benzene}} \times \frac{1 \text{ mol } C_6H_5NO_2}{123.0 \text{ g } C_6H_5NO_2} \times \frac{1000 \text{ g benzene}}{1 \text{ kg benzene}}$$

$$= \frac{0.453 \text{ mol } C_6H_5NO_2}{1 \text{ kg benzene}} = 0.453 \, m$$

The boiling-point elevation is calculated as follows from Equation 14-1:

$$\Delta T_b = \text{molality } (m) \cdot K_b = 0.453 \, m \times \frac{2.53°C}{1 \, m} = 1.15°C$$

(The value of K_b for benzene is found in Table 14-3.) Therefore, the boiling point of the benzene solution (1 atm) is 81.25°C (80.10°C + 1.15°C = 81.25°C). *Answer*

The freezing-point depression is calculated as follows from Equation 14-2:

$$\Delta T_f = \text{molality } (m) \cdot K_f = 0.453 \, m \times \frac{5.12°C}{1 \, m} = 2.32°C$$

(The value of K_f for benzene is found in Table 14-3). Therefore, the freezing point of the benzene solution is 3.18°C (5.50°C $-$ 2.32°C $=$ 3.18°C). *Answer*

(d) Calculating the molality of the urea (CH_4N_2O, molecular mass = 60.0 amu) solution as follows:

$$\frac{9.75 \text{ g } CH_4N_2O}{250 \text{ g ethyl alcohol}} \times \frac{1 \text{ mol } CH_4N_2O}{60.0 \text{ g } CH_4N_2O} \times \frac{1000 \text{ g ethyl alcohol}}{1 \text{ kg ethyl alcohol}}$$

$$= \frac{0.650 \text{ mol } CH_4N_2O}{1 \text{ kg ethyl alcohol}} = 0.650 \, m$$

The boiling-point elevation is calculated as follows from Equation 14-1:

$$\Delta T_b = \text{molality } (m) \cdot K_b = 0.650 \, m \times \frac{1.22°C}{1 \, m} = 0.793°C$$

(The value of K_b for ethyl alcohol is found in Table 14-3.) Therefore, the boiling point of this solution (1 atm) is 79.29°C (78.50°C + 0.793°C = 79.29°C, expressed to the hundredth place). *Answer*

The freezing-point depression is calculated as follows from Equation 14-2:

$$\Delta T_f = \text{molality } (m) \cdot K_f = 0.650 \, m \times \frac{1.99°C}{1 \, m} = 1.29°C$$

Work Problems 37 and 38.

(The value of K_f for ethyl alcohol is found in Table 14-3.) Therefore, the freezing point of this solution is $-118.59°C$ ($-117.30°C - 1.29°C = -118.59°C$). *Answer*

The boiling-point elevation and freezing-point depression can also be used to calculate the molecular mass of a compound. Consider the following problem example:

Problem Example 14-21

Calculate the molecular mass of a nonvolatile nonionized unknown given the following data:

(a) 4.35 g of the unknown was dissolved in $\overline{100}$ g of water with the resulting solution having a freezing point of $-1.31°C$.

(b) 5.65 g of the unknown was dissolved in 125 g of benzene with the resulting solution having a freezing point of 4.25°C.

(c) 6.50 g of the unknown was dissolved in 95.0 g of ethyl alcohol with the resulting solution having a boiling point of 79.55°C.

SOLUTION

(a) Since the freezing point of the solution is $-1.31°C$, the $\Delta T_f = 1.31$ [freezing point of water (Table 14-3) $-$ freezing point of the solution $= 0.00°C - (-1.31°C) = 1.31°C$] and the molality of the solution may be calculated from Equation 14-2 as follows:

$$\text{molality } (m) = \frac{\Delta T_f}{K_f} = 1.31°C \times \frac{1 \, m}{1.86°C} = 0.704 \, m = \frac{0.704 \text{ mol of unknown}}{1 \text{ kg water}}$$

(The value of K_f for water is found in Table 14-3.) Using the solution of the unknown in water and the molality (0.704 mol of unknown/kg of water), the molecular mass is calculated solving for *grams per mole* (g/mol) as follows:

$$\frac{4.35 \text{ g unknown}}{100 \text{ g } H_2O} \times \frac{1000 \text{ g } H_2O}{1 \text{ kg } H_2O} \times \frac{1 \text{ kg } H_2O}{0.704 \text{ mol}} = 61.8 \text{ g/mol, } 61.8 \text{ amu}   \textit{Answer}$$

(b) Calculating the $\Delta T_f = 1.25°C$ [freezing point of benzene $-$ freezing point of solution $= 5.50°C - 4.25°C = 1.25°C$] and the molality of the solution may be calculated from Equation 14-2 as follows:

$$\text{molality } (m) = \frac{\Delta T_f}{K_f} = 1.25°C \times \frac{1 \, m}{5.12°C} = 0.244 \, m = \frac{0.244 \text{ mol unknown}}{1 \text{ kg benzene}}$$

(The value of K_f for benzene is found in Table 14-3.) Using the solution of the unknown in benzene and the molality (0.244 mol of unknown/kg of benzene), the molecular mass is calculated solving for *grams per mole* (g/mol) as follows:

$$\frac{5.65 \text{ g unknown}}{125 \text{ g benzene}} \times \frac{1000 \text{ g benzene}}{1 \text{ kg benzene}} \times \frac{1 \text{ kg benzene}}{0.244 \text{ mol unknown}}$$

$$= 185 \text{ g/mol, } 185 \text{ amu} \quad \textit{Answer}$$

(c) The boiling point elevation (ΔT_b) is calculated as follows:

[boiling point of solution − boiling point of ethyl alcohol

$$= 79.55°C - 78.50°C = 1.05°C] = \Delta T_b$$

(Note that the boiling point of the solution is subtracted from that of the ethyl alcohol since we are calculating a boiling point elevation.) The molality of the solution may be calculated from Equation 14-1 as follows:

$$\text{molality } (m) = \frac{\Delta T_b}{K_b} = 1.05°C \times \frac{1 \text{ } m}{1.22°C} = 0.861 \text{ } m = \frac{0.861 \text{ mol unknown}}{1 \text{ kg ethyl alcohol}}$$

(The value of K_b for ethyl alcohol is found in Table 14-3.) Using the solution of the unknown in ethyl alcohol and the molality (0.861 mol of unknown/kg of ethyl alcohol), the molecular mass is calculated solving for *grams per mole* (g/mol) as follows:

Work Problems 39 and 40.

$$\frac{6.50 \text{ g unknown}}{95.0 \text{ g ethyl alcohol}} \times \frac{1000 \text{ g ethyl alcohol}}{1 \text{ kg ethyl alcohol}} \times \frac{1 \text{ kg ethyl alcohol}}{0.861 \text{ mol unknown}}$$

$$= 79.5 \text{ g/mol, } 79.5 \text{ amu} \quad \textit{Answer}$$

14-13

Colloids

Earlier in this chapter, we stated that colloids bridged the gap between solutions and suspensions. In **colloids,** the particles are dispersed without appreciable bonding to solvent molecules, and they do not settle out on standing. In this discussion we shall relate colloids to solutions.

In discussing solutions, we spoke of solutes and solvents, but in discussing colloids, we use the terms "dispersed particles" (dispersed phase) and "dispers*ing* medium" (dispers*ing* phase). **Dispersed particles** (dispersed phase) are the colloidal particles, comparable to the solute in a solution. **Dispersing medium** (dispers*ing* phase) is the substance in which the colloidal particles are distributed, comparable to the solvent in a solution. Milk is an example of a colloid; butterfat constitutes the dispersed particles, and water is the dispersing medium.

In a solution, the particles of solute and solvent are molecules, or ions with one or more solvent molecules bound to each solute particle. In colloids, the suspended particles are molecules or aggregates of molecules, *larger in size* than those found in *solutions* but *smaller* than those found in *suspensions*, and on the average with less than one molecule of dispersing medium bound per molecule of the dispersed particles. The colloidal particles are in various shapes and are in the range from 10 up to 2000 Å ($1Å = 10^{-10}$ m) in diameter. They can consist of one huge molecule, such as a starch molecule, or aggregates of *many* molecules.

As with solutions, colloids may exist in any one of the three physical states of matter. These different types of colloids have different names: a *foam* is a gas dispersed in a liquid or solid; a *liquid aerosol* is a liquid dispersed in a gas; a *solid aerosol* or a *smoke* is a solid dispersed in a gas; an *emulsion* is a liquid dispersed in a liquid; a *gel* is a liquid dispersed in a solid; and a *sol* is a solid dispersed in a liquid or solid. Table 14-4 lists these various types of colloids and gives some examples of each.

TABLE 14-4   Types of Colloids

Dispersed Particles	Dispersing Medium	Name	Example
Gas	Liquid	Foam	Foaming shaving cream
Gas	Solid	Solid foam	Styrofoam
Liquid	Gas	Liquid aerosol	Fog; mist
Liquid	Liquid	Emulsion	Milk; some pharmaceutical preparations, such as liniments and mineral oil emulsion in water; mayonnaise; butter; hand cream
Liquid	Solid	Gel	Jell-O; certain hair gels
Solid	Gas	Solid aerosol or smoke	Dust; smoke
Solid	Liquid	Sol	Latex paints
Solid	Solid	Solid sol	Gems, such as ruby, turquoise, garnet

Besides particle size, colloids have other identifying properties. Properties peculiar to colloids are (1) optical effect, (2) motion effect, (3) electrical charge effect, and (4) adsorption effect.

Optical Effect

When a relatively narrow beam of light is passed through a colloid, such as *dust* particles in the air (Table 14-3), the light is scattered by the dust particles and they appear in the beam as bright, tiny specks of light (see Figure 14-10). In a solution, the appearance is different. The scattering of light in a colloid is due to the reflection of the light by the larger colloidal particles producing a *visible* beam of light. No observable reflection is caused by the smaller solute

FIGURE 14-10

Optical effect (Tyndall effect) of colloids.

particles in a solution. Hence, a beam of light passing through a solution is *invisible*. This optical effect is named the **Tyndall effect,** after British physicist John Tyndall (1820–1893), who critically investigated it in 1869.

Motion Effect

If a colloid is viewed with a special microscope, the dispersed colloidal particle appears to move in a zigzag, random motion through the dispersing medium. The special microscope consists of a regular microscope focused on the colloid with a strong light source at right angles to the colloid. It is viewed against a black background. What are actually seen are reflections from the colloidal particles. The erratic, random motion (a "jittering dance") of the dispersed colloidal particles in the dispersing medium is due to bombardment of the dispersed colloidal particles by the medium, creating a zigzag, random motion, as shown in Figure 14-11. Since this reflection of light is due to the colloidal particle size, no such motion is seen in a solution; however, the solute and solvent particles are in continuous random motion. This motion effect of col-

FIGURE 14-11

Motion effect (Brownian movement) of colloids.

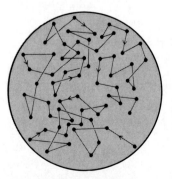

loids is one reason why colloidal particles do not settle out on standing. The constant bombardment by the dispersing medium particles on the dispersed colloidal particles keeps the dispersed colloidal particles suspended indefinitely. This effect, named after the English botanist Robert Brown (1773–1858), who first studied it with cytoplasmic granules in pollen grains in the summer of 1827, is called **Brownian movement.**

Electrical Charge Effect

A dispersed colloidal particle can *adsorb* (or bind) electrically charged particles (ions) on its *surface*. The charged species adsorbed on the surface of a given kind of colloidal particle may be either positive or negative. Colloidal particles of a given kind will *all* have the *same* sign of excess charge. Since like charges repel each other, the dispersed colloidal particles repel each other. In addition to the random motion of the particles (Brownian movement), this electrical charge effect also prevents the dispersed colloidal particles from coagulating and precipitating.

In colloidal dispersions of hydrated metal oxides, such as iron(III) hydroxide, the adsorbed ions are positive and the colloidal particles acquire a positive charge. Conversely, in the case of certain sulfides, such as arsenic(III) sulfide sol, the colloidal particles adsorb negative ions and thus acquire a negative charge. Dirt in river beds adsorbs negative ions and acquires a negative charge.

If a colloid of one charge comes in contact with a colloid of another charge or an ion of opposite charge, then the dispersed colloidal particles generally precipitate from the dispersing medium. Such is the case in the mixing of the two colloids, iron(III) hydroxide and arsenic(III) sulfide. When river water with negatively charged particles meets the ocean with positively charged particles (ionic salts), dirt is deposited, forming a fertile river delta.

The coagulation effect of an electrical charge on colloids is used in removing suspended particles from the effluent gases in industrial smokestacks. The device used in this operation is called a *Cottrell precipitator*, and smaller models of it are used as electronic air cleaners in homes. Airborne colloidal particles (dust, pollen, etc.) are effectively removed from the atmosphere in the home as follows. These air-borne particles pass through an "ionizing" section of the device where they are given an intense electrical charge. Next, the air carries them into the collecting section where they are attracted to metal plates under the influence of a powerful electrical field, as shown in Figure 14-12. The particles cling to these metal plates and the clean air is returned to the ventilation system.

Adsorption Effect

Colloids, due to their great surface area, have great adsorbing power. The property of adsorption is not limited to colloids, for other substances—such as charcoal, filter paper (cellulose), aluminum oxide, and bentonite—also have

FIGURE 14-12

An electronic air cleaner for homes (* are large airborne particles; dots are smaller airborne particles).

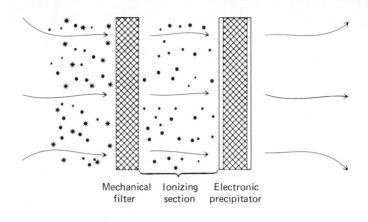

Mechanical filter — Ionizing section — Electronic precipitator

great adsorbing power. Charcoal has been used as an adsorbing agent in the brewing of beer in three different ways. In the first way, it is used to purify the water and to remove any odors or taste in the water. In the second way, it is used to clarify the beer and to adsorb the cloudy colloids that are formed during the fermentation process. And, in the third way, it is used to purify the carbon dioxide that is added to carbonate the beer. The carbon dioxide gives the beer a tang and its presence causes the beer to form a "head" and makes it foamy.

Chapter Summary

A *solution* is homogeneous throughout and is composed of two or more pure substances; its composition can be varied usually within certain limits. The *solute* is usually the component in the lesser quantity and the *solvent* is the component in the greater quantity. In this chapter, factors that affect the solubility and rate of solution of a solute in a solvent, including the reading of a graph of the solubility of a salt vs temperature and the solubility of a gas in a liquid with pressure (*Henry's Law*), were discussed. The terms saturated, unsaturated, and supersaturated solutions were distinguished. Application of these solutions to purification by crystallization of a solid was mentioned.

Five methods of expressing concentration of solutions are: (1) percent by mass, (2) parts per million, (3) molarity, (4) normality, and (5) molality. Problem examples involving each of these types were solved. In addition, problems involving the interconversion of concentrations of percent by mass, molarity, normality, and molality were given. The effects of a nonvolatile-nonionized solute on the boiling and freezing points of a volatile solvent were considered. This solute *raises* the boiling point and *lowers* the freezing point of the solvent. Using the effects of boiling-point raising and freezing-point lowering on a particular volatile solvent, the molecular mass of an unknown substance can be calculated.

Last, the types of colloids and their properties and the comparison of colloids with solutions and suspension were described.

EXERCISES

1. Define or explain
   (a) homogeneous matter (b) heterogeneous matter
   (c) solution (d) pure substance
   (e) colloid (f) solute
   (g) solvent (h) Henry's Law
   (i) saturated solution (j) unsaturated solution
   (k) supersaturated solution (l) percent by mass
   (m) parts per million (n) molality
   (o) molarity (p) normality
   (q) equivalent of an acid (r) equivalent of a base
   (s) equivalent of a salt (t) dispersed particles
   (u) dispersing medium (v) foam
   (w) liquid aerosol (x) solid aerosol or smoke
   (y) emulsion (z) gel
   (aa) sol (bb) Tyndall effect
   (cc) Brownian movement (dd) Cottrell precipitator

2. Distinguish between
   (a) homogeneous and heterogeneous matter
   (b) a solution and a pure substance
   (c) a colloid and a suspension
   (d) a colloid and a solution
   (e) solute and solvent
   (f) solubility and rate of solution
   (g) unsaturated and saturated solution
   (h) saturated and supersaturated solution
   (i) percent by mass and parts per million
   (j) molality and molarity
   (k) molarity and normality
   (l) molality and normality
   (m) dispersed particles and dispersing medium
   (n) foam and liquid aerosol
   (o) emulsion and sol
   (p) Tyndall effect and Brownian movement

Types of Solutions   (See Section 14-2)

3. Classify the following solutions according to states of matter involved:
   (a) salt in water
   (b) ammonia water
   (c) instant coffee in hot water
   (d) Prestone added to the cooling system of your car
   (e) glucose (a sugar) in the blood
   (f) alcohol in water
   (g) chlorine added to water for purification
   (h) oxygen dissolved in water
   (i) sulfur dioxide in the fluid surrounding mucous membranes, as in your nose and throat
   (j) sugar in water

Factors Affecting Solubility and Rate (See Section 14-3)

4. Classify the following compounds regarding relative solubility in water vs solubility in carbon tetrachloride; list the better solvent:
   (a) KCl
   (b) NaF
   (c) gasoline
   (d) $Pb(NO_3)_2$

Saturated, Unsaturated, and Supersaturated Solutions (See Section 14-4)

5. Describe in your own words the process of purification of a solid by crystallization.

6. How would you determine if a given solution were saturated, unsaturated, or supersaturated? Explain.

Colloids (See Section 14-13)

7. Classify the following colloids according to possible types and give the name of the colloid:
   (a) gold in water
   (b) soap foam
   (c) clouds
   (d) mayonnaise

8. Two solids, **A** and **B**, are dispersed in pure water. When a narrow beam of light is passed through **A** in aqueous medium, no visible light path is observed, but with **B** in aqueous medium, the path of the beam of light is visible. Which one of the two, **A** or **B**, forms a colloid and which one forms a solution?

9. Approximately 1000 years ago, a river delta was formed where the city of New Orleans is now located. This delta consists of deposited silt formerly held in colloidal dispersion in the Mississippi River. Explain why this silt from the Mississippi River was deposited at the point where the river empties into the salt-laden Gulf of Mexico.

PROBLEMS

Factors Affecting Solubility and Rate (See Section 14-3)

10. From Figure 14-3, determine the solubility of the following in grams per $\overline{100}$ g of water:
    (a) $CaCrO_4$ at $1\overline{0}°C$
    (b) $CaCrO_4$ at $8\overline{0}°C$
    (c) NaCl at $2\overline{0}°C$
    (d) KCl at $2\overline{0}°C$
    (e) KCl at $75°C$
    (f) KNO_3 at $3\overline{0}°C$

11. Calculate the solubility in grams per liter of oxygen in water at a partial pressure of 3.00 atm and 5°C. The solubility is 0.0132 g of oxygen per liter of water at 5°C and a *total* pressure of 1.000 atmosphere and 5°C. (See Appendix V.)

12. Calculate the solubility in grams per liter of nitrogen in water at a partial pressure of $76\overline{00}$ torr and 0°C. The solubility is 0.0292 g of nitrogen per liter of water at 0°C and a *total* pressure of $76\overline{0}$ of torr. (See Appendix V.)

Percent by Mass   (See Section 14-6)

13. Calculate the percent of the solute in each of the following aqueous solutions:
     (a)   8.50 g of sodium chloride in 95.0 g of solution
     (b)   25.2 g of potassium carbonate in 100.0 g of water
     (c)   3.88 g of calcium chloride in 78.50 g of water

14. Calculate the percent of the solute in each of the following aqueous solutions:
     (a)   13.7 g of sodium chloride in $11\overline{0}$ g of solution
     (b)   12.4 g of barium chloride in 80.7 g of water
     (c)   0.155 g of phenol (C_6H_6O) in 15.000 g of glycerol

15. Calculate the grams of solute that must be dissolved in
     (a) $35\overline{0}$ g of water in the preparation of a 17.0 percent potassium sulfate solu-
            tion
     (b)   15.0 g of water in the preparation of a 12.0 percent sodium chloride solution

16. Calculate the grams of water that must be added to
     (a)   16.0 g of sugar $(C_{12}H_{22}O_{11})$ in the preparation of a 23.0 percent sugar solu-
            tion
     (b)   4.00 g of potassium iodide in the preparation of a 1.90 percent potassium
            iodide solution

17. Calculate the number of grams of solution necessary to provide the following:
     (a)   68.3 g of sodium chloride from a 12.0 percent aqueous sodium chloride
            solution
     (b)   1.20 g of sodium bicarbonate from a 6.00 percent aqueous sodium bicar-
            bonate solution

Parts per Million   (See Section 14-7)

18. Calculate the parts per million (ppm) of the solute in each of the following
     aqueous solutions. (Assume that the density of the very dilute water sample is
     1.00 g/mL.)
     (a)   128 mg of sodium (Na^{1+}) ions in $75\overline{0}$ mL of a water sample
     (b)   172 mg of potassium (K^{1+}) ions in $85\overline{0}$ mL of a water sample
     (c)   2.5 mg of aluminum (Al^{3+}) ions in 1.5 L of ocean water

19. Calculate the parts per million (ppm) of the solute in each of the following
     aqueous solutions. (Assume that the density of the very dilute water sample is
     1.00 g/mL.)
     (a)   195 mg of sodium chloride (NaCl) in $30\overline{0}$ mL of a water sample
     (b)   6.5 mg of potassium (K^{1+}) in $5\overline{0}$ mL of a water sample
     (c) 2.7×10^{-3} mg of gold (Au) in 450 L of ocean water

20. Calculate the number of mg of the solute dissolved in the following. (Assume that
     the density of the very dilute water sample is 1.00 g/mL.)
     (a)   3.20 L of a water sample having 15 ppm strontium (Sr^{2+}) ions
     (b)   9.80 L of ocean water having 65 ppm bromide (Br^{1-}) ions
     (c)   15.0 L of ocean water having 3.0×10^{-4} ppm silver (Ag)

Molarity   (See Section 14-8)

21. Calculate the molarity of each of the following aqueous solutions:
    (a)   75.5 g of ethyl alcohol (C_2H_6O) in $45\overline{0}$ mL of solution
    (b)   2.65 g of sodium chloride (NaCl) in 40.0 mL of solution. Also calculate the molarity of the chloride (Cl^{1-}) ion.
    (c)   20.8 g of sugar ($C_{12}H_{22}O_{11}$) in 275 mL of solution

22. Calculate the molarity of each of the following solutions:
    (a)   22.0 g of sodium bromide (NaBr) in 850 mL of solution. Also calculate the molarity of the bromide (Br^{1-}) ion.
    (b)   12.0 g of calcium chloride ($CaCl_2$) in $64\overline{0}$ mL of solution. Also calculate the molarity of the chloride (Cl^{1-}) ion.
    (c)   15.0 g of barium bromide in 1150 mL of solution. Also calculate the molarity of the bromide (Br^{1-}) ion.

23. Calculate the number of grams of solute necessary to prepare the following aqueous solutions. Explain how each solution would be prepared.
    (a) $50\overline{0}$ mL of a 0.110 M sodium hydroxide (NaOH) solution
    (b) $25\overline{0}$ mL of a 0.220 M calcium chloride ($CaCl_2$) solution
    (c) $10\overline{0}$ mL of a 0.155 M sodium sulfate solution

24. Calculate the number of milliliters of aqueous solution required to provide the following:
    (a)   5.10 g of sodium bromide (NaBr) from a 0.100 M solution
    (b)   7.65 g of calcium chloride ($CaCl_2$) from a 1.40 M solution
    (c)   1.20 mol of H_2SO_4 from a 6.00 M solution

Normality   (See Section 14-9)

25. Calculate the normality of each of the following aqueous solutions:
    (a)   8.85 g of sodium hydroxide (NaOH) in $45\overline{0}$ mL of solution
    (b)   2.10 g of barium hydroxide [$Ba(OH)_2$] in $50\overline{0}$ mL of solution in reactions that replace both hydroxide ions
    (c)   65.5 g of H_3PO_4 in $25\overline{0}$ mL of solution in reactions that replace all three hydrogen ions

26. Calculate the normality of each of the following aqueous solutions:
    (a)   12.1 g of H_2SO_4 in $75\overline{0}$ mL of solution in reactions that replace both hydrogen ions
    (b)   14.1 g of sodium sulfate in 625 mL of solution in reactions that replace both sodium ions
    (c)   11.2 g of calcium chloride in $70\overline{0}$ mL of solution in reactions that replace both chloride ions

27. Calculate the number of grams of solute necessary to prepare the following aqueous solutions:
    (a) $26\overline{0}$ mL of a 0.0100 N sulfuric acid solution in reactions that replace both hydrogen ions
    (b)   145 mL of a 0.800 N phosphoric acid solution in reactions that replace all three hydrogen ions
    (c)   255 mL of a 0.0500 N calcium chloride solution in reactions that replace both chloride ions

28. Calculate the number of milliliters of aqueous solution required to provide the following:
    (a) 62.0 g of H_2SO_4 from a 4.00 N solution in reactions that replace both hydrogen ions
    (b) 78.5 g of calcium chloride from a 2.00 N solution in reactions that replace both chloride ions
    (c) 1.85 g of barium hydroxide from a 0.0400 N solution in reactions that replace both hydroxide ions

Molality (See Section 14-10)

29. Calculate the molality of each of the following solutions:
    (a) $17\overline{0}$ g of ethyl alcohol (C_2H_6O) in $65\overline{0}$ g of water
    (b) 3.50 g of H_2SO_4 in 12.0 g of water
    (c) 2.60 g of glucose ($C_6H_{12}O_6$) in $11\overline{0}$ g of water

30. Calculate the molality of each of the following solutions:
    (a) 12.6 g of ethylene glycol ($C_2H_6O_2$, Prestone) in 485 g of water
    (b) 28.0 g of calcium chloride ($CaCl_2$) in $62\overline{0}$ g of water
    (c) 2.40 mol of sugar ($C_{12}H_{22}O_{11}$) in $86\overline{0}$ g of water

31. Calculate the number of grams of solute necessary to prepare the following aqueous solutions:
    (a) $40\overline{0}$ g of a 0.400 m solution of ethyl alcohol (C_2H_6O)
    (b) $70\overline{0}$ g of a 0.500 m solution of sulfuric acid (H_2SO_4)
    (c) 425 g of a 3.20 m solution of ethylene glycol ($C_2H_6O_2$)

32. Calculate the number of grams of water that must be added to
    (a) 65.0 g of glucose ($C_6H_{12}O_6$) in the preparation of a 2.00 m solution
    (b) 95.0 g of sugar ($C_{12}H_{22}O_{11}$) in the preparation of an 8.00 m solution
    (c) 4.10 mol of H_2SO_4 in the preparation of a 12.0 m solution

Conversion of Concentration of Solutions (See Section 14-11)

33. Calculate the percent by mass of the following aqueous solutions:
    (a) 2.00 M sodium chloride (NaCl) solution (density of solution = 1.08 g/mL)
    (b) 12.0 N sulfuric acid (H_2SO_4) solution in reactions that replace both hydrogen ions (density of solution = 1.34 g/mL)
    (c) 2.75 m nitric acid solution

34. Calculate the molarity of each of the following aqueous solutions:
    (a) 25.0 percent calcium nitrate [$Ca(NO_3)_2$] solution (density of solution = 1.21 g/mL)
    (b) 0.200 N sulfuric acid solution in reactions that replace both hydrogen ions
    (c) 1.00 m sugar ($C_{12}H_{22}O_{11}$) solution (density of solution = 1.11 g/mL)

35. Calculate the normality of each of the following aqueous solutions:
    (a) 20.0 percent sodium hydroxide solution (density of solution = 1.22 g/mL)
    (b) 0.500 M phosphoric acid solution in reactions that replace all three hydrogen ions
    (c) 4.00 m sulfuric acid solution (density of solution = 1.20 g/mL) in reactions that replace both hydrogen ions

36. Calculate the molality of each of the following aqueous solutions:
    (a) 20.0 percent hydrochloric acid solution
    (b) 8.00 M ethyl alcohol (C_2H_6O) solution (density of solution = 0.941 g/mL)
    (c) 10.0 N sulfuric acid solution (density of solution = 1.282 g/mL) in reactions that replace both hydrogen ions

Colligative Properties of Solutions (See Section 14-12 and Table 14-3 for additional data)

37. Calculate the boiling point (1 atm) and freezing point of the following solutions:
    (a) a 1.65 m aqueous glucose ($C_6H_{12}O_6$) solution
    (b) a glucose solution containing 5.35 g of glucose ($C_6H_{12}O_6$) in 75.0 g of water
    (c) a glycerol solution containing 7.65 g of glycerol ($C_3H_8O_3$) in 125 g of ethyl alcohol

38. Calculate the boiling point (1 atm) and freezing point of the following solutions:
    (a) a 0.750 m aqueous urea (CH_4N_2O) solution
    (b) a urea solution containing 4.35 g of urea (CH_4N_2O) in $1\overline{1}0$ g of water
    (c) a urea solution containing 8.50 g of urea (CH_4N_2O) in $\overline{2}00$ g of ethyl alcohol

39. Calculate the molecular mass of a nonvolatile-nonionized unknown given the following data:
    (a) 4.20 g of the unknown was dissolved in 80.0 g of water with the resulting solution having a freezing point of −1.00°C
    (b) 3.20 g of the unknown was dissolved in 125 g of water with the resulting solution having a freezing point of −1.10°C
    (c) 3.65 g of the unknown was dissolved in $1\overline{2}0$ g of benzene with the freezing point of 4.10°C

40. Calculate the molecular mass of a nonvolatile-nonionized known given the following data:
    (a) 6.80 g of the unknown was dissolved in 135 g of water with the resulting solution having a freezing point of -1.25°C
    (b) 7.30 g of the unknown was dissolved in $\overline{2}00$ g of water with the resulting solution having a freezing point of −1.65°C
    (c) 4.32 g of the unknown was dissolved in 75.0 g of benzene with the resulting solution having a boiling point of 81.20°C (1 atm)

General Problems

41. Laboratory concentrated sulfuric acid is approximately 98.0 percent H_2SO_4 and its density is 1.83 g/mL. Calculate the molality, molarity, and normality (where both hydrogen ions react) of the sulfuric acid. Calculate the molarity of the hydrogen ions if both hydrogens in H_2SO_4 are completely ionized.

42. Concentrated hydrochloric acid is approximately 37.0 percent HCl and its density is 1.18 g/mL. Calculate the molality, molarity, and normality of the hydrochloric acid.

43. A nonvolatile-nonionized unknown gave on analysis: 40.0 percent carbon, 6.7 percent hydrogen, and 53.3 percent oxygen. A solution of this compound (10.2 g

in $1\overline{0}0$ g of water) had a freezing point of $-1.05°C$. Calculate the molecular formula of the nonvolatile unionized unknown. (*Hint:* See 8-5.)

Readings

Myers, Joel N., "Fog." *Sci. Am.*, Dec. 1968, v. 219, p. 74. An interesting article discussing the interrelation of fog and air pollution and methods used to dissipate fogs and inhibit fog formation, some of which are based on the properties of fog as a colloid.

Wilson, J. N., "Colloid and Surface Chemistry in Industrial Research." *J. Chem. Educ.*, 1962, v. 39, p. 187. A discussion of the industrial chemical problems encountered in colloids and surface; includes the colloidal problems encountered with latex paints.

Acids, Bases, and Ionic Equations

Acids and bases found in the home and kitchen: Amphojel (an antacid containing aluminum hydroxide); Drano (a drain opener containing sodium hydroxide); root beer (a soft drink containing carbonic acid); household ammonia, vinegar (a dilute solution of acetic acid); baking soda (sodium bicarbonate); boric acid; and lemon juice (contains citric acid).

TASKS

1. Memorize the Arrhenius and Brønsted-Lowry definitions for an acid and a base.

2. Memorize the properties of acids and bases in aqueous solutions.

3. (a) Memorize the definitions of pH, pOH, and their relations to each other.
   (b) Memorize the pH range for acidity, for basicity, and the neutral pH point.

4. Memorize the list of strong and weak electrolytes and nonelectrolytes as given in Table 15-4.

OBJECTIVES

1. Given the following terms, define each term and describe the distinguishing characteristic of each:
   (a) Arrhenius acid (Section 15-1)
   (b) Arrhenius base (Section 15-1)
   (c) ionization (Section 15-1)
   (d) dissociation (Section 15-1)
   (e) Brønsted-Lowry acid (Section 15-1)
   (f) Brønsted-Lowry base (Section 15-1)
   (g) amphiprotic (Section 15-1)
   (h) indicators (Section 15-1)
   (i) titration (Section 15-2)
   (j) electrolytes (Section 15-5)
   (k) nonelectrolytes (Section 15-5)
   (l) strong electrolytes (Section 15-5)
   (m) weak electrolytes (Section 15-5)
   (n) ionic equations (Section 15-6)
   (o) net ionic equation (Section 15-6)

2. Given the Arrhenius definitions and the formulas of various substances, classify them as acids or bases (Section 15-1, Problem 6).

3. Given the Brønsted-Lowry definitions and the formulas of various substances, classify them as acids, bases, or amphiprotic substances (Section 15-1, Problem 7).

4. (a) Given the results of a titration of a definite volume of an unknown concentration of an acid or base solution with a volume of another solution of known concentration, calculate the unknown

concentration. Given the results of a titration of a solution of unknown concentration of an acid or base solution with a known *mass* of a base or acid, calculate the unknown concentration. This calculated concentration may be expressed in molarity, percent by mass if the density of the solution is known, or normality if the use of the acid or base in the reaction is known.

(b) Given the volume and normality concentration of an acid or a base, the normality of another base or acid, the formulas of the compounds, and the use of the acid or base in the reaction, calculate the volume of the other base or acid needed in the reaction.

[Problem Examples 15-1, 15-2, 15-3, 15-4, 15-5, and 15-6, Problems 8, 9, 10, 11, 12, 13, 14, 15, and 16]

5. Given the hydrogen ion concentration of a solution in moles per liter and the Table of Logarithms, calculate the pH and pOH of the solution (Problem Examples 15-7, 15-8, and 15-9, Problems 17, 18, 19, and 20).

6. Given the pH or pOH of a solution and the Table of Logarithms, calculate either the hydrogen ion or the hydroxide ion concentration of the solution in moles per liter (Problem Examples 15-10, 15-11, and 15-12, Problems 21, 22, and 23).

7. Given the formulas of reactants and their states, complete and balance the equations, expressing them as total ionic equations and as net ionic equations. Indicate any precipitate by (s) and any gas by (g) (Problem Examples 15-13, 15-14, 15-15, 15-16, and 15-17, Problems 24 and 25).

In Chapter 7 (7-6), we introduced the terms "acid" and "base" in discussing nomenclature. Later, in Chapter 9 (9-11), we considered the reaction of an acid with a base as an example of a neutralization reaction. In this chapter, we shall define an acid and a base in more explicit terms and consider their properties more fully. The properties of acids and bases are important to your everyday life—even to your survival. A majority of the food you eat is acidic, but your blood is slightly basic. If the acid-base balance in the blood changes slightly in either direction, death may result.

After we have considered acids and bases and their properties, we shall consider **ionic equations.** In Chapter 9, we stated that equations may be written as molecular equations or as ionic equations, and we demonstrated how to balance and complete molecular equations. In this chapter, you will learn how to convert molecular equations to ionic equations.

15-1

Definitions and Properties of Acids and Bases

Two definitions of acids and bases we will consider in this text are:

1. Arrhenius definition

2. Brønsted-Lowry definition

Arrhenius Definition

The Arrhenius definition of acids and bases was proposed in 1884 by Svante August Arrhenius (1859–1927), a Swedish physicist and chemist. Arrhenius defined an **acid** as a substance that yields hydrogen ions (H^{1+}) when dissolved in *water*. The hydrogen ion is a bare proton, but in water solution it is hydrated and exists as a hydronium ion (H_3O^{1+}), as shown in the following equation:[1]

$$H^{1+}) \quad + \ \overset{..}{H_2O}: \quad \rightarrow \quad \left[\begin{array}{c} H \\ \diagdown \\ \diagup \overset{..}{O}-H \\ H \end{array} \right]^{1+} \tag{15-1}$$

hydrogen ion   water   hydronium ion (H_3O^{1+})

Some common examples of acids, according to Arrhenius' definition, are nitric acid (HNO_3) and acetic acid ($HC_2H_3O_2$).

Arrhenius proposed that a **base** is a substance that yields hydroxide ions (OH^{1-}) when dissolved in *water*. According to Arrhenius' definition, some common examples of bases are sodium hydroxide (NaOH) and calcium hydroxide [Ca(OH)$_2$)].

Work Problem 6.

Acids in the pure anhydrous state exist as highly polar covalent molecules. For example, hydrogen chloride is a gas, pure hydrogen sulfate is a viscous liquid, and pure hydrogen nitrate and glacial acetic acid are liquids. None of these common acids yield more than a few ionic species in the pure state but are composed nearly entirely of molecules that are highly polar in nature. The change of such substances into *ions* is accomplished by *dissolving* them in *water*. In the dissolving process, the highly polar water molecules react with the molecules of acid to form hydronium ions and the corresponding hydrated anion from the acid. This process is called **ionization** and refers to the *formation* of *ions* from atoms or molecules by the transfer of electrons. The following equations illustrate this process of *ionization*:

---

[1] Some investigations of the hydrated proton reveals that in moderate concentrations it also exists as the $H_5O_2^{1+}$ ion and as the $H_9O_4^{1+}$ ion in dilute solutions. This can be represented best as $H^{1+}_{(aq)}$.

$$HCl_{(g)} + H_2O_{(\ell)} \longrightarrow H_3O^{1+}_{(aq)} + Cl^{1-}_{(aq)} \qquad (15\text{-}2)$$

$$H_2SO_{4(\ell)} + H_2O_{(\ell)} \longrightarrow H_3O^{1+}_{(aq)} + HSO_4^{1-}_{(aq)} \qquad (15\text{-}3)$$

$$HNO_{3(\ell)} + H_2O_{(\ell)} \longrightarrow H_3O^{1+}_{(aq)} + NO_3^{1-}_{(aq)} \qquad (15\text{-}4)$$

$$HC_2H_3O_{2(\ell)} + H_2O_{(\ell)} \overset{\rightarrow}{\longleftarrow} H_3O^{1+}_{(aq)} + C_2H_3O_2^{1-}_{(aq)} \qquad (15\text{-}5)$$

Hydrochloric, sulfuric, and nitric acids are all nearly completely ionized in this process, but acetic acid undergoes only partial dissociation into ions and a dynamic equilibrium ($\rightleftarrows$) is established between the ionic species and the undissociated acetic acid molecules. (When the longer arrow is in one direction, it means that the dynamic equilibrium is displaced in that direction.)

Certain compounds of metals with nonmetals also exist as highly polar covalent molecules in the anhydrous condition. Examples of such substances are anhydrous aluminum chloride, iron(III) chloride, and tin(IV) chloride.[2] These compounds also react with water to yield hydrated cations and anions, as illustrated by the following equations:

$$Al_2Cl_{6(s)} + 12\ H_2O_{(\ell)} \rightarrow 2\ Al(H_2O)_6^{3+}_{(aq)} + 6\ Cl^{1-}_{(aq)} \qquad (15\text{-}6)$$

$$Fe_2Cl_{6(s)} + 12\ H_2O_{(\ell)} \rightarrow 2\ Fe(H_2O)_6^{3+}_{(aq)} + 6\ Cl^{1-}_{(aq)} \qquad (15\text{-}7)$$

$$SnCl_4 + 4\ H_2O \rightarrow Sn(H_2O)_4^{4+}_{(aq)} + 4\ Cl^{1-}_{(aq)} \qquad (15\text{-}8)$$

Basic compounds existing as covalent molecules that react with water to form ions in solution are also known. The most common of these is ammonia gas. Equation 15-9 illustrates its reaction with water. Note that this reaction gives an equilibrium similar to Equation 15-5.

$$NH_{3(g)} + H_2O \overset{\rightarrow}{\longleftarrow} NH_4^{1+}_{(aq)} + OH^{1-}_{(aq)} \qquad (15\text{-}9)$$

The substance, NH_4OH, does not exist. Ammonia (NH_3) in water gives a few ammonium ions (NH_4^{1+}) and hydroxide ions (OH^{1-}) but *no NH_4OH molecules*.

Compounds such as salts and alkali metal hydroxides, which are *ionic in the pure anhydrous state*, also dissolve in water to give ions in solution. In these cases, however, the anydrous compound exists as *ions in the crystalline state*, and the highly polar *water molecules* simply help *to break down the crystal structure by hydrating the ions present*. This process is called **dissociation** and refers to the *separation of ionic* substances into ions by the actions of the solvent. Equations 15-10 and 15-11 illustrate this process:

---

[2]At room temperature, aluminum chloride and iron(III) chloride exist as *dimers*—that is, $(AlCl_3)_2$ = Al_2Cl_6, and $(FeCl_3)_2$ = Fe_2Cl_6.

$$Na^{1+}OH^{1-}_{(s)} \xrightarrow{\text{in } H_2O_{(\ell)}} Na^{1+}_{(aq)} + OH^{1-}_{(aq)} \qquad (15\text{-}10)$$

$$Na^{1+}Cl^{1-}_{(s)} \xrightarrow{\text{in } H_2O_{(\ell)}} Na^{1+}_{(aq)} + Cl^{1-}_{(aq)} \qquad (15\text{-}11)$$

Brønsted-Lowry Definition

Arrhenius defined acids and bases in the specific case using *water* as the solvent for preparing the solutions. In 1923, Johannes N. Brønsted (1879–1947), a Danish chemist, and Thomas M. Lowry (1874–1936), an English chemist, independently proposed a more general definition of acids and bases. The Brønsted-Lowry concept defines an **acid** as any substance that can give or donate a proton (H^{1+}) to some other substance. A **base** is defined as any substance capable of receiving or accepting a proton from some other substance. In simple terms, an **acid** is a *proton donor*, whereas a **base** is a *proton acceptor*. Note that the definitions made here are independent of the solvent medium used in preparing a solution of the particular acid or base. According to the Brønsted-Lowry idea, *ions*, as well as *uncharged molecules*, may be acids or bases. Indeed, any substance that is an Arrhenius acid or base will also be a Brønsted-Lowry acid or base.

The use of a few examples may be helpful to illustrate this concept. In Equations 15-2 through 15-5, the molecules of HCl, H_2SO_4, HNO_3, and $HC_2H_3O_2$ are behaving as Brønsted-Lowry acids by donating protons to the *base, water,* which accepted them. In Equation 15-9, however, the water is behaving as an *acid* as it donates a proton to a molecule of ammonia, a base, in the forward reaction (reaction to the right). Ammonium ion behaves as an acid whereas hydroxide ion is a base in the reverse reaction (reaction to the left) of the equilibrium.

Other examples are shown in Table 15-1. Note that both *ions* and *molecules* are capable of acid and base behavior; furthermore, the reverse reaction is also an acid-base reaction.

Substances capable of behaving either as a Brønsted-Lowry acid or base are said to be **amphiprotic** (**amphi-** means "of *both* kinds"). Water, for exam-

TABLE 15-1 Brønsted-Lowry Acids and Bases[a]

Acid	Base	Acid	Base	
H_2SO_4 + H_2O	$\rightarrow$	H_3O^{1+}	+ HSO_4^{1-}	**(15-12)**
HSO_4^{1-} + H_2O	$\rightleftharpoons$	H_3O^{1+}	+ SO_4^{2-}	**(15-13)**
H_2SO_4 + Cl^{1-}	$\rightleftharpoons$	HCl	+ HSO_4^{1-}	**(15-14)**
H_3O^{1+} + OH^{1-}	$\rightleftharpoons$	H_2O	+ H_2O	**(15-15)**
NH_4^{1+} + Cl^{1-}	$\rightleftharpoons$	HCl	+ NH_3	**(15-16)**
NH_3 + NH_3	$\rightleftharpoons$	NH_4^{1+}	+ NH_2^{1-} (amide ion)	**(15-17)**

[a]The pair of arrows indicates the presence of a dynamic equilibrium. The longer arrow in one direction means that the dynamic equilibrium is displaced in that direction.

ple, behaves as a base (proton acceptor) toward hydrogen chloride (Equation 15-2) and as an acid toward ammonia (Equation 15-9). Certain ions, HSO_4^{1-} (Equations 15-12 and 15-13) and hydrogen carbonate (HCO_3^{1-}), are also amphiprotic. In the self-ionization of water (see 15-3) and ammonia (Equation 15-17), *one molecule* of the substance acts as a *base toward another* acting as an *acid*. When solid ammonium chloride is heated to form the vapor phase, an acid-base reaction takes place, as Equation 15-16 in Table 15-1 illustrates.

Work Problem 7.

Now, let us consider some properties of acids and bases in aqueous solutions, which you have probably encountered previously. Aqueous solutions of acids generally have the following properties:

1. *Tastes sour.* Citrus fruits (lemons, limes, etc.) have a sour taste due to the citric acid found in them. Dill pickles are sour due to the presence of vinegar (acetic acid). Sour milk contains lactic acid, which gives it its sour taste. A small amount of sodium hydrogen carbonate ($NaHCO_3$) is sometimes added to "sweeten" slightly sour milk. (In regard to tasting chemicals in general, do not do so unless you have previously been advised by a qualified chemist to the contrary. In any event, *never* taste concentrated acids.)

2. *Turns blue litmus paper red.* Litmus is an indicator. **Indicators** are compounds the color of which is affected by acid and base. Aqueous acid turns blue litmus paper red, whereas aqueous base turns red litmus blue.

3. *Neutralizes bases.* Acids react with bases, as was mentioned in 9-11.

Aqueous solutions of bases in general have the following properties:

1. *Tastes bitter.* Milk of magnesia [$Mg(OH)_2$] has a relatively bitter taste, which is sometimes masked with a mint flavor.

2. *Feels soapy or slick.* A dilute solution of sodium hydroxide has a soapy or slick feeling on the skin.

3. *Turns red litmus paper blue.*

4. *Neutralizes acids.*

15-2

Titration

In 9-11, we discussed a type of reaction called neutralization. In *neutralization*, an acid (or an acid oxide, 9-7) reacts with a base (or basic oxide, 9-7), with the formation of water in most cases being the driving force behind the reaction. This neutralization can be represented by a general equation,

$$HX + MOH \rightarrow MX + HOH \qquad (15\text{-}18)$$

where *HX* is an acid and *MOH* is a base. Water is one of the products.

Neutralization reactions are often carried out by a procedure called titration. **Titration** (with reference to neutralization) is a procedure for determining the *concentration* of an acid or base in a solution through the addition of a base or an acid of *known concentration* until the neutralization point or end point is reached, as shown by an indicator.[3] Indicators often used in titrations are methyl orange (red in acid and yellow in base), methyl red (red in acid and yellow in base), phenolphthalein (colorless in acid and red in base), and bromthymol blue (yellow in acid and blue in base), as shown in Table 15-2.

TABLE 15-2 Color Changes of Indicators

Indicator	Approximate pH at Which Color Changes[a]	Color in Acid	Color in Base
Methyl orange	4	Red	Yellow
Methyl red	5	Red	Yellow
Phenolphthalein	9	Colorless	Red
Bromthymol blue	7	Yellow	Blue

[a]The choice of an indicator depends on the pH (see 15-4) of the aqueous solution of the salt formed when the acid or base is neutralized.

In a titration, a measured amount of the acid or base of unknown concentration is placed in an Erlenmeyer flask and a drop or two of an appropriate indicator is added. To this solution, a solution of an acid or base (whichever is appropriate) of known concentration is slowly added from a buret until the color of the indicator just changes. This is called the *end point*. For example, we may want to determine the concentration of an aqueous solution of sodium hydroxide. A measured volume or mass of this solution is placed in an Erlenmeyer flask, a drop or two of the indicator—phenolphthalein—is added, and the solution turns red. Hydrochloric acid of known concentration is added slowly from a buret until the red color of the indicator *just* fades. Finally, a nearly colorless solution results at the end point, as shown in Figure 15-1. By measuring the volume of the hydrochloric acid solution used, we can determine the amount of sodium hydroxide in the unknown sample.

Consider the following problem examples:

Problem Example 15-1

In the titration of 30.0 mL, of sodium hydroxide solution of unknown concentration, 45.2 mL of 0.100 M hydrochloric acid was required to neutralize the sodium hydroxide solution to a phenolphthalein end point. Calculate the molarity of the sodium hydroxide solution.

---

[3]Instead of an indicator, an instrument called the pH meter can be used, as will be mentioned in 15-4. The pH meter is more accurate than the indicator.

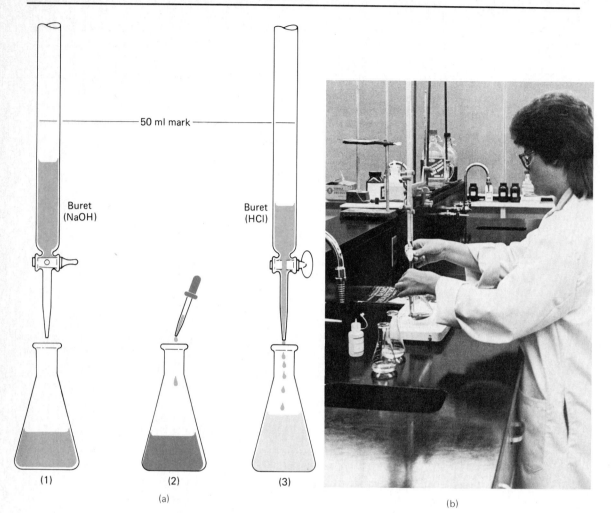

FIGURE 15-1

Titration of a sodium hydroxide solution of unknown concentration. (a) (1) Measure an exact amount of the sodium hydroxide solution of unknown concentration into an Erlenmeyer flask. (2) Add to this solution one or two drops of phenolphthalein (an indicator) to give a red solution. (3) From a buret add hydrochloric acid solution of known concentration until the red color just fades and a nearly colorless solution appears at the end point. (b) The actual titration. (Courtesty of David S. Seese)

SOLUTION: As with stoichiometry problems, the first thing we must know is the balanced equation:

$$NaOH + HCl \rightarrow NaCl + H_2O$$

Next, apply the stoichiometry procedure (Chapter 10) and the units of molarity to calculate the moles of sodium hydroxide neutralized with 45.2 mL of 0.100 M hydrochloric acid solution:

$$45.2 \text{ mL solution} \times \frac{1 \text{L}}{1000 \text{ mL}} \times \frac{0.100 \text{ mol HCl}}{1 \text{L solution}} \times \frac{1 \text{ mol NaOH}}{1 \text{ mol HCl}}$$

Refer to the
balanced equation

$$= 0.00452 \text{ mol NaOH}$$

Finally, calculate the concentration of the sodium hydroxide solution in moles per liter from the 30.0 mL of the sodium hydroxide solution that was used:

$$\frac{0.00452 \text{ mol NaOH}}{30.0 \text{ mL solution}} \times \frac{1000 \text{ mL}}{1 \text{ L}} = \frac{0.151 \text{ mol NaOH}}{1 \text{ L solution}} = 0.151 \ M \quad Answer$$

ALTERNATE SOLUTION: In working neutralization problems it is not necessary to work with the unit, moles. It is sometimes more convenient to work with millimoles (mmol, see 8-2) as shown below:

$$45.2 \text{ mL solution} \times \frac{0.100 \text{ mmol HCl}}{1 \text{ mL solution}} \times \frac{1 \text{ mmol NaOH}}{1 \text{ mmol HCl}} = 4.52 \text{ mmol NaOH}$$

$$\left[\text{The units, } \frac{\text{millimole (mmol) of solute}}{\text{milliliter (mL) of solution}}, \text{ are also equal to molarity } (M). \text{ Prove this!} \right]$$

Next, calculate the concentration of the sodium hydroxide in mmol/mL from the 30.0 mL of the sodium hydroxide solution that was used:

$$\frac{4.52 \text{ mmol NaOH}}{30.0 \text{ mL solution}} = \frac{0.151 \text{ mmol NaOH}}{1 \text{ mL solution}} = 0.151 \ M \quad Answer$$

Problem Example 15-2

Household ammonia is a dilute solution of ammonia in water. In the titration of 2.00 mL of household ammonia, 34.9 mL of 0.110 M hydrochloric acid solution was required to neutralize this solution to a methyl red end point. Calculate (a) the molarity and (b) the percent (density of solution = 0.985 g/mL) of the dilute ammonia solution.

SOLUTION

(a) Calculate the molarity of the NH_3. The equation is

$$NH_3 + HCl \rightarrow NH_4Cl$$

The moles of NH_3 are

$$34.9 \text{ mL solution} \times \frac{1 \text{L}}{1000 \text{ mL}} \times \frac{0.110 \text{ mol HCl}}{1 \text{ L solution}} \times \frac{1 \text{ mol NH}_3}{1 \text{ mol HCl}}$$

$$= 0.00384 \text{ mol NH}_3$$

and molarity of NH_3 is

$$\frac{0.00384 \text{ mol NH}_3}{2.00 \text{ mL solution}} \times \frac{1000 \text{ mL}}{1 \text{ L}} = \frac{1.92 \text{ mol NH}_3}{1 \text{ L solution}} = 1.92 \text{ M} \quad Answer$$

(b) Calculate the percent of the ammonia, using the density of the solution as 0.985 g/mL and the molecular mass of ammonia as 17.0 amu.

$$\frac{1.92 \text{ mol NH}_3}{1 \text{ L solution}} \times \frac{17.0 \text{ g NH}_3}{1 \text{ mol NH}_3} \times \frac{1 \text{ L}}{1000 \text{ mL}} \times \frac{1 \text{ mL solution}}{0.985 \text{ g solution}} \times 100$$

$$= 3.31\% \text{ NH}_3 \quad Answer$$

ALTERNATE SOLUTION

(a) Millimoles of ammonia:

$$34.9 \text{ mL solution} \times \frac{0.110 \text{ mmol HCl}}{1 \text{ mL solution}} \times \frac{1 \text{ mmol NH}_3}{1 \text{ mmol HCl}} = 3.84 \text{ mmol NH}_3$$

Molarity of ammonia:

$$\frac{3.84 \text{ mmol NH}_3}{2.00 \text{ mL solution}} = \frac{1.92 \text{ mmol NH}_3}{1 \text{ mL solution}} = 1.92 \text{ M} \quad Answer$$

(b) Calculate the percent of ammonia, using the density of the solution as 0.985 g/mL and the molecular mass of ammonia as 17.0 amu, but remembering that 1 *mmol* of NH_3 = 17.0 *mg* of NH_3.

$$\frac{1.92 \text{ mmol NH}_3}{1 \text{ mL solution}} \times \frac{17.0 \text{ mg NH}_3}{1 \text{ mmol NH}_3} \times \frac{1 \text{ mL solution}}{0.985 \text{ g solution}} \times \frac{1 \text{ g solution}}{1000 \text{ mg solution}}$$

$$\times 100 = 3.31\% \quad Answer$$

Problem Example 15-3

Pure sodium carbonate is used as a standard in determining the molarity of an acid. If 0.875 g of pure sodium carbonate was dissolved in water and the solution was titrated with 35.6 mL of hydrochloric acid to a methyl orange end point, calculate the molarity of the hydrochloric acid solution.

SOLUTION

Equation: $Na_2CO_3 + 2 HCl \rightarrow 2 NaCl + CO_2 + H_2O$

Moles of HCl (the formula mass of Na_2CO_3 is 106.0 amu):

$$0.875 \text{ g Na}_2\text{CO}_3 \times \frac{1 \text{ mol Na}_2\text{CO}_3}{106.0 \text{ g Na}_2\text{CO}_3} \times \frac{2 \text{ mol HCl}}{1 \text{ mol Na}_2\text{CO}_3} = 0.0165 \text{ mol HCl}$$

Refer to the
balanced equation

Molarity of HCl:

$$\frac{0.0165 \text{ mol HCl}}{35.6 \text{ mL solution}} \times \frac{1000 \text{ mL}}{1 \text{ L}} = \frac{0.463 \text{ mol HCl}}{1 \text{ L solution}} = 0.463 \text{ } M \quad Answer$$

ALTERNATE SOLUTION

Millimoles of HCl:

$$0.875 \text{ g Na}_2\text{CO}_3 \times \frac{1000 \text{ mg Na}_2\text{CO}_3}{1 \text{ g Na}_2\text{CO}_3} \times \frac{1 \text{ mmol Na}_2\text{CO}_3}{106.0 \text{ mg Na}_2\text{CO}_3} \times \frac{2 \text{ mmol HCl}}{1 \text{ mmol Na}_2\text{CO}_3} = 16.5 \text{ mmol HCl}$$

Refer to the
balanced equation

(Remember that 1 *mmol* of $Na_2CO_3 = 106.0$ *mg* of Na_2CO_3.)

Molarity of HCl:

$$\frac{16.5 \text{ mmol HCl}}{35.6 \text{ mL solution}} = \frac{0.463 \text{ mmol HCl}}{1 \text{ mL solution}} = 0.463 \text{ } M \quad Answer$$

Problem Example 15-4

If 1.51 g of potassium carbonate (100.0 percent pure) is titrated with 15.4 mL of hydrochloric acid to a methyl orange end point, calculate (a) the molarity and (b) the percent by mass (density of solution = 1.023 g/mL) of the hydrochloric acid solution.

SOLUTION

(a) Calculate the molarity of HCl. The balanced equation is

$$K_2CO_3 + 2 \text{ HCl} \rightarrow 2 \text{ KCl} + CO_2 + H_2O$$

The number of moles of HCl is

$$1.51 \text{ g K}_2\text{CO}_3 \times \frac{1 \text{ mol K}_2\text{CO}_3}{138.2 \text{ g K}_2\text{CO}_3} \times \frac{2 \text{ mol HCl}}{1 \text{ mol K}_2\text{CO}_3} = 0.0219 \text{ mol HCl}$$

Refer to the
balanced equation

(The formula mass of K_2CO_3 is 138.2 amu.)

The molarity is

$$\frac{0.0219 \text{ mol HCl}}{15.4 \text{ mL solution}} \times \frac{1000 \text{ mL}}{1 \text{ L}} = \frac{1.42 \text{ mol HCl}}{1 \text{ L solution}} = 1.42 \text{ } M \quad Answer$$

(b) Calculate the percent by mass of the hydrogen chloride, using the density of the solution as 1.023 g/mL and the molecular mass of HCl as 36.5 amu.

$$\frac{1.42 \text{ mol HCl}}{1 \text{ L solution}} \times \frac{36.5 \text{ g HCl}}{1 \text{ mol HCl}} \times \frac{1 \text{ L}}{1000 \text{ mL}} \times \frac{1 \text{ mL solution}}{1.023 \text{ g solution}} \times 100$$

$$= 5.07\% \quad \textit{Answer}$$

ALTERNATE SOLUTION

(a) Calculate the molarity of HCl. The number of mmol of HCl is calculated as follows:

$$1.51 \text{ g K}_2\text{CO}_3 \times \frac{1000 \text{ mg K}_2\text{CO}_3}{1 \text{ g K}_2\text{CO}_3} \times \frac{1 \text{ mmol K}_2\text{CO}_3}{138.2 \text{ mg K}_2\text{CO}_3} \times \frac{2 \text{ mmol HCl}}{1 \text{ mmol K}_2\text{CO}_3}$$

$$= 21.9 \text{ mmole HCl}$$

(Remember that 1 *mmol* of K_2CO_3 is equal to 138.2 *mg* of K_2CO_3.)

The molarity of HCl is calculated as follows:

$$\frac{21.9 \text{ mmol HCl}}{15.4 \text{ mL solution}} = \frac{1.42 \text{ mmol HCl}}{1 \text{ mL solution}} = 1.42 \ M \quad \textit{Answer}$$

(b) Calculate the percent by mass of the hydrogen chloride.

Work Problems 8, 9, 10, 11, 12, 13, and 14.

$$\frac{1.42 \text{ mmol HCl}}{1 \text{ mL solution}} \times \frac{36.5 \text{ mg HCl}}{1 \text{ mmol HCl}} \times \frac{1 \text{ mL solution}}{1.023 \text{ g solution}} \times \frac{1 \text{ g solution}}{1000 \text{ mg solution}}$$

$$\times 100 = 5.07\% \quad \textit{Answer}$$

In acid-base titrations the number of equivalents of acid is equal to the number of equivalents of base. This statement follows from our definition of *one* equivalent (see 14-9) of acid or base supplying 6.02×10^{23} hydrogen ions (if an acid) or 6.02×10^{23} hydroxide ions (if a base). Thus, *one equivalent of any acid will exactly combine with one equivalent of any base.* We can expand on this by using milliequivalents (meq), since a milliequivalent is $\frac{1}{1000}$ of an equivalent. Therefore, we have the following equation for the neutralization of an acid (A) by a base (B) or vice versa:

$$\text{equivalents (eq) A} = \text{equivalents (eq) B} \quad \text{or} \qquad (15\text{-}19)$$

$$\text{milliequivalents (meq) A} = \text{milliequivalents (meq) B} \qquad (15\text{-}20)$$

Using the definition of normality as $\dfrac{\text{equivalents}}{\text{liter of solution}}$, which is also equal to $\dfrac{\text{milliequivalents}}{\text{milliliter of solution}}$, we can write equation 15-20 as follows:

$$meq\ A = mL_A \cdot N_A\ (meq_A/mL_A)$$

$$meq\ B = mL_B \cdot N_B\ (meq_B/mL_B)$$

$$meq\ A = meq\ B$$

$$mL_A \cdot N_A = mL_B \cdot N_B \qquad (15\text{-}21)$$

Equation 15-21 can then be used to solve titration problems involving acids and bases.

Problem Example 15-5

In the titration of 34.5 mL of sodium solution hydroxide solution of unknown concentration, 27.5 mL of 0.100 N sulfuric acid solution was required to neutralize the sodium hydroxide in reactions where both hydrogen ions of the sulfuric acid react. Calculate the normality of the sodium hydroxide solution.

SOLUTION: From Equation 15-21, mL_A = 27.5 mL of H_2SO_4 solution, N_A = $\dfrac{0.100\ meq\ H_2SO_4}{mL\ H_2SO_4\ solution}$, and mL_B = 34.4 mL of NaOH solution, the normality of the sodium hydroxide (N_B) is calculated as follows:

$$27.5\ \cancel{mL\ H_2SO_4\ solution} \times \frac{0.100\ meq\ H_2SO_4}{\cancel{mL\ H_2SO_4\ solution}} = 34.5\ mL\ NaOH\ solution \times N_B$$

$$N_B = \frac{(27.5)(0.100)}{34.5} = \frac{0.0797\ meq\ NaOH}{mL\ NaOH\ solution} = 0.0797\ N \quad Answer$$

Problem Example 15-6

In the titration of a 0.160 N sodium hydroxide solution, 36.5 mL of 0.120 N phosphoric acid solution was required to neutralize the sodium hydroxide in reactions that replace all three hydrogen ions of the phosphoric acid. Calculate the number of mL of the sodium hydroxide solution needed in the reaction.

SOLUTION: From equation 15-21, mL_A = 35.6 mL of H_3PO_4 solution,

$$N_A = \frac{0.120\ meq\ H_3PO_4}{mL\ H_3PO_4\ solution} \quad and \quad N_B = \frac{0.160\ meq\ NaOH}{mL\ NaOH\ solution},$$

The number of mL of sodium hydroxide (mL_B) is calculated as follows:

$$35.6\ \cancel{mL\ H_3PO_4\ solution} \times \frac{0.120\ meq\ H_3PO_4}{\cancel{mL\ H_3PO_4\ solution}} = mL_B \times \frac{0.160\ meq\ NaOH}{mL\ NaOH\ solution}$$

Work Problems 15
and 16.

$$mL_B = \frac{(35.6)(0.120)}{0.160} = 26.7\ mL\ NaOH\ solution \quad Answer$$

15-3

Ionization of Water

In 15-2, we mentioned that the formation of water in most cases was the driving force behind a neutralization reaction. The reason for this is that water is *only slightly* ionized, as shown by its very slight conduction of an electric current using sensitive *instruments* (see 15-4). The following equations illustrate this slight ionization:

$$\text{HOH} + \text{H}_2\ddot{\text{O}}: \;\rightleftharpoons\; \underset{\substack{\text{hydronium}\\\text{ion}}}{\text{H}_3\text{O}^{1+}} + \underset{\substack{\text{hydroxide}\\\text{ion}}}{\text{OH}^{1-}} \tag{15-22}$$

or, simply,

$$\text{HOH} \rightleftharpoons \underset{\substack{\text{hydrogen}\\\text{ion}}}{\text{H}^{1+}_{\ (aq)}} + \underset{\substack{\text{hydroxide}\\\text{ion}}}{\text{OH}^{1-}_{\ (aq)}} \tag{15-23}$$

Equations 15-22 and 15-23 represent the equilibrium of water with its respective ions, H_3O^{1+} or $\text{H}^{1+}_{(aq)}$ and $\text{OH}^{1-}_{(aq)}$. Equilibrium systems will be discussed in more detail in Chapter 17.

In 1 L of pure water at 25°C, there are 1×10^{-7} mol/L of hydrogen ion (hydronium ion) and 1×10^{-7} mol/L of hydroxide ion. In any aqueous solution at 25°C, the product of the concentration of the hydrogen ion in moles per liter and the concentration of the hydroxide ion in moles per liter is *equal* to a *constant*, K_w. This constant K_w is called the *ion product constant* for water and includes the concentration of water. It is equal to 1×10^{-14}. Hence, for *any* aqueous solution at 25°C,

$$[\text{H}^{1+}]\,[\text{OH}^{1-}] = K_w = 1 \times 10^{-14} \;(\text{mol}^2/\text{L}^2) \tag{15-24}$$

where the brackets represent the concentration in moles per liter of the substance whose formula is enclosed in the brackets. Changes in the hydrogen ion concentration will result in a corresponding change in the hydroxide ion concentration, such that the product of these concentrations in any aqueous solution at 25°C will always be equal to the constant K_w, 1×10^{-14} (mol^2/L^2).

If a solution has a *hydrogen ion* concentration *larger* than 1×10^{-7} mol/L (that is, a smaller negative exponent), such as 1.0×10^{-5} mol/L, the solution is termed "acidic." If the *hydroxide ion* concentration is *larger* than 1×10^{-7} mol/L, such as 1.0×10^{-5} mol/L, or the *hydrogen ion* concentration is *less* than 1×10^{-7} mol/L liter, such as 1.0×10^{-9} mol/L, the solution is called "basic." If the product of the hydrogen ion concentration in moles per liter $[\text{H}^{1+}]$ and the hydroxide ion concentration in moles per liter $[\text{OH}^{1-}]$ is assumed to equal 1×10^{-14}, a neutral solution can occur only when the hydrogen ion concentration is **equal** to the hydroxide ion concentration—that is, when *each* is equal to 1×10^{-7} mol/L. Hence, using the assumption above, we have

$$[H^{1+}] > 1 \times 10^{-7} \text{ mol/L} = \textbf{acidic} \quad (> = \text{greater than})$$
$$[OH^{1-}] > 1 \times 10^{-7} \text{ mol/L or}$$
$$[H^{1+}] < 1 \times 10^{-7} \text{ mol/L} = \textbf{basic} \quad (< = \text{less than})$$
$$[H^{1+}] = [OH^{1-}] = 1 \times 10^{-7} \text{ mol/L} = \textbf{neutral}$$

Carbonated soft drinks have a hydrogen ion concentration of approximately 1×10^{-4} mol/L; therefore, they are acidic.

15-4

pH and pOH

For convenience, hydrogen ion concentration is expressed in terms of pH, which is defined as the negative logarithm of the hydrogen ion concentration:[4]

$$\textbf{pH} = -\log [H^{1+}] \tag{15-25}$$

The hydroxide ion concentration may be expressed in terms of pOH, which is defined as the negative logarithm of the hydroxide ion concentration:

$$\textbf{pOH} = -\log [OH^{1-}] \tag{15-26}$$

The sum of the pH and pOH for any solution will equal 14 (see Equation 15-28). This expression may be derived as follows. For an aqueous solution at 25°C from Equation 15-24,

$$[H^{1+}] [OH^{1-}] = K_w = 1 \times 10^{-14} \tag{15-27}$$

---

[4]The *common logarithm of a number* is defined as the **exponent** to which 10 must be raised to give the number (N), as $N = 10^{\text{exponent} = (\log N)}$. Thus, $\log 10^3 = 3$, $\log 10^{-4} = -4$, $\log 1 = 0$ (since 1 $= 10^0$ or any nonzero number raised to the 0 power). A number that cannot be represented as 10 raised to an integer power will have a logarithm which is not an integer. Logarithms of such numbers may be obtained from Logarithms of Numbers, Appendix VI. For example, the logarithm of 4.40 is given in Appendix VI as 0.6435. Read down the N column to 4.4, and then go the 0 column. Therefore, $4.40 = 10^{0.6435}$. Consider another example: the logarithm of 4.45 is 0.6484. Again, read down the N column to 4.4, and then go to the 5 column. Therefore, $4.45 = 10^{0.6484}$. Consider one more example: the logarithm of 8.62 is 0.9355. Again, read down the N column to 8.6, and then go to the 2 column. Therefore, $8.62 = 10^{0.9355}$. The logarithm of a number can also be obtained from many electronic calculators. For calculators with a logarithm key (log), enter the number and then press the log key, but be sure to round off your number to the appropriate number of significant digits.

The logarithm of the product of two numbers is equal to the **sum** of the logarithms of the numbers; therefore, from Equation 15-27,

$$\log [H^{1+}] + \log [OH^{1-}] = \log K_w = \log 1 \times 10^{-14}$$
$$= \log 1 + \log 10^{-14}$$
$$= 0 + (-14) = -14$$

Therefore,

$$\log [H^{1+}] + \log [OH^{1-}] = \log K_w = -14$$

Multiplying both sides of the equation by -1 gives

$$(-1)\{\log [H^{1+}] + \log [OH^{1-}]\} = (-1)(\log K_w) = (-1)(-14)$$
$$(-\log [H^{1+}]) + (-\log [OH^{1-}]) = (-\log K_w) = 14$$

Hence, from our definitions of pH and pOH (see Equations 15-25 and 15-26), and a corresponding definition for $pK_w = -\log K_w$,

$$\mathbf{pH + pOH = pK}_w = 14 \qquad (15\text{-}28)$$

The relationship between $[H^{1+}]$, $[OH^{1-}]$, pH, and pOH for aqueous solutions is illustrated in Table 15-3, along with the pH of some common examples. From this table you should note that the pH range for acidic and basic solutions is

$$0 \longleftrightarrow 7 \longleftrightarrow 14$$
$$\uparrow$$
$$\textbf{Acidic} \quad \textbf{Neutral} \quad \textbf{Basic}$$

In the table, you will also notice that the pH of the blood is slightly *basic*, with a pH range of a mere 0.20 pH unit (7.3 to 7.5). If the pH of the blood goes much below 7.3, acidosis occurs; if it falls below 7.0, death may ensue. If the pH goes above 7.5, alkalosis occurs; if it goes above 7.8, death may result. The pH of the blood is maintained in this narrow range by *buffers*, which are solutions of substances preventing a rapid change in pH. Three main types of buffers in the blood are

1. carbonic acid (H_2CO_3) and sodium hydrogen carbonate ($NaHCO_3$)

2. sodium dihydrogen phosphate (NaH_2PO_4) and disodium hydrogen phosphate (Na_2HPO_4)

3. certain proteins

TABLE 15-3 Relationship Between $[H^{1+}]$, $[OH^{1+}]$, pH, and pOH; and pH of Some Common Examples

	$[H^{1+}]$ (mol/L)	$[OH^{1-}]$ (mol/L)	pH^a	pOH	Acid or Base Strength	Common Examples (Approximate pH Range)
ACIDIC	10^0 (1)	10^{-14}	0	14	Strongly acidic	1 *M* HCl (0)
	10^{-1}	10^{-13}	1	13		Gastric juice (1-3)
	10^{-2}	10^{-12}	2	12		Limes (1.8-2.0)
						Soft drinks (2.0-4.0)
						Lemons (2.2-2.4)
	10^{-3}	10^{-11}	3	11	Weakly acidic	Dill pickles (3.2-3.6)
						Acid rain (3-4)
	10^{-4}	10^{-10}	4	10		
	10^{-5}	10^{-9}	5	9		Urine (4.5-8.0)
	10^{-6}	10^{-8}	6	8		Sour milk (6.0-6.2)
						Milk (6.5-6.7)
						Saliva (6.5-7.5)
	10^{-7}	**10^{-7}**	**7**	**7**	**Neutral**	Blood (7.3-7.5)
	10^{-8}	10^{-6}	8	6	Weakly basic	
	10^{-9}	10^{-5}	9	5		
BASIC	10^{-10}	10^{-4}	10	4		Milk of magnesia (9.9-10.1)
	10^{-11}	10^{-3}	11	3		Household ammonia (11.5-12.0)
	10^{-12}	10^{-2}	12	2	Strongly basic	
	10^{-13}	10^{-1}	13	1		
	10^{-14}	10^0 (1)	14	0		1 *M* NaOH (14)

^{a}The *normal* pH range is from 0 to 14, although solutions with negative pH (to -2) exist, as do solutions having a pH greater than 14 (to 16).

Acid rain (pH 3–4) has been blamed for destroying the forests in northeastern United States and southeastern Canada (see Figure 15-2). This acid rain is a dilute solution of acids, including sulfuric and nitric acids. The reaction of sulfur oxides and nitrogen oxides with water produces the oxyacids (see 9-7). These sulfur oxides and nitrogen oxides are produced from industrial pollution and automobiles, respectively.

Now, with some knowledge of pH and its applications, let us consider the calculation of pH and pOH, given the hydrogen ion concentration in moles per liter.

FIGURE 15-2

Dead spruce trees on Camel Hump in Vermont's Green Mountains. Evidence suggests that acid rain may be the cause of the destruction of our forests in northeastern United States and southeastern Canada (Hubert W. Vogelmann, *Natural History*, November 1982). Courtesy David Like, Botany Department, University of Vermont.

Problem Example 15-7

Calculate the pH and pOH of a solution whose hydrogen ion concentration is 4.6×10^{-6} mol/L.

SOLUTION

$$pH = -\log [H^{1+}] = -\log [4.6 \times 10^{-6}]$$

From Appendix VI, find log 4.6 by reading down the N column to 4.6; then look to the 0 column, and the log 4.6 = 0.6628. Hence,

$$pH = -[0.6628 + (-6.0000)] = -0.6628 + 6.0000 = 5.3372$$

(*Note:* You **add** the **logarithms** of **two numbers** to obtain the **logarithm** of their **product:** log $(x \cdot y) = \log x + \log y$.) In this text, we shall express our answers for pH and pOH problems to the *tenths' decimal place* − hence, 5.3. *Answer*

$$pOH = 14 - pH = 14.0 - 5.3 = 8.7 \quad Answer$$

Calculator:

$$pH = -\log [4.6 \times 10^{-6}]$$

Enter the decimal (4.6), then press the EE ↓ key, then the positive value of the exponent (6), the change sign key, $+/-$, to make the exponent negative, and finally the log key and the change sign key, $+/-$.

$$pH = -[-5.3372] = 5.3372, 5.3 \quad Answer$$

$$pOH = 14.0 - 5.3 = 8.7 \quad Answer$$

Problem Example 15-8

Gatorade, a popular antithirst drink, has a hydrogen ion concentration of 8.0×10^{-4} mol/L. Calculate its pH and pOH.

SOLUTION

$$pH = -\log [H^{1+}] = -\log [8.0 \times 10^{-4}]$$

From Appendix VI, find log 8.0 by reading down the N solumn to 8.0; then look to the 0 column, and the log 8.0 = 0.9031.

$$pH = -[0.9031 + (-4.0000)] = -0.9031 + 4.0000 = 3.0969, 3.1 \quad Answer$$

$$pOH = 14 - pH = 14.0 - 3.1 = 10.9 \quad Answer$$

Calculator:

$$pH = -\log [8.0 \times 10^{-4}]$$

Enter the decimal (8.0), then press the EE ↓ key, then the positive value of the exponent (4), the change sign key, $+/-$, to make the exponent negative, and finally the log key and the change sign key, $+/-$.

$$pH = -[-3.0969] = 3.0969, 3.1 \quad Answer$$

$$pOH = 14.0 - 3.1 = 10.9 \quad Answer$$

Problem Example 15-9

A commercial tomato juice has a hydrogen ion concentration of 25×10^{-6} mol/L. Calculate its pH and pOH.

SOLUTION: The hydrogen ion concentration should first be expressed in scientific notation (see 2-4); the hydrogen ion concentration is 2.5×10^{-5} mol/L.

$$pH = -\log [H^{1+}] = -\log [2.5 \times 10^{-5}]$$

From Appendix VI, log 2.5 = 0.3979, and

$$pH = -[0.3979 + (-5.0000)] = -0.3979 + 5.0000 = 4.6021, 4.6 \quad Answer$$

$$pOH = 14 - pH = 14.0 - 4.6 = 9.4 \quad Answer$$

Calculator:

$$pH = -\log [25 \times 10^{-6}]$$

Enter 25 by pressing the EE ↓ key, then the positive value of the exponent (6), the change sign button, $+/-$, to make the exponent negative, and finally the log key and the change sign key, $+/-$.

Work Problems 17, 18, 19, and 20.

$$pH = -[-4.6021] = 4.6021, 4.6 \quad \textit{Answer}$$

$$pOH = 14.0 - 4.6 = 9.4 \quad \textit{Answer}$$

The pH of a solution can be determined directly with a pH meter, as shown in Figure 15-3, where the pH is usually read by *approximating* to the *hundredths'* place.

The pH of a solution as determined with a pH meter may be used to calculate the hydrogen ion or hydroxide ion concentrations of the solution, as illustrated in Problem Examples 15-10, 15-11, and 15-12.

FIGURE 15-3

A pH meter. (Courtesy E. H. Sargent and Company.)

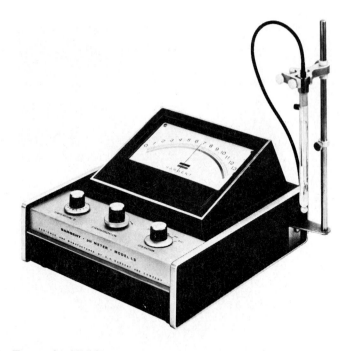

Problem Example 15-10

The pH of a urine sample is measured as 5.4 using a pH meter. Calculate the hydrogen ion concentration in the sample.

SOLUTION: Since the pH of the sample is 5.4, $\log [H^{1+}] = -5.4$ and therefore $[H^{1+}] = 10^{-5.4}$; however, this must be converted to scientific notation to be useful. A technique for doing this is to convert this number to the product of two numbers, one of which is 10 raised to a *positive* decimal power (the logarithms in Appendix VI are positive decimals) and the other is 10 raised to a negative integer power. To do this, multiply $10^{-5.4}$ by $10^6 \times 10^{-6}$ ($10^6 \times 10^{-6} = 1$) and combine the first two factors: $10^{-5.4} \times 10^6 \times 10^{-6} = 10^{0.6} \times 10^{-6}$ (see 2-3). From Appendix VI, we find that log 3.98 = 0.5999 (~0.6), and, therefore, $10^{0.6} \times 10^{-6} = 3.98 \times 10^{-6}$. Rounding off to two significant digits gives $[H^{1+}] = 4.0 \times 10^{-6}$ mol/L. *Answer*

Calculator:

$$[H^{1+}] = 10^{-5.4}$$

Enter the base (10) by pressing the EE ↓ key, then the y^x key, the positive value of the exponent (5.4), the change sign key, $+/-$, to make the exponent negative, and finally the = key.

$$[H^{1+}] = 3.9811 - 06 = 3.9811 \times 10^{-6} \text{ mol/L},$$

$$4.0 \times 10^{-6} \text{ mol/L} \quad \textit{Answer}$$

Problem Example 15-11

The pH of a solution is found to be 9.8. Calculate the hydroxide ion concentration in the solution.

SOLUTION: The **pOH** of the solution is readily calculated as **14.0 − 9.8 = 4.2**, and the $[\mathbf{OH^{1-}}] = \mathbf{10^{-4.2}}$. This is converted to scientific notation as in Problem Example 15-10: $10^{-4.2} \times 10^5 \times 10^{-5} = 10^{0.8} \times 10^{-5}$. From Appendix VI, log 6.31 = 0.8000, and $10^{0.8} \times 10^{-5} = 6.31 \times 10^{-5}$. Rounding off gives $[OH^{1-}] = 6.3 \times 10^{-5}$ mol/L. *Answer*

Calculator:

$$pOH = 14.0 - 9.8 = 4.2, [OH^{1-}] = 10^{-4.2}$$

Enter the base (10) by pressing the EE ↓ key, then the y^x key, the positive value of the exponent (4.2), the change sign key, $+/-$, to make the exponent negative, and finally the = key.

$$[OH^{1-}] = 6.3096 - 05 = 6.3096 \times 10^{-5} \text{ mol/L},$$

$$6.3 \times 10^{-5} \text{ mol/L} \quad \textit{Answer}$$

Problem Example 15-12

The pH of the rain in Pitlochry, Scotland, on April 10, 1974, was 2.4. This is the lowest pH value for rain ever recorded. Calculate the hydrogen ion concentration of the rain.

SOLUTION: Since the pH is 2.4, the $[H^{1+}] = 10^{-2.4}$. This is converted to scientific notation as in Problem Example 15-10: $10^{-2.4} \times 10^3 \times 10^{-3} = 10^{0.6} \times 10^{-3}$. From

Appendix VI, log 3.98 = 0.5999 (~0.6) and $10^{0.6} \times 10^{-3} = 3.98 \times 10^{-3}$. Rounding off to two significant digits gives $[H^{1+}] = 4.0 \times 10^{-3}$ mol/L. *Answer*

Calculator:

$$[H^{1+}] = 10^{-2.4}$$

Enter the base (10) by pressing the EE ↓ key, then the y^x key, the positive value of the exponent (2.4), the change sign key, $+/-$, to make the exponent negative, and finally the = key.

Work Problems 21, 22, and 23.

$$[H^{1+}] = 3.9811 - 03 = 3.9811 \times 10^{-3} \text{ mol/L},$$

$$4.0 \times 10^{-3} \text{ mol/L} \quad \textit{Answer}$$

15-5

Electrolytes vs Nonelectrolytes

Substances whose aqueous solutions or melted salts conduct an electric current are called **electrolytes,** and those substances whose aqueous solutions do not conduct an electric current are referred to as **nonelectrolytes.** To determine whether a substance is an electrolyte or a nonelectrolyte, an aqueous solution of the substance is prepared, and the solution is tested with two electrodes connected to a source of electric current (direct or alternating)[5] with a standard light bulb in the circuit, as shown in Figure 15-4. If the *bulb glows,* the substance is an *electrolyte;* and if it does *not glow,* it is a *non*electrolyte.

The reason for the conduction of electric current by electrolytes was explained in 1884 by the Swedish physicist and chemist Svante August Arrhenius (15-1).[6] His explanation was that ions exist in aqueous solutions of electro-

FIGURE 15-4

An apparatus for determining conduction of an electric current in an aqueous solution of a substance: (a) An electrolyte. (b) A nonelectrolyte.

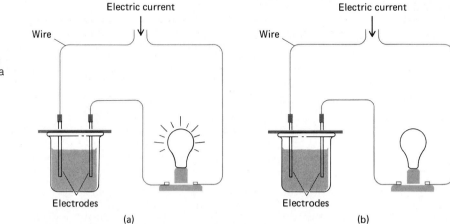

Electric current Wire Electrodes (a) Electric current Wire Electrodes (b)

---

[5]A source of direct current is an automobile battery; a source of alternating current is the electric outlet in your home.

[6]At the time Arrhenius proposed his explanation, he was a graduate student and only 25 years old. At first, his explanation was not widely accepted by his more seasoned colleagues, but as time passed it received wide acclaim, for which he received the Nobel Prize in chemistry in 1903.

lytes and that no ions are present in aqueous solutions of nonelectrolytes. If a *direct* electric current (battery) is used, one of the electrodes becomes positively charged and the other negatively charged. The negative electrode is called the *cathode,* and the positive electrode is the *anode.* The positive ions in the solution migrate to the cathode (negative electrode) and are called *cations,* whereas the negative ions in the solution migrate to the anode (positive electrode) and hence are called *anions.*

Acids, bases, and salts are *electrolytes,* due to the presence of ions in their aqueous solutions, and these solutions **do** conduct electric current. Examples of *nonelectrolytes* are *sugar* (sucrose, $C_{12}H_{22}O_{11}$), *ethyl alcohol* (C_2H_6O), and *glycerine* $C_3H_8O_3$), whose aqueous solutions **do not** conduct electric current; these nonelectroytes exist as molecules rather than as ions in aqueous solution. *Pure water* is also shown to be essentially a nonelectrolyte by the electrode method, since the standard bulb *does* **not** *glow.* There do not appear to be *sufficient* ions present in pure water to conduct electric current under these conditions.[7]

Electrolytes can be further subdivided into *strong* and weak electrolytes. For **strong electrolytes** the standard light bulb glows *brightly,* but for **weak electrolytes** the standard bulb has only a *dull* glow. *Most* soluble *salts,* such as sodium chloride (NaCl), some acids, such as *sulfuric acid* (H_2SO_4), *hydrochloric acid* (HCl), *nitric acid* (HNO_3), and *perchloric acid* ($HClO_4$), some bases, such as *group IA hydroxides*—sodium and potassium hydroxides (NaOH and KOH)—and other bases, such as *barium, strontium,* and *calcium hydroxides* [$Ba(OH)_2$, $Sr(OH)_2$, and $Ca(OH)_2$], are classed as **strong** electrolytes. These strong electrolytes (acids and bases) are considered to be strong acids and bases. Some **weak** electrolytes are *most acids* and *bases,* such as acetic acid ($HC_2H_3O_2$), hydrocyanic acid or hydrogen cyanide (HCN), hydrosulfuric acid or hydrogen sulfide (H_2S), ammonia water [$NH_{3(aq)}$], and the salts *lead(II) acetate* [$Pb(C_2H_3O_2)_2$] and *mercury(II)* or *mercuric chloride* ($HgCl_2$). Strong electrolytes are considered to be 75 to 100 percent dissociated or ionized into ions in aqueous solutions, whereas weak electrolytes are considered to be only a few percent ionized. Strong electrolytes are present in solution mostly as ions, whereas weak electrolytes consist mostly of molecules containing covalent bonds in equilibrium with a few ions in aqueous solution.

Acetic acid is an example of a weak electrolyte. Pure acetic acid (glacial acetic acid) does not conduct an electric current; that is, the standard bulb does not glow. Dilution of the acetic acid with water (a nonelectrolyte) gives a solution that conducts an electric current slightly; that is, the standard bulb has a dull glow. This conduction of electric current on dilution with water is a result of the partial ionization (15-1) of the acetic acid. Pure acetic acid acts as a nonelectrolyte, like water with very few ions being present. However, on dilution with water, more ions are formed; hence, there is greater conduction of electric current, as shown by the following equations:

---

[7]Tap water contains salts which produce ions and conduct an electric current. This is one reason it is very dangerous to touch electrical objects while taking a bath. Also, body fluids contain salts which give the bath water increased conductivity.

$$HC_2H_3O_{2(\ell)} \rightleftharpoons H^{1+}{}_{(\ell)} + C_2H_3O_2{}^{1-}{}_{(\ell)} \qquad (15\text{-}29)$$

DOES NOT GLOW (not enough ions)

$$HC_2H_3O_{2(\ell)} + H_2O_{(\ell)} \rightleftharpoons H_3O^{1+}{}_{(aq)} + C_2H_3O_2{}^{1-}{}_{(aq)} \qquad (15\text{-}30)$$

GLOWS

Equation 15-30 represents an equilibrium involving the weak electrolyte aqueous acetic acid. This equilibrium will be treated quantitatively in 17-4.

Table 15-4 summarizes strong and weak electrolytes and nonelectrolytes. Knowing the examples of weak and strong electrolytes and nonelectrolytes given in this table will help in writing ionic equations.

TABLE 15-4 Summary of Strong and Weak Electrolytes and Nonelectrolytes[a]

Strong Electrolytes	Weak Electrolytes	Nonelectrolytes
Most soluble salts	Most acids and bases	$C_{12}H_{22}O_{11}$ (sugar or sucrose)
H_2SO_4 (sulfuric acid)	$Pb(C_2H_3O_2)_2$ [lead(II) acetate]	C_2H_6O (ethyl alcohol)
HCl (hydrochloric acid)	$HgCl_2$ [mercury(II) or mercuric chloride]	$C_3H_8O_3$ (glycerine) H_2O (water)
HNO_3 (nitric acid)		
$HClO_4$ (perchloric acid)		
Group IA hydroxides, as NaOH (sodium hydroxide) KOH (potassium hydroxide)		
$Ba(OH)_2$ (barium hydroxide)		
$Sr(OH)_2$ (strontium hydroxide)		
$Ca(OH)_2$ (calcium hydroxide)		

[a]You may find it helpful to make "flash cards" for these strong electrolytes, weak electrolytes, and nonelectrolytes, for we shall refer to them again in writing ionic equations (15-6).

15-6

Guidelines for Writing Ionic Equations

In our discussion of chemical equations (9-1), we stated that equations may be written in two general ways: as **molecular equations** and as **ionic equations**. We have considered molecular equations in detail in Chapter 9 and are now prepared to consider ionic equations. **Ionic equations** express a chemical change (reaction) in terms of ions for those compounds existing mostly in ionic form in aqueous solution. For *ionic compounds*, the reacting particles are actually *ions*; hence, in ionic equations the ions are written as they *actually* exist in the solution. Therefore, ionic equations give a better representation of a chemical change in aqueous solution than do molecular equations.

In the discussion on balancing molecular equations (9-3), we suggested a few guidelines to help you balance equations by inspection. We suggest the following guidelines for writing ionic equations:

1. Complete and balance an equation in the form in which it is given to you. If it is given in the molecular form, then complete the equation in the molecular form, balance it, and *then* change it to the ionic form. If it is given in ionic form, complete it, and balance it in the ionic form. To complete chemical equations see Chapter 9.

2. The formulas for compounds written in **molecular form** are
   a. *nonelectrolytes*, such as those listed in Table 15-4;
   b. *weak electrolytes*, such as those listed in Table 15-4;
   c. *solids* and *precipitates*[8] from aqueous solutions (see solubility rules in 9-10, page 216, or inside back cover of this text), such as $CaCO_{3(s)}$; and $AgCl_{(s)}$; and
   d. *gases*, such as H_2, N_2, and O_2, written as the diatomic gas, $H_{2(g)}$, etc. All *strong electrolytes*, such as those listed in Table 15-4, are written in **ionic** form. (As you may have noticed, Table 15-4 is very important in writing ionic equations.)

3. When you write compounds in ionic form, use subscripts only to express polyatomic ions. For example, **1** mol of sulfuric acid (H_2SO_4) in ionic form is written as $2\ H^{1+}\ +\ SO_4^{2-}$ (use subscript 4, since SO_4^{2-} is a polyatomic ion). Write **3** mol of sodium sulfate (3 Na_2SO_4) as $6\ Na^{1+}\ +\ 3\ SO_4^{2-}$.

4. Each ion (monatomic or polyatomic ion) and each atom should be checked (✔) to make sure it is balanced on both sides of the equation. The net charge on each side of the equation must be the **same.**

5. The **net ionic equation** shows only those ions that have actually undergone a chemical change. The ions appearing on *both sides* of the equation that have *not* undergone a change are crossed out and are not included in the net ionic equation. These unaltered ions are included in the *total ionic equation*, but *not* in the net ionic equation. They just go along for the ride! Finally, the net ionic equations should be checked (✔) for ions, atoms, and charge and to see that the coefficients are in the lowest possible integral ratio.

---

[8]A solid and a precipitate most often exist as ions, but both are customarily written in the molecular form. The reason is that the ions in the solid are *not* bound to solvent molecules (see 14-1), and they are *not* separated from ions of the opposite charge in the manner that ions from the dissociation of soluble ionic crystals are separated in aqueous solutions.

15-7

**Examples
of Ionic
Equations**

Now let us apply these guidelines to writing ionic equations:

Problem Example 15-13

$$AgNO_{3(aq)} + HCl_{(aq)} \rightarrow$$

Completing and balancing the equation according to guideline 1 gives

$$AgNO_{3(aq)} + HCl_{(aq)} \rightarrow AgCl_{(s)} + HNO_{3(aq)}$$

Consider each of the reactants and products in regard to molecular or ionic form:

Formula	Identification	Conclusion
$AgNO_3$	Salt, strong electrolyte	Ionic form
HCl	Acid, strong electrolyte	Ionic form
AgCl	Salt, precipitate (see solubility rules-back of book)	Molecular form
HNO_3	Acid, strong electrolyte	Ionic form

Write the total ionic equation by applying guidelines 2 and 3. All compounds here are written in ionic form, except AgCl, which is a precipitate (see Table 15-4 and solubility rules, 9-10). The total ionic equation is

$$Ag^{1+}_{(aq)} + NO_3^{1-}_{(aq)} + H^{1+}_{(aq)} + Cl^{1-}_{(aq)} \rightarrow AgCl_{(s)} + H^{1+}_{(aq)} + NO_3^{1-}_{(aq)}$$

Check each ion, atom, and charge according to guideline 4. The total ionic equation is

$$\overset{\smile}{Ag}^{1+}_{(aq)} + \overset{\smile}{NO}_3^{1-}_{(aq)} + \overset{\smile}{H}^{1+}_{(aq)} + \overset{\smile}{Cl}^{1-}_{(aq)} \rightarrow$$

Charges: 1^+ $+ 1^-$ $+ 1^+$ $+ 1^-$ $= 0$

$$\overset{\smile}{Ag}\overset{\smile}{Cl}_{(s)} + \overset{\smile}{H}^{1+}_{(aq)} + \overset{\smile}{NO}_3^{1-}_{(aq)}$$

$$= 0 + 1^+ \quad + 1^- = 0$$

Write the net ionic equation by crossing out ions that appear on both sides of the equation, according to guideline 5. Check the final net ionic equation for ions, atoms, charge, and lowest possible ratio of coefficients.

$$Ag^{1+}_{(aq)} + \cancel{NO_3^{1-}}_{(aq)} + \cancel{H^{1+}}_{(aq)} + Cl^{1-}_{(aq)} \rightarrow AgCl_{(s)} + \cancel{H^{1+}}_{(aq)} + \cancel{NO_3^{1-}}_{(aq)}$$

The net ionic equation is

$$\overset{\smile}{Ag}^{1+}_{(aq)} + \overset{\smile}{Cl}^{1-}_{(aq)} \rightarrow \overset{\smile}{Ag}\overset{\smile}{Cl}_{(s)}$$

Charges: 1^+ $+ 1^- = 0$ $= 0$

From the net ionic equation, you should note that this reaction is the reaction of any soluble ionic silver salt with a soluble strongly ionic chloride compound.

Problem Example 15-14

$$NaOH_{(aq)} + H_2SO_{4(aq)} \rightarrow$$

Completing and balancing the equation according to guideline 1 gives

$$2\ NaOH_{(aq)} + H_2SO_{4(aq)} \rightarrow Na_2SO_{4(aq)} + 2\ H_2O_{(\ell)}$$

Consider each of the reactants and products in regard to molecular or ionic form:

Formula	Identification	Conclusion
NaOH	Base, strong electrolyte	Ionic form
H_2SO_4	Acid, strong electrolyte	Ionic form
Na_2SO_4	Salt, soluble (see solubility rules), strong electrolyte	Ionic form
H_2O	Nonelectrolyte	Molecular form

Write the total ionic equation by applying guidelines 2 and 3. All compounds here are written in ionic form, except H_2O (a nonelectrolyte, see Table 15-4). The total ionic equation is

$$2\ Na^{1+}_{(aq)} + 2\ OH^{1-}_{(aq)} + 2\ H^{1+}_{(aq)} + SO_4^{2-}_{(aq)} \rightarrow 2\ Na^{1+}_{(aq)} + SO_4^{2-}_{(aq)} + 2\ H_2O_{(\ell)}$$

Check each ion, atom, and charge according to guideline 4.

$$2\ Na^{1+}_{(aq)} + 2\ OH^{1-}_{(aq)} + 2H^{1+}_{(aq)} + SO_4^{2-}_{(aq)} \rightarrow$$

Charges: $2(1^+)$ $+ 2(1^-)$ $+ 2(1^+)$ $+ 2^- = 0$

$$2\ Na^{1+}_{(aq)} + SO_4^{2-}_{(aq)} + 2\ H_2O_{(\ell)}$$

$$= 2(1^+) \qquad + 2^- + 0 = 0$$

Crossing out the ions that appear on both sides of the equation according to guideline 5 gives the net ionic equation. Check the net ionic equation for ions, atoms, charge, and lowest possible ratio of coefficients:

$$2\ \cancel{Na^{1+}}_{(aq)} + 2\ OH^{1-}_{(aq)} + 2\ H^{1+}_{(aq)} + \cancel{SO_4^{2-}}_{(aq)} \rightarrow 2\ \cancel{Na^{1+}}_{(aq)} + \cancel{SO_4^{2-}}_{(aq)} + 2\ H_2O_{(\ell)}$$

$$2\ OH^{1-}_{(aq)} + 2\ H^{1+}_{(aq)} \rightarrow 2\ H_2O_{(\ell)}$$

Dividing both sides of the equation by 2, the net ionic equation is

$$OH^{1-}_{(aq)} + H^{1+}_{(aq)} \rightarrow H_2O_{(\ell)}$$

Charges: 1^- $+ 1^+$ $= 0 = 0$

This reaction is a *neutralization* reaction (9-11), and as a net ionic equation, it is simply the reaction of a hydroxide ion with a hydrogen ion to form water. This, then, is the reaction of any *strong acid* with a *strong base*.

Problem Example 15-15

$$Al_{(s)} + H_2SO_{4(aq)} \rightarrow$$

This is a single replacement reaction involving the electromotive or activity series (see 9-9). According to guideline 1, completing and balancing the equation gives

$$2\,Al_{(s)} + 3\,H_2SO_{4(aq)} \rightarrow Al_2(SO_4)_{3(aq)} + 3\,H_{2(g)}$$

Consider each of the reactants and products in regard to molecular or ionic form:

Formula	Identification	Conclusion
Al	Metal, solid	Molecular form
H_2SO_4	Acid, strong electrolyte	Ionic form
$Al_2(SO_4)_3$	Salt, soluble (see solubility rules), strong electrolyte	Ionic form
H_2	Gas	Molecular form

Write the total ionic equation by applying guidelines 2 and 3. All substances here are written in ionic form, except Al, a solid and a free metal (*not* an ion), and H_2, a gas. Checking as in guideline 4 gives the following total ionic equation:

$$2\,Al_{(s)} + 6\,H^{1+}_{(aq)} + 3\,SO_4^{2-}_{(aq)} \rightarrow 2\,Al^{3+}_{(aq)} + 3\,SO_4^{2-}_{(aq)} + 3\,H_{2(g)}$$

Charges: 0   + 6(1⁺)   + 3(2⁻)   = 0 = 2(3⁺) + 3(2⁻)   + 0 = 0

Crossing out the ions that appear on both sides of the equation and checking again as in guideline 5, we have the following net ionic equation:

$$2\,Al_{(s)} + 6\,H^{1+}_{(aq)} + \cancel{3\,SO_4^{2-}}_{(aq)} \rightarrow 2\,Al^{3+}_{(aq)} + \cancel{3\,SO_4^{2-}}_{(aq)} + 3\,H_{2(g)}$$

(Note that neither Al nor H^{1+} ion can be crossed out since they appear in the products as Al^{3+} and H_2, respectively.)

$$2\,Al_{(s)} + 6\,H^{1+}_{(aq)} \rightarrow 2\,Al^{3+}_{(aq)} + 3\,H_{2(g)}$$

Charges: 0   + 6(1⁺) = 6⁺ = 2(3⁺)   + 0 = 6⁺

Problem Example 15-16

$$NH_{3(aq)} + H_2O_{(\ell)} + Al_2(SO_4)_{3(aq)} \rightarrow$$

Completing and balancing the equation according to guideline 1, with $NH_3 + H_2O$ acting as $NH_4^{1+} + OH^{1-}$ (see Equation 15-9), gives

$$6 \ NH_{3(aq)} + 6 \ H_2O + Al_2(SO_4)_{3(aq)} \rightarrow 3 \ (NH_4)_2SO_{4(aq)} + 2 \ Al(OH)_{3(s)}$$

Consider each of the reactants and products in regard to molecular or ionic form:

Formula	Identification	Conclusion
NH_3	Base, weak electrolyte	Molecular form
H_2O	Nonelectrolyte	Molecular form
$Al_2(SO_4)_3$	Salt, soluble (see solubility rules), strong electrolyte	Ionic form
$(NH_4)_2SO_4$	Salt, soluble (see solubility rules), strong electrolyte	Ionic form
$Al(OH)_3$	Base, weak electrolyte, and precipitate (see solubility rules)	Molecular form

Write the total ionic equation by applying 2 and 3. All compounds here are written in ionic form, except $NH_{3(aq)}$, a weak electrolyte, H_2O, a nonelectrolyte, and $Al(OH)_3$, a precipitate (see Table 15-4 and solubility rules, 9-10). Checking as in guideline 4 gives the following total ionic equation:

$$6 \ \overset{\smile\smile}{N}\overset{\smile}{H}_{3(aq)} + 6 \ \overset{\smile\smile}{H}_2\overset{\smile}{O}_{(\ell)} + 2 \ \overset{\smile}{Al}^{3+}{}_{(aq)} + 3 \ \overset{\smile}{S}\overset{\smile}{O}_4{}^{2-}{}_{(aq)} \rightarrow$$

Charges: 0   + 0   + 2(3$^+$)   + 3(2$^-$)   = 0

$$6 \ \overset{\smile\smile}{N}\overset{\smile}{H}_4{}^{1+}{}_{(aq)} + 3 \ \overset{\smile}{S}\overset{\smile}{O}_4{}^{2-}{}_{(aq)} + 2 \ \overset{\smile}{Al}(\overset{\smile\smile}{OH})_{3(s)}$$

= 6(1$^+$)   + 3(2$^-$)   + 0 = 0

Crossing out the ions that appear on both sides of the equation and checking again as in guideline 5, we have the following net ionic equation:

$$6 \ NH_{3(aq)} + 6 \ H_2O_{(\ell)} + 2 \ Al^{3+}{}_{(aq)} + \cancel{3 \ SO_4{}^{2-}{}_{(aq)}} \rightarrow$$
$$6 \ NH_4{}^{1+}{}_{(aq)} + \cancel{3 \ SO_4{}^{2-}{}_{(aq)}} + 2 \ Al(OH)_{3(s)}$$

Dividing both sides of the equation by 2:

$$3 \ \overset{\smile\smile}{N}\overset{\smile}{H}_{3(aq)} + 3 \ \overset{\smile\smile}{H}_2\overset{\smile}{O}_{(\ell)} + \overset{\smile}{Al}^{3+}{}_{(aq)} \rightarrow 3 \ \overset{\smile\smile}{N}\overset{\smile}{H}_4{}^{1+}{}_{(aq)} + \overset{\smile}{Al}(\overset{\smile\smile}{OH})_{3(s)}$$

Charges: 0   + 0   + 3$^+$ = 3$^+$ = 3(1$^+$)   + 0 = 3$^+$

Problem Example 15-17

$$Ag^{1+}{}_{(aq)} + H_2S_{(aq)} \rightarrow$$

This is a double replacement reaction (see 9-10). The anion bound to the Ag^{1+} is any anion that produces a soluble silver salt in water, such as acetate or nitrate. Completing and balancing the equation according to guideline 1 gives the following ionic equation:

$$2 \ Ag^{1+}{}_{(aq)} + H_2S_{(aq)} \rightarrow Ag_2S_{(s)} + 2 \ H^{1+}{}_{(aq)}$$

Consider each of the reactants and products in regard to molecular or ionic form:

Formula	Identification	Conclusion
Ag^{1+}	Ion	Ionic form
H_2S	Acid, weak electrolyte	Molecular form
Ag_2S	Salt, precipitate	Molecular form
H^{1+}	Ion	Ionic form

Write the precipitate, Ag_2S, and the weak electrolyte, H_2S, in molecular form, according to guidelines 2 and 3 (see Table 15-4 and solubility rules, 9-10). Check the net ionic equation for ions, atoms, and charge, according to guideline 4.

$$2\ Ag^{1+}_{(aq)} + H_2S_{(aq)} \rightarrow Ag_2S_{(s)} + 2\ H^{1+}_{(aq)}$$

Charges: $2(1^+)$ $+ 0 = 2^+ = 0 + 2(1^+) = 2^+$

Work Problems 24 and 25.

The net ionic equation is the same as this ionic equation, since the same ions do not appear on *both* sides of the equation.

Chapter Summary

In this chapter, acids and bases and their properties were considered in more detail than in Chapters 7 and 9. The *Arrhenius* definition states that an *acid* is a substance that yields hydrogen ions (H^{1+}) when dissolved in water and that a *base* is a substance that yields hydroxide (OH^{1-}) ions in water. The *Brønsted-Lowry* definition describes an *acid* as any substance that can give or donate a proton (H^{1+}) to some other substance and a *base* as any substance capable of receiving or accepting a proton. A substance that is capable of behaving as either a Brønsted-Lowry acid or base is *amphiprotic*.

Titration of acids and bases is a procedure for determining their concentration in a solution. Various titration problems were solved where the concentration of the acid or base was determined in molarity, percent by mass, or normality; the volume of the acid or base consumed in the titration was also calculated.

pH is defined as the negative logarithm of the hydrogen ion concentration and *pOH* as the negative logarithm of the hydroxide ion concentration. The sum of the pH and the pOH is equal to 14. Various problems calculating pH or pOH were solved, given the hydrogen ion concentration, and calculating the hydrogen or hydroxide ion concentration, given the pH or pOH.

Electrolytes are substances whose aqueous solutions conduct an electric current, while *nonelectrolytes* are substances whose aqueous solutions do not conduct an electric current. Electrolytes are divided into *strong* or *weak electrolytes*, depending on their ability to conduct electricity. The more ions present in the solution, the better the conduction. Table 15-4 summarizes strong and weak electrolytes and nonelectrolytes.

Ionic equations express a chemical change (reaction) in terms of ions for those compounds existing mostly in ionic form in aqueous solutions. Guidelines for writing ionic equations were presented. These guidelines are strongly dependent on (1) completing equations (Chapter 9), (2) the list of strong and weak electrolytes and nonelectrolytes (Table 15-4), and (3) the solubility rules (9-10 or inside back cover of this text). Problem examples of ionic and *net* ionic equations were given along with detailed solutions.

EXERCISES

1. Define or explain
   (a) Arrhenius definition of an acid
   (b) Arrhenius definition of a base
   (c) Brønsted-Lowry definition of an acid
   (d) Brønsted-Lowry definition of a base
   (e) amphiprotic substance   (f) hydronium ion
   (g) ionization   (h) dissociation
   (i) indicator   (j) titration
   (k) ion product constant for   (l) pH
        water, K_w
   (m) pOH   (n) pK_w
   (o) electrolyte   (p) nonelectrolyte
   (q) cathode   (r) cation
   (s) anode   (t) anion
   (u) strong electrolyte   (v) weak electrolyte
   (w) ionic equation   (x) net ionic equation

2. Distinguish between
   (a) an acid by the Arrhenius definition and by the Brønsted-Lowry definition
   (b) a base by the Arrhenius definition and by the Brønsted-Lowry definition
   (c) hydronium ion and hydroxide ion
   (d) hydronium ion and a proton
   (e) ionization and dissociation
   (f) pH and pOH

Acids and Bases (See Section 15-1)

3. List at least three general characteristics each for an acid and for a base.

Electrolytes vs Nonelectrolytes (See Section 15-5)

4. List at least four strong electrolytes, four weak electrolytes, and four nonelectrolytes.

5. In your own words, explain why pure acetic acid does not light the standard light bulb, whereas dilute acetic acid does.

PROBLEMS

Acids and Bases   (See Section 15-1)

6. Classify the following as acids or bases according to the Arrhenius theory:
    (a) $HNO_{3(aq)}$ (b) $KOH_{(aq)}$ (c) $Ca(OH)_{2(aq)}$
    (d) $HCl_{(aq)}$ (e) $H_2SO_{4(aq)}$ (f) $NaOH_{(aq)}$

7. Classify the following as acids, bases, or amphiprotic substances according to the Brønsted-Lowry theory:
    (a) NH_4^{1+} (b) HCO_3^{1-} (c) H_2O
    (d) H_3O^{1+} (e) H_2SO_4 (f) HSO_4^{1-}

Titration   (See Section 15-2)

8. If 25.0 mL of 0.100 M hydrochloric acid solution is necessary to neutralize 55.0 mL of a solution of sodium hydroxide to a phenolphthalein end point, calculate the molarity of the sodium hydroxide solution.

9. If 32.0 mL of a dilute solution of lime water (calcium hydroxide) required 12.4 mL of 0.100 M hydrochloric acid solution for neutralization to a methyl red end point, calculate the molarity of the lime water.

10. If 20.0 mL of potassium hydroxide solution required 16.4 mL of 0.150 M hydrochloric acid solution for neutralization to a bromthymol blue end point, calculate the molarity of the potassium hydroxide solution.

11. If 37.5 mL of 0.500 M sodium hydroxide solution is necessary to neutralize 25.0 mL of a hydrochloric acid solution to a phenolphthalein end point, calculate (a) the molarity, and (b) the percent by mass (density of solution = 1.013 g/mL) of the hydrochloric acid solution.

12. Vinegar is a dilute solution of acetic acid. In the titration of 5.00 mL of vinegar, 37.7 mL of 0.105 M sodium hydroxide solution was required to neutralize the vinegar to a phenolphthalein end point. Calculate (a) the molarity, and (b) the percent by mass (density of solution = 1.007 g/mL) of the vinegar.

13. If 0.625 g of pure sodium carbonate was dissolved in water and the solution titrated with 30.8 mL of hydrochloric acid to a methyl orange end point, calculate the molarity of the hydrochloric acid solution.

14. If 0.200 g of pure sodium hydroxide is titrated with 5.70 mL of hydrochloric acid solution to a phenolphthalein end point, calculate (a) the molarity, and (b) the percent by mass (density of solution = 1.016 g/mL) of the hydrochloric acid solution.

15. In the titration of 24.5 mL of potassium hydroxide solution of unknown concentration, 35.7 mL of 0.110 N sulfuric acid was required to neutralize the potassium hydroxide in reactions where both hydrogen ions of the sulfuric acid react. Calculate the normality of the potassium hydroxide solution.

16. In the titration of a 0.125 N sodium hydroxide solution, 28.5 mL of 0.155 N sulfuric acid solution was required to neutralize the sodium hydroxide in reactions

that replace both hydrogen ions of the sulfuric acid. Calculate the number of mL of the sodium hydroxide solution needed in the reaction.

pH and pOH   (See Section 15-4)

17. Calculate the pH and pOH of the following solutions:
    (a) hydrogen ion concentration is 1.0×10^{-9} mol/L
    (b) hydrogen ion concentration in household ammonia is 2.0×10^{-12} mol/L
    (c) hydrogen ion concentration in commercial milk is 2.0×10^{-7} mol/L
    (d) hydrogen ion concentration in sour milk is 6.3×10^{-7} mol/L
    (e) hydrogen ion concentration in 7-Up is 25×10^{-5} mol/L

18. Calculate the pH and pOH of the following solutions:
    (a) hydrogen ion concentration is 2.4×10^{-5} mol/L
    (b) hydrogen ion concentration in vinegar is 7.9×10^{-4} mol/L
    (c) hydrogen ion concentration in a dilute solution (0.133%) of citric acid found in various fruits is 1.4×10^{-3} mol/L
    (d) hydrogen ion concentration in seawater is 5.3×10^{-9} mol/L
    (e) hydrogen ion concentration in a given sample of human urine is 8.6×10^{-6} mol/L

19. In general, the flavor of devil's-food cake is best if the hydrogen ion concentration is between 1.0×10^{-8} and 3.2×10^{-8} mol/L. (a) Calculate this range on the pH scale. (b) With a hydrogen ion concentration of less than 1.0×10^{-8} mol/L, the cake has a bitter soapy taste. Explain.

20. The hydrogen ion concentration of soils varies considerably. In forest soils the hydrogen ion concentration is about 3.2×10^{-5} mol/L, while in desert soils the hydrogen ion concentration is about 1.0×10^{-10} mol/L. The higher hydrogen ion concentration in forest soils is due to the decomposition of organic matter resulting in the production of carbon dioxide. Calculate the pH and pOH for (a) forest and (b) desert soils.

21. Calculate the hydrogen ion concentration in moles per liter for each of the following solutions:
    (a) a solution whose pH is 5.2
    (b) a solution whose pH is 9.6
    (c) a solution whose pOH is 2.5
    (d) a solution whose pOH is 12.0
    (e) a solution whose hydroxide ion concentration is 1.7×10^{-5} mol/L
    (f) a solution whose hydroxide ion concentration is 3.4×10^{-5} g/L

22. Calculate the hydroxide ion concentration in moles per liter for each of the following solutions:
    (a) a solution whose pOH is 5.0
    (b) a solution whose pOH is 9.6
    (c) a solution whose pH is 5.7
    (d) a solution whose pH is 8.2
    (e) a solution whose hydrogen ion concentration is 4.0×10^{-6} mol/L

23. Rainwater has a pH of 4.0. Calculate its hydrogen ion concentration in moles per liter. Explain why the water is acidic instead of neutral.

Ionic Equations (See Sections 15-6 and 15-7). *Hint:* First, review Chapter 9 so you will be able to complete and balance these equations.

24. Complete and balance the following equations, writing them as total ionic equations and as net ionic equations; indicate any precipitate by (s) and any gas by (g):

(a) $BaCl_{2(aq)} + (NH_4)_2CO_{3(aq)} \rightarrow$
(b) $Fe(NO_3)_{3(aq)} + NH_{3(aq)} + H_2O_{(\ell)} \rightarrow$
(c) $SrCl_{2(aq)} + K_2CO_{3(aq)} \rightarrow$
(d) $Na_2CO_{3(aq)} + HCl_{(aq)} \rightarrow$
(e) $KCl_{(aq)} + AgNO_{3(aq)} \rightarrow$
(f) $Al_{(s)} + HCl_{(aq)} \rightarrow$
(g) $CO_{2(g)} + Ca(OH)_{2(aq)} \rightarrow$
(h) $SrCO_{3(s)} + HC_2H_3O_{2(aq)} \rightarrow$
(i) $Fe_{(s)} + CuSO_{4(aq)} \rightarrow$
(j) $CdCl_{2(aq)} + H_2S_{(aq)} \rightarrow$

25. Complete and balance the following equations, writing them as total ionic equations and as net ionic equations. Indicate any precipitate by (s) and any gas by (g). Assume that all reactants are in water solution, unless otherwise indicated.

(a) $HgCl_2 + H_2S \rightarrow$ (b) $MgSO_4 + NaOH \rightarrow$
(c) $CaO_{(s)} + HCl \rightarrow$ (d) $FeCl_3 + NH_3 + H_2O \rightarrow$
(e) $FeSO_4 + (NH_4)_2S \rightarrow$ (f) $Al(OH)_{3(s)} + HCl \rightarrow$
(g) $H_3PO_4 + KOH \rightarrow$ (h) $MgCl_2 + Na_2CO_3 \rightarrow$
(i) $Cl_{2(g)} + NaBr \rightarrow$ (j) $C_{12}H_{22}O_{11} \xrightarrow{\Delta}$

General Problems

26. If 1.50 g of 80.0 percent by mass potassium hydroxide is titrated with 10.0 mL of hydrochloric acid solution to a phenolphthalein end point, calculate (a) the molarity, and (b) the percent by mass (density of solution = 1.037 g/mL) of the hydrochloric acid solution.

27. A theory used to explain how acid rain destroys forests is that the acid makes soluble the *insoluble* aluminum salts in the soil. These *soluble* aluminum salts are taken up by the tree roots and are toxic to the tree. Assuming that the insoluble salt in the soil is AlZ_3 (Z = polyatomic organic anion) and the acid rain is nitric acid (dilute aqueous solution), write a balanced equation for this reaction. Write the ionic and net ionic equation for the reaction.

Reading

King, Edward J., *Ionic Reactions and Separations; Experiments in Qualitative Analysis.* New York: Harcourt Brace Jovanovich, 1973. This book includes units on ionic reactions and ionic equations (Chapter 1, Sections 1.1 to 1.7) and may be helpful at this time.

16

Oxidation-Reduction Equations and Electrochemistry

The lead storage battery or other source of direct current may be used to silver-plate eating utensils or serving pieces.

TASK

Memorize what to add when H and O atoms are needed in acidic or basic reactions in balancing oxidation-reduction equations by the ion electron method (see Section 16-3).

OBJECTIVES

1. Given the following terms, define each term and describe the distinguishing characteristics of each:
   (a) oxidation number (Section 16-1)
   (b) oxidation (Section 16-1)
   (c) reduction (Section 16-1)
   (d) oxidizing agent (Section 16-1)
   (e) reducing agent (Section 16-1)

2. Given the formulas of the reactants and products in oxidation-reduction equations, balance these equations by the oxidation number method (Problem Examples 16-1, 16-2, 16-3, and 16-4, Problems 8, 9, 10, and 11).

3. Given the formulas of the reactants and products in oxidation-reduction equations, balance these equations by the ion electron method (Problem Examples 16-5, 16-6, 16-7, and 16-8, Problems 12, 13, 14, and 15).

4. Given the standard reduction potentials and equations consisting of metals and metal ions, determine if these reactions will occur (Section 16-6, Problem 16).

5. Given the standard reduction potentials and various half-cell oxidation or reduction equations, give the E^0 value for these reactions (Section 16-6, Problem 17).

In Chapter 9 (9-6), we mentioned that there is a type of chemical reaction called an *oxidation-reduction reaction,* and that to balance the complex form of *oxidation-reduction equations* (also called *redox equations*) we need to use special techniques. In general, we cannot balance these equations "by inspection" as we balanced the five simple types of reactions (combination, decomposition, single replacement, double replacement, and neutralization) in Chapter 9. The equations of three of these types of reactions (combination, decomposition, and single replacement) are simple cases of oxidation-reduction equations;

therefore, we are able to balance them "by inspection." In this chapter, we shall consider the special techniques for balancing *complex* oxidation-reduction equations.

In addition, we shall now consider electrochemistry, involving oxidation-reduction equations. **Electrochemistry** is concerned with the relationship of electrical energy and chemical energy. We shall study the use of electrical energy to produce certain chemical reactions that would otherwise not occur; an example is the case of electrolytic cells used in electroplating, such as silverplating on flatware. We shall also consider the basis of the electromotive series mentioned in 9-9 and 13-6, as well as cells that produce electricity from chemical reactions, such as the dry cell (a flashlight battery), the lead storage battery of your car, and the fuel cell, one type of which was used in the Apollo space missions. Corrosion of metals will also be considered.

16-1

Definitions of Oxidation and Reduction. Oxidizing and Reducing Agents

Before we consider the definitions of oxidation and reduction, we need to review oxidation numbers as outlined in 6-1. In that section, we defined **oxidation number** (ox no) as a positive or negative whole number used to describe the combining capacity of an element in a compound.[1] The change in oxidation number from one state to another implies the number of *electrons lost* or *gained*. An example of a **positive change** would be in going from the free state (zero oxidation number) to the combined state with a *positive* oxidation number. *Electrons are lost* (remember, electrons are negatively charged). An example of a **negative change** would be in going from the free state to the combined state with a *negative* oxidation number. *Electrons are gained.*

Oxidation at one time referred only to the combination of an element with oxygen, but the term has been expanded, and **oxidation** is now defined as a chemical change in which a substance *loses electrons*, or one or more elements in it *increase in oxidation number.* If electrons (negative) are lost from an element, then the resulting element will have an increase in oxidation number.

Reduction is a chemical change in which a substance *gains electrons*, or one or more elements in it *decrease in oxidation number.* If an element gains electrons (negative), then the resulting element will have a decrease (algebraic) in oxidation number. Figure 16-1 may help you differentiate these two terms. A useful relationship to memorize is that **reduction** involves a **decrease in oxidation number.** The opposite, then, would be true for oxidation, or an increase in oxidation number is involved in oxidation. These can also be translated into gaining or losing electrons, since a gain of negative electrons will result in a decrease in oxidation number (reduction), and a loss of electrons will result in a gain in oxidation number (oxidation).

In a given reaction, whenever a substance is oxidized, it loses electrons to another substance, which is thereby reduced; hence, *oxidation accompanies re-*

---

[1]Fractional average oxidation numbers of atoms in compounds do exist, as was mentioned in Chapter 6 (footnote 1).

FIGURE 16-1

Oxidation and reduction, change in oxidation number, and electron transfer.

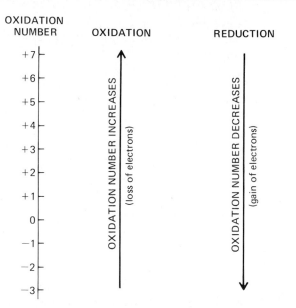

duction and reduction accompanies oxidation. The equation is therefore called an oxidation-reduction equation (or "**redox**" equation).

In an oxidation-reduction equation the substance *oxidized* is called the **reducing agent** (reductant), since it produces reduction in another substance. The substance being *reduced* is called the **oxidizing agent** (oxidant), since it produces oxidation in another substance.

A simple example of a combination reaction will illustrate this point:

$$Ca_{(s)} + S_{(s)} \xrightarrow{\Delta} CaS_{(s)} \tag{16-1}$$

Calcium metal (zero oxidation number) combines with sulfur (zero oxidation number) to form calcium sulfide (2^+ oxidation number for calcium and 2^- oxidation number for sulfur). The **calcium** therefore has **increased in oxidation number** (or lost electrons) and is **oxidized**. The **sulfur** has **decreased in oxidation number** (or gained electrons) and is **reduced**. Since calcium has been *oxidized*, it is called the *reducing agent;* since sulfur has been *reduced*, it is called the *oxidizing agent.* This, then, is an example of an oxidation-reduction equation.

16-2

Balancing Oxidation-Reduction Equations. The Oxidation Number Method

Oxidation-reduction equations can be balanced by two methods: the **oxidation number method** or the **ion electron method**. The oxidation number method can be used for both molecular and ionic equations, and both types will be illustrated as examples.

As we did in balancing molecular equations (9-3) and in writing ionic equations (15-6), we shall suggest a few guidelines to follow in balancing oxidation-reduction equations by the **oxidation number method**.

1. Determine by inspection which elements undergo a change in oxidation number. At first, this may involve calculating the oxidation number for each element in the equation (see 6-1), but after balancing a few oxidation-reduction equations you should readily be able to recognize which elements undergo a change in oxidation number.

2. Above each element oxidized or reduced write its oxidation number.[2]

3. Determine the *change in oxidation number* for each element undergoing a change in oxidation number. This is best shown by drawing an arrow from the element in the reactant to the element in the product and indicating the *increase or decrease of oxidation number* above the arrow.

4. Balance the increase and decrease of oxidation number by coefficients placed in front of the increase or decrease of oxidation number. *The total increase must equal the total decrease.*

5. Place these coefficients in front of the corresponding formula of the reactants and complete the balancing of the equation by inspection. If it becomes necessary to double the value of a coefficient in the oxidation or reduction, you must carry out the corresponding change with the other substances in the oxidation or reduction. This must be done to keep the *total increase* in oxidation number *equal* to that *total decrease* in oxidation number.

6. Place a check (✔) above each atom on both sides of the equation to ensure that the equation is balanced. Also check the equation to see that the coefficients are the lowest possible ratio. For ionic equations, the net *charges on both sides* of the equation must be *equal*.

Now, let us consider some examples. In this text, the products of the oxidation-reduction reaction will be given (due to the often complicated nature of these products). Thus, you will be asked only to balance the equations. Also, to simplify writing the equation, the physical states will normally not be included in the equation.

Problem Example 16-1

$$C + H_2SO_{4(conc)} \xrightarrow{\Delta} CO_2 + SO_2 + H_2O$$

According to guideline 1, determine by inspection the elements undergoing a change in oxidation number. Carbon in the free state in the reactants must change, since in the products it appears in the combined state as CO_2. In the reactants, S in H_2SO_4 must also change, because in the products it appears as SO_2. According to guideline 2,

---

[2]The oxidation numbers can also be written below each element, so as to avoid confusion with ionic charges on polyatomic ions. An oxidation number is an arbitrary assignment based on certain rules (see 6-1), while an ion has an actual charge on a particle.

write the oxidation number above the element oxidized or reduced. (See 6-1 for a review of calculation of oxidation numbers.)

$$\overset{0}{C} + \overset{6^+}{H_2SO_4} \xrightarrow{\Delta} \overset{4^+}{CO_2} + \overset{4^+}{SO_2} + H_2O$$

Determine the change in oxidation number for each element that undergoes a change in oxidation number, according to guideline 3. Then balance the increase and decrease in oxidation numbers by placing a coefficient in front of the increase and decrease of oxidation numbers, according to guideline 4.

$$\overset{0}{C} + \overset{6^+}{H_2SO_4} \xrightarrow{\Delta} \overset{4^+}{CO_2} + \overset{4^+}{SO_2} + H_2O$$

1 (inc 4 ox no)

2 (dec 2 ox no)

Place these coefficients in front of the corresponding formulas and balance the equation by inspection, according to guideline 5. The following equation results:

$$\overset{0}{C} + 2 \overset{6^+}{H_2SO_4} \xrightarrow{\Delta} \overset{4^+}{CO_2} + 2 \overset{4^+}{SO_2} + 2 H_2O$$

1 (inc 4 ox no)

2 (dec 2 ox no)

(Since the 1 is understood in front of the carbon, it can be deleted.) Check each atom on both sides of the equation according to guideline 6. The following balanced equation results:

$$C + 2 H_2SO_4 \xrightarrow{\Delta} CO_2 + 2 SO_2 + 2 H_2O$$

Since *carbon* increases in oxidation number and hence loses electrons, it is oxidized and is the *reducing agent*, whereas S in H_2SO_4 decreases in oxidation number and thus gains electrons. Therefore, H_2SO_4 is reduced and is the *oxidizing agent*.

Problem Example 16-2

$$KMnO_4 + HCl \rightarrow MnCl_2 + Cl_2 + H_2O + KCl$$

Apply guidelines 1 and 2 to this equation and place the oxidation numbers above the elements that change in oxidation number. The following equation results:

$$\overset{7+}{K}MnO_4 + \overset{1^-}{H}Cl \rightarrow \overset{2+}{M}nCl_2 + \overset{0}{Cl_2} + H_2O + KCl$$

According to guideline 3, determine the change in oxidation number for each element undergoing a change in oxidation number.

$$\overset{7^+}{K}MnO_4 + \overset{1^-}{H}Cl \rightarrow \overset{2^+}{Mn}Cl_2 + \overset{0}{Cl_2} + H_2O + KCl$$

(inc 1 ox no)

(dec 5 ox no)

Quick examination would seem to indicate that we need a 5 in front of the increase of oxidation number, but this is not the case. Since we need an *even* number of free Cl atoms on the right to enable us to place a whole number in front of the Cl_2 molecule, we must use 10 HCl and 2 $KMnO_4$ on the left, according to guidelines 4 and 5:

$$2\,\overset{7^+}{K}MnO_4 + \boxed{10}\,\overset{1^-}{H}Cl \rightarrow 2\,\overset{2^+}{Mn}Cl_2 + \overset{0}{Cl_2} + H_2O + KCl$$

10 (inc 1 ox no)

2 (dec 5 ox no)

Continue with guideline 5 and balance the K, Mn, O, and Cl_2 (molecule). The following equation results:

$$2\,KMnO_4 + 10\,HCl \rightarrow 2\,MnCl_2 + 5\,Cl_2 + 8\,H_2O + 2\,KCl$$

We must add six (6) *more* HCl molecules to the left to supply enough Cl^{1-} ions for the 2 $MnCl_2$ and 2 KCl on the right and enough hydrogens for the oxygens. Note that these Cl^{1-} ions do not undergo a change in oxidation number and therefore are not involved in the oxidation process. Balance out the hydrogens and check each atom according to guideline 6. The following balanced equation results:

$$2\,KMnO_4 + 16\,HCl \rightarrow 2\,MnCl_2 + 5\,Cl_2 + 8\,H_2O + 2\,KCl$$

The Cl in HCl increases in oxidation number and hence loses electrons; therefore, *HCl* is oxidized and is the *reducing agent*. In $KMnO_4$, Mn decreases in oxidation number and hence gains electrons; therefore, *$KMnO_4$* is reduced and is the *oxidizing agent*.

Problem Example 16-3

$$Ag + H^{1+} + NO_3^{1-} \rightarrow Ag^{1+} + NO + H_2O$$

This is an ionic equation, and it, too, can be balanced by the oxidation number method. According to guidelines 1 and 2, place the oxidation numbers of elements that change in oxidation number above the elements.

$$\overset{0}{Ag} + H^{1+} + \overset{5^+}{N}O_3^{1-} \rightarrow \overset{2^+}{Ag}^{1+} + NO + H_2O$$

Determine the change in oxidation number for each element undergoing a change in oxidation number, according to guideline 3. Then, balance the increase and decrease in oxidation number by placing a coefficient in front of the increase or decrease, according to guideline 4.

$$\overset{0}{Ag} + \overset{1+}{H} + \overset{5+}{NO_3{}^{1-}} \rightarrow \overset{2+}{Ag^{1+}} + NO + H_2O$$

3 (inc 1 ox no)

1 (dec 3 ox no)

Place these coefficients in front of the formulas and balance the equation by inspection, according to guideline 5. The following equation results:

$$3\ \overset{0}{Ag} + 4\ \overset{1+}{H} + \overset{5+}{NO_3{}^{1-}} \rightarrow 3\ \overset{2+}{Ag^{1+}} + NO + 2\ H_2O$$

3 (inc 1 ox no)

1 (dec 3 ox no)

Check each atom on both sides of the equation, and check the *charge* on both sides, according to guideline 6. The following balanced equation results:

$$3\ \overset{\nearrow}{Ag} + 4\ \overset{\nearrow}{H^{1+}} + \overset{\nearrow\nearrow}{NO_3{}^{1-}} \rightarrow 3\ \overset{\nearrow}{Ag^{1+}} + \overset{\nearrow\nearrow}{NO} + 2\ \overset{\nearrow\nearrow}{H_2O}$$

Charges: $4^+ + \quad 1^- = \quad 3^+ \quad = 3^+$

Since the *Ag* increases its oxidation number and hence loses electrons, it is oxidized and thus is the *reducing agent*. In $NO_3{}^{1-}$, N decreases in oxidation number and hence gains electrons; therefore, $NO_3{}^{1-}$ is reduced and is the *oxidizing agent*.

Problem Example 16-4

$$I^{1-} + Cr_2O_7{}^{2-} + H^{1+} \rightarrow Cr^{3+} + I_2 + H_2O$$

Apply guidelines 1 and 2 to this ionic equation as follows:

$$I^{1-} + \overset{6+}{Cr_2O_7{}^{2-}} + H^{1+} \rightarrow Cr^{3+} + \overset{0}{I_2} + H_2O$$

Apply guideline 3 for an increase of 1 in oxidation number for the I and a decrease of 3 in oxidation number for the Cr. Since there are 2 Cr atoms changing in oxidation number, the total decrease in oxidation number is 6. Therefore, we must balance with a coefficient of 6, according to guideline 4.

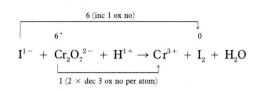

$$I^{1-} + Cr_2O_7^{2-} + H^{1+} \rightarrow Cr^{3+} + I_2 + H_2O$$

Place the coefficients in front of the formulas and balance the equation by inspection, according to guideline 5. The following balanced equation results:

$$6\ I^{1-} + Cr_2O_7^{2-} + 14\ H^{1+} \rightarrow 2\ Cr^{3+} + 3\ I_2 + 7\ H_2O$$

Check each atom on both sides of the equation, and check the charge on both sides, according to guideline 6. The following balanced equation results:

$$6\ I^{1-} + Cr_2O_7^{2-} + 14\ H^{1+} \rightarrow 2\ Cr^{3+} + 3\ I_2 + 7\ H_2O$$

Charges: $6^- \quad + 2^- \quad\quad + 14^+ = 6^+ = 2(3^+) = 6^+$

Work Problems 8, 9, 10, and 11.

Since the I^{1-} increases in oxidation number and hence loses electrons, it is oxidized and is the *reducing agent*. In $Cr_2O_7^{2-}$, Cr decreases in oxidation number and hence gains electrons; therefore, $Cr_2O_7^{2-}$ is reduced and is the *oxidizing agent*.

16-3

Balancing Oxidation-Reduction Equations. The Ion Electron Method

The ion electron method is a second method used to balance oxidation-reduction equations. The technique used in this method involves *two partial equations* representing *half-reactions:* one equation describes the *oxidation*, and the other equation describes the *reduction*. The two partial equations are then added to produce a final balanced equation. Although we artificially divide the original reaction into two partial equations, these partial equations do not take place alone, and whenever oxidation occurs, so does reduction. These partial equations are very useful in electrochemistry, as we shall learn in the next sections.

As with the oxidation number method, we shall also suggest a few guidelines to balance oxidation-reduction equations by the **ion electron method.**

1.   Write the equation in *net ionic* form (see 15-6) *without* attempting to balance it.

2.   Determine by inspection the elements that undergo a change in oxidation number and then write two partial equations: an *oxidation half-reaction* and a *reduction half-reaction*.

3. Balance the atoms on each side of the partial equations. In *acid* solution, H^{1+} ions and H_2O molecules may be added. For each hydrogen atom (H) needed, a H^{1+} ion is added. For each oxygen atom (O) needed, a H_2O molecule is added, with 2 H^{1+} ions being shown on the other side of the partial equation. In *basic* solution, OH^{1-} ions and H_2O molecules may be added. For each hydrogen atom (H) needed, an H_2O molecule is added with an OH^{1-} ion written on the other side of the partial equation. For **each** oxygen atom (O) needed, *two* OH^{1-} ions are added with *one* H_2O molecule written on the other side of the partial equation. The following summarizes these additions in acid and base:

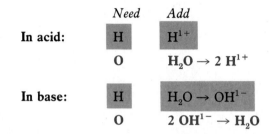

	Need	*Add*
In acid:	H	H^{1+}
	O	$H_2O \rightarrow 2\ H^{1+}$
In base:	H	$H_2O \rightarrow OH^{1-}$
	O	$2\ OH^{1-} \rightarrow H_2O$

(To remember what to add in a given type of solution, you may find it helpful to make "flash cards" of the preceding summary.)

4. Balance the partial equation electrically by adding electrons to the appropriate side of the equation so that the *charges* on both sides of the partial equations are *equal*. These two partial equations are defined as follows: The *oxidation* half-reaction equation, in which the reactant loses electrons, is written with the electrons on the *products* side; the *reduction* half-reaction equation in which the reactant *gains* electrons is written with the electrons on the reactant side.

$$\text{Oxidation:}\quad M \rightarrow M^{1+} + 1e^-$$
$$\text{Reduction:}\quad A + 1e^- \rightarrow A^{1-}$$

5. Multiply each **entire** partial equation by an appropriate number so that the *electrons lost in one partial equation* (oxidation half-reaction) *are equal to the electrons gained in the other partial equation* (reduction half-reaction).

6. Add the two partial equations and eliminate those electrons, ions, or water molecules that are on both sides of the equation.

7. Place a check (✔) above each atom on both sides of the equation to ensure that the equation is balanced. Also check the net charges on both sides of the equation to see that they are equal. Check the equation to see that the coefficients are the lowest possible ratios.

Problem Example 16-5

$$Fe^{2+} + MnO_4^{1-} \rightarrow Fe^{3+} + Mn^{2+} \text{ in acid solution}$$

Guideline 1 does not apply, since the equation is already in ionic form, so use guideline 2. The following two partial equations for the half-reactions result:

(1) $$Fe^{2+} \rightarrow Fe^{3+}$$

(2) $$MnO_4^{1-} \rightarrow Mn^{2+}$$

Balance the atoms for each partial equation, according to guideline 3. In partial equation (2), add 4 H_2O to the products to balance the O atoms in MnO_4^{1-} and then add 8 H^{1+} to the reactants, since the reaction is carried out in acid.

(1) $$Fe^{2+} \rightarrow Fe^{3+}$$

(2) $$8\ H^{1+} + MnO_4^{1-} \rightarrow Mn^{2+} + 4\ H_2O$$

Then, balance the two partial equations electrically by adding electrons (negatively charged) to the appropriate sides, according to guideline 4. In partial equation (1), electrons are lost; hence, this equation represents the *oxidation* half-reaction. In partial equation (2), electrons are gained; hence, this equation represents the *reduction* half-reaction.

(1) *Oxidation:* $$Fe^{2+} \rightarrow Fe^{3+} + 1e^-$$

Charges: $$2^+ = 3^+ \quad + 1^- = 2^+$$

(2) *Reduction:* $$8\ H^{1+} + MnO_4^{1-} + 5e^- \rightarrow Mn^{2+} + 4\ H_2O$$

Charges: $$8^+ + 1^- \quad + 5^- = 2^+ \quad = 2^+$$

Multiply the *entire* partial equation (1) by 5, so that the gain of electrons in partial equation (2) is equal to the loss, according to the guideline 5, giving the following partial equations:

(1) *Oxidation:* $$5\ Fe^{2+} \rightarrow 5\ Fe^{3+} + 5e^-$$

(2) *Reduction:* $$8\ H^{1+} + MnO_4^{1-} + 5e^- \rightarrow Mn^{2+} + 4\ H_2O$$

(A common error is to forget to multiply the **entire** partial equation by the appropriate factor.) Add the two partial equations, and eliminate those electrons on both sides of the equation, according to guideline 6. The following equation results:

$$5\ Fe^{2+} + 8\ H^{1+} + MnO_4^{1-} + \cancel{5e^-} \rightarrow 5\ Fe^{3+} + \cancel{5e^-} + Mn^{2+} + 4\ H_2O$$

Check each atom and the charges on both sides of the equation to obtain the final balanced equation, according to guideline 7:

$$5\overset{\rightthreetimes}{Fe}^{2+} + 8\overset{\rightthreetimes}{H}^{1+} + \overset{\rightthreetimes}{MnO_4}^{1-} \rightarrow 5\overset{\rightthreetimes}{Fe}^{3+} + \overset{\rightthreetimes}{Mn}^{2+} + 4\overset{\rightthreetimes}{H_2O}$$

Charges: $5\,(2^+) \ + \ 8^+ \ + \ 1^- \ = \ 17^+ \ = \ 5(3^+) + \ 2^+ \ = \ 17^+$

In partial equation (1), since the Fe^{2+} is oxidized, it is the *reducing agent;* in partial equation (2), since MnO_4^{1-} is reduced, it is the *oxidizing agent.*

Problem Example 16-6

$$Zn \ + \ HgO \rightarrow ZnO_2^{2-} \ + \ Hg \quad \text{in basic solution}$$

Excluding guideline 1, since the equation is already in ionic form, and proceeding to guideline 2, we can write the following two partial equations for the half-reactions:

(1) $\qquad\qquad\qquad\qquad\qquad\qquad Zn \rightarrow ZnO_2^{2-}$

(2) $\qquad\qquad\qquad\qquad\qquad\qquad HgO \rightarrow Hg$

Balance the atoms for each partial equation, according to guideline 3. In partial equation (1), two oxygen atoms are required in the reactants, and, since the solution is basic, add $4\,OH^{1-}$ ions to the reactants and $2\,H_2O$ molecules to the products to balance the atoms. In partial equation (2), one oxygen atom is required in the products, so add $2\,OH^{1-}$ ions to the products and one H_2O molecule to the reactants to balance the atoms.

(1) $\qquad\qquad\qquad\qquad Zn \ + \ 4\,OH^1 \rightarrow ZnO_2^{2-} \ + \ 2\,H_2O$

(2) $\qquad\qquad\qquad\qquad HgO \ + \ H_2O \rightarrow Hg \ + \ 2\,OH^{1-}$

According to guideline 4, balance the two partial equations electrically by adding electrons to the appropriate side. (Remember that a free metal has zero charge; that is, Zn and Hg are neutral, zero oxidation number.) In partial equation (1), the electrons are lost; hence, this equation represents the *oxidation* half-reaction. In partial equation (2), electrons are gained; hence, this equation represents the *reduction* half-reaction.

(1) *Oxidation:* $\ Zn \ + \ 4\,OH^{1-} \rightarrow ZnO_2^{2-} \ + \ 2\,H_2O \ + \ 2e^-$

$\qquad$ Charges: $\qquad\quad 4^- \ = \quad 2^- \qquad\qquad + \qquad 2^- = 4^-$

(2) *Reduction:* $\ HgO \ + \ H_2O \ + \ 2e^- \rightarrow Hg \ + \ 2\,OH^{1-}$

$\qquad$ Charges: $\qquad\qquad\qquad\quad 2^- \ = \qquad\quad 2^-$

In the two partial equations, the number of electrons lost is equal to the number of electrons gained, according to guideline 5. Therefore, add the two partial equations and eliminate those electrons and ions on both sides of the equation, according to guideline 6, to obtain the following equation:

$$Zn \ + \ 4\,OH^{1-} \ + \cancel{2e^-} \ + \ HgO \ + \ H_2O \rightarrow ZnO_2^{2-} \ + \ 2\,H_2O \ + \cancel{2e^-} \ + \ Hg \ + \ 2\,OH^{1-}$$

The resulting OH^{1-} ions present on the left side are 2 OH^{1-} (4 OH^{1-} on the left minus 2 OH^{1-} on the right), and the resulting H_2O molecules on the right side are 1 H_2O (2 H_2O on the right minus 1 H_2O on the left). Hence, the following equation results:

$$Zn + 2\ OH^{1-} + HgO \rightarrow ZnO_2^{2-} + H_2O + Hg$$

Check each atom and the charge on both sides of the equation to obtain the final balanced equation, according to guideline 7:

$$\overset{\smile}{Z}n + 2\ \overset{\smile\smile}{O}H^{1-} + \overset{\smile\smile}{H}gO \rightarrow \overset{\smile\smile}{Z}nO_2^{2-} + \overset{\smile\smile}{H}_2O + \overset{\smile}{H}g$$

Charges: 2^- $= 2^-$

In partial equation (1), since the *Zn* is oxidized, it is the *reducing agent,* and in partial equation (2), since *HgO* is reduced, it is the *oxidizing agent.*

Problem Example 16-7

$$NaI + Fe_2(SO_4)_3 \rightarrow I_2 + FeSO_4 + Na_2SO_4 \text{ in aqueous solution}$$

Apply guideline 1 by writing the equation in *net ionic* form *without* attempting to balance it. Refer to the solubility rules of inorganic substances in water in the back of the book. The following net ionic equation results:

$$\cancel{Na^{+}} + I^{1-} + Fe^{3+} + \cancel{SO_4^{2-}} \rightarrow I_2 + Fe^{2+} + \cancel{SO_4^{2-}} + \cancel{Na^{+}} + \cancel{SO_4^{2-}}$$

Net Ionic: $I^{1-} + Fe^{3+} \rightarrow I_2 + Fe^{2+}$

(Note that no attempt is made to balance the ions.) Write two partial equations according to guideline 2:

(1) $I^{1-} \rightarrow I_2$

(2) $Fe^{3+} \rightarrow Fe^{2+}$

Balance the atoms for each partial equation, according to guideline 3:

(1) $2\ I^{1-} \rightarrow I_2$

(2) $Fe^{3+} \rightarrow Fe^{2+}$

Then balance these two equations electrically, according to guideline 4. In partial equation (1), electrons are lost; hence, this equation represents the *oxidation* half-reaction. In partial equation (2), electrons are gained; hence, this equation represents the *reduction* half-reaction.

(1) *Oxidation:* $2\ I^{1-} \quad \rightarrow I_2 + 2e^-$

Charges: 2^- $=$ 2^-

(2) *Reduction:*

$$Fe^{3+} + 1e^- \rightarrow Fe^{2+}$$

$$\text{Charges: } 3^+ + 1^- = 2^+$$

In the two partial equations, the number of electrons lost must be equal to the number of electrons gained, according to guideline 5; therefore, multiply partial equation (2) by 2.

(1) *Oxidation:* $\quad\quad\quad 2\, I^{1-} \rightarrow I_2 + 2e^-$

(2) *Reduction:* $\quad\quad\quad 2\, Fe^{3+} + 2e^- \rightarrow 2\, Fe^{2+}$

Add the two partial equations and eliminate the electrons on opposite sides of the equation, according to guideline 6, to obtain the following equation:

$$2\, I^{1-} + 2\, Fe^{3+} + \cancel{2e^-} \rightarrow I_2 + \cancel{2e^-} + 2\, Fe^{2+}$$

$$2\, I^{1-} + 2\, Fe^{3+} \rightarrow I_2 + 2\, Fe^{2+}$$

Check each atom and the charge on both sides of the equation to obtain the final balanced equation, according to guideline 7:

$$2\, I^{1-} + 2\, Fe^{3+} \rightarrow I_2 + 2\, Fe^{2+}$$

$$\text{Charges: } 2^- + 2(3^+) = 4^+ = 2(2^+) = 4^+$$

In partial equation (1), since the I^{1-} is oxidized, it is the *reducing agent;* and in partial equation (2), since Fe^{3+} is reduced, it is the *oxidizing agent.*

Problem Example 16-8

$$CrO_2 + ClO^{1-} \rightarrow CrO_4^{2-} + Cl^{1-} \text{ in basic solution}$$

Excluding guideline 1, since the equation is already in ionic form, and proceeding to guideline 2, we can write the following two partial equations for the half-reactions:

(1) $\quad\quad\quad\quad\quad CrO_2 \rightarrow CrO_4^{2-}$

(2) $\quad\quad\quad\quad\quad ClO^{1-} \rightarrow Cl^{1-}$

Balance the atoms for each partial equation, according to guideline 3:

(1) $\quad\quad\quad CrO_2 + 4\, OH^{1-} \rightarrow CrO_4^{2-} + 2\, H_2O$

(2) $\quad\quad\quad ClO^{1-} + H_2O \rightarrow Cl^{1-} + 2\, OH^{1-}$

Balance these two equations electrically, according to guideline 4. In partial equation (1), electrons are lost; hence, this equation represents the *oxidation* half-reaction. In partial equation (2), electrons are gained; therefore, this equation represents the *reduction* half-reaction.

(1) *Oxidation:* $CrO_2 + 4\,OH^{1-} \rightarrow CrO_4^{2-} + 2\,H_2O + 2e^-$

Charges: $4^- \quad = 2^- \quad + \quad 2^- = 4^-$

(2) *Reduction:* $ClO^{1-} + H_2O + 2e^- \rightarrow Cl^{1-} + 2\,OH^{1-}$

Charges: $1^- \quad + \quad 2^- = 3^- = 1^- \quad + 2^- = 3^-$

According to guideline 5, the number of electrons lost must equal the number of electrons gained. Now, add the two partial equations and eliminate these electrons and ions on both sides of the equation, according to guideline 6, to obtain the following equation:

$$CrO_2 + 4\,OH^{1-} + ClO^{1-} + H_2O + \cancel{2e^-} \rightarrow CrO_4^{2-} + 2\,H_2O + \cancel{2e^-} + Cl^{1-} + 2\,OH^{1-}$$

The resulting OH^{1-} ions present on the left side are $2\,OH^{1-}$ ($4\,OH^{1-}$ on the left minus $2\,OH^{1-}$ on the right), and the resulting H_2O molecules on the right side are 1 H_2O ($2\,H_2O$ on the right minus $1\,H_2O$ on the left). Hence, the following equation results:

$$CrO_2 + 2\,OH^{1-} + ClO^{1-} \rightarrow CrO_4^{2-} + H_2O + Cl^{1-}$$

Check each atom and the charge on both sides of the equation to obtain the final balanced equation, according to guideline 7:

$$CrO_2 + 2\,OH^{1-} + ClO^{1-} \rightarrow CrO_4^{2-} + H_2O + Cl^{1-}$$

Charges: $2^- \quad + 1^- \quad = 3^- = 2^- \quad + \quad 1^- = 3^-$

Work Problems 12, 13, 14, and 15.

In partial equation (1), since the CrO_2 is oxidized, it is the *reducing agent*, and in partial equation (2), since the ClO^{1-} is reduced, it is the *oxidizing agent*.

16-4

Electrolytic Cells

Electrolytic cells *use electrical energy to produce certain chemical reactions* that would otherwise not occur at an appreciable rate. The apparatus used in an electrolytic cell is similar to that used to determine whether a substance is an electrolyte or nonelectrolyte (15-5, see Figure 15-4). The electric current in an electrolytic cell must be a direct current, as in a car battery. Two electrodes, a cathode (C) and an anode (A), as shown in Figure 16-2, are in the cell. The solution contains an *electrolyte*, XY.

Electricity is the flow of electrons, and the direct-current source (battery or generator) withdraws electrons from one electrode and pushes them to the other. One electrode becomes negatively charged in this process and is called the *cathode* (C). The positive ions of the electrolyte (X^{1+}) migrate to this cathode and accept electrons from it. Thus **reduction occurs at the cathode** ($X^{1+} + e^- \rightarrow X^0$). The other electrode, which becomes positively charged by the action of the direct current source, is called the *anode* (A). The negative ions

FIGURE 16-2

A simple electrolytic cell.

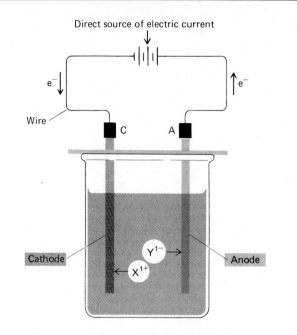

of the electrolyte (Y^{1-}) migrate to this anode and transfer electrons to the anode, thus completing the electrical circuit. Hence, **oxidation occurs at the anode** ($Y^{1-} \rightarrow Y^0 + e^-$).

We now represent the two half-cell reactions that occur in the electrolytic cell in Figure 16-2; electrons flow through the wire (external circuit) from the anode (A) to the cathode (C):

Cathode: $\quad\quad X^{1+} + 1e^- \rightarrow \quad X^0 \quad$ (reduction) $\quad\quad\quad$ (16-2)

deposited at cathode (C)

Anode: $\quad\quad Y^{1-} \quad\quad\quad \rightarrow \quad Y^0 \; + \; 1e^- \quad$ (oxidation) $\quad\quad\quad$ (16-3)

deposited at anode (A)

One application of an electrolytic cell is the electroplating of silver on utensils, such as knives, forks, spoons, etc., as shown in Figure 16-3.[3] The utensil acts as the cathode with pure silver as the anode, and the electrolyte is a soluble silver salt. The two half-reactions are illustrated by the following equations:

---

[3]Silverplate utensils should not be confused with sterling silver utensils. The plate utensils are made now of mostly nickel (copper was previously used) plated with silver. Sterling silver utensils are made primarily of silver with a slight amount of copper or nickel added for hardness. No plating is placed or needed on sterling silver.

FIGURE 16-3

Electroplating a spoon. The spoon acts as the cathode and pure silver acts as the anode. Reduction occurs at the *cathode,* and oxidation at the *anode.*

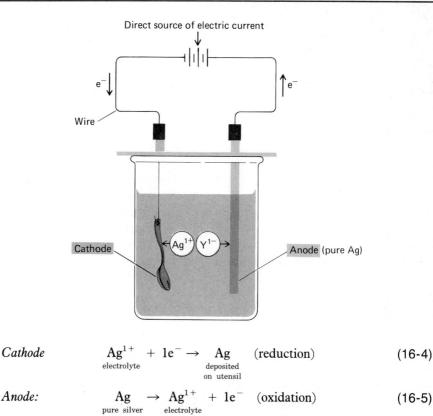

$$\text{\textit{Cathode}} \qquad \underset{\text{electrolyte}}{Ag^{1+}} + 1e^- \rightarrow \underset{\substack{\text{deposited} \\ \text{on utensil}}}{Ag} \quad \text{(reduction)} \qquad (16\text{-}4)$$

$$\text{\textit{Anode:}} \qquad \underset{\text{pure silver}}{Ag} \rightarrow \underset{\text{electrolyte}}{Ag^{1+}} + 1e^- \quad \text{(oxidation)} \qquad (16\text{-}5)$$

(Note that reduction occurs at the cathode, and oxidation occurs at the anode.) The silver ions produced at the anode (lose electrons–oxidation) migrate to the cathode (gain electrons–reduction), forming a deposit of silver on the utensil. The negative ions (Y^{1-}) of the electrolyte (soluble silver salt) migrate to the anode, accompanying the production of the silver ions.

A second application is in the final purification to obtain pure copper.[4] In this purification, an electrolytic cell is used. The impure copper acts as the anode and a thin plate of pure copper acts as the cathode, with the electrolyte being copper(II) sulfate. The pure copper forms on the cathode and is better than 99.9 percent copper. Other metals found in copper ores, such as silver, gold, and platinum, are deposited as a "sludge" at the bottom of the cell. These metals are further refined, and their value makes the entire operation quite economical. Because this final step in purification requires electricity, cheap electrical power should be available at or near the electrolytic cell plant site.

A third application of an electrolytic cell is in making "tin" cans. The cans

---

[4]Other metals besides copper which are prepared in nearly pure form using an electrolytic cell are aluminum, sodium, potassium, magnesium, and calcium.

can be made by electroplating a *thin* film of tin on a steel can or by dipping the steel can in molten tin. Hence, "tin" cans consist of *very little* tin.

16-5

Voltaic or Galvanic Cells

Voltaic or **galvanic cells** *produce electrical energy from certain chemical reactions.* These cells were named after two Italian physicists, Count Alessandro Volta (1745–1827) and Luigi (or Aloisio) Galvani (1737–1798), who contributed to the basic development of the cell. The cell consists of two electrodes as in an electrolytic cell, with **reduction** occurring at the **cathode** and **oxidation** at the **anode,** *as before.* The reaction is carried out in such a manner that electrons are permitted to flow along a wire *from the anode to the cathode.* Thus, the direction of the current (electron) flow is the same as in the electrolytic cell.

The reaction of zinc metal with copper(II) ions is a simple example of a voltaic or galvanic cell used to produce electricity. If a piece of zinc metal is immersed in a solution of copper(II) ions, a reaction takes place immediately, as shown in Equation 16-6. The *energy* released as this reaction takes place is evolved as *heat,* the *transfer of electrons* from zinc to copper(II) ions taking place *directly:*

$$Zn + Cu^{2+} \rightarrow Zn^{2+} + Cu \qquad (16\text{-}6)$$

If the reaction is carried out by *not* allowing the *zinc metal* to come in *direct contact* with the *copper(II) ions,* then the energy evolved can be obtained as electrical energy–electricity. This can be done with a cell, as Figure 16-4a shows. In this cell, a solution of zinc ions is *separated* from a solution containing copper(II) ions by a *porous partition* through which ions may diffuse. These two solutions can also be connected through a *salt bridge,* consisting of a solution of some electrolyte, such as potassium sulfate, as shown in Figure 16-4b.

No reaction takes place until the switch in the external circuit is closed. The electricity produced is enough to make a light bulb glow. Then Zn will go into solution as Zn^{2+} ions, and the electrons deposited on the zinc electrode will travel along the wire to the copper electrode. There, electrons are accepted by Cu^{2+} ions at the surface of this electrode to deposit copper metal on the copper electrode. At the zinc electrode, electrons are lost *(oxidation),* and hence the zinc electrode is the *anode.* At the copper electrode, electrons are gained *(reduction),* and hence the copper electrode is the *cathode.* The two half-reactions are given in the following equations:

Anode: $\qquad \underset{\text{metal}}{Zn} \rightarrow \underset{\text{electrolyte}}{Zn^{2+}} + 2e^- \quad \text{(oxidation)} \qquad (16\text{-}7)$

Cathode: $\qquad \underset{\text{electrolyte}}{Cu^{2+}} + 2e^- \rightarrow \underset{\substack{\text{metal} \\ \text{deposited}}}{Cu} \quad \text{(reduction)} \qquad (16\text{-}8)$

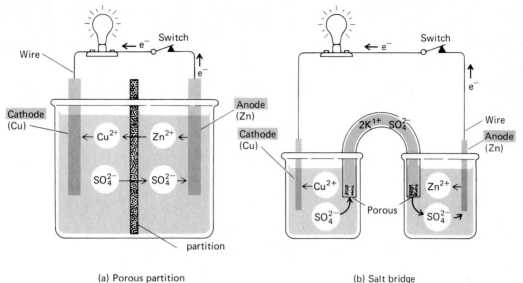

(a) Porous partition   (b) Salt bridge

FIGURE 16-4

A simple voltaic or galvanic cell. (a) A porous partition separates the two solutions of zinc sulfate and copper(II) sulfate. (b) A salt bridge of potassium sulfate connects the two solutions of zinc sulfate and copper(II) sulfate.

The sum of these two half-cell equations represents the overall cell reaction, as given in Equation 16-6.

Since the zinc is going into solution and the copper is coming out, there should be as few zinc ions and as many copper(II) ions in solution as possible to prevent the cell reaction from slowing down, due to the abundance or scarcity of certain ions. The porous partition or salt bridge allows the excess negative ions accumulating in the copper compartment to neutralize the excess positive ions accumulating in the zinc compartment. The electrons flow through the wire from the zinc electrode (anode) to the copper electrode (cathode). The positive ions (cations) migrate to the cathode (copper electrode), and the negative ions (anions) migrate to the anode compartment (zinc electrode), the same as in an electrolytic cell.

16-6

Standard Reduction Potentials

In Chapter 9 (9-9), we discussed the *electromotive* or *activity series* in regard to replacement reactions. We stated that it is possible to arrange the metals in an order called the electromotive or activity series, so that each element in the series will displace any of those following it in an aqueous solution of its salt. The basis of this electromotive or activity series is the **standard reduction potentials** for the *ions* of the metals forming the *free metals*, as listed in Table 16-1. These metals are listed in order of *energy* or *potential* in which the ion

TABLE 16-1   Standard Reduction Potentials for Various Metals[a]

Reducation Half-Cell Equation[b]	E^0(V)
$Li^{1+} + 1e^- \rightleftharpoons Li$	-3.04
$K^{1+} + 1e^- \rightleftharpoons K$	-2.92
$Ba^{2+} + 2e^- \rightleftharpoons Ba$	-2.91
$Ca^{2+} + 2e^- \rightleftharpoons Ca$	-2.87
$Na^{1+} + 1e^- \rightleftharpoons Na$	-2.71
$Mg^{2+} + 2e^- \rightleftharpoons Mg$	-2.36
$Al^{3+} + 3e^- \rightleftharpoons Al$	-1.66
$Zn^{2+} + 2e^- \rightleftharpoons Zn$	-0.76
$Fe^{2+} + 2e^- \rightleftharpoons Fe$	-0.44
$Cd^{2+} + 2e^- \rightleftharpoons Cd$	-0.40
$Ni^{2+} + 2e^- \rightleftharpoons Ni$	-0.25
$Sn^{2+} + 2e^- \rightleftharpoons Sn$	-0.14
$Pb^{2+} + 2e^- \rightleftharpoons Pb$	-0.13
2 H^{1+} + 2e^- $\rightleftharpoons$ H_2	**0.00**
$Cu^{2+} + 2e^- \rightleftharpoons Cu$	$+0.34$
$Hg_2^{2+} + 2e^- \rightleftharpoons 2\,Hg$	$+0.79$
$Ag^{1+} + 1e^- \rightleftharpoons Ag$	$+0.80$
$Hg^{2+} + 2e^- \rightleftharpoons Hg$	$+0.85$
$Au^{3+} + 3e^- \rightleftharpoons Au$	$+1.42$

[a]Although hydrogen is not a metal, it is included in this series for comparison.

[b]Aqueous solutions of ions at 1.0 M concentration at 25°C and hydrogen gas at 1.0 atm pressure and 25°C.

gains electrons (reduction). This potential is expressed in volts (V), and is called *standard electrode potential, E^0* (read ee, zero).[5] Since the reaction involves the gain of electrons (reduction), it is more precisely called *standard reduction potential*. These standard reduction potentials are obtained by immersing the metal in a solution of its ions at a 1.0 M (**molar;** see 14-8) concentration and 25°C and measuring the potential of such a half-cell in comparison with a hydrogen electrode at a pressure of 1.0 atm and 25°C, which is arbitrarily assigned a value of 0.00 V for its standard electrode potential, E^0. The hydrogen electrode is constructed by immersing a small sheet of platinum metal in a solution 1.0 M in hydrogen ions (25°C). Hydrogen gas is bubbled over the platinum metal at 1.0 atm pressure (25°C), as shown in Figure 16-5

---

[5]The IUPAC (see 7-1) has recommended the use of standard reduction potentials (instead of standard oxidation potentials) for *standard electrode potentials*.

FIGURE 16-5

Measurement of E^0 (volt) of a zinc half-cell with a salt bridge. Solutions 1.0 M at 25°C. $E^0 = -0.76$ V.

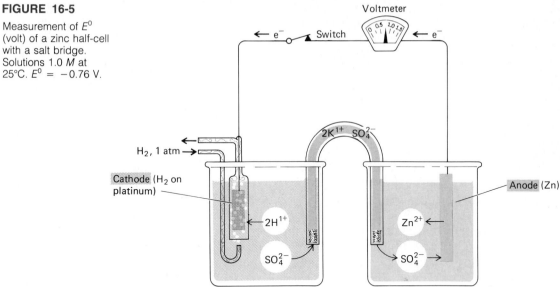

in the measurement of the standard reduction potential of zinc. The half-cell reactions are given in the following two equations:

Anode: $\qquad\qquad$ $Zn \rightarrow Zn^{2+} + 2e^-$ (oxidation) $\qquad$ (16-9)

Cathode: $\qquad\qquad$ $2\,H^{1+} + 2e^- \rightarrow H_2$ (reduction) $\qquad$ (16-10)

The overall cell equation is obtained by adding the two half-cell equations:

Overall cell equation: $Zn + 2\,H^{1+} \rightarrow Zn^{2+} + H_2$ $\qquad$ (16-11)

These standard reduction potentials are assigned positive or negative values, depending upon their ability to force hydrogen ions to accept an electron. Those elements capable of forcing hydrogen ions to *accept an electron* are assigned *negative* standard *reduction* potentials (negative E^0 values), whereas those elements *unable* to force hydrogen ions to *accept an electron* [but instead the ion of the metal must accept electrons from the hydrogen (H_2) molecules] are assigned *positive* standard *reduction* potentials (positive E^0 values). **The more negative is the E^0 value, the more active is the metal.** Hence, such an active metal will replace an ion of a metal with an algebraically greater E^0 value from its salt by transferring electrons to the ion. For example, Mg ($E^0 = -2.36$ V) will replace Zn ($E^0 = -0.76$ V) from its salt, as $Mg + Zn^{2+} \rightarrow Mg^{2+} + Zn$. The E^0 values from negative to positive follow the *same* order as the electromotive or activity series and are the basis of this series.

Work Problem 16.

You may understand this order better if you consider *oxidation*. All the half-cell equations considered so far were written in the direction of reduction occurring from left to right in the reaction; that is, the electrons were written on the reactants side. If the *complete reaction* is *reversed*—that is, if the electrons are written on the products side—the *sign* is *reversed* and the half-cell equation is an oxidation. For example, the E^0 for the oxidation half-cell equation $Li \leftrightarrows Li^{1+} + le^-$ is $+3.04$ V, the **standard oxidation potential,** and this reaction **readily** occurs, placing lithium at the top of the electromotive or activity series. Therefore, by writing the *oxidation* half-cell equation, we see that those metals above hydrogen in the electromotive or activity series have a *positive* E^0 value for the standard **oxidation** potential. Those metals below hydrogen in the electromotive or activity series have a *negative* E^0 value for the standard **oxidation** potential, with hydrogen having a value of 0.00 V. This order, from positive to negative values of E^0 for the standard *oxidation* potentials, also follows the order of the electromotive or activity series.

Work Problem 17.

Also in Chapter 9 (9-9), we listed another series similar to the electromotive or activity series for the halogens. In this series, fluorine will replace chlorine from an aqueous solution of its salt; chlorine will replace bromine; and bromine will replace iodine. Again, the basis of this series is the *standard reduction potential,* as shown in Table 16-2, with F_2 having the greatest tendency (most positive E^0 value) to **gain** electrons (reduction) of the halogens. Therefore, in the halogens, the order of activity could be listed in order of the decreasing magnitude of the *positive* standard reduction potential, such as $F_2 > Cl_2 > Br_2 > I_2$.

TABLE 16-2   Standard Reduction Potentials for the Halogens

Reduction Half-Cell Equation[a]	E^0 (V)
$I_2 + 2e^- \rightleftarrows 2\,I^{1-}$	$+0.54$
$Br_2 + 2e^- \rightleftarrows 2\,Br^{1-}$	$+1.06$
$Cl_2 + 2e^- \rightleftarrows 2\,Cl^{1-}$	$+1.36$
$F_2 + 2e^- \rightleftarrows 2\,F^{1-}$	$+2.87$

[a]Aqueous solutions of ions at a 1.0 *M* concentration at 25°C and gases at 1.0 atm pressure at 25°C.

16-7

Practical Cells

The practical production of electrical energy from certain chemical reactions is used in the *dry cell*—such as a flashlight battery—the *lead storage battery* of a car, and the fuel cell, one type of which was used in the Apollo space missions.

The **dry cell** (potential, 1.5 V), as shown in Figure 16-6, consists of a zinc

FIGURE 16-6

The dry cell.

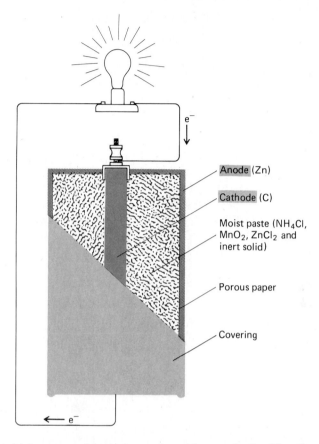

Anode (Zn)

Cathode (C)

Moist paste (NH$_4$Cl, MnO$_2$, ZnCl$_2$ and inert solid)

Porous paper

Covering

container, which acts as the anode, and a carbon cathode. The electrolyte is a moist paste of ammonium chloride, manganese(IV) oxide, and zinc chloride. The following two equations give the half-cell reactions:

Anode: $\text{Zn} \rightarrow \text{Zn}^{2+} + 2e^-$ (oxidation)   (16-12)

Cathode: $2\,\text{MnO}_2 + 2\,\text{NH}_4^{1+} + 2e^- \rightarrow$
$$\text{H}_2\text{O} + \text{Mn}_2\text{O}_3 + 2\,\text{NH}_3 \text{ (reduction)} \quad \text{(16-13)}$$

The anode is separated from the moist paste of the electrolyte by a porous paper or plastic film, which prevents the electrons given off at the anode from entering the moist paste directly. Hence, they must pass through the external circuit and reenter at the carbon cathode. If the manganese(IV) oxide were absent, the cathode reaction would be

$$2\,\text{NH}_4^{1+} + 2e^- \rightarrow 2\,\text{NH}_3 + \text{H}_2 \quad \text{(16-14)}$$

The hydrogen liberated would cover and insulate the carbon cathode from the mixture of ammonium chloride and zinc chloride, and the reaction would soon stop. But the presence of manganese(IV) oxide prevents the formation of hydrogen, making the dry cell a practical, long-lived source of uninterrupted direct-current electricity. The overall cell reaction for the dry cell is given in the following equation:

Overall cell equation:

$$Zn + 2\ MnO_2 + 2\ NH_4^{1+} \rightarrow$$
$$H_2O + Zn^{2+} + Mn_2O_3 + 2\ NH_3 \quad (16\text{-}15)$$

The **lead storage battery** (potential, approximately 2 V per cell), as shown in Figure 16-7, is another practical chemical system that produces electricity. Its advantage over the dry cell is that it can be readily recharged after the electrode reactions have depleted the electrolyte (dilute sulfuric acid). The alternator of a car acts to recharge the battery by converting some of the mechanical energy from the engine to electrical energy. This forces electrons through the battery in the reverse direction in order to charge the battery. The electrodes in this battery are lead and lead(IV) oxide. The lead is the anode and the lead(IV) oxide is the cathode, with the electrolyte being dilute sulfuric acid (battery acid).[6] The half-cell reactions are given in the following two equations:

FIGURE 16-7

One cell of a lead storage battery.

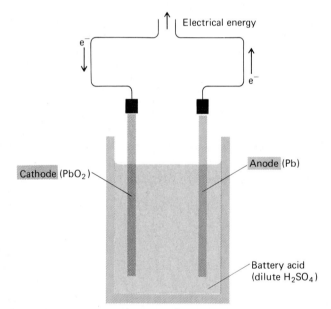

Electrical energy

e^-

e^-

Cathode (PbO_2)

Anode (Pb)

Battery acid (dilute H_2SO_4)

[6]Once the battery is charged with battery acid (dilute sulfuric acid, density of 1.25 to 1.30 g/mL), only water is added to replace water lost by evaporation. As the sulfuric acid is used up and water is formed in the discharge reaction (Equation 16-18), the density of the dilute acid decreases. When the density of the solution falls to about 1.2 g/mL the battery needs to be recharged. To prevent this acid from "eating away" at the battery terminals, a slurry of baking soda ($NaHCO_3$) in water, automatic transmission fluid, or Karo syrup is often added to or painted on the battery terminals. This baking soda neutralizes the acid. Write the balanced equation for the reaction. Many cars do not start simply because there is a poor connection at the battery terminals.

Anode: $Pb_{(s)} + SO_4^{2-} \rightarrow PbSO_{4(s)} + 2e^-$ (oxidation)   (16-16)

Cathode: $PbO_{2(s)} + 4 H^{1+} + SO_4^{2-} + 2e^- \rightarrow$

$$PbSO_{4(s)} + 2 H_2O \text{ (reduction)} \quad (16\text{-}17)$$

The sulfate ions decrease during discharge. Lead(II) sulfate forms at both electrodes. The overall cell reaction is given in the following equation:

Overall cell equation: $Pb_{(s)} + PbO_{2(s)} + 4 H^{1+} + 2 SO_4^{2-} \underset{\text{charge}}{\overset{\text{discharge}}{\rightleftharpoons}}$

$$2 PbSO_{4(s)} + 2 H_2O \quad (16\text{-}18)$$

For a lead storage battery to be *charged*, it must be converted to an electrolytic cell with an outside source of energy (such as a battery charger), which reverses the discharge reaction.

New batteries are being developed to power electric cars. One such battery that is in the experimental stage now is the aluminum-air battery. This battery is refueled with tap water (about 6 gallons) every 250 to 300 miles. Every 1000 to 3000 miles the aluminum plates are replaced. A five-passenger electric car requires 60 of these batteries, about the size of a regular lead storage battery, with a weight of 970 pounds for these 60 batteries. The car would be able to travel at around 55 miles per hour. A prototype car powered by these aluminum-air batteries is planned for 1989.

The **fuel cell** is another type of practical cell that converts chemical energy into electrical energy. The primary difference between the fuel cell and the dry cell or the lead storage battery is that in the fuel cell the *reactants* are *continuously supplied* and the *products continuously removed*. This is not the case in the dry cell and the lead storage battery. One type of fuel cell used in the Apollo space missions involves the reaction of hydrogen with oxygen to form water. This cell (potential, slightly less than 1 V) consists of a hydrogen electrode of nickel metal and an oxygen electrode of nickel-nickel oxide. The electrolyte is potassium hydroxide. The hydrogen electrode is the anode, and the oxygen electrode is the cathode. These half-cell reactions are given in the following two equations:

Anode: $2 H_2 + 4 OH^{1-} \rightarrow 4 H_2O + 4e^-$ (oxidation)   (16-19)

Cathode: $O_2 + 4e^- + 2 H_2O \rightarrow 4 OH^{1-}$ (reduction)   (16-20)

The overall cell reaction is given in the following equation:

Overall cell reaction: $2 H_2 + O_2 \rightarrow 2 H_2O$ (16-21)

A single power plant called a Powercel,[7] as shown in Figure 16-8, consists of 31 of these fuel cells. Three Powercels were used in the Apollo space mis-

---

[7]Powercel is the trademark for Pratt & Whitney Aircraft fuel cell power plants.

FIGURE 16-8

The Powercel, a fuel cell power plant. (Courtesy Pratt & Whitney Aircraft.)

sions, each having a mass of approximately 225 pounds. The Powercel operates best at about 205°C, but can operate between 196°C and 260°C with heat being liberated during the course of the reaction. This heat is removed from the fuel cell through heat exchangers, and the excess heat is either dissipated into space or also used as a source of electrical energy by means of thermoelectric devices. The water produced in this reaction is used as drinkable water for the astronauts.

Fuel cells such as the hydrogen and oxygen fuel cell just described show promise as an energy source for automobiles. This cell does not pollute the atmosphere with carbon monoxide and nitrogen oxides as the present internal-combustion engine does, but rather produces drinkable water. It does produce some heat energy, but it is less than that emitted in the present internal-combustion engines to produce the same power.

16-8

Corrosion

Corrosion is the process in which a metal is eaten into or worn away by some substance exposed to the metal to form an undesirable substance lacking metallic properties. In the rusting of iron, oxygen, and water eat away at the iron

FIGURE 16-9

Even Volkswagens rust. One of the many Volkswagens owned by one of the authors (WSS).

to form hydrated iron(III) oxide. We are aware how our cars seem to rust away after a few years, unless great care is taken to prevent this (see Figure 16-9).

The rusting of iron is electrochemical and involves an oxidation-reduction reaction.

Anode: $Fe \longrightarrow Fe^{2+} + 2e^-$ (oxidation)   (16-22)

Cathode: $O_2 + 4 H^{1+} + 4e^- \rightarrow 2 H_2O$ (reduction)   (16-23)

The iron forms iron(II) ions with the loss of two electrons (oxidation) at the anode. These electrons are conducted in the metal to another area where the rust appears and there is an ample supply of oxygen. At the cathode the oxygen and acid accept the electrons to form water (reduction). The iron(II) ions are further oxidized to iron(III) ions (Fe^{3+}). The iron(III) ions form the rust, a hydrated iron(III) oxide, $Fe_2O_3 \cdot x\,H_2O$, with a variable amount of water, x H_2O. The reaction is as follows:

$$x\,H_2O + 4\,Fe^{2+} + 4\,H_2O + O_2 \rightarrow 2\,Fe_2O_3 \cdot x\,H_2O + 8\,H^{1+} \qquad (16\text{-}24)$$

In removing rust from car bodies, the rust must not only be removed but the "pit" in the metal must be smooth, for here is where the oxidation occurs. The rusting process appears to start with (1) impurities in the steel, or (2) strain on the steel. Salts (NaCl or $CaCl_2$) used on highways in the winter to melt the ice act as electrolytes to conduct the electrons from the anode to the cathode.

To protect the steel or iron against corrosion various protective metals are used to cover the steel. "Tin" cans have a thin film of tin on the steel and "galvanized" iron have a thin film of zinc on the steel.

Chapter Summary

Oxidation is defined as a chemical change in which a substance loses electrons, or one or more elements in it increase in oxidation number, while *reduction* is a chemical change in which a substance gains electrons, or one or more elements in it decrease in oxidation number. In a given reaction, oxidation accompanies reduction and reduction accompanies oxidation. The equation is therefore called an oxidation-reduction equation. The substance that is oxidized is called the *reducing agent*, and the substance that is reduced is called the *oxidizing agent*. In this chapter, oxidation-reduction equations were balanced by (1) the oxidation number method and (2) the ion electron method. Guidelines were given for balancing these equations by both methods and four examples of balancing equations by each method were also given.

Electrolytic cells use electrical energy to produce certain chemical reactions that would otherwise not occur at an appreciable rate. They are used in electroplating and the purification of various metals. *Voltaic* or *galvanic cells* produce electrical energy from certain chemical reactions. The order of the *electromotive series* (see 9-9) in terms of standard reduction potentials was discussed. Based on the E^0 value in volts of the standard reduction potential, it was determined whether or not a given reaction will occur. Values of E^0 in volts for a given oxidation or reduction equation were determined from the standard reduction potentials.

Finally, practical cells, such as a dry cell, a lead storage battery, and a fuel cell, which produce electrical energy, were discussed. The corrosion of metals was also briefly considered.

EXERCISES

1. Define or explain
   - (a) oxidation number
   - (b) valence
   - (c) oxidation
   - (d) reduction
   - (e) oxidizing agent (oxidant)
   - (f) reducing agent (reductant)
   - (g) oxidation half-reaction
   - (h) reduction half-reaction
   - (i) electrolytic cells
   - (j) cathode in regard to oxidation and reduction
   - (k) anode in regard to oxidation and reduction
   - (l) voltaic or galvanic cells
   - (m) porous partition
   - (n) salt bridge
   - (o) E^0
   - (p) negative E^0 value for standard *reduction* potential

(q) positive E^0 value for stan-   (r) corrosion
     dard *reduction* potential

2. Distinguish between
(a) oxidation number and valence
(b) oxidation and reduction
(c) oxidizing agent (oxidant) and reducing agent (reductant)
(d) oxidation half-reaction and reduction half-reaction
(e) cathode and anode, in regard to oxidation and reduction
(f) electrolytic cells and voltaic or galvanic cells
(g) negative E^0 values and positive E^0 values for standard *reduction* potentials
(h) fuel cell and dry cell

Electrolytic Cells (See Section 16-4)

3. In your own words, describe an electrolytic cell. Include in your description a drawing and the reactions that occur at the anode and cathode.

Voltaic or Galvanic Cells (See Section 16-5)

4. In your own words, describe a voltaic or galvanic cell. Clearly point out the difference between an electrolytic cell and a voltaic or galvanic cell. Include a drawing of both cells.

Practical Cell (See Section 16-7)

5. In your own words, describe the following cells; include in your description the balanced equations for the half-reactions that occur at the anode and cathode.
(a) dry cell   (b) lead storage battery   (c) fuel cell

6. Discuss the positive and negative effects on the environment of the use of a fuel cell such as a hydrogen and oxygen fuel cell in powering automobiles.

Corrosion (See Section 16-8)

7. In your own words, describe corrosion and include in your description the balanced equations for the half-reactions that occur at the anode and cathode.

PROBLEMS

Oxidation-Reduction Equation—Oxidation Number Method (See Sections 16-1 and 16-2)

8. Balance the following oxidation-reduction equations by the oxidation number method:
(a) $HNO_3 + HI \rightarrow NO + I_2 + H_2O$
(b) $KI + H_2SO_{4(conc)} \rightarrow H_2S + H_2O + I_2 + K_2SO_4$
(c) $Cu + HNO_{3(dil)} \rightarrow Cu(NO_3)_2 + NO + H_2O$

(d) $KIO_4 + KI + HCl \rightarrow KCl + I_2 + H_2O$

(e) $HNO_3 + I_2 \rightarrow NO_2 + H_2O + HIO_3$

(f) $Ag + H_2SO_{4(conc)} \rightarrow Ag_2SO_4 + SO_2 + H_2O$

(g) $Cr_2O_7^{2-} + Fe^{2+} + H^{1+} \rightarrow Cr^{3+} + Fe^{3+} + H_2O$

(h) $Cu + H^{1+} + SO_4^{2-} \rightarrow Cu^{2+} + SO_2 + H_2O$

(i) $I_2 + CdS + H^{1+} \rightarrow HI + S + Cd^{2+}$

(j) $Zn + Cr_2O_7^{2-} + H^{1+} \rightarrow Zn^{2+} + Cr^{3+} + H_2O$

9. In the equations in Problem 8. indicate the substances that are oxidized and reduced and the oxidizing and reducing agent.

10. Balance the following oxidation-reduction equations by the oxidation number method:

(a) $Bi(OH)_3 + K_2SnO_2 \rightarrow Bi + K_2SnO_3 + H_2O$

(b) $Sb + HNO_{3(dil)} \rightarrow Sb_2O_5 + NO + H_2O$

(c) $Na_2TeO_3 + NaI + HCl \rightarrow NaCl + Te + H_2O + I_2$

(d) $Mn(NO_3)_2 + NaBiO_3 + HNO_3 \rightarrow HMnO_4 + Bi(NO_3)_3 + NaNO_3 + H_2O$

(e) $CoSO_4 + KI + KIO_3 + H_2O \rightarrow Co(OH)_2 + K_2SO_4 + I_2$

(f) $I_2O_5 + CO \rightarrow I_2 + CO_2$

(g) $Cr_2O_7^{2-} + H^{1+} + Cl^{1-} \rightarrow Cr^{3+} + Cl_2 + H_2O$

(h) $I^{1-} + MnO_4^{1-} + H^{1+} \rightarrow Mn^{2+} + I_2 + H_2O$

(i) $MnO_4^{1-} + SO_2 + H_2O \rightarrow Mn^{2+} + SO_4^{2-} + H^{1+}$

(j) $H_2S + Cr_2O_7^{2-} + H^{1+} \rightarrow S + Cr^{3+} + H_2O$

11. In the equations in Problem 10, indicate the substances that are oxidized and reduced and the oxidizing and reducing agents.

Oxidation-Reduction Equations—Ion Electron Method (See Sections 16-1 and 16-3)

12. Balance the following oxidation-reduction equations by the ion electron method:

(a) $Sn^{2+} + IO_3^{1-} \rightarrow Sn^{4+} + I^{1-} + H_2O$ in acid solution

(b) $AsO_2^{1-} + MnO_4^{1-} \rightarrow AsO_3^{1-} + Mn^{2+} + H_2O$ in acid solution

(c) $C_2O_4^{2-} + MnO_4^{1-} \rightarrow CO_2 + Mn^{2+} + H_2O$ in acid solution

(d) $Mn^{2+} + BiO_3^{1-} \rightarrow MnO_4^{1-} + Bi^{3+} + H_2O$ in acid solution

(e) $MnO_4^{1-} + H_2O_2 \rightarrow Mn^{2+} + H_2O + O_2$ in acid solution

(f) $Fe + NO_3^{1-} \rightarrow Fe^{3+} + NO + H_2O$ in acid solution

(g) $Cl_2 \rightarrow ClO_3^{1-} + Cl^{1-} + H_2O$ in basic solution

(h) $Cl_2 \rightarrow ClO^{1-} + Cl^{1-} + H_2O$ in basic solution (cold)

(i) $MnO_{2(s)} + O_2 \rightarrow MnO_4^{2-} + H_2O$ in basic solution

(j) $PbS_{(s)} + H_2O_2 \rightarrow PbSO_{4(s)} + H_2O$ in acid solution
(*Hint:* H_2O_2 is a nonelectrolyte.)

13. In the equations in Problem 12, indicate the substances that are oxidized and reduced and the oxidizing and reducing agents.

14. Balance the following oxidation-reduction equations by the ion electron method:
    (a) $Cr_2O_7^{2-} + C_2O_4^{2-} \rightarrow Cr^{3+} + CO_2 + H_2O$ in acid solution
    (b) $S^{2-} + NO_3^{1-} \rightarrow S + NO + H_2O$ in acid solution
    (c) $SO_3^{2-} + MnO_4^{1-} \rightarrow SO_4^{2-} + Mn^{2+} + H_2O$ in acid solution
    (d) $AsO_3^{3-} + Br_2 + H_2O \rightarrow AsO_4^{3-} + Br^{1-}$ in acid solution
    (e) $AsO_4^{3-} + I^{1-} \rightarrow AsO_3^{3-} + I_2$ in acid solution
    (f) $Bi(OH)_3 + SnO_2^{2-} \rightarrow SnO_3^{2-} + Bi + H_2O$ in basic solution
    (g) $Mn^{2+} + H_2O_2 \rightarrow MnO_{2(s)} + H_2O$ in basic solution
    (h) $MnO_4^{1-} + ClO_2^{1-} + H_2O \rightarrow MnO_{2(s)} + ClO_4^{1-}$ in basic solution
    (i) $NiS_{(s)} + HCl + HNO_3 \rightarrow NiCl_2 + NO_{(g)} + S_{(s)} + H_2O$ in acid solution
    (j) $HSbCl_6 + H_2S_{(g)} \rightarrow H_3SbCl_6 + S_{(s)}$ in acid solution
    (*Hint:* $HSbCl_6$ ionizes as H^{1+} and $SbCl_6^{1-}$, and H_3SbCl_6 ionizes as 3 H^{1+} and $SbCl_6^{3-}$.)

15. In the equations in Problem 14, indicate the substances that are oxidized and reduced and the oxidizing and reducing agents.

Standard Reduction Potentials   (See Section 16-6)

16. From the standard reduction potentials for metals (Table 16-1), determine whether the following reactions will occur:
    (a) $Al + Cd^{2+} \rightarrow Al^{3+} + Cd$
    (b) $Zn + Ni^{2+} \rightarrow Zn^{2+} + Ni$
    (c) $Pb + Zn^{2+} \rightarrow Pb^{2+} + Zn$
    (d) $Ni + Ag^{1+} \rightarrow Ni^{2+} + Ag$
    (e) $Hg + Pb^{2+} \rightarrow Hg^{2+} + Pb$

17. From the standard reduction potentials for the metals (Table 16-1) and for halogens (Table 16-2), give the E^0 value for the following reactions:
    (a) $Mg \rightleftarrows Mg^{2+} + 2e^-$
    (b) $Sn^{2+} + 2e^- \rightleftarrows Sn$
    (c) $Fe^{2+} + 2e^- \rightleftarrows Fe$
    (d) $2 Cl^{1-} \rightleftarrows Cl_2 + 2e^-$
    (e) $Br_2 + 2e^- \rightleftarrows 2 Br^{1-}$

General Problems

18. Using Tables 16-1 and 16-2, determine (a) the best oxidizing agent and (b) the best reducing agent of the following: Hg, Zn^{2+}, Ca, F_2, Br^{1-}.

19. "Tin" cans consist of a thin coating of tin on steel, which is primarily iron. In time, iron rusts in the atmosphere to form hydrated iron(III) oxide, which is then washed into the soil. Based on the standard oxidation potentials of tin and iron, why is tin used to coat these cans?

20. Refer to Problem 8, equation (a).
    (a) How many grams of nitrogen oxide (nitric oxide) can be produced from 75.0 g of pure nitric acid?

(b) How many liters of nitrogen oxide at STP can be produced from 75.0 g of pure nitric acid?

(c) How many liters of nitrogen oxide at 27°C and $\overline{700}$ torr pressure can be produced from 75.0 g of pure nitric acid?

(d) How many grams of nitrogen oxide can be produced from 45.0 mL of concentrated nitric acid? (Concentrated nitric acid is 72.0 percent pure nitric acid and has a density of 1.42 g/mL.)

21. In Problem 10, equation (a), how many grams of bismuth hydroxide are required to produce 6.25 g of precipitated bismuth metal?

22. Consider the following equation:

$$KMnO_4 + FeSO_4 + H_2SO_4 \rightarrow MnSO_4 + Fe_2(SO_4)_3 + K_2SO_4 + H_2O$$

(a) Balance this oxidation-reduction equation by the oxidation number method.

(b) Calculate the number of moles of potassium permanganate required to oxidize 3.20 g of iron(II) sulfate.

(c) How many mL of a 0.100 M potassium permanganate solution would be required to oxidize 3.20 g of iron(II) sulfate?

23. In one of the tests to determine alcohol content in the breath, the police officer uses a chemical test. The equation is as follows:

$$\underset{\text{ethyl alcohol}}{C_2H_6O} + K_2Cr_2O_7 + H_2SO_4 \rightarrow \underset{\text{acetic acid}}{C_2H_4O_2} + Cr_2(SO_4)_3 + K_2SO_4 + H_2O$$

The basis of the test is the change of the red-orange potassium dichromate ($K_2Cr_2O_7$) to the blue chromium(III) sulfate [$Cr_2(SO_4)_3$]. (a) Balance the equation by the oxidation number method. (b) What is the oxidizing agent?

Readings

Hoffman, George A., "The Electric Automobile." *Sci. Am.*, Oct. 1966, v. 215, p. 34. A timely article comparing the electric car with the internal-combustion car. Discusses the battery systems available and briefly mentions the possibility of using hydrogen-air fuel cells.

Kalb, Doris, "The Chemical Equation. Part II: Oxidation-Reduction Reaction." *J. Chem. Educ.*, 1978, v. 55, p. 326. Reviews balancing oxidation-reduction equations by the methods given in your text.

Treptow, Richard S., "Determination of Alcohol in Breath for Law Enforcement. *J. Chem. Educ.*, 1974, v. 51, p. 651. Discusses in detail the on-the-spot chemical test for alcohol from a person's breath.

17

Reaction Rates
and Chemical Equilibria

In nature physical equilibrium exists between the cloud cover and evaporation of water from the earth's surface to keep everything in balance. Chemical equilibria involve dynamic balance between chemical reactions and other chemical phenomena.

OBJECTIVES

1. Given the following terms, define each term and describe the distinguishing characteristic of each:
   (a) reaction rate (Introduction to Chapter 17)
   (b) chemical kinetics (Introduction to Chapter 17)
   (c) Law of Mass Action (Section 17-1)
   (d) Le Châtelier's Principle (Section 17-3)

2. Given the chemical equation and the rate equation for various reactions, determine the reaction order of each reactant and the overall reaction order (Problem Example 17-1, Problem 7).

3. Given the chemical equation and the reaction rate equation for various reactions, determine the effect a change in concentration of reactants would have on the reaction rate (Problem Example 17-2, Problem 8).

4. Given the values of the ionization constants (K) at a given temperature in water for various weak acids or bases of the same type, list these acids or bases in order of decreasing strength (Section 17-2, Problems 9 and 10).

5. Given the equation of a reaction along with the states of the reactants and products, write the expression for the equilibrium constant [(K), Problem Example 17-3, Problems 11 and 12].

6. Given the equation of a reaction along with the states of the reactants and products and the heat energy of the reaction, predict the direction the equilibrium will be shifted by a change in concentration of reactants and products, a temperature change, and a pressure change (Le Châtelier's Principle, Section 17-3, Problems 13, 14, 15, and 16).

7. Given the molar concentration of a weak, singly ionized acid or base and the percent ionization, calculate the equilibrium constant (K) for the acid or base (Problem Example 17-4, Problems 17 and 18).

8. Given the equilibrium constant and the molar concentration of a weak, singly ionized acid or base, calculate the molar concentration of the hydrogen ion or hydroxide ion and the percent ionization for the acid or base (Problem Examples 17-5 and 17-6, Problems 19 and 20).

9. Given the values of the solubility product constants (K_{sp}) at a given temperature and in water for several slightly soluble electrolytes of the same type, determine their orders of solubilities in water (Section 17-5, Problem 21).

10. Given the formula of a slightly soluble electrolyte, write the expression for the solubility product constant [(K_{sp}), Problem Example 17-7, Problems 22 and 23].

11. Given the concentration of a slightly soluble electrolyte in grams per liter, its formula or name (all of the *AB* type), and the Table of Approximate Atomic Masses, calculate the solubility product constant (K_{sp}) for the electrolyte (Problem Example 17-8, Problems 24 and 25).

12. (a) Given the solubility product constant (K_{sp}) for a slightly soluble electrolyte, its formula or name (all of the *AB* type), and the Table of Approximate Atomic Masses, calculate its concentration in moles per liter (molarity) and grams per liter.

(b) Given the solubility product constant (K_{sp}) for a slightly soluble electrolyte, its formula or name (all of the *AB* type), and the molar concentration of the electrolyte, predict if precipitation would occur.

(Problem Examples 17-9 and 17-10, Problems 26 and 27)

In addition to finding the products for a given reaction and its stoichiometry by balancing the equation, we must consider two other factors in a chemical reaction: reaction rates and chemical equilibria. The **reaction rate** is the rate or speed at which the products are produced or the reactants consumed. An explosion is an example of a fast reaction, while the formation of oil from decayed organic matter is an example of a slow reaction.

The study of reaction rates and the mechanism or path of a reaction from its reactants to its products constitutes a special subdivision of most chemistry branches (physical, inorganic, organic, biochemistry, etc.) called **chemical kinetics**. We have previously encountered dynamic equilibria many times: with vapor pressure (12-3); in the process of melting or freezing—going from the solid to the liquid states or vice versa (12-9); with saturated solutions in the rate of dissolution and crystallization (14-4); and, more recently, with standard reduction potential half-cell equations (16-6), to name just a few. In the last case, these are *chemical* equilibria, since they involve chemical reactions and not just changes of state or condition. *Most* chemical reactions are *practically* reversible; *all* chemical reactions are *theoretically* reversible. The reactants react to form products and the products react to form reactants, until a state of dynamic equilibrium is reached, which is indicated by a double arrow ($\rightleftharpoons$) when the rates of both reactions become the same. Therefore, reactions "go both ways."

Reaction rates and chemical equilibria are related, in that in the course of a reaction the rate of the forward reaction *decreases* and the rate of the reverse reaction *increases* until equilibrium is reached.

17-1

Reaction Rates   In 1864, Cato M. Guldberg (1836–1902) and Peter Waage (1833–1900), Norwegian chemists, proposed the Law of Mass Action, describing the rate of a chemical reaction in relation to the concentration of the reactants. The **Law of Mass Action** states that the rate of a chemical reaction is proportional to the "active masses" of the reactants. The "active masses" have been found related to the relative molar concentration (see 14-8) of the reactants in moles per liter for solutions or for gases.

For a general reaction,

$$aA \;+\; bB \rightarrow cC \;+\; dD \tag{17-1}$$

the rate of the *overall reaction*[1] according to the Law of Mass Action is proportional to the concentrations of the reactants in moles per liter raised to certain powers:

$$\text{Rate} \propto [A]^x[B]^y \tag{17-2}$$

where

$$[A] = \text{concentration of } A \text{ in moles per liter}$$

$$[B] = \text{concentration of } B \text{ in moles per liter}$$

$$x \text{ and } y = \text{whole numbers, fractional numbers, negative numbers, or}$$
$$\text{zero, as determined by } \textit{experimentation}$$

The equation for the rate expression is, therefore,

$$\text{Rate} = k[A]^x[B]^y \tag{17-3}$$

where

$$k = \text{a proportionality constant, called the } \textit{specific rate constant}$$

The values of x and y are *sometimes* equal to the coefficients of the balanced equation—that is, a and b—but they do not have to be; in fact, they may be fractional numbers, negative numbers, or zero. The actual values for x and y must be determined by *experimentation* followed by considerable calculations.

---

[1]The overall reaction may involve many steps; rates for each step may be written. The rate equations given here are for the *overall* reaction. As in *any sequential* operation, the overall rate is the same as the *rate of the* **slowest** *step* in the sequence.

The value of x and y in Equation 17-3 is the **reaction order** of each **reactant**. The sum of x and y is the **overall reaction order** for the reaction. If x is equal to **1** and y is equal to **2**, then the reaction order is *first* order in A and *second* order in B, with the overall reaction order being *three* (**1** + **2** = 3) for this reaction.

For example, consider the decomposition of two different nitrogen oxide compounds:

1. decomposition of nitrogen dioxide

$$2 \, NO_{2(g)} \rightarrow 2 \, NO_{(g)} + O_{2(g)} \qquad (17\text{-}4)$$

2. decomposition of dinitrogen pentoxide

$$2 \, N_2O_{5(g)} \rightarrow 4 \, NO_{2(g)} + O_{2(g)} \qquad (17\text{-}5)$$

In the decomposition of nitrogen dioxide (Equation 17-4), the rate of the overall reaction has been found by experimentation to be

$$\text{Rate} = k[NO_2]^2 \qquad (17\text{-}6)$$

Here, note that the superscript is the same as the coefficient of NO_2 in the balanced equation. The reaction is second order in NO_2 and the overall reaction is also second order. In the decomposition of dinitrogen pentoxide (Equation 17-5), the rate of the overall reaction has been found to be

$$\text{Rate} = k[N_2O_5] \qquad (17\text{-}7)$$

In the rate for this reaction, the superscript is not the same as the coefficient of N_2O_5 in the balanced equation. In the balanced equation, the coefficient in front of N_2O_5 is 2, while in the rate equation the superscript is 1. These superscripts must be determined by *experimentation*. The reaction is first order in N_2O_5 and the overall reaction is also first order.

The rate of a given chemical reaction is often followed (observed or traced) by a change in color or pressure from reactants to products that accompanies the reaction. In the preceding decompositions, the rate of the reaction can be observed in Equation 17-4 by the disappearance of the brown gas of nitrogen dioxide and in Equation 17-5 by the appearance of the brown gas.

Problem Example 17-1

Given the following chemical equations and rate equations, determine the reaction order of each reactant and the overall reaction order.

(a) $Cl_{2(g)} + \underset{\text{chloroform}}{CHCl_{3(g)}} \rightarrow HCl_{(g)} + \underset{\substack{\text{carbon} \\ \text{tetrachloride}}}{CCl_4}$

$\text{rate} = k[Cl_2]^{1/2}[CHCl_3]$

(b) $A + B \rightarrow C$
$\text{rate} = k[A]^2[B]^3$

SOLUTION

Work Problem 7.

(a) The reaction is half order in chlorine and first order in chloroform, with the overall reaction order being 1.5. *Answer*

(b) The reaction is second order in A and third order in B with the overall reaction order being 5. *Answer*

Let us now consider the four factors that influence the rate of a given chemical reaction: (1) nature of reactants, (2) concentrations of reactants, (3) temperature, and (4) catalyst.

Nature of Reactants

The rate of a chemical reaction varies depending on the reactivity of the reactants, that is, their ability to break old bonds and form new ones. Some chemical reactions are fast and some are slow.

The decomposition of nitrogen dioxide (Equation 17-4) is an example of a relatively fast reaction that is complete in a few minutes. The decomposition of dinitrogen pentoxide (Equation 17-5), with comparable conditions to Equation 17-4, requires over 100 minutes to go to completion.

Concentration of Reactants

From the rate of a reaction determined experimentally, as expressed in Equation 17-3, you would expect the rate to increase with the concentration of the reactants, since they appear in the rate equation. This occurs. As the concentration increases, more reactant molecules collide with other reactant molecules (picture a number of people in a small room), and more product molecules are formed. Increasing the pressure on a gaseous reaction system would decrease the volume and would consequently increase the concentrations of the reactants, and hence increase the rate. The effect on the rate will depend on the power to which the reactant is raised in the rate equation. In the decomposition of nitrogen dioxide, **doubling** the concentration of nitrogen dioxide would cause a *fourfold* increase [$(1 \times 2)^2 = 4$] in the rate, since the rate $= k[NO_2]^2$. In the decomposition of dinitrogen pentoxide, **doubling** the concentration of nitrogen pentoxide would cause a *twofold* increase [$(1 \times 2)^1 = 2$] in the rate, since the rate $= k[N_2O_5]$.

Problem Example 17-2

Given the following chemical equation:

$$A + B \rightarrow C$$

with the reaction rate $= k[A][B]^2$. Calculate the effect on the reaction rate if the following occurs:

(a) The concentration of A is doubled.
(b) The concentration of B is doubled.
(c) The concentration of B is tripled.
(d) The concentration of B is increased 1.50 times.

SOLUTION

Work Problem 8.

(a) $A = (1 \times 2)^1 = 2$ times the rate   *Answer*
(b) $B = (1 \times 2)^2 = 4$ times the rate   *Answer*
(c) $B = (1 \times 3)^2 = 9$ times the rate   *Answer*
(d) $B = (1 \times 1.50)^2 = 1.50 \times 1.50 = 2.25$ times the rate   *Answer*

Temperature

As the temperature increases for a given reaction, the number of molecules of reactants with sufficient energy to react increases, and hence the rate of the reaction also increases. The reactant molecules need a sufficient amount of energy to react and get over a "reaction *energy* barrier" (activation energy). This increase in rate with increased temperature applies to *both* exothermic and endothermic reactions (10-7) and is somewhat analogous to giving the riders on a toboggan a push to get them started down a hill.

In the decomposition of nitrogen dioxide (Equation 17-4), from a temperature of 320°C to 330°C (10°C rise) the rate increases by a factor of 1.5. In the decomposition of dinitrogen pentoxide (Equation 17-5), from a temperature of 45°C to 55°C (10°C rise) the rate increases by a factor of 3.0. Therefore, an increase in temperature increases the rate, but the exact amount of rate increase varies with the reaction and with the nature of the reactants. As a rule, for *each 10°C* rise in temperature, the rate roughly *doubles* or *triples*, as you can see from the two cases above.

Catalyst

We have previously (9-2) defined a *catalyst* as a substance that speeds up a chemical reaction but is recovered without appreciable change at the end of the reaction. The catalyst affects the rate of a reaction by increasing it. It affects the time but *not* the point at which equilibrium is reached (see 17-2). It shortens the time required for a reaction to reach equilibrium. Various types of catalysts are enzymes, chlorophyll in photosynthesis reactions, and manganese(IV) oxide in the decomposition of potassium chlorate (9-4).

17-2

Irreversible and Reversible Reactions. Chemical Equilibria

Some reactions are irreversible *in practice;* that is, a chemical equilibrium is apparently not established and the reactions essentially go to completion. This occurs when the products are removed or the reverse reaction rate is so slow as to be negligible. The frying of a hamburger or an egg is an example of an "irreversible" reaction. The formation of any one of the following substances as products acts as the driving force for a reaction to go "irreversibly," that is, to go toward completion: (1) gas, (2) precipitate, or (3) a nonionized or only partially ionized substance, such as water.

The formation of a gas, **which is removed as soon as it forms,** drives to completion one type of double replacement reaction (9-10). If the gas remains in contact with the reactants, as in a closed container, then a reversible reaction occurs and an equilibrium is established. An example of the production of a gas that is removed as it is formed is

$$MgCO_{3(s)} + 2\ HCl_{(aq)} \rightarrow MgCl_{2(aq)} + H_2O_{(\ell)} + CO_{2(g)} \qquad (17\text{-}8)$$

or, as a net ionic equation (see 15-7),

$$MgCO_{3(s)} + 2\ H^{1+}{}_{(aq)} \rightarrow Mg^{2+}{}_{(aq)} + H_2O_{(\ell)} + CO_{2(g)} \qquad (17\text{-}9)$$

The formation of a precipitate from a solution acts to drive toward completion other types of double replacement reactions. The precipitation of the substance acts essentially to remove it from the reaction. (Actually, the reaction is still reversible, as long as the solid remains in contact with the solution, but the equilibrium strongly favors the products.) An example is

$$AgNO_{3(aq)} + HCl_{(aq)} \rightleftharpoons AgCl_{(s)} + HNO_{3(aq)} \qquad (17\text{-}10)$$

or

$$Ag^{1+}{}_{(aq)} + Cl^{1-}{}_{(aq)} \rightleftharpoons AgCl_{(s)} \qquad (17\text{-}11)$$

Water, which we mentioned in 15-3 as being only slightly ionized, acts to drive toward completion some neutralization reactions (9-11). (Again, the equilibrium is established but strongly favors the products.) An example is

$$NaOH_{(aq)} + HBr_{(aq)} \rightleftharpoons NaBr_{(aq)} + H_2O_{(\ell)} \qquad (17\text{-}12)$$

or

$$OH^{1-}{}_{(aq)} + H^{1+}{}_{(aq)} \rightleftharpoons H_2O_{(\ell)} \qquad (17\text{-}13)$$

Now, let us consider in detail a general case of a reversible reaction:

$$A + B \rightleftharpoons C + D \qquad (17\text{-}14)$$

The reaction $A + B \rightarrow C + D$ is considered the forward reaction, and $C + D \rightarrow A + B$ the reverse reaction. If we start with a mixture of A and B at a given temperature and pressure, A and B will react with each other to form C and D at a rate dependent on the initial (starting) concentrations. Initially, the reverse reaction has a rate of 0, since no C and D molecules are present; however, as their concentrations increase, the rate of the reverse reaction will increase. During this time, the rate of the forward reaction decreases as the concentration of A and B decreases. Eventually, the rate at which the C and D molecules react to form A and B molecules *will be equal* to the rate at which the A and B molecules form C and D molecules. *When this point is reached, the system is in chemical equilibrium.* This is a dynamic equilibrium, because these two opposite reactions are continually taking place. Once equilibrium is achieved, *no* net increase or decrease of reactants and products is observed if the conditions remain the same. The graph of this general reaction is depicted in Figure 17-1.

FIGURE 17-1

Chemical equilibrium is reached when the rate of the forward reaction is equal to the rate of the reverse reaction.

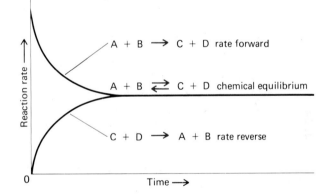

For any equilibrium reaction, a constant known as the equilibrium constant (K) can be obtained if all the quantities in the expression can be determined experimentally. Consider the reaction in Equation 17-14 and assume it to occur in either direction in a single step. For a single-step reaction, the rate of the *forward* reaction is directly proportional to the concentration of A and B from the Law of Mass Action. The rate of the forward reaction may be expressed as

$$\text{rate forward} \propto [A][B] \qquad (17\text{-}15)$$

$$\text{rate forward} = k_f[A][B] \qquad (17\text{-}16)$$

and the rate of the reverse reaction as

$$\text{rate reverse} \propto [C][D] \tag{17-17}$$

$$\text{rate reverse} = k_r[C][D] \tag{17-18}$$

where k_f and k_r are the specific rate constants for the forward and reverse reactions, respectively. The superscripts in the equilibrium expression are the same as in the balanced equation. The total forward rate and the total reverse rate in an equilibrium system are related to the concentrations of the reactants or products raised to appropriate powers.[2]

*At equilibrium, the rate of the forward reaction will be **equal** to the rate of the reverse reaction.* **Or,**

$$k_f[A][B] = k_r[C][D] \tag{17-19}$$

which can be arranged to

$$\frac{k_f}{k_r} = \frac{[C][D]}{[A][B]} \tag{17-20}$$

Since k_f and k_r are constants, k_f/k_r is a constant and is given by the symbol K. This is the equilibrium constant for the reaction in Equation 17-14, and it will have a certain value which is constant at a *fixed temperature* for a given reaction.

$$K = \frac{[C][D]}{[A][B]} \tag{17-21}$$

For the very general reaction

$$aA + bB \rightleftarrows cC + dD \tag{17-22}$$

where A, B, C, and D represent different molecular species and a, b, c, and d are coefficients in the balanced equation, we can obtain the following expression for the equilibrium constant:

$$K = \frac{[C]^c[D]^d}{[A]^a[B]^b} \tag{17-23}$$

Thus, the **coefficients** of the substance in the chemical equation become **exponents** for the relative concentrations in the expression for the equilibrium constant, K. At a *given* temperature, K is essentially constant.

---

[2] These powers may be equal to the coefficients of the substances in the balanced equation; however, they could be different depending on the mechanism of the reaction.

The *larger* is the value of K, the *greater is the concentration of product,* assuming the reactions are all of the same type, that is, an HX acid. For example, the constant for the aqueous ionization of acetic acid ($HC_2H_3O_2$) is 1.76×10^{-5} (mol/L) at 25°C, whereas that for the aqueous ionization of lactic acid ($HC_3H_5O_3$) is 1.37×10^{-4} or 13.7×10^{-5} (mol/L) at 25°C. Therefore, since the K value for lactic acid is larger (smaller negative exponent) than it is for acetic acid, more ions are produced in a given volume of solution per mole of lactic acid than per mole of acetic acid; hence, lactic acid is a stronger acid than acetic acid.

Work Problems 9
and 10.

$$HC_2H_3O_{2(aq)} \rightleftharpoons H^{1+}_{(aq)} + C_2H_3O_2^{1-}_{(aq)} \qquad (17\text{-}24)$$
$$\text{acetic acid}$$

$$HC_3H_5O_{3(aq)} \rightleftharpoons H^{1+}_{(aq)} + C_3H_5O_3^{1-}_{(aq)} \qquad (17\text{-}25)$$
$$\text{lactic acid}$$

The constant K may not have units. In Equation 17-21, if all the concentrations are measured in moles per liter, then K would have no units, as

$$K = \frac{[\text{mol/L}][\text{mol/L}]}{[\text{mol/L}][\text{mol/L}]}$$

If the concentration of C had to be raised to the second (2nd) power (see Equations 17-22 and 17-23), then the units of K would be moles per liter, as

$$K = \frac{[\text{mol/L}]^2[\text{mol/L}]}{[\text{mol/L}][\text{mol/L}]} = (\text{mol/L})$$

In this text, we shall place the units of K in parentheses immediately following the numerical value for K, if such units exist, as was done for K_w (see 15-3) and just now. When there are no units, we shall give only the numerical value of K.

In the equilibrium expression for an equation that contains a solid, the **solid** is *not considered* in the *expression*, since at constant temperature the concentration in moles per liter of the solid in the solid phase is constant and will not change, and this value is automatically included in K.

An equilibrium in which at least one reactant or product is heterogeneously dispersed in the other is classed as a *heterogeneous equilibrium*. Examples of this are a dense solid and a gas in equilibrium in a closed container or two insoluble liquids (immiscible). If all reactants and products are homogeneously dispersed, as in the gaseous state, the equilibrium is classed as a *homogeneous equilibrium*.

The commercial production of ammonia by the Haber process is a typical example of a reversible reaction that exists in *homogeneous* chemical equilibrium. The Haber process, developed in 1914 by Fritz Haber[3] (1869–1934), a German chemist, consists of the reaction of nitrogen with hydrogen to form ammonia, which is in equilibrium with the reactants. The reaction is exothermic, with 2.2×10^4 cal (9.2×10^4 J) of heat energy evolved per mole of nitrogen used.

$$N_{2(g)} + 3\ H_{2(g)} \xrightarrow{\text{catalyst}} 2\ NH_{3(g)} + 2.2 \times 10^4\ \text{cal}\ (9.2 \times 10^4\ \text{J}) \quad (17\text{-}26)$$

The reaction is carried out at approximately 500°C, with a pressure of about 200 to 600 atm, in the presence of a catalyst such as magnetic iron oxide (Fe_3O_4) and a potassium oxide-aluminum oxide mixture, $K_2O \cdot Al_2O_3$. After leaving the catalyst chamber, the ammonia is liquefied and separated from the nitrogen and hydrogen gases, which are recycled in the process.

Referring to Equation 17-23 for the general reaction, we can write the equilibrium expression for this reaction as follows:

$$K = \frac{[NH_3]^2}{[N_2][H_2]^3}$$

The coefficients of the substances in the chemical equation become exponents for the concentrations in the expression for the equilibrium constant K. To simplify the equilibrium expression, we do not write the state of the reactants and products.

Consider the following problem example for expressing the equilibrium constant.

Problem Example 17-3

Write the expression for the equilibrium constant for each of the following reactions:

(a)
$$CO_{(g)} + Cl_{2(g)} \rightleftarrows COCl_{2(g)}$$

(b)
$$H_{2(g)} + I_{2(g)} \rightleftarrows 2\ HI_{(g)}$$

(c)
$$CaCO_{3(s)} \rightleftarrows CaO_{(s)} + CO_{2(g)}$$

SOLUTION: From Equation 17-23, K is as follows for the preceding reactions:

(a)
$$K = \frac{[COCl_2]}{[CO][Cl_2]} \quad \textit{Answer}$$

(b)
$$K = \frac{[HI]^2}{[H_2][I_2]} \quad \textit{Answer}$$

---

[3]Haber received the Nobel Prize in chemistry in 1918.

(c) $$K = [CO_2]   \textit{Answer}$$

Work Problems 11 and 12.

Both $CaCO_3$ and CaO are solids, and hence their concentrations are constant at a given temperature. They are, therefore, not considered in the equilibrium expression because they are included in the value for the constant K.

17-3

Le Châtelier's Principle

In 1888, Henry Louis Le Châtelier (le·shä′ te·lyā′) (1850–1926), French chemist, formulated a principle governing equilibrium. **Le Châtelier's Principle** states that if an *equilibrium system* is subject to a change in conditions of *concentration, temperature,* or *pressure,* the system will change to a new equilibrium position, where possible, in a direction that will tend to *restore* the original conditions. Now let us consider these changes in conditions, the effect of (1) concentration, (2) temperature, and (3) pressure.

Concentration

When the concentration of one of the substances in a system at equilibrium is increased, Le Châtelier's Principle predicts that the *equilibrium will shift so as to use up partially the added substance.* Decreasing the concentration of one of the substances in a system at equilibrium will cause the equilibrium to shift so as to replenish partially the substance removed. In all cases, the equilibrium constant K will remain essentially *constant,* with the concentration of the reactants or products varying.

In the Haber process for ammonia (17-2; see Equation 17-26), an increase in concentration of either nitrogen or hydrogen will shift the equilibrium to the right (the products side) and will decrease the concentration of the other reactant. Increasing the concentration of ammonia will shift the equilibrium to the left (the reactants side). Conversely, decreasing the concentration of either nitrogen or hydrogen will shift the equilibrium to the left (the reactants side), and decreasing the concentration of ammonia will shift the equilibrium to the right (the products side). Figure 17-2a summarizes the effect of an increase in concentration of various substances on the equilibrium in the Haber process. Therefore, to obtain a maximum yield of ammonia by the Haber process, either nitrogen or hydrogen is used in *excess,* with the excess being recycled and the ammonia gas that is formed being constantly removed.

Temperature

Le Châtelier's Principle, as applied to the temperature effect on an equilibrium system, may be stated as follows: If the temperature of a system *at equilibrium* is changed, the equilibrium will shift so as to change the temperature toward its original value. If the reaction is *exothermic,* as in the following equation,

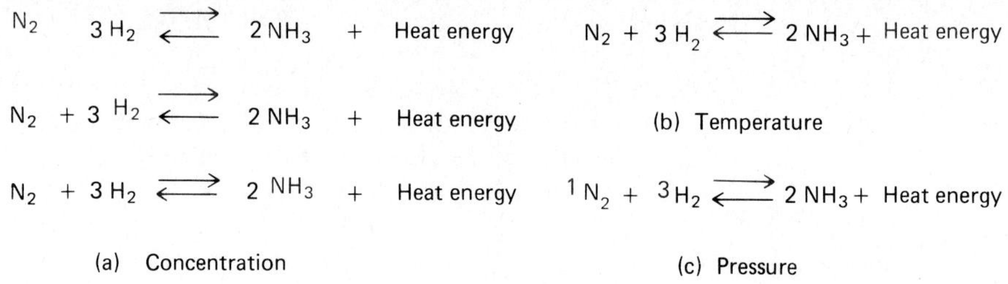

$$N_2 \quad 3\,H_2 \; \overset{\longrightarrow}{\longleftarrow} \; 2\,NH_3 \;+\; \text{Heat energy} \qquad N_2 + 3\,H_2 \; \overset{\longrightarrow}{\longleftarrow} \; 2\,NH_3 + \text{Heat energy}$$

$$N_2 + 3\,H_2 \; \overset{\longrightarrow}{\longleftarrow} \; 2\,NH_3 \;+\; \text{Heat energy} \qquad \text{(b) Temperature}$$

$$N_2 + 3\,H_2 \; \overset{\longrightarrow}{\longleftarrow} \; 2\,NH_3 \;+\; \text{Heat energy} \qquad {}^1N_2 + {}^3H_2 \; \overset{\longrightarrow}{\longleftarrow} \; 2\,NH_3 + \text{Heat energy}$$

(a) Concentration   (c) Pressure

FIGURE 17-2

Summary of the effects of an increase of various factors on the equilibrium for the Haber process for the production of ammonia. (a) Concentration. (b) Temperature. (c) Pressure. (Note the relationship of the **colored** symbols and/or words with the **colored** arrows.)

the heat resulting from an increased temperature will act *as if it were one of the products*.

$$A + B \rightleftarrows C + D + \text{heat energy} \tag{17-27}$$

An increase in temperature will shift the equilibrium to the left (the reactants side), whereas a decrease in temperature will shift the equilibrium to the right (the products side). If the reaction is *endothermic*, as in the following equation, the heat resulting from an increased temperature will act *as if it were one of the reactants*.

$$A + B \rightleftarrows C + D - \text{heat energy} \tag{17-28}$$

or

$$A + B + \text{heat energy} \rightleftarrows C + D$$

An increase in temperature will shift the equilibrium to the right, whereas a decrease in temperature will shift the equilibrium to the left. As well as shifting the equilibrium with a change in temperature, the equilibrium constant *also* changes when the temperature is changed.

The Haber process for ammonia (see Equation 17-26) is *exothermic*, and hence an increase in temperature will shift the equilibrium to the left, decreasing the production of ammonia and also decreasing the value of the equilibrium constant. But, at a very low temperature, the time to obtain the maximum equilibrium yield may be too long. However, at higher temperature the time to reach equilibrium is shortened, but the point of equilibrium is unfavorable to the product. Compensation for this is made by increasing the pressure of the system, which improves the product yield as described below. Figure 17-2b summarizes the effect of an increase in temperature on equilibrium for the Haber process.

Pressure

As applied to the effect of pressure on an equilibrium system, involving one or more gaseous substances, Le Châtelier's Principle would predict that *increasing* the *pressure* on a system at equilibrium will shift the equilibrium in that direction which will *decrease the volume*. Decreasing the pressure will have the opposite effect. If there is *no volume change* in going from reactants to products, pressure will have *no effect* on the equilibrium. The equilibrium constant does not change with pressure.

In the Haber process for ammonia (see Equation 17-26), one volume of nitrogen gas reacts with three volumes of hydrogen gas to form two volumes of ammonia gas. If pressure is applied, the system reacts to offset the strain by forming ammonia, since **four** volumes of *gas* are on the left side of the equation $(1 + 3 = 4)$ and only **two** volumes of *gas* are on the right. Therefore, an increased yield of ammonia requires high pressures. A pressure of about 200 to 600 atm has been found to be satisfactory. Higher pressure could also be used, but the cost of equipment for handling it would offset the value of the increased yield. Figure 17-2c summarizes the effect of an increase in pressure on equilibrium for the Haber process. Using a pressure of 600 atm and 500°C, a 42.1 percent yield of ammonia by volume is obtained in the equilibrium mixture.

Consider as another example the formation of nitrogen oxide according to the following equation:

$$N_{2(g)} + O_{2(g)} \rightleftarrows 2\,NO_{(g)} \qquad (17\text{-}29)$$

Pressure would have *no* effect on the equilibrium, since **two** volumes of reactants (*one* volume of nitrogen and *one* volume of oxygen) react to give **two** volumes of product (nitrogen oxide). Also, as we would expect, a solid would not be affected by a pressure change.

In 17-1, we mentioned that a catalyst acts to increase the rate of a reaction. A catalyst does *not affect the position* of the equilibrium, since it increases both the forward and the reverse reactions to the same degree. Hence, it *will not* increase the yield of the desired product, but it will *decrease the time* required for equilibrium to be established. In the Haber process, the catalysts are magnetic iron oxide (Fe_3O_4) and a potassium oxide-aluminum oxide mixture $(K_2O \cdot Al_2O_3)$. They make the process economically feasible, for without them, the reaction would take considerable time to reach equilibrium.

Work Problems 13, 14, 15, and 16.

17-4

Weak Electrolyte Equilibria

As we mentioned in 15-5, weak electrolytes are ionized only a few percent. Therefore, solutions of weak electrolytes contain both the *un-ionized substance* and the *ions* resulting from the ionization of the weak electrolyte. The partial ionization in water of a weak electrolyte reaches equilibrium. It may therefore be treated as are equilibrium reactions, discussed in 17-2, with the same form as was used in expressing the equilibrium constant (K) in Problem Example 17-3.

A solution of acetic acid in water is a simple equilibrium involving a weak electrolyte. The following equation represents the equilibrium between the nonionized acetic acid ($HC_2H_3O_2$) and its ions in solution:

$$HC_2H_3O_{2(aq)} \rightleftharpoons H^{1+}_{(aq)} + C_2H_3O_2^{1-}_{(aq)} \qquad (17\text{-}30)$$

The equilibrium constant, now called K_a for an acid, is given by the following expression:

$$K_a = \frac{[H^{1+}][C_2H_3O_2^{1-}]}{[HC_2H_3O_2]} \qquad (17\text{-}31)$$

The $[H^{1+}]$ and $[C_2H_3O_2^{1-}]$ are the concentrations of hydrogen ions and acetate ions expressed in moles per liter, and $[HC_2H_3O_2]$ is the concentration of acetic molecules (*at equilibrium*) in moles per liter. The constant must be determined experimentally, and for acetic acid at 25°C it is 1.76×10^{-5} (mol/L). For simplicity, the state of reactants and products are not written. The concentration of the water and its slight ionization is included in the K_a constant.

A similar equilibrium is found in aqueous ammonia solution. The following equation represents the equilibrium between the nonionized aqueous ammonia and its ions in solution:

$$NH_{3(aq)} + HOH_{(\ell)} \rightleftharpoons NH_4^{1+}_{(aq)} + OH^{1-}_{(aq)} \qquad (17\text{-}32)$$

The equilibrium constant, now called K_b for a base, is given by the following expression:

$$K_b = \frac{[NH_4^{1+}][OH^{1-}]}{[NH_3]} \qquad (17\text{-}33)$$

The $[NH_4^{1+}]$ and $[OH^{1-}]$ are the concentrations of ammonium ions and hydroxide ions expressed in moles per liter, and $[NH_3]$ is the concentration of aqueous ammonia (*at equilibrium*) in moles per liter. The experimentally determined constant for aqueous ammonia at 25°C is 1.79×10^{-5} (mol/L). Again, for simplicity, we do not write the state of the reactants and products. The concentration of the water and its slight ionization is included in the K_b constant.

Now, let us consider some calculations involving weak electrolytes in equilibrium.

Problem Example 17-4

The hydrofluoric acid in a 0.0400 M HF solution is 13.4 percent ionized. Calculate K_a for HF.

SOLUTION: The equation representing the ionization is

$$HF_{(aq)} \rightleftharpoons H^{1+}_{(aq)} + F^{1-}_{(aq)}$$

and the expression for K_a would be

$$K_a = \frac{[H^{1+}][F^{1-}]}{[HF]}$$

To evaluate the K_a, determine the concentrations of H^{1+} and F^{1-} and HF (*at equilibrium*) from the data given. If the HF is 13.4 percent ionized, the percent *nonionized* at equilibrium would be 86.6 percent (100.0 − 13.4), and the concentrations of H^{1+}, F^{1-}, and HF in solution are calculated as follows:

$$[H^{1+}] = 0.0400 \text{ mol/L} \times 0.134 = 0.00536 \text{ mol/L}$$

(The $[F^{1-}]$ is equal to the $[H^{1+}]$, since in the balanced equation **1 mol** of HF ionizes to form **1 mol** of H^{1+} and **1 mol** of F^{1-}; therefore, **0.00536 mol** of HF would ionize to form **0.00536 mol** of H^{1+} and **0.00536 mol** of F^{1-}.)

$$[F^{1-}] = 0.00536 \text{ mol/L}$$

$$[HF] = 0.0400 \text{ mol/L} \times 0.866 = 0.0346 \text{ mol/L } at\ equilibrium$$

We can also obtain the [HF] by subtracting the mol/L of ionized HF from the mol/L of the initial HF, as 0.0400 mol/L (initially) − 0.00536 mol/L (ionized from the mole relationship in the balanced equation) = 0.0346 mol/L. The value of K_a is calculated as follows:

$$K_a = \frac{[H^{1+}][F^{1-}]}{[HF]} = \frac{[0.00536 \text{ mol/L}][0.00536 \text{ mol/L}]}{[0.0346 \text{ mol/L}]}$$

$$= \frac{[5.36 \times 10^{-3}][5.36 \times 10^{-3}](mol^2/L^2)}{[3.46 \times 10^{-2}](mol/L)}$$

Work Problems 17 and 18.

$$= 8.30 \times 10^{-4} \text{ (mol/L)} \quad Answer$$

Problem Example 17-5

Calculate the hydrogen ion concentration and the percent ionization at 25°C in a 0.102 M acetic acid ($HC_2H_3O_2$) solution. The K_a for $HC_2H_3O_2 = 1.76 \times 10^{-5}$ (mol/L) at 25°C.

SOLUTION: The equation representing the ionization is

$$HC_2H_3O_{2(aq)} \rightleftharpoons H^{1+}_{(aq)} + C_2H_3O_2^{1-}_{(aq)}$$

and the expression for K_a is

$$K_a = \frac{[H^{1+}][C_2H_3O_2^{1-}]}{[HC_2H_3O_2]}$$

In the balanced equation, **1 mol** of $HC_2H_3O_2$ on *ionizing* forms **1 mol** of H^{1+} and **1 mol** of $C_2H_3O_2^{1-}$. If **5 mol** of $HC_2H_3O_2$ ionize, we would have **5 mol** of H^{1+} and **5 mol** of $C_2H_3O_2^{1-}$, and if x **mol** of $HC_2H_3O_2$ ionize, this would form **x mol** of H^{1+} and x **mol** of $C_2H_3O_2^{1-}$. Therefore, let $x = [H^{1+}]$ in mol/L, and x mol/L also equals $[C_2H_3O_2^{1-}]$. The concentration of the nonionized $HC_2H_3O_2$ remaining *at equilibrium* is $[HC_2H_3O_2] = 0.102$ mol/L $- x$ mol/L. The value of 0.102 mol/L of acetic acid *initially* and the value of 0.102 mol/L $- x$ mol/L of acetic acid *at equilibrium* may need some further explanation. This new value at equilibrium is what is left *after* some of the acetic acid has ionized. The following illustration may be helpful:

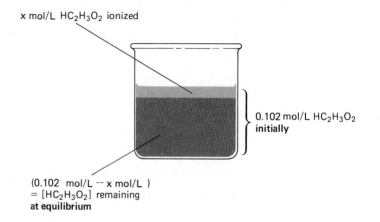

x mol/L $HC_2H_3O_2$ ionized

0.102 mol/L $HC_2H_3O_2$
initially

(0.102 mol/L — x mol/L)
= $[HC_2H_3O_2]$ remaining
at equilibrium

Substituting the values for $[H^{1+}]$, $[C_2H_3O_2^{1-}]$, and $[HC_2H_3O_2]$ into the equilibrium expression just given, and dropping the units of x for simplicity, we have the following:

$$1.76 \times 10^{-5} \text{ (mol/L)} = \frac{[x][x]}{[0.102 \text{ mol/L} - x]}$$

$$1.76 \times 10^{-5} \text{ (mol/L)} = \frac{[x]^2}{[0.102 \text{ mol/L} - x]}$$

To solve this equation for x, we must use the quadratic equation; however, we can simplify by assuming that x is very small in relation to the original concentration of $HC_2H_3O_2$ (0.102 M). Therefore, the quantity 0.102 mol/L $- x$ will have a value not much different from 0.102 mol/L. Hence, x in the *denominator* may be neglected to simplify solving the equation without greatly affecting the answer. Dropping the x is analogous to taking a bucket of water from the ocean. The loss of such a small amount has relatively negligible effect on the overall volume of water in the ocean. In some cases, such as very low concentrations and relatively large K_a values, the x cannot be neglected and the quadratic equation must be solved. For our purposes in this text, we shall consider only cases where we can neglect the x in the denominator. Neglecting the x gives the following expression:

$$1.76 \times 10^{-5} \text{ (mol/L)} = \frac{[x]^2}{0.102 \text{ mol/L}}$$

and

$$x^2 = 1.76 \times 0.102 \times 10^{-5} \frac{\text{mol}^2}{\text{L}^2} = 1.80 \times 10^{-6} \frac{\text{mol}^2}{\text{L}^2}$$

$$x = \sqrt{1.80 \times 10^{-6} \frac{\text{mol}^2}{\text{L}^2}} = 1.34 \times 10^{-3} \frac{\text{mol}}{\text{L}} = [\text{H}^{1+}] \quad \textit{Answer}$$

You can obtain the square root of 1.80 from a calculator. To obtain the square root of 10^{-6}, divide the exponent by 2 to give 10^{-3}. Always adjust the power of 10 by exponential notation (see 2-3) to give an *even*-numbered exponent. Calculate the percent ionization as follows:

$$\frac{\text{mol/L of acid ionized}}{\text{mol/L of acid initially}} \times 100$$

The moles per liter of acid ionized is the same as the moles per liter of hydrogen ions, since in the balanced equation the mole relationship is $1:1$. Hence,

$$\frac{1.34 \times 10^{-3} \, \cancel{\text{mol/L}}}{0.102 \, \cancel{\text{mol/L}}} \times 100 = \frac{1.34 \times 10^{-3} \times 10^2}{1.02 \times 10^{-1}} = 1.31\% \quad \textit{Answer}$$

Problem Example 17-6

Calculate the hydroxide ion concentration and the percent ionization at $20°C$ in a 0.100 M aqueous ammonia (NH_3) solution. The K_b for aqueous NH_3 = 1.70×10^{-5} (mol/L) at $20°C$.

SOLUTION: The equation representing the ionization is

$$NH_{3(aq)} + HOH_{(aq)} \rightleftharpoons NH_4{}^{1+}{}_{(aq)} + OH^{1-}{}_{(aq)}$$

and the expression for K_b is

$$K_b = \frac{[NH_4{}^{1+}][OH^{1-}]}{[NH_3]}$$

Let $x = [OH^{1-}]$ in mol/L; therefore, x in mol/L also equals $[NH_4{}^{1+}]$, since in the balanced equation **1 mol** of aqueous NH_3 ionizes to form **1 mol** of $NH_4{}^{1+}$ and **1 mol** of OH^{1-}. The concentration of the nonionized NH_3 *at equilibrium* is

$$[NH_3] = 0.100 \text{ mol/L} - x \text{ mol/L}$$

again from the mole relationship in the balanced equation being $1:1$. Substituting these values into the equilibrium expression and dropping the units of x for simplicity, we obtain

$$1.70 \times 10^{-5} \text{ (mol/L)} = \frac{[x][x]}{[0.100 \text{ mol/L} - x]}$$

$$1.70 \times 10^{-5} \text{ (mol/L)} = \frac{[x]^2}{[0.100 \text{ mol/L} - x]}$$

Neglecting the x in $0.100 \text{ mol/L} - x$ and further solving the equation, we have

$$1.70 \times 10^{-5} \text{ (mol/L)} = \frac{[x]^2}{0.100 \text{ mol/L}}$$

$$x^2 = 0.100 \text{ mol/L} \times 1.70 \times 10^{-5} \text{ mol/L}$$

$$= 0.170 \times 10^{-5} \frac{\text{mol}^2}{\text{L}^2}$$

$$x^2 = 1.70 \times 10^{-6} \frac{\text{mol}^2}{\text{L}^2}$$

$$x = \sqrt{1.70 \times 10^{-6} \frac{\text{mol}^2}{\text{L}^2}} = 1.30 \times 10^{-3} \frac{\text{mol}}{\text{L}} \quad Answer$$

The percent ionization is calculated as follows:

Work Problems 19 and 20.

$$\frac{1.30 \times 10^{-3} \text{ mol/L}}{0.100 \text{ mol/L}} \times 100 = \frac{1.30 \times 10^{-3} \times 10^2}{1.00 \times 10^{-1}} = 1.30\% \quad Answer$$

17-5

Solubility Product Equilibria

The equilibrium between a slightly soluble electrolyte and its ions in solution is a dynamic equilibrium involving constant *dissolving* of the electrolyte and *reprecipitation* of it at the surface of the solid crystals. At equilibrium, the solution is *saturated*—the *rate of solution* **equals** the *rate of precipitation*.

Consider a slightly soluble electrolyte of the formula A_2B_3 added to pure water. The A_2B_3 will dissolve until a saturated solution is obtained. The equation for this equilibrium is expressed as follows:

$$A_2B_{3(s)} \rightleftharpoons 2A^{3+}{}_{(aq)} + 3B^{2-}{}_{(aq)} \qquad (17\text{-}34)$$
$$\text{(in saturated solution)}$$

The equilibrium constant for the equation could be expressed as

$$K = \frac{[A^{3+}]^2[B^{2-}]^3}{[A_2B_3]} \qquad (17\text{-}35)$$

However, since A_2B_3 is in the solid state and its concentration in the solid phase is constant, the term A_2B_3 is incorporated into the constant K, and the expression becomes

$$K_{sp} = [A^{3+}]^2[B^{2-}]^3 \qquad (17\text{-}36)$$

The constant K_{sp} is called the *solubility product constant* and is defined for A_2B_3 by Equation 17-36, where $[A^{3+}]$ and $[B^{2-}]$ represent the concentrations of these ions in moles per liter in a solution *at equilibrium with solid A_2B_3* (at saturation). Thus, the **coefficients** of the ions of a slightly soluble electrolyte in equilibrium with its ions become **exponents** for the concentrations of the ions in solution in the expression for the solubility product constant, K_{sp}. Figure 17-3 depicts the solubility product equilibrium for A_2B_3.

FIGURE 17-3

Solubility product equilibrium for A_2B_3.

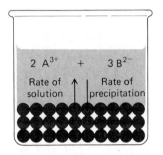

The solubility product constant K_{sp} is essentially constant at a *given* temperature for a particular slightly soluble electrolyte. This value will change with a change in temperature, as did the equilibrium constant K.

In comparing values of K_{sp}, the electrolytes must all be of the *same* type, such as all AB, or all A_2B_3. The larger is the K_{sp} value of an electrolyte of a *given* type, the more soluble is the electrolyte in water. For example, the solubility product constant for silver chloride (AgCl) is 1.56×10^{-10} (mol^2/L^2) at 25°C, whereas that of silver bromide (AgBr) is 7.00×10^{-13} or 0.00700×10^{-10} (mol^2/L^2) at 25°C. Both are AB-type electrolytes. Therefore, since the K_{sp} value for silver chloride is larger (smaller negative exponent), the solubility of silver chloride in water is greater than that of silver bromide.

Work Problem 21.

$$AgCl_{(s)} \rightleftharpoons Ag^{1+}{}_{(aq)} + Cl^{1-}{}_{(aq)}$$
$$\text{(in saturated solution)}$$

$$AgBr_{(s)} \rightleftharpoons Ag^{1+}{}_{(aq)} + Br^{1-}{}_{(aq)}$$
$$\text{(in saturated solution)}$$

When the product of the concentrations of the ions each raised to their respective powers is *equal* to the value of the solubility product constant, then a *saturated* solution exists. When the product of these concentrations of the ions each raised to their respective powers is *less* than the value of the solubility product constant, then an *unsaturated* solution exists, and when this product is *greater* than the value of the solubility product constant, a *supersaturated* solution exists and precipitation normally occurs. Therefore, the solubility product constant gives the limit to the solubility of the electrolyte in water at a given temperature.

The solubility product constant has units. In Equation 17-36, the units would be

$$K_{sp} = [mol/L]^2[mol/L]^3 = \frac{mol^2}{L^2} \times \frac{mol^3}{L^3} = \left(\frac{mol^5}{L^5}\right) \qquad (17\text{-}37)$$

If the coefficients were one for each of the ions, then the units would be

$$K_{sp} = [mol/L][mol/L] = \left(\frac{mol^2}{L^2}\right) \qquad (17\text{-}38)$$

as was shown for the K_{sp} for AgCl and AgBr. In this text, we shall place the units of K_{sp} in parentheses immediately following the numerical value for K_{sp}, as we did for the equilibrium constant K.

Now, let us write the solubility product constant for a specific slightly soluble electrolyte, silver chromate (Ag_2CrO_4). The K_{sp} expression for silver chromate is obtained as follows:

$$Ag_2CrO_{4(s)} \rightleftarrows 2\ Ag^{1+}_{(aq)} + CrO_4^{2-}_{(aq)} \qquad (17\text{-}39)$$
$$\text{(in saturated solution)}$$

$$K_{sp} = [Ag^{1+}]^2[CrO_4^{2-}]$$

The coefficients of the ions become exponents for the concentrations of the ions in the expression for K_{sp}.

Consider some more K_{sp} expressions for slightly soluble electrolytes.

Problem Example 17-7

Write the expression for the K_{sp} for each of the following slightly soluble electrolytes:

(a) $Ba_3(PO_4)_2$

(b) $Mn(OH)_2$

SOLUTION

(a) The K_{sp} for $Ba_3(PO_4)_2$ is obtained as follows:

$$Ba_3(PO_4)_{2(s)} \rightleftarrows 3\ Ba^{2+}_{(aq)} + 2\ PO_4^{3-}_{(aq)}$$
$$\text{(in saturated solution)}$$

$$K_{sp} = [Ba^{2+}]^3[PO_4^{3-}]^2 \quad Answer$$

(b) The K_{sp} for $Mn(OH)_2$ is obtained as follows:

Work Problems 22
and 23.

$$Mn(OH)_{2(s)} \rightleftarrows Mn^{2+}_{(aq)} + 2\ OH^{1-}_{(aq)}$$
$$\text{(in saturated solution)}$$

$$K_{sp} = [Mn^{2+}]\ [OH^{1-}]^2\quad \textit{Answer}$$

The following problem examples illustrate the application of solubility product equilibria to quantitative calculations.

Problem Example 17-8

The solubility of silver chloride in water at 25°C is 0.00179 g/L. Calculate the K_{sp} for AgCl at 25°C.

SOLUTION: The equation for the equilibrium is

$$AgCl_{(s)} \rightleftarrows Ag^{1+}_{(aq)} + Cl^{1-}_{(aq)}$$
$$\text{(in saturated solution)}$$

and

$$K_{sp} = [Ag^{1+}]\ [Cl^{1-}]$$

The K_{sp} expression gives the concentration of the ions in moles per liter; hence, the solubility must be expressed in *moles per liter*. The formula mass for AgCl is 143.4 amu, and the concentration in moles per liter (molarity) is

$$\frac{0.00179\ \text{g AgCl}}{1\ L} \times \frac{1\ \text{mol AgCl}}{143.4\ \text{g AgCl}} = \frac{1.79 \times 10^{-3}}{1.434 \times 10^{2}} = 1.25 \times 10^{-5}\ \text{mol AgCl/L}$$

The concentrations of both Ag^{1+} and Cl^{1-} in the solution would be 1.25×10^{-5} mol/L each, since **1 mol** of AgCl *in solution* would produce **1 mol each** of Ag^{1+} and Cl^{1-} from the balanced equation. The solubility product constant for AgCl is, therefore,

Work Problems 24
and 25.

$$K_{sp} = [Ag^{1+}][Cl^{1-}] = [1.25 \times 10^{-5}\ \text{mol/L}][1.25 \times 10^{-5}\ \text{mol/L}]$$

$$= 1.56 \times 10^{-10}\ (mol^2/L^2)\quad \textit{Answer}$$

Problem Example 17-9

The solubility product constant for barium chromate is $2.00 \times 10^{-10}\ (mol^2/L^2)$ at $\overline{2}0°C$. (a) Calculate the molarity of a saturated solution of $BaCrO_4$ at $\overline{2}0°C$. (b) What is its concentration in grams $BaCrO_4$ per liter at $\overline{2}0°C$?

SOLUTION

(a) The equation for the equilibrium is

$$BaCrO_{4(s)} \rightleftharpoons Ba^{2+}_{(aq)} + CrO_4^{2-}_{(aq)}$$
$$\text{(in saturated solution)}$$

and

$$K_{sp} = 2.00 \times 10^{-10} (mol^2/L^2) = [Ba^{2+}][CrO_4^{2-}]$$

Let x = mol of $BaCrO_4$/L of saturated solution. The concentration of $[Ba^{2+}]$ and $[CrO_4^{2-}]$ is also equal to x mol/L, since from the balanced equation **1 mol** of $BaCrO_4$ *in solution* yields **1 mol** of Ba^{2+} and **1 mol** of CrO_4^{2-}. Hence,

$$x \text{ mol/L} = [Ba^{2+}] = [CrO_4^{2-}]$$

and K_{sp} is expressed as follows:

$$2.00 \times 10^{-10} \text{ mol}^2/L^2 = [Ba^{2+}][CrO_4^{2-}] = [x][x] = [x^2]$$

$$x = \sqrt{2.00 \times 10^{-10} \frac{mol^2}{L^2}} = 1.41 \times 10^{-5} \frac{mol}{L}$$

As with weak electrolytes in equilibrium (see Problem Example 17-5), you can obtain the square root of 2.00 from a calculator. To obtain the square root of 10^{-10}, divide the exponent by 2 to give 10^{-5}. Therefore, the solution is 1.41×10^{-5} mol of $BaCrO_4$/L, or 1.41×10^{-5} M $BaCrO_4$.   *Answer*

(b) The formula mass of $BaCrO_4$ is 253.3 amu; hence, the solubility in grams per liter is calculated

$$\frac{1.41 \times 10^{-5} \text{ mol } BaCrO_4}{1 \text{ L}} \times \frac{253.3 \text{ g } BaCrO_4}{1 \text{ mol } BaCrO_4}$$

$$= 3.57 \times 10^{-3} \text{ g } \frac{BaCrO_4}{L} \quad Answer^4$$

Problem Example 17-10

The K_{sp} for a slightly soluble strong electrolyte (AB) is 9.00×10^{-12} (mol^2/L^2) at 25°C. (a) What is the solubility of AB in moles per liter? (b) If the concentration of AB in a solution reaches 3.00×10^{-5} mol/L, would precipitation occur?

---

[4]For more complicated compounds, such as AB_2 example $Fe(OH)_2$, or A_2B, example Ag_2CrO_4, the same method can be used. For both cases (AB_2 and A_2B) the final equation would be the same, that is, $4x^3 = K_{sp}$. The solution of this equation would involve solving a cubic equation.

SOLUTION

(a) The equation for the equilibrium is

$$AB_{(s)} \rightleftharpoons A^{1+}_{(aq)} + B^{1-}_{(aq)}$$
$$\text{(in saturated solution)}$$

and

$$K_{sp} = 9.00 \times 10^{-12} \,(\text{mol}^2/\text{L}^2) = [A^{1+}][B^{1-}]$$

Let x = mol of AB/L of saturated solution. The concentration of $[A^{1+}]$ and $[B^{1-}]$ is also equal to x mol/L, since from the balanced equation **1 mol** of AB *in solution* yields **1 mol** of A^{1+} and **1 mol** of B^{1-}. Hence,

$$x \text{ mol/L} = [A^{1+}] = [B^{1-}]$$

and K_{sp} is expressed as follows:

$$9.00 \times 10^{-12} \,(\text{mol}^2/\text{L}^2) = [A^{1+}][B^{1-}] = [x][x] = [x]^2$$

$$x = \sqrt{9.00 \times 10^{-12} \,\frac{\text{mol}^2}{\text{L}^2}} = 3.00 \times 10^{-6} \,\frac{\text{mol}}{\text{L}}$$

Therefore, the solution is

$$3.00 \times 10^{-6} \,\frac{\text{mol } AB}{\text{L}} \quad Answer$$

(b) If the concentration of $A\overset{\cdot}{B}$ is 3.00×10^{-5} mol/L, then the concentrations of A^{1+} and B^{1-} must also be 3.00×10^{-5} mol/L, since from the balanced equation the mole relationship is **1 mol** of AB yields **1 mol** of A^{1+} and **1 mol** of B^{1-}. Hence,

$$[A^{1+}] = 3.00 \times 10^{-5} \text{ mol/L}$$

$$[B^{1-}] = 3.00 \times 10^{-5} \text{ mol/L}$$

and the product of the concentrations of the ions raised to their respective powers is expressed as follows:

$$[A^{1+}][B^{1-}] = \left[3.00 \times 10^{-5} \,\frac{\text{mol}}{\text{L}} \right]\left[3.00 \times 10^{-5} \,\frac{\text{mol}}{\text{L}} \right] = 9.00 \times 10^{-10} \,\frac{\text{mol}^2}{\text{L}^2}$$

Work Problems 26 and 27.

This value (9.00×10^{-10} mol^2/L^2) is *greater* than the value of the K_{sp} (9.00×10^{-12} or 0.0900×10^{-10} mol^2/L^2), and hence precipitation would occur.

Answer

Chapter Summary

The *reaction rate* of a chemical reaction is the rate or speed at which the products are formed or the reactants consumed. The reaction order of reactants and overall reaction order for various reactions were considered. Factors that influence the rate of a chemical reaction are (1) the nature of the reactants, (2) the concentration of the reactants, (3) the temperature, and (4) a catalyst. Reaction rates and chemical equilibrium are related; in the course of a reaction, the rate of the forward reaction decreases and the rate of the reverse reaction increases until equilibrium is reached. At equilibrium, the rate of the forward reaction will equal the rate of the reverse reaction.

Some chemical reactions essentially go to completion and are considered irreversible. These irreversible reactions frequently are dependent on the formation of the products in the reaction such as (1) a gas, which is removed as soon as it is formed, (2) a precipitate, and (3) a nonionized or only partially ionized substance, such as water.

The equilibrium expression was developed for the equilibrium constant (K). The value of K was compared for different reactions that were of the same type. Examples of expressing this constant (K) for different chemical reactions were given.

Le Châtelier's Principle states that if an equilibrium system is subjected to a change in conditions of concentration, temperature, or pressure, the system will change in a direction that will tend to restore the original conditions. Changes in these conditions were discussed for given reactions.

Weak electrolyte equilibria were considered. Problem examples involving the calculation of the equilibrium constant (K_a for acid and K_b for base) and the calculation of the hydrogen ion or hydroxide ion concentration along with the percent ionization were given. These examples were limited to the types HX for an acid and MOH for a base.

Solubility product equilibria were discussed. The solubility product expression was developed for the solubility product constant (K_{sp}) for slightly soluble electrolytes and the value of K_{sp} was compared for different electrolytes of the same type. Examples of expressing this constant for different electrolytes were given. Problem examples involving the calculation of K_{sp}, the calculation of the solubility of the slightly soluble electrolyte (limited to the type AB), and the precipitation of the electrolyte given its concentration were solved.

EXERCISES

1. Define or explain
   - (a) reaction rate
   - (b) chemical kinetics
   - (c) Law of Mass Action
   - (d) specific rate constant
   - (e) reaction order
   - (f) overall reaction order
   - (g) equilibrium constant
   - (h) Le Châtelier's Principle
   - (i) K_a
   - (j) K_b
   - (k) K_{sp}

2. Distinguish between
   (a) reaction rate and equilibrium
   (b) specific rate constant and equilibrium constant
   (c) K_a and K_b
   (d) K_a and K_{sp}

Reaction Rate (See Section 17-1)

3. List and describe, in your own words, the four factors that influence the rate of a reaction.

4. The yield in a certain industrial preparation in equilibrium with its reactants appeared to be progressively decreasing in a certain chemical plant. One suggested solution to the problem was that a study be made of various catalysts that could increase the yield. What do you think about this solution to the problem?

Irreversible and Reversible Reactions (See Section 17-2)

5. List three classes of products that act as a driving force for a reaction to go toward completion, with the point of equilibrium shifted far toward the products side.

Le Châtelier's Principle (See Section 17-3)

6. List and describe, in your own words, the three changes in conditions that affect the equilibrium, according to Le Châtelier's Principle.

PROBLEMS

Reaction Rates (see Section 17-1)

7. Given the following chemical equations and rate equations, determine the reaction order for each reactant and the overall reaction order.

   (a) $2\,NO_{(g)} + Br_{2(g)} \rightarrow 2\,NOBr_{(g)}$

   rate $= k[NO]^2[Br_2]$

   (b) $A + B \rightarrow C$

   rate $= k[A]^2$

8. Given the following reaction:

$$2\,NO_{(g)} + 2\,H_{2(g)} \rightarrow N_{2(g)} + 2\,H_2O_{(g)}$$

with the reaction rate $= k[NO]^2[H_2]$. Calculate the effect on the reaction rate if the following occurs:
   (a) The concentration of NO is doubled.
   (b) The concentration of NO is tripled.

    (c) The concentration of H_2 is tripled.

    (d) The concentration of NO is increased by 2.50 times.

Chemical Equilibria   (See Section 17-2)

9. The following are ionization constants for the monoionization of various acids (K_a) at 25°C. List them in order of decreasing acid strength by their K_a values.

Acid (in approximately 0.1 N aqueous solutions)	K_a at 25°C (mol/L)
Acetic acid	1.76×10^{-5}
Barbituric acid	$9.8 \ \times 10^{-5}$
Chloroacetic acid	1.40×10^{-3}
Formic acid	1.77×10^{-4}
Lactic acid	1.37×10^{-4}
Sulfurous acid	1.72×10^{-2}

10. The following are ionization constants for the monoionization of various bases (K_b) at 25°C. List them in order of decreasing basic strength by their K_b values.

Base (in approximately 0.1 N aqueous solutions)	K_b at 25°C (mol/L)
Ammonia	1.79×10^{-5}
Codeine	$9 \ \ \times 10^{-7}$
Nicotine	$7 \ \ \times 10^{-7}$
Novocain	$7 \ \ \times 10^{-6}$
Silver hydroxide	$1.1 \ \times 10^{-4}$
Urea	$1.5 \ \times 10^{-14}$

11. Write the expression for the equilibrium constant for each of the following reactions:

    (a) $CH_{4(g)} + Cl_{2(g)} \rightleftarrows CH_3Cl_{(g)} + HCl_{(g)}$

    (b) $SO_{2(g)} + NO_{2(g)} \rightleftarrows SO_{3(g)} + NO_{(g)}$

    (c) $2\ NO_{(g)} + O_{2(g)} \rightleftarrows 2\ NO_{2(g)}$

    (d) $NH_4Cl_{(s)} \rightleftarrows NH_{3(g)} + HCl_{(g)}$

    (e) $4\ H_2O_{(g)} + 3\ Fe_{(s)} \rightleftarrows Fe_3O_{4(s)} + 4\ H_{2(g)}$

    (f) $3\ O_{2(g)} \rightleftarrows 2\ O_{3(g)}$

    (g) $SO_2Cl_{2(g)} \rightleftarrows SO_{2(g)} + Cl_{2(g)}$

    (h) $4\ NH_{3(g)} + 5\ O_{2(g)} \rightleftarrows 4\ NO_{(g)} + 6\ H_2O_{(g)}$

    (i) $2\ Pb(NO_3)_{2(s)} \rightleftarrows 2\ PbO_{(s)} + 4\ NO_{2(g)} + O_{2(g)}$

    (j) $2\ H_{2(g)} + O_{2(g)} \rightleftarrows 2\ H_2O_{(g)}$

12. Write the expression for the equilibrium constant for each of the following reactions:

    (a) $N_{2(g)} + O_{2(g)} \rightleftarrows 2\ NO_{(g)}$

    (b) $2\ SO_{2(g)} + O_{2(g)} \rightleftarrows 2\ SO_{3(g)}$

    (c) $BaSO_{3(s)} \rightleftarrows BaO_{(s)} + SO_{2(g)}$

(d) $C_{(s)} + H_2O_{(g)} \rightleftarrows CO_{(g)} + H_{2(g)}$

(e) $2 NOCl_{(g)} \rightleftarrows 2 NO_{(g)} + Cl_{2(g)}$

(f) $PCl_{5(g)} \rightleftarrows PCl_{3(g)} + Cl_{2(g)}$

(g) $2 HgO_{(s)} \rightleftarrows 2 Hg_{(g)} + O_{2(g)}$

(h) $COBr_{2(g)} \rightleftarrows CO_{(g)} + Br_{2(g)}$

(i) $CH_{4(g)} + 2 O_{2(g)} \rightleftarrows CO_{2(g)} + 2 H_2O_{(g)}$

(j) $2 Ag_{(s)} + Cl_{2(g)} \rightleftarrows 2 AgCl_{(s)}$

Le Châtelier's Principle   (See Section 17-3)

13. One of the preparations of chlorine gas is the Deacon process, which involves the following equilibrium:

$$4 HCl_{(g)} + O_{2(g)} \rightleftarrows 2 Cl_{2(g)} + 2 H_2O_{(g)}$$

In which direction will the equilibrium be shifted by each of the following changes?

(a) increasing the concentration of HCl

(b) decreasing the concentration of HCl

(c) increasing the concentration of O_2

(d) decreasing the concentration of O_2

(e) increasing the concentration of Cl_2

(f) decreasing the concentration of Cl_2

(g) increasing the concentration of H_2O

(h) decreasing the concentration of H_2O

(i) increasing the pressure

(j) decreasing the pressure

14. One of the commercial preparations of hydrogen gas involves the following equilibrium:

$$CO_{(g)} + H_2O_{(g)} \rightleftarrows CO_{2(g)} + H_{2(g)}$$

In which direction will the equilibrium be shifted by each of the following changes?

(a) increasing the concentration of CO

(b) decreasing the concentration of CO

(c) increasing the concentration of H_2O

(d) decreasing the concentration of H_2O

(e) increasing the concentration of CO_2

(f) decreasing the concentration of CO_2

(g) increasing the concentration of H_2

(h) decreasing the concentration of H_2

(i) increasing the pressure

(j) decreasing the pressure

15. Predict the effect on equilibrium of the following when (1) the temperature is increased, (2) the temperature is decreased, (3) the pressure is increased, and (4) the pressure is decreased:

(a) $2 H_{2(g)} + O_{2(g)} \rightleftarrows 2 H_2O_{(g)} + 115.6$ kcal

(b) $H_{2(g)} + Cl_{2(g)} \rightleftarrows 2 HCl_{(g)} + 185$ kJ

(c) $H_{2(g)} + I_{2(g)} \rightleftarrows 2 HI_{(g)} - 51.9$ kJ

(d) $2 F_{2(g)} + O_{2(g)} \rightleftarrows 2 OF_{2(g)} - 46.0$ kJ

(e) $4 Al_{(s)} + 3 O_{2(g)} \rightleftarrows 2 Al_2O_{3(s)} + 798.2$ kcal

16. Predict the effect on equilibrium of the following when (1) the temperature is increased, (2) the temperature is decreased, (3) the pressure is increased, and (4) the pressure is decreased:

(a) $C_6H_{6(g)} + 3 H_{2(g)} \rightleftarrows C_6H_{12(g)} + 206$ kJ

(b) $2 NO_{(g)} \rightleftarrows N_{2(g)} + O_{2(g)} + 43.2$ kcal

(c) $2 CO_{(g)} + O_{2(g)} \rightleftarrows 2 CO_{2(g)} + 135.2$ kcal

(d) $N_{2(g)} + 2 O_{2(g)} \rightleftarrows 2 NO_{2(g)} - 16.2$ kcal

(e) $C_{(s)} + O_{2(g)} \rightleftarrows CO_{2(g)} + 393$ kJ

Weak Electrolyte Equilibria   (See Section 17-4)

17. Calculate the ionization constants for each of the following weak electrolytes from the percent ionization at the concentrations given:

(a) A 0.500 M solution of aqueous NH_3 is 0.600 percent ionized.

(b) A 0.100 M solution of HF is 8.23 percent ionized.

(c) A 0.800 M solution of HCN is 0.00300 percent ionized.

(d) A 0.400 M solution of HA is 1.50 percent ionized.

18. Calculate the ionization constants for each of the following weak electrolytes from the percent ionization or appropriate data at the concentration given:

(a) A 2.00 M solution of $HC_2H_3O_2$ is 0.300 percent ionized.

(b) A 0.0100 M solution of formic acid ($HCHO_2$) is 13.1 percent ionized. [*Hint:*
$HCHO_{2(aq)} \rightleftarrows H^{1+}_{(aq)} + CHO_2^{1-}_{(aq)}$.]

(c) A 0.0300 M solution of MOH is 4.00 percent ionized.

(d) A 0.0200 M solution of HA, whose $[H^{1+}] = 1.80 \times 10^{-3}$ mol/L.

19. From the ionization constants of each of the following weak electrolytes, calculate the hydrogen ion concentration (for acids) or the hydroxide ion concentration (for bases) in moles per liter and the percent ionization of the weak electrolyte in each of the following solutions:

(a) 0.200 M acetic acid at 5°C. The K_a for $HC_2H_3O_2 = 1.70 \times 10^{-5}$ (mol/L) at 5°C.

(b) 0.500 M aqueous ammonia at $20̄$°C. The K_b for aqueous $NH_3 = 1.70 \times 10^{-5}$ (mol/L) at $20̄$°C.

(c) 2.00 M formic acid at $50̄$°C. The K_a for $HCHO_2 = 1.65 \times 10^{-4}$ (mol/L) at 50°C.

(d) 0.250 M MOH. The K_b for MOH $= 6.40 \times 10^{-7}$ (mol/L).

20. From the ionization constants of each of the following weak electrolytes, calculate the hydrogen ion concentration (for acids) or the hydroxide ion concentration (for bases) in moles per liter and the percent ionization of the weak electrolyte in each of the following solutions:

(a) 0.100 M hypochlorous acid at 25°C. The K_a for HClO $= 3.50 \times 10^{-9}$ (mol/L) at 25°C.

(b) 0.0278 M hypoiodous acid at 25°C. The K_a for HIO $= 2.30 \times 10^{-11}$ (mol/L) at 25°C.

(c) 0.0500 M MOH. The K_b for MOH = 1.10×10^{-6} (mol/L).

(d) 0.300 M MOH. The K_b for MOH = 3.00×10^{-8} (mol/L).

Solubility Product Equilibria   (See Section 17-5)

21. The following are solubility product constants (K_{sp}) for various slightly soluble electrolytes in water at 20°C. List them in order of decreasing solubility in water.

Slightly Soluble Electrolyte	K_{sp} at $\overline{2}0$°C (mol^2/L^2)
Silver acetate	4.0×10^{-3}
Silver iodide	8.5×10^{-17}
Silver bromate	6.0×10^{-5}
Silver bromide	5.0×10^{-13}
Silver chloride	1.8×10^{-10}

22. Write the expression for the solubility product constant (K_{sp}) for each of the following slightly soluble electrolytes:
(a) $AgCl$ (b) Bi_2S_3
(c) SnS_2 (d) $Pb_3(AsO_4)_2$
(e) As_2S_5

23. Write the expression for the solubility product constant (K_{sp}) for each of the following slightly soluble electrolytes:
(a) HgS (b) BaF_2
(c) $Fe(OH)_3$ (d) $Pb(IO_3)_2$
(e) Cu_2S

24. From the solubility of each of the following compounds in pure water at a given temperature, calculate the solubility product constant (K_{sp}) for the compound at that temperature:
(a) silver bromide: 0.000165 g of AgBr/L at 25°C
(b) barium carbonate: 9.00×10^{-5} mol of $BaCO_3$/L at 18°C
(c) lead(II) sulfide: 2.64×10^{-15} mol of PbS/L at 25°C
(d) AB: 0.00135 g of AB/L (formula mass AB = 94.0 amu)

25. From the solubility of each of the following compounds in pure water at a given temperature, calculate the solubility product constant (K_{sp}) for the compound at that temperature:
(a) strontium carbonate: 0.00590 g of $SrCO_3$/L at 25°C
(b) cobalt sulfide: 1.57×10^{-11} g of CoS/L at 18°C
(c) iron(II) sulfide: 5.34×10^{-8} g of FeS/L at 18°C
(d) AB: 0.000755 mol of AB/L

26. From the solubility product constant for each of the following salts, (1) calculate the molarity of a saturated solution of the salt at the given temperature, and (2) calculate the solubility of the salt in grams per liter at a given temperature:
(a) cadmium sulfide at 18°C; the K_{sp} for CdS = 3.60×10^{-29} (mol^2/L^2) at 18°C.
(b) barium carbonate at 16°C; the K_{sp} for $BaCO_3$ = 7.00×10^{-9} (mol^2/L^2) at 16°C.
(c) thallium(I) bromide at 25°C; the K_{sp} for TlBr = 4.00×10^{-6} (mol^2/L^2) at 25°C.

(d) nickel sulfide at 18°C; the K_{sp} for NiS $= 1.40 \times 10^{-24}$ (mol²/L²) at 18°C. If the concentration of NiS in a solution reaches 2.00×10^{-11} mol/L, would precipitation occur?

27. From the solubility product constants for each of the following salts, (1) calculate the molarity of a saturated solution of the salt at the given temperature, and (2) calculate the solubility of the salt in grams per liter at the given temperature:
    (a) silver iodide at 25°C; the K_{sp} for AgI $= 1.50 \times 10^{-16}$ (mol²/L²) at 25°C.
    (b) lead carbonate at 18°C; the K_{sp} for PbCO$_3$ $= 3.30 \times 10^{-14}$ (mol²/L²) at 18°C.
    (c) calcium chromate at 18°C; the K_{sp} for CaCrO$_4$ $= 2.30 \times 10^{-2}$ (mol²/L²) at 18°C.
    (d) barium chromate at 28°C; the K_{sp} for BaCrO$_4$ $= 2.40 \times 10^{-10}$ (mol²/L²) at 28°C. If the concentration of BaCrO$_4$ in a solution reaches 3.60×10^{-6} mol/L, would precipitation occur?

General Problems

28. Calculate the pH and pOH of each of the following solutions:
    (a) a 0.530 M aqueous solution of acetic acid at 5°C; the K_a for HC$_2$H$_3$O$_2$ $= 1.70 \times 10^{-5}$ (mol/L) at 5°C.
    (b) a 0.100 M aqueous solution of aqueous ammonia at 20°C; the K_b for aqueous NH$_3$ $= 1.70 \times 10^{-5}$ (mol/L) at 20°C.

29. Calculate the pH and pOH of each of the following solutions:
    (a) a 0.0100 M aqueous solution of acetic acid at 20°C; HC$_2$H$_3$O$_2$ is 4.20 percent ionized at 20°C.
    (b) a 0.500 M aqueous solution of aqueous ammonia at 27°C; aqueous NH$_3$ is 0.600 percent ionized at 27°C.

30. Swimmers often complain that the chlorine in the swimming pool "burns" their eyes when the pH is low. The ionic equation for the reaction of chlorine with water is as follows:

$$Cl_2 + H_2O \rightleftharpoons HClO + H^{1+} + Cl^{1-}$$

Explain this fact in terms of pH and chlorine formation.

Readings

Eyring, Henry, and Edward M. Eyring, *Modern Chemical Kinetics*. New York: Reinhold Publishing Corp., 1963. A classic in the field of chemical kinetics. Chapter 1 may be interesting after you have studied the material in your text.

Feldman, Martin R., "Fritz Haber," *J. Chem. Educ.*, 1983, v. 60, p. 463. Discusses the life of Fritz Haber, known for his unique synthesis of ammonia.

House, J. E., Jr., "Chemical Queries." *J. Chem. Educ.*, 1969, v. 46, p. 674. Discusses the effect of a rise of 10°C on reaction rate.

Organic Chemistry I: Hydrocarbons

Polymer products used in the home: food storage bags (polyethylene); Saran Wrap (a chlorinated polyethylene); Teflon (fluorinated polyethylene) coated fry pan; spatula with plastic blade; polyethylene storage container; and polyethylene sauce dispenser.

TASKS

1. Memorize the names and structural formulas of the alkanes given in Table 18-1.

2. Memorize the names and structural formulas of the alkyl groups given in Table 18-2.

3. Memorize the structural formulas for (1) benzene, (2) toluene, and (3) o-, m-, and p-xylene, as given in Section 18-7.

OBJECTIVES

1. Define each of the following terms and describe the distinguishing characteristics of each:
   (a) organic chemistry (Introduction to Chapter 18)
   (b) hydrocarbons (Introduction to Chapter 18)
   (c) sigma (σ) bond (Section 18-1)
   (d) alkanes (Section 18-3)
   (e) homologous series (Section 18-3)
   (f) constitutional (structural) isomers (Section 18-3)
   (g) alkenes (Section 18-4)
   (h) pi (π) bond (Section 18-4)
   (i) addition polymers (Section 18-5)
   (j) alkynes (Section 18-6)
   (k) aromatic hydrocarbons (Section 18-7)

2. Given the general molecular formula for an open-chain alkane (C_nH_{2n+2}), write condensed structural formulas for all of the constitutional isomers for a given number of carbons from C_1 to C_7 (Problem Example 18-1, Problem 3).

3. Given the structural formulas for the following types of organic compounds, write an acceptable IUPAC name for each:
   (a) alkanes (Problem Example 18-2, Problem 4)
   (b) alkenes (Problem Example 18-4, Problem 7)
   (c) alkynes (Problem Example 18-6, Problem 11)
   (d) aromatic hydrocarbons (Problem Example 18-8, Problem 14)

4. Given the IUPAC names or other acceptable names for the following types of organic compounds, write the structural formula for each:
   (a) alkanes (Problem Example 18-3, Problem 5)
   (b) alkenes (Problem Example 18-5, Problem 8)
   (c) alkynes (Problem Example 18-7, Problem 12)
   (d) aromatic hydrocarbons (Problem Example 18-9, Problem 15)

5. Given the structural formula of a specific compound from the following types of organic compounds, complete and balance the equation for its reaction with chlorine or with bromine, giving appropriate conditions or suitable solvents:
   (a) alkanes (Section 18-3, Problem 6)
   (b) alkenes (Section 18-4, Problem 9)
   (c) alkynes (Section 18-6, Problem 13)

6. Given the following addition polymers; give the structural formula for the monomer and the polymer;
   (a) polyethylene (Section 18-5, Problem 10)
   (b) Teflon (Section 18-5, Problem 10)
   (c) polyvinyl chloride (Section 18-5, Problem 10)
   (d) Orlon (Section 18-5, Problem 10)

7. Given benzene (as an example of an aromatic hydrocarbon), complete and balance the equation for its reaction with chlorine (iron catalyst), bromine (iron catalyst), and nitric acid (in sulfuric acid) to give the appropriate substitution product (Section 18-7, Problem 16).

Organic chemistry is the study of compounds containing the element carbon; inorganic chemistry is the study of all other elements and compounds. The number of known organic compounds, about 4 million, far exceeds the number of known inorganic compounds, approximately 100,000. A typical organic compound is aspirin ($C_9H_8O_4$), and a typical inorganic compound is sodium chloride (salt, $NaCl$). Some compounds that contain carbon, such as those containing the polyatomic ions (see Table 6-4) cyanide (CN^{1-}), hydrogen carbonate or bicarbonate (HCO_2^{1-}), and carbonate (CO_3^{2-}), are considered inorganic compounds because their properties resemble inorganic compounds and not organic compounds.

The two primary sources of organic compounds are oil and coal. Oil is the chief source. We burn much of our available oil to produce energy for automobiles, and we burn coal and oil to produce energy for electric power plants.

In an attempt to solve our energy problems, some energy experts have proposed developing synthetic fuels on a large scale. Actually, "synthetic fuels" are not artificial products produced in the laboratory. They are the results of converting organic compounds from sources other than oil into petroleum products. For example, coal, shale rock, and tar sands, which are relatively abundant, can be converted to oil; also, coal may be converted to natural gas. If a synthetic fuel industry is developed in the United States, it would probably not be in operation until the 1990s. At present with an oil glut and the falling price of oil, the development of a synthetic fuel industry in the United States in the near future seems unlikely. The chief problem with the production of synthetic fuels is that it also produces considerable pollution.

Today's petroleum industry provides us with many useful organic materials used to synthesize other organic compounds. Various organic reaction sequences can be performed on simple organic compounds to obtain other valuable organic compounds; in addition, many useful organic substances are obtained from other natural sources such as plants, animals, and microorganisms. Useful organic substances that should be of interest to you are textiles derived from natural fibers (cotton, wool, rayons) and synthetic polyamide and polyester fibers (nylon and Dacron, respectively); vitamins (A, B_1, B_2, B_6, B_{12}, C, D, E, K); hormones (estrone, progesterone, testosterone, insulin, corticosterone, adrenalin, etc.); and medicinals such as aspirin, caffeine, antihistamine drugs, and antibiotics (penicillins, streptomycin, tetracyclines, etc.).

We will divide the discussion of organic chemistry into two parts: (1) the **hydrocarbons** and (2) the **derivatives of the hydrocarbons.** In this chapter, we will consider the hydrocarbons. In Chapter 19, we will consider derivatives of the hydrocarbons. The **hydrocarbons** are organic compounds that contain only the elements carbon and hydrogen. The organic compounds in oil and coal are largely hydrocarbons.[1] The simplest hydrocarbon is methane (CH_4); we will begin our discussion of the hydrocarbons with this simple molecule.

18-1

Methane and Some Larger Molecules

A methane molecule (CH_4) is composed of four hydrogen atoms and one carbon atom. But, before we consider the methane molecule, we must first consider the electronic structure of these atoms. In sublevels, the electronic structures of hydrogen and carbon atoms are

$$^{1}_{1}H \ = \ 1s^1$$

$$^{12}_{6}C \ = \ 1s^2, \ 2s^2 \ 2p^2$$

Hydrogen has *one* valence electron and carbon has *four* valence electrons (see 4-7). The electron configuration of the carbon atom in orbitals is: $1s^2$; $2s^2$ $2p_x^1 \ 2p_y^1 \ 2p_z^0$, since according to Hund's rule (see 13-1) one electron is placed in each p orbital (p_x, p_y, p_z), before pairing, to the maximum of two electrons per orbital. In almost all of its covalent compounds, carbon forms *four* bonds; however, as shown above, in its ground state the carbon atom has only *two* unpaired electrons (p_x^1 and p_y^1). If carbon used only these unpaired electrons for bonding, it would be expected to form only *two* covalent bonds, but carbon uses all *four* of its valence electrons to form *four* covalent bonds. We can look for ways in which the valence electrons of carbon can be redistributed to give *four* unpaired electrons. The redistribution is the basis of the concept of *hybridized orbitals* or mixed orbitals, discussed earlier in 13-1.

---

[1]Crude oil also contains sulfur compounds. The high-grade oil contains less sulfur (less than 0.5 percent by mass) and is called "sweet crude," while the low-grade oil contains more sulfur (0.5 percent by mass and greater) and is called "sour crude." The fields in Saudi Arabia consist of "sour crude." The sulfur content in the oil is one of the factors that determines the varied price of a barrel of oil from the many oil fields around the world.

Suppose we promoted one of the $2s$ electrons to the $2p_z$ orbital and thus had an electron configuration as follows:

$$1s^2,\ 2s^1\ 2p_x^1\ 2p_y^1\ 2p_z^1$$

Furthermore let us assume that these *four* electrons could occupy four *equivalent* **hybrid orbitals** which were *one* part s and *three* parts p character. Such sp^3 hybrid orbitals would have bond angles of 109.5° (see Figure 13-2) or would be *tetrahedral* (similar to a three-sided-base pyramid). Physical measurements have shown that compounds such as methane and carbon tetrachloride are tetrahedral. Carbon uses these sp^3 orbitals to form molecular orbitals when carbon is bonded by **single** bonds to other atoms.

In methane, the four sp^3 orbital electrons in carbon form covalent bonds with each of the $1s$ orbital electrons of the four hydrogen atoms. Figure 18-1 depicts the methane molecule showing the overlap of each of the four sp^3 orbital electrons in carbon with the four $1s$ orbital electrons of hydrogen to form *four* covalent sigma (σ) bonds. A **sigma (σ) bond** is cylindrically symmetrical about the line joining the nuclei—in this case carbon and hydrogen atoms. Figure 6-17 shows molecular models of methane.

For convenience, a molecule of methane is generally written as an electron-dot formula or an expanded or condensed structural formula, as follows:

$$\begin{array}{c} H \\[-2pt] H\!:\!\overset{\cdot\cdot}{\underset{\cdot\cdot}{C}}\!:\!H \\[-2pt] H \end{array}$$	$$\begin{array}{c} H \\ | \\ H\!-\!C\!-\!H \\ | \\ H \end{array}$$	CH_4
electron-dot formula	expanded structural formula	condensed structural formula[2]

FIGURE 18-1

The methane (CH_4) molecule is formed from the four sp^3 hybrid orbital electrons of carbon and four $1s$-orbital electrons of hydrogen. All four bonds in methane are sigma (σ) bonds: C-nucleus •; H-nuclei •

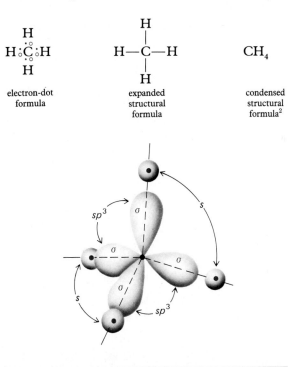

[2]In the condensed structural formula, the hydrogen atoms are written collectively next to the carbon atom to which they are attached.

In methane, carbon bonds only to hydrogen atoms, but carbon can bond to other carbon atoms to form *chains*, *branched chains*, and *rings* of carbon atoms. Examples of compounds having chains of carbon atoms are ethane (C_2H_6) and propane (C_3H_8); their expanded structural formulas and condensed structural formulas are as follows:

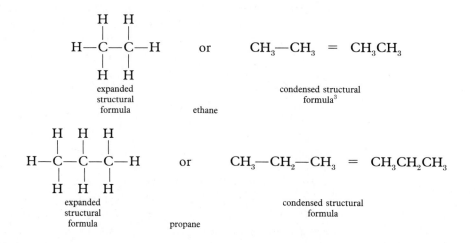

The structural formula for propane appears as a "straight chain," but this is not the case, since the bond angle between all atoms is approximately 109.5°. So, we consider a "straight chain" to be a continuous chain of carbon atoms as shown in Figure 18-2.

Examples of compounds containing rings of carbon atoms are cyclopentane (C_5H_{10}) and cyclohexane (C_6H_{12}). Their structural formulas are shown below, both as full and as convenient skeletal formulas with each corner of the ring representing a —CH_2—.

FIGURE 18-2

Molecular models of propane (C_3H_8). (a) Prentice-Hall models; (b) Stuart-Briegleb models.

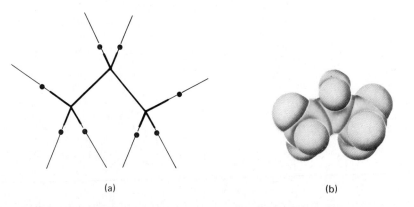

(a)   (b)

---

[3]A dash (—) may or may not be used to represent the bond between the carbon atoms.

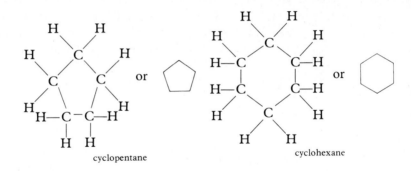

cyclopentane

cyclohexane

In the structural formulas above, carbon has four *lines* or *bonds* extending from it. These bonds are sigma bonds. They are tetrahedrally directed and result from the overlap of the sp^3 orbitals with the orbitals of the atoms bonded to the carbon atom.

18-2

Classification of the Hydrocarbons

The **hydrocarbons** are organic compounds consisting of only the elements carbon and hydrogen. Hydrocarbons are divided into two groups based on structure: the **aliphatic** hydrocarbons and the **aromatic** hydrocarbons. The aliphatic hydrocarbons are further divided into three general groups: the alkanes, the alkenes, and the alkynes. In 18-3 to 18-6, we will consider the aliphatic hydrocarbons, in 18-7 we will consider the aromatic hydrocarbons. Figure 18-3 summarizes the classification of the hydrocarbons.

FIGURE 18-3

Classification of the hydrocarbons.

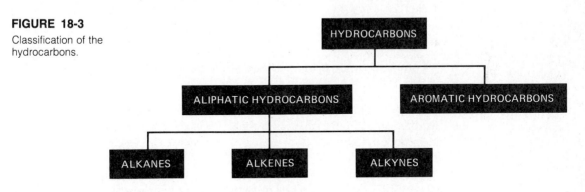

18-3

Alkanes

The **alkanes** (al'kāns), also called the *saturated hydrocarbons* (paraffins), are aliphatic hydrocarbons that have the general molecular formula C_nH_{2n+2} for open-chain systems.[4] The simplest alkane is methane (natural gas, CH_4),

---

[4]The general molecular formula for *cyclo*alkanes is C_nH_{2n}.

which follows the general molecular formula, with $n = 1$. Gasoline consists of a mixture of alkanes containing five to ten carbon atoms.

We have already mentioned (18-1) the first three hydrocarbons in the alkane series (methane, ethane, and propane) and have shown their structural formulas. Each of these hydrocarbons has the general molecular formula C_nH_{2n+2}. The first 10 continuous chain alkanes are shown in Table 18-1. The prefix in each of these hydrocarbons is characteristic of the number of carbon atoms in the chain and, as you will see later, these prefixes are used in the nomenclature of many classes of organic compounds. Note that the prefixes used from pentane to decane are nearly the same prefixes that are used in inorganic nomenclature (see Table 7-1) and indicate the number of carbon atoms. You must memorize the names and structural formulas of those alkanes in Table 18-1.

TABLE 18-1 The Alkanes, Methane to Decane

Name[a]	Molecular Formula	n	Condensed Structural Formula
Methane	CH_4	1	CH_4
Ethane	C_2H_6	2	CH_3CH_3
Propane	C_3H_8	3	$CH_3CH_2CH_3$
Butane	C_4H_{10}	4	$CH_3CH_2CH_2CH_3$
Pentane	C_5H_{12}	5	$CH_3CH_2CH_2CH_2CH_3$
Hexane	C_6H_{14}	6	$CH_3CH_2CH_2CH_2CH_2CH_3$
Heptane	C_7H_{16}	7	$CH_3CH_2CH_2CH_2CH_2CH_2CH_3$
Octane	C_8H_{18}	8	$CH_3CH_2CH_2CH_2CH_2CH_2CH_2CH_3$
Nonane	C_9H_{20}	9	$CH_3CH_2CH_2CH_2CH_2CH_2CH_2CH_2CH_3$
Decane	$C_{10}H_{22}$	10	$CH_3CH_2CH_2CH_2CH_2CH_2CH_2CH_2CH_2CH_3$

[a]The hydrocarbons from butane to decane were formerly named n-butane to n-decane, the n standing for normal, which meant that the carbon atoms were in a continuous chain.

If we examine the molecular formulas of the series above, we notice that they differ from each other by one carbon and two hydrogen atoms, a CH_2, or *methylene*, group. For example, propane (CH_3—CH_2—CH_3) differs from ethane (CH_3—CH_3) by one —CH_2— group. Butane (CH_3—CH_2—CH_2—CH_3) differs from propane (CH_3—CH_2—CH_3) also by one —CH_2— group. A series of compounds in which each compound in the series differs from the next compounded by a *multiple,* such as a —CH_2—, is called a **homologous series.** Therefore, the series methane to decane is a homologous series.

Constitutional (Structural) Isomers

In the previous paragraph, we considered only those alkanes whose carbon atoms form a continuous chain and for which we write only one structural formula. But, once the number of carbon atoms in the molecular formula for an alkane reaches four or more, it is possible to write more than one structural formula for a given molecular formula. Compounds that have the *same* molecular formula but *different* structural formulas are called **constitutional isomers (structural isomers)**. One constitutional isomer of a given molecular formula has *different* properties than another constitutional isomer of the *same* molecular formula.

Simple examples of constitutional isomers are the butanes, C_4H_{10}, and pentanes, C_5H_{12}. There are two isomeric butanes and three isomeric pentanes:

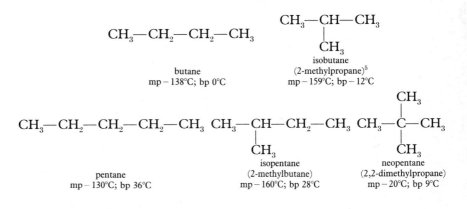

butane
mp – 138°C; bp 0°C

isobutane
(2-methylpropane)[5]
mp – 159°C; bp – 12°C

pentane
mp – 130°C; bp 36°C

isopentane
(2-methylbutane)
mp – 160°C; bp 28°C

neopentane
(2,2-dimethylpropane)
mp – 20°C; bp 9°C

Each of the isomers is a distinct chemical compound having physical properties different from the others. As we continue to increase the number of carbons in the homologous series of the alkanes, we increase the number of constitutional isomers. For example, there are 5 hexanes, 9 heptanes, 18 octanes, 35 nonanes, and 75 decanes. All the constitutional isomers in each case are entirely different compounds and have different properties. For example, notice the different melting and boiling points of the butanes and pentanes.

In drawing constitutional isomers of alkanes the following guidelines may be helpful:

1. Draw a carbon skeleton (no hydrogen atoms) using all of the carbons in a **continuous** chain.

2. Remove **one** carbon (C) atom from the end of the chain and place it on another carbon atom so that the new skeleton differs from the previous carbon skeleton. Repeat this procedure until you exhaust all possibilities of relocating *one* carbon atom.

3. Next, if necessary, remove **two** carbon atoms from the continuous chain skeleton in (1) and relocate them either as *single* carbon atoms or as a *two-carbon* fragment on other carbons in the chain. Write all possible *different* skeletons relocating two carbon atoms. Continue this procedure, if needed, for **three** carbon atoms until you have the number of constitu-

---

[5]The names in parentheses are the IUPAC names. We will explain them later in this section.

tional isomers as given in the problem. Check all skeletons to be sure that they are all different.

4.   Place H atoms on the C atoms in each skeleton in (1), (2), and (3), remembering that there are **four bonds to each C atom.**

Problem Example 18-1

Write condensed formulas for the constitutional (structural) isomers of hexane (C_6H_{14}). There are five such isomers.

SOLUTION

Guideline 1:
(1)   C—C—C—C—C—C

Guideline 2:

(2) 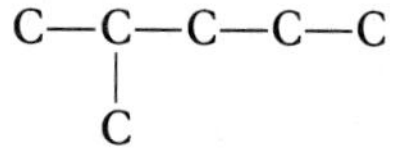   C—C—C—C—C   [If a C were placed on the end, such as
          | C—C—C—C—C, this form would have been the
          C |
                                               C

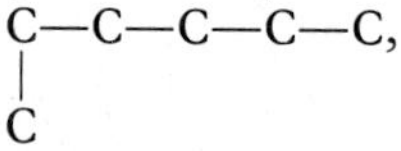

                                     same as (1).]

(3) 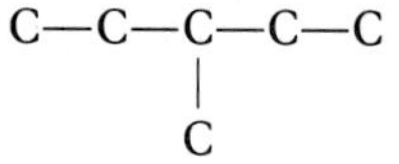   C—C—C—C—C
              |
              C

Guideline 3:

```
              C
              |
(4)   C—C—C—C        (5)   C—C—C—C
          |                    |  |
          C                    C  C
```

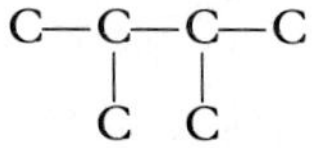

Guideline 4 (Remember that there are four bonds to each carbon atom):

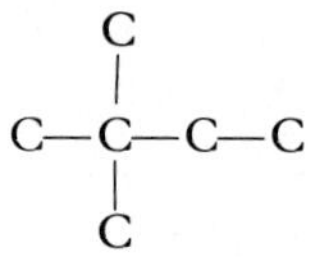

(1) CH_3—CH_2—CH_2—CH_2—CH_2—CH_3 (2) 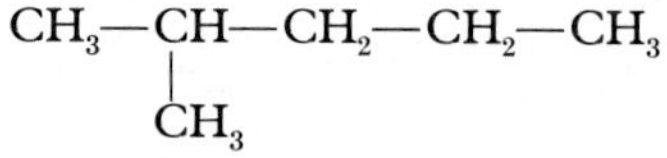 CH_3—CH—CH_2—CH_2—CH_3
                                                                    |
                                                                    CH_3

                                                                    CH_3
                                                                    |
(3) CH_3—CH_2—CH—CH_2—CH_3 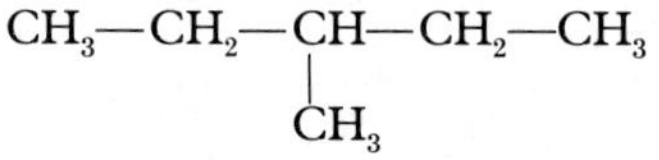   (4) 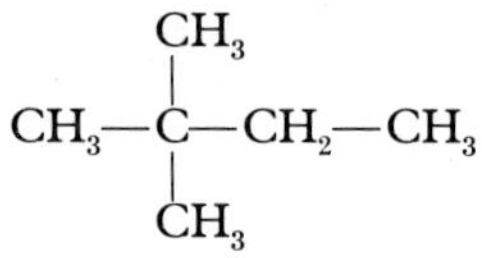 CH_3—C—CH_2—CH_3
                      | |
                      CH_3 CH_3

(5) 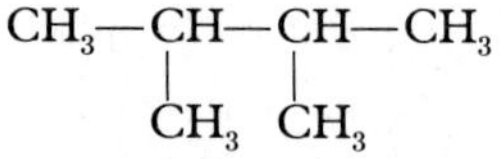 CH_3—CH—CH—CH_3
              | |
              CH_3 CH_3

Work Problem 3.   Now, following these guidelines, draw the condensed structural formulas for the two constitutional isomers of butane (C_4H_{10}) and the three constitutional isomers of pentane (C_5H_{12}).

Nomenclature

In naming the isomeric butanes and pentanes, we used prefixes such as *iso-* and *neo-;* we call these *trivial* prefixes. The nomenclature of the compounds becomes awkward if we use such trivial prefixes in naming isomeric hexanes and higher hydrocarbons in order to distinguish among their isomers. So, a systematic method of nomenclature is necessary, and the IUPAC (International Union of Pure and Applied Chemistry) system that has been developed over the years is now the preferred method of naming organic compounds.[6] Basically, this system uses names composed of two parts. The **terminal** portion names the longest continuous chain in the molecule, the *parent chain;* the first portion names the **substituent groups** attached to the parent chain.

Before discussing this system in more detail, we must consider the nomenclature of alkyl groups, which are frequently substituents on "parent chains." **Alkyl groups** are derived by removal of one hydrogen atom from an alkane; they are named generally by replacing the **-ane** ending of the alkane by **-yl** in the case of the simpler hydrocarbons. In the continuous chain hydrocarbons higher than ethane, "alkyl" is reserved only for the alkyl group obtained by removal of a hydrogen atom from the terminal carbon atoms.[7] Thus:

CH_4 is meth**ane** CH_3— is meth**yl**

CH_3—CH_3 is eth**ane** CH_3—CH_2— is eth**yl**

CH_3—CH_2—CH_3 is prop**ane** CH_3—CH_2—CH_2— is prop**yl**

CH_3—CH_2—CH_2—CH_3 is but**ane** CH_3—CH_2—CH_2—CH_2— is but**yl**

CH_3—CH_2—CH_2—CH_2—CH_3 is pent**ane** CH_3—CH_2—CH_2—CH_2—CH_2— is pent**yl**

CH_3—CH_2—CH_2—CH_2—CH_2—CH_3 is hex**ane** CH_3—CH_2—CH_2—CH_2—CH_2—CH_2— is hex**yl**

The nomenclature of the alkyl groups that are isomeric to propyl, butyl, pentyl, etc., is more difficult, but a suitable trivial method of naming these groups has been developed. The groups are isopropyl, *sec*-butyl, isobutyl, *tert*-butyl, isopentyl, and isohexyl.

The two isomeric propyl groups can be derived from propane by removing a hydrogen atom from the end of the chain (propyl) *or* from the middle carbon atom (isopropyl):

$$CH_3—CH_2—CH_2— \quad CH_3—CH—CH_3 \; = \; \overset{\displaystyle CH_3}{\underset{\displaystyle CH_3}{CH—}} \; = \; (CH_3)_2CH—$$

propyl   isopropyl

---

[6]This system began in 1892 at a conference held in Geneva, Switzerland, and through the IUPAC has been revised periodically since that time.

[7]Alkyls were formerly named as the *n*-alkyl groups (normal alkyl groups), but this is no longer done. Thus, *n*-butyl is butyl, *n*-pentyl is pentyl, *n*-hexyl is hexyl.

The four isomeric butyl groups are readily derived from the two butanes, butane and isobutane, by removing an appropriate hydrogen atom as follows:

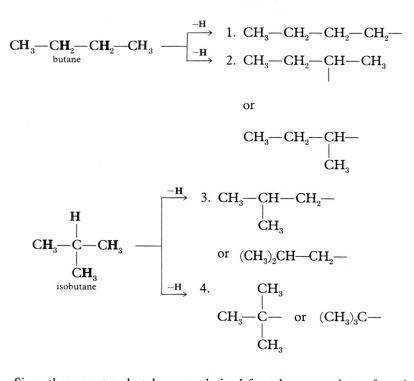

$$CH_3—CH_2—CH_2—CH_3$$
butane

1. $CH_3—CH_2—CH_2—CH_2—$

2. $CH_3—CH_2—CH—CH_3$

or

$CH_3—CH_2—CH—$
$\qquad\qquad\quad CH_3$

$$CH_3—C(H)(CH_3)—CH_3$$
isobutane

3. $CH_3—CH—CH_2—$
$\qquad\quad CH_3$

or $(CH_3)_2CH—CH_2—$

4. $CH_3—C(CH_3)(CH_3)—$ or $(CH_3)_3C—$

Since there are two butyl groups derived from butane and two from isobutane, it is necessary to differentiate further between these groups in order to name them adequately. Alkyl groups can be classified into three different types:

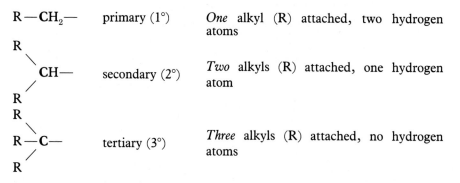

$R—CH_2—$ primary (1°) *One* alkyl (R) attached, two hydrogen atoms

$\overset{R}{\underset{R}{\diagdown}}CH—$ secondary (2°) *Two* alkyls (R) attached, one hydrogen atom

$R—\overset{R}{\underset{R}{C}}—$ tertiary (3°) *Three* alkyls (R) attached, no hydrogen atoms

It is now possible to use the following terms to describe the four different butyl groups mentioned above:

1.   primary butyl
2.   *sec*-butyl

—simplified to—

3.   primary isobutyl

4.   *tert*-isobutyl

butyl = $CH_3CH_2CH_2CH_2$—

sec-butyl = $CH_3CH_2\overset{|}{C}H$
$\underset{CH_3}{|}$

(sec- = secondary)

isobutyl = $CH_3\overset{|}{C}HCH_2$—
$\underset{CH_3}{|}$

$tert$-butyl = $\overset{\overset{CH_3}{|}}{CH_3\underset{\underset{CH_3}{|}}{C}}$—

($tert$- = tertiary)

In general, for any alkyl chain that has up to six carbon atoms, the group having the structure

$$CH_3-\underset{\underset{CH_3}{|}}{C}H-(CH_2)_n-$$

is called the *iso-alkyl group;* thus:

$$CH_3-\underset{\underset{CH_3}{|}}{C}H-CH_2-CH_2- \quad \text{is isopentyl}$$

and

$$CH_3-\underset{\underset{CH_3}{|}}{C}H-CH_2-CH_2-CH_2- \quad \text{is isohexyl}$$

You must learn the trivial names of the simple alkyl groups. Table 18-2 summarizes them for you.

The IUPAC names for the alkanes are obtained by using the following rules:

1.   The alkane hydrocarbons all have the ending **-ane.**

2.   The longest continuous chain of carbons is determined and this chain is used as the parent structure. For example, if the longest continuous chain of carbon atoms is **5**, then the parent structure is called a *pent*ane. Consider the following example:

TABLE 18-2 Summary of Simple Alkyl Groups and Their Trivial Names

Alkyl Group	Trivial Name
CH_3-	Methyl
CH_3-CH_2-	Ethyl
$CH_3-CH_2-CH_2-$	Propyl
$CH_3-CH- \\ \quad\quad | \\ \quad\quad CH_3$	Isopropyl
$CH_3-CH_2-CH_2-CH_2-$	Butyl
$CH_3-CH_2-CH- \\ \quad\quad\quad\quad | \\ \quad\quad\quad\quad CH_3$	*sec*-Butyl
$\quad\quad CH_3 \\ \quad\quad | \\ CH_3-CH-CH_2-$	Isobutyl
$\quad\quad CH_3 \\ \quad\quad | \\ CH_3-C- \\ \quad\quad | \\ \quad\quad CH_3$	*tert*-Butyl
$CH_3-CH-CH_2-CH_2- \\ \quad\quad | \\ \quad\quad CH_3$	Isopentyl
$CH_3-CH-CH_2-CH_2-CH_2- \\ \quad\quad | \\ \quad\quad CH_3$	Isohexyl

$$\overset{1}{CH_3}-\overset{2}{CH}-\overset{3}{CH}-\overset{4}{CH_2}-\overset{5}{CH_3}$$
$$\underset{5}{}\;\underset{4}{CH_3}\;\underset{3}{CH_3}\;\underset{2}{}\;\underset{1}{}$$

The name of the parent structure is *pentane*. This longest continuous chain may be written as a "straight" chain, but may also be "bent," as follows:

$$\overset{1}{CH_3}-\overset{2}{CH}-\overset{3}{CH}-\overset{4}{CH_2}$$
$$\quad\quad CH_3\;\;CH_3\;\;CH_3\;\;{}_5$$

The name of the parent structure is still *pentane*. **If the chain is in the form of a ring, then the prefix *cyclo-* is used.**

3. The carbon atoms in this chain are numbered by starting at the end that would give the *lowest numbers* to the group or groups attached to the parent structure. In the above example, the lowest numbers would be obtained by numbering from the left side (**2** and **3** are lower than **3** and **4**).

4. The group attached to the parent structure, other than hydrogen, is given both *name* and *number*. Halogens attached to the parent structure are named *halo*, such as -Cl is *chloro*, -Br is *bromo-*, and -I is *iodo-*. The $-NO_2$ group is named *nitro-*. Alkyl groups are given their accepted IUPAC names.

5. If more than one of the *same* group appears as a substituent in a given molecule, then the prefixes *di, tri, tetra-*, and *penta-* are used to indicate the number of times this group appears—two, three, four, or five times, respectively—and the position of these groups on the numbered parent structure is indicated in *increasing numerical order. Each group* must have a *number* to indicate its position on the parent structure. In the case above, two methyl groups appear on the pentane and hence a *di-* is used, in the name *dimethyl*. These methyl groups are attached at the **2** and **3** positions, **2,3**-dimethyl. Notice that a comma (,) is used between the numbers and a hyphen (-) between the number and the name. Therefore, the correct IUPAC name for the compound above is 2,3-dimethylpentane.

6. If more than *one type of group* is attached to the parent structure, these groups are placed in alphabetical order in the name. However, in determining such alphabetical order, prefixes such as *di-, tri, tetra-*, denoting the number of groups, and hyphenated prefixes such as *sec-* and *tert-* are ignored. The actual name of the group following such prefixes is used to determine alphabetical order. Note that the prefix *iso-* is not hyphenated when used and is used in the determination of the alphabetical sequence. In the following list of groups, the letter in **bold** print is used to determine the alphabetical order: di**c**hloro, tri**m**ethyl, **i**sopropyl, sec-**b**utyl, tert-**b**utyl, tetra**e**thyl.

Problem Example 18-2

Give the IUPAC name for each of the following compounds:

Answers

(a) CH₃—CH—CH₃
            |
           CH₃

2-methylpropane

(b) CH₃—CH₂—CH₂—Cl

1-chloropropane

(c) CH₃—CH—CH—CH₂—CH₃
            | |
           Cl   CH₃

2-chloro-3-methylpentane
(Note the alphabetical order of the substituents.)

(d)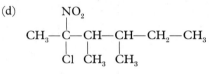

2-chloro-3,4-dimethyl-2-nitrohexane

Work Problem 4.

(e)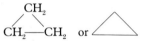

cyclopropane
(Note that each corner represents a
—CH_2—.)

Now that we can name alkanes, we must be able to write the structural formula for a compound given its name. In doing this we shall follow three steps:

1. Write the chain or ring of carbon atoms for the parent structure.

2. Add the groups to the parent structure.

3. Place the necessary hydrogen atoms on the carbon atoms such that each carbon has *four* covalent bonds.

Problem Example 18-3

Write the structural formula for each of the following compounds:

Answers

(a) 2-methylpentane

$$CH_3-CH-CH_2-CH_2-CH_3$$
$$\qquad\quad |$$
$$\qquad\;\; CH_3$$

(b) 2-bromobutane

$$CH_3CH_2CHCH_3$$
$$\qquad\qquad |$$
$$\qquad\qquad Br$$

(c) 1,3,5-trichlorohexane

$$ClCH_2CH_2CHCH_2CHCH_3$$
$$\qquad\qquad\; |\qquad\;\; |$$
$$\qquad\qquad\; Cl\qquad\; Cl$$

(c) 3-chloro-2,4-dimethylheptane

$$CH_3-CH-CH-CH-CH_2-CH_2-CH_3$$
$$\qquad\quad |\qquad |\qquad |$$
$$\qquad\;\; CH_3\; Cl\;\; CH_3$$

Work Problem 5.

(e) methylcyclopentane

(All positions on the ring are equivalent, so the —CH_3 can be attached to any one of the carbons in the cyclopentane ring.)

Reactions

There are only a few useful reactions of the alkanes, because they are *relatively inert*. Alkanes do not react under ordinary conditions with acids, such as hydrochloric acid; bases, such as sodium hydroxide; oxidizing agents (see 16-1), such as potassium permanganate; or reducing agents (see 16-1), such as sodium metal. They do react with oxygen to produce, primarily, heat energy which heats homes and powers automobiles (see Problem Examples 9-3 and 9-10). Another reaction of alkanes is *halogenation*.

In halogenation, a halogen atom, chlorine or bromine, replaces a hydrogen atom. This reaction is classed in a general group called **substitution reactions.** Hence, alkanes react by substitution when they are halogenated, as illustrated in the following general equation for monohalogenation:

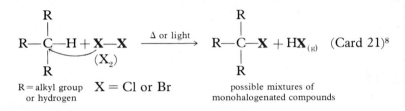

R = alkyl group or hydrogen   X = Cl or Br   possible mixtures of monohalogenated compounds

Heat (Δ) or light is required for the reaction to proceed. The reaction forms two products: the *organic monohalogenated compound* and the *hydrogen halide gas,* such as hydrogen chloride or hydrogen bromide gas:

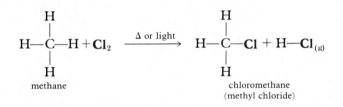

methane   chloromethane (methyl chloride)

In larger alkane molecules there is the possibility of the halogen attacking different carbon atoms in the molecule, producing a mixture of monohalogenated organic compounds. If the reaction is allowed to continue, di-, tri-, tetra-, etc. halogenated organic compounds can be formed. In the case of chlorination of methane, these compounds would be dichloromethane or methylene chloride ($ClCH_2Cl$), trichloromethane or chloroform ($HCCl_3$), and tetrachloromethane or carbon tetrachloride (CCl_4). To obtain primarily monohalogenated compounds, a large excess of methane is used and the methane is recycled through the reaction mixture.

Work Problem 6.

---

[8]In Chapter 9, we suggested that you make flash cards, with the reactants on one side, including catalysts and conditions, and products on the other side to help you remember the reactions. In this chapter, continue the same procedure for the general equation.

18-4

Alkenes

The **alkenes** (al′kēns) are aliphatic hydrocarbons that have the general molecular formula C_nH_{2n} for open-chain systems. The simplest alkene is ethylene (C_2H_4), which follows the general molecular formula with $n = 2$. The alkenes and alkynes (see 18-6) are both referred to as **unsaturated hydrocarbons,** since they contain fewer than the maximum number of hydrogen atoms in their general molecular formula. The alkenes contain a *double* bond.

Ethylene

Physical measurements on ethylene have shown the molecule to be planar, with the bond angles between any two bonds being 120°, as shown below:

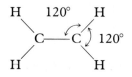

Suppose we assume that one of the $2s$ electrons of carbon is promoted to the vacant $2p_z$ orbital (see 18-1). Furthermore, let us consider that the $2s^1$, $2p_x^{\ 1}$ and $2p_y^{\ 1}$ electrons could occupy equivalent *hybrid orbitals* that are one part s and only *two* parts p character; the fourth valence electron remains in the $2p_z$ orbital. It can be shown that such sp^2 hybrid orbitals would have bond angles of 120° and would be lying in the XY plane; the axis of the nonhybridized $2p_z$ orbital is perpendicular to this plane:

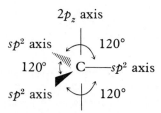

In the above illustration, the three sp^2 axes lie in a plane perpendicular to the page with the ▶ bond coming out of the page, the ⦚⦚⦚▸ bond going into the page, and the —— bond in the plane of the page.

We consider that a molecule of ethylene is composed of two such carbon atoms bonded together by a σ bond formed by the overlap of two sp^2 hybrid orbitals, with the remaining four sp^2 orbitals forming σ bonds by overlapping with the s orbitals of the hydrogen atoms. The fourth valence electron of each carbon atom lies in the remaining p orbital whose axis is perpendicular to the plane of the sp^2 axes:

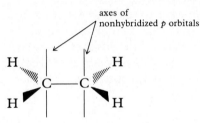

The two electrons in the two nonhybridized p orbitals[9] form the second bond between the carbon atoms, but this bond is *not* a σ bond. It is formed by a side-to-side overlap of the p orbitals and has its maximum strength when the axes of the p orbitals are parallel. It is called a **pi (π) bond.** See Figures 18-4 and 18-5. A π bond differs from a sigma (σ) bond in that it is *not* cylin-

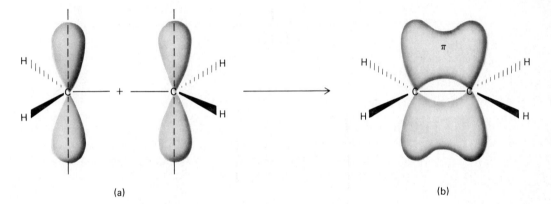

(a)   (b)

FIGURE 18-4

The overlap of p-orbital electrons in ethylene (C_2H_4) to form a π bond with lobes of electron clouds *above* and *below* the plane of carbon atoms. (a) Two p orbitals; (b) a π bond.

FIGURE 18-5

Molecular models of ethylene (C_2H_4). (a) Prentice-Hall model; (b) Stuart-Briegleb model.

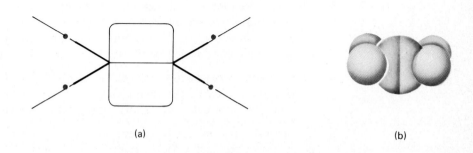

(a)   (b)

[9]A $2p$ orbital is more correctly represented as shown in Figure 4-5. For convenience, the representation shown in Figures 18-4a, 18-6a, and 18-8a is more generally used and is used in this chapter.

drically symmetrical about the bond joining the two nuclei. It is also a *weaker* bond than the σ bond. Thus, the carbon-carbon double bond in ethylene is considered to consist of *one* σ bond and *one* π bond.

For convenience, a molecule of ethylene is usually represented by either its expanded structural formula or its condensed structural formula:

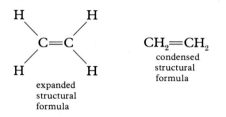

Nomenclature

"Ethylene" is a trivial name and is not used in the IUPAC system. The correct IUPAC name is "ethene." We can use trivial names for a few alkenes, such as ethylene, propylene, and isobutylene, but now we will consider only the IUPAC nomenclature. The following rules govern the IUPAC nomenclature of alkenes:

1. The alkene hydrocarbons all have the ending **-ene.**

2. The longest continuous chain of carbons that *contains the double bond* is used as the parent structure. For example, if the longest continuous chain of carbon atoms that contains the double bond is 5 carbon atoms, then the parent structure is named as a *pent*ene. Consider the following example:

$$\overset{1}{C}H_3-\overset{2}{C}=\overset{3}{C}H-\overset{4}{C}H-\overset{5}{C}H_3$$
$$\underset{CH_3}{|} \qquad \underset{CH_3}{|}$$

3. The carbon atoms in this chain are numbered so as to give the *lowest* possible number to the *doubled bond* regardless of the groups attached to the parent structure. In the example above, the lowest possible number for the double bond would be obtained by numbering from the left side.

4. The position of the double bond is indicated by placing the lower numbered carbon atom of the double bond before the name of the parent structure; this procedure would make the pentene above a 2-pentene.

5. The groups attached to the parent structure, other than hydrogen, are given both *name* and *number*, as was done in naming the alkanes, and these groups are placed in alphabetical order. For the compound above, the correct IUPAC name is 2,4-dimethyl-2-pentene.

Problem Example 18-4

Give the IUPAC name for each of the following compounds:

Answers

(a) $CH_2=CH-CH_2-CH_3$

1-butene

(b) $CH_3-CH=CH-CH_2-CH_3$

2-pentene

(c) $CH_3-C=CH-CH-CH_3$
         | |
       CH_3 Cl

4-chloro-2-methyl-2-pentene

(d) $CH_3-C==C-CH_2-CH_2-CH-CH_3$
       | | |
     CH_3 Br CH_3

3-bromo-2,6-dimethyl-2-heptene

(e) CH_3
             |
$CH_3-CH-C-CH-CH_2-CH=CH_2$
        | | |
      NO_2 Cl CH_3

Work Problem 7.

5-chloro-4,5-dimethyl-6-nitro-1-heptene

In writing structures of alkenes from the name, we follow the same steps that we took with alkanes except that we add the *double bond* in the *correct* position in the parent structure.

Problem Example 18-5

Write the structural formula for each of the following compounds:

Answers

(a) propene[10] $CH_2=CH-CH_3$

(b) 1-pentene $CH_2=CH-CH_2-CH_2-CH_3$

(c) 2-hexene $CH_3-CH=CH-CH_2-CH_2-CH_3$

(d) 4-methyl-1-pentene $CH_2=CH-CH_2-CH-CH_3$
                                     |
                                   CH_3

Work Problem 8.   (e) 4,5,5-trimethyl-2-hexene

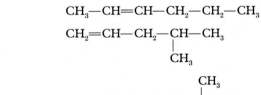

[10]A number is not needed in the name, as only *one* structure can be written.

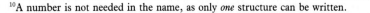

Reactions

As we discussed in 18-3, alkanes react with chlorine or bromine by a substitution reaction under the influence of *heat* or *light*. Alkenes also react with chlorine or bromine (halogenation), but generally the reaction is an **addition reaction**. The reaction occurs readily at room temperature. The reactive site in an alkene is the π bond, and the halogen atoms add across this bond to form two carbon-halogen σ bonds, as shown in the following general equation. Notice that the π bond is broken and is not present in the product.

$$R—CH\!\!=\!\!CH_2 + X—X \longrightarrow R—CH—CH_2 \quad \text{(Card 22)}$$
$$\underset{(X_2)}{} \qquad\qquad\qquad \overset{|}{X} \quad \overset{|}{X}$$

R = alkyl or H   X = Cl or Br

$$CH_3—CH\!\!=\!\!CH_2 + Br_2 \longrightarrow CH_3—CH—CH_2$$
$$\underset{\text{propene}}{} \quad \underset{\substack{\text{in}\\ CCl_4}}{} \qquad\qquad \overset{|}{Br} \quad \overset{|}{Br}$$
$$\text{1,2-dibromopropane}$$

Bromine dissolves in carbon tetrachloride to yield a reddish-brown solution, and as the bromine reacts with the alkene, the reddish-brown color of the bromine disappears and a colorless solution is generally obtained. This reaction is usually carried out at room temperature and the presence of an alkene double bond is noted by the disappearance of the reddish-brown color.

Work Problem 9.

Alkenes also add other reagents besides chlorine or bromine across the π bond. These include hydrogen—when platinum is used as a catalyst—hydrogen chloride, hydrogen bromide, hydrogen iodide, water, and many others. These addition reactions are illustrated with ethylene as the alkene:

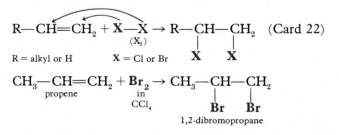

Hydrogenation reactions are carried out on vegetable oils to form higher-melting solids which are sold as semisolid oleomargarine (see 19-8).

18-5

Addition Polymers

Alkenes, beside undergoing simple addition reactions, also react with themselves to form "giant" molecules called *polymers* or "plastic." *Poly* means "many" and *mer* comes from the Greek meaning parts, so **polymers** by definition are molecules that have many parts. They are formed from small molecules called *monomers* (one part or molecule). These monomers react with each other to form long-chain molecules with high molecular masses. **Addition polymers** are polymers formed from alkene monomers which react with each other by "addition" across the double bond breaking the pi (π) bond.

In this section, we will consider four polymers that may be familiar to you. They are: (1) polyethylene, (2) Teflon, (3) polyvinyl chloride (PVC), and (4) Orlon.

Polyethylene is formed from the monomer, ethylene, using an organic peroxide catalyst and heat and pressure. The equation for the reaction is as follows:

$$n \text{ CH}_2{=}\text{CH}_2 \rightarrow \left[\text{CH}_2{-}\text{CH}_2 \right]_n$$

ethylene   polyethylene

In the formula of polyethylene the n denotes a large number of the monomer units with a molecular mass for polyethylene of about 1,000,000 amu. Polyethylene is used in making plastic bags, trash cans, milk containers, and squeeze bottles.

Teflon is formed from the monomer, tetrafluoroethylene, using a catalyst of hydrogen peroxide and iron(III) ions. The equation for the reaction is as follows:

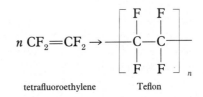

$$n \text{ CF}_2{=}\text{CF}_2 \rightarrow$$

tetrafluoroethylene   Teflon

Teflon is resistant to all types of chemical action and can withstand high temperatures. It is used on nonstick cooking utensils, in greaseless bearings, and as an insulator.

Polyvinyl chloride (PVC) is formed from the monomer, vinyl chloride, using an organic peroxide catalyst and heat. The equation for the reaction is as follows:

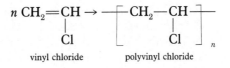

$$n \text{ CH}_2{=}\text{CH} \rightarrow$$

vinyl chloride   polyvinyl chloride

Polyvinyl chloride has a molecular mass of about 1,500,000 amu. It is used to make vinyl tile floors, plumbing pipes, phonograph records, rainware, shower curtains, and garden hoses.

Orlon is formed from the monomer, acrylonitrile, and a catalyst of hydrogen peroxide and iron(II) sulfate. The equation for the reaction is as follows:

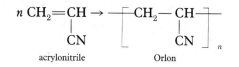

$$n\,CH_2{=}CH \rightarrow \left[CH_2{-}CH \right]_n$$

acrylonitrile   Orlon

Work Problem 10.   Orlon is used to make carpets and clothing, such as sweaters.

18-6

Alkynes

The **alkynes** (al'kīns) are aliphatic hydrocarbons that have the general molecular formula C_nH_{2n-2} for the open-chain systems. Like the alkenes, they are also *unsaturated hydrocarbons*. The simplest alkyne is acetylene (C_2H_2), which follows the general molecular formula with $n = 2$. The alkynes contain a *triple* bond.

Acetylene

Because acetylene is a linear molecule, its carbon and hydrogen atoms all lie in a straight line, which means that its C—C—H bond angle is 180°. If one of the $2s$ electrons of carbon is promoted to the vacant $2p_z$ orbital, the electron configuration of carbon becomes $1s^2, 2s^1\,2p_x{}^1\,2p_y{}^1\,2p_z{}^1$. Now, let us mix the $2s$ and $2p_x$ orbitals to obtain two *hybrid orbitals* that are one part s and one part p. These sp orbitals would have bond angles of 180° and would lie along the p_x axis. The axes of the nonhybridized p_y and p_z orbitals would be perpendicular to each other and would lie in a plane perpendicular to the axis of the sp hybridized orbitals:

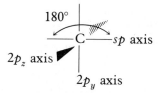

A molecule of acetylene is formed by two such carbon atoms which are bonded together by a σ bond through the overlap of two sp hybrid orbitals; the remaining two sp hybrid orbitals form σ bonds with hydrogen. The third and fourth valence electrons of each carbon atom lie in the remaining p orbitals which are nonhybridized. These four electrons form the second and third bonds between the carbon atoms by π bond formation. Thus, the *triple bond* in acetylene consists of **one** σ bond and **two** π bonds. Figure 18-6 depicts this overlap of p orbitals to form two π bonds, and Figure 18-7 shows molecular models of acetylene.

Generally, expanded structural formulas or condensed structural formulas are used to represent a molecule of acetylene:

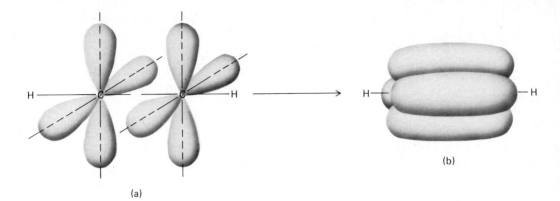

(a)

FIGURE 18-6

The overlap of *p*-orbital electrons in acetylene (C_2H_2) to form two π bonds. (a) Four *p* orbitals; (b) two π bonds.

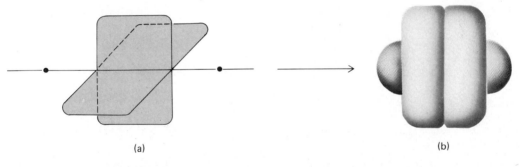

(a)   (b)

FIGURE 18-7

Molecular models of acetylene (C_2H_2). (a) Prentice-Hall model; (b) Stuart-Briegleb model.

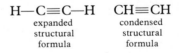

H—C≡C—H	CH≡CH
expanded	condensed
structural	structural
formula	formula

Nomenclature

When alkynes are named by the IUPAC system, the following rules are used:

1. The alkyne hydrocarbons all have the ending **-yne.** So, acetylene has the IUPAC name "ethyne."

2. The longest continuous chain of carbon atoms that *contain the triple bond* is used as the parent structure.

3. This chain of carbon atoms is numbered such that the triple bond has the lowest possible number.

4. The position of the triple bond is indicated by placing the lowest number of the carbon atom of the triple bond before the name of the parent structure.

5. The groups attached to the parent structure other than hydrogen are given both *name* and *number* and are placed in alphabetical order.

Problem Example 18-6

Give the IUPAC name for each of the following compounds:

Answers

(a) $CH{\equiv}C{-}CH_2{-}CH_3$ 1-butyne

(b) 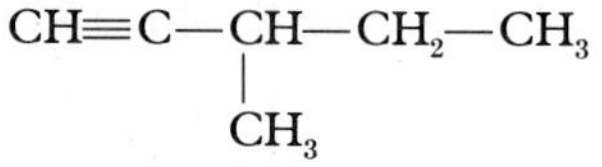 $CH{\equiv}C{-}CH{-}CH_2{-}CH_3$ 3-methyl-1-pentyne
$\qquad\qquad\quad |$
$\qquad\qquad CH_3$

(c) 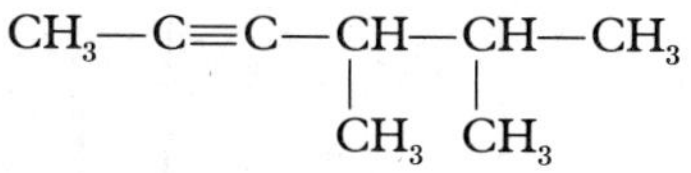 $CH_3{-}C{\equiv}C{-}CH{-}CH{-}CH_3$ 4,5-dimethyl-2-hexyne
$\qquad\qquad\qquad\quad |\quad\ |$
$\qquad\qquad\qquad CH_3\ CH_3$

(d) 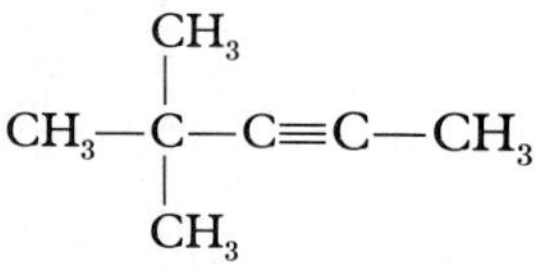 $\qquad\quad CH_3$
$\qquad\qquad |$
$\quad CH_3{-}C{-}C{\equiv}C{-}CH_3$ 4,4-dimethyl-2-pentyne
$\qquad\qquad |$
$\qquad\qquad CH_3$

(e) 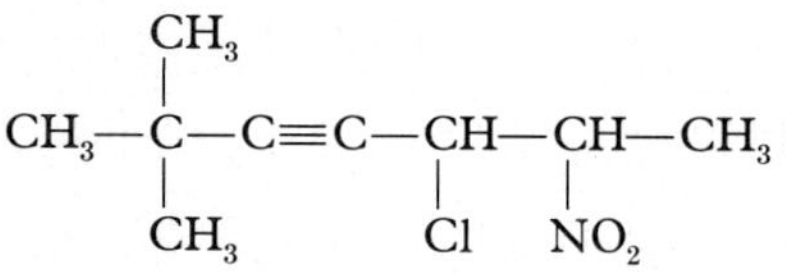 $\qquad\quad CH_3$
$\qquad\qquad |$
$\quad CH_3{-}C{-}C{\equiv}C{-}CH{-}CH{-}CH_3$ 5-chloro-2,2-dimethyl-6-nitro-3-heptyne
Work Problem 11. $\qquad\ |\qquad\qquad |\quad\ |$
$\qquad\qquad CH_3\qquad\ Cl\ \ NO_2$

When we write structures of alkynes from the names of the alkynes, we follow the steps that we took with alkenes, except that we place the *triple bond* in the *correct* position in the parent structure.

Problem Example 18-7

Write the structural formula for each of the following compounds:

Answers

(a) 3-methyl-1-butyne $CH_3{-}CH{-}C{\equiv}CH$
$\qquad\qquad\qquad\qquad\qquad |$
$\qquad\qquad\qquad\qquad\ CH_3$

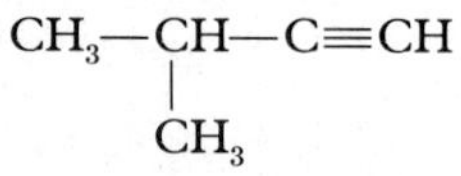

(b) 2,2,5-trimethyl-3-heptyne $\qquad\ CH_3$
$\qquad\qquad\qquad\qquad\qquad\qquad\quad |$
$\qquad\qquad\qquad\qquad CH_3{-}C{-}C{\equiv}C{-}CH{-}CH_2{-}CH_3$
$\qquad\qquad\qquad\qquad\qquad\quad |\qquad\qquad |$
$\qquad\qquad\qquad\qquad\qquad CH_3\qquad\quad CH_3$

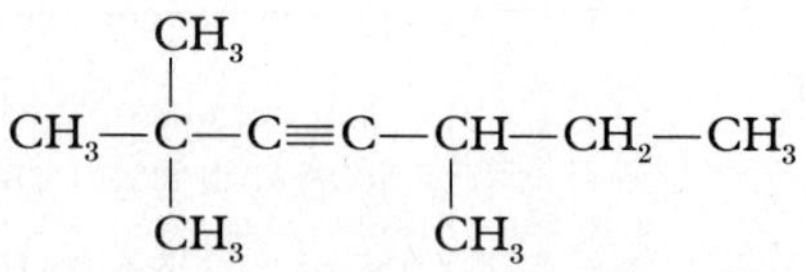

(c) 2-pentyne

$$CH_3—C≡C—CH_2—CH_3$$

(d) 4,4-dimethyl-2-hexyne

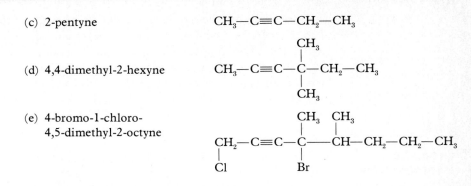

(e) 4-bromo-1-chloro-
     4,5-dimethyl-2-octyne

Work Problem 12.

Reactions

Alkynes undergo **addition reactions** similar to the reactions of alkenes. They react readily with chlorine and bromine at room temperature. The reaction sites in an alkyne are the two π bonds, and the halogen atoms add across these bonds to form carbon-halogen σ bonds. The reaction can occur in steps, as shown in the following general equation:

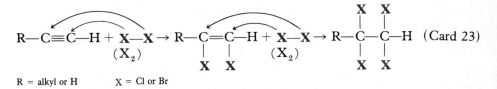

R = alkyl or H   X = Cl or Br

Note that a π bond is broken each time an addition occurs and that *two* moles of the halogen (X) are used.

$$CH_3—C≡C—H + 2\ Br_2 \underset{\substack{\text{excess}\\ \text{in } CCl_4}}{\longrightarrow} CH_3—\overset{\displaystyle Br}{\underset{\displaystyle Br}{C}}—\overset{\displaystyle Br}{\underset{\displaystyle Br}{C}}—H$$

1,1,2,2,-tetrabromopropane

Work Problem 13.

Alkynes can add other reagents across the π bonds. These include hydrogen—when a platinum catalyst is used—hydrogen chloride, hydrogen bromide, and hydrogen iodide. These reactions can occur in steps, as shown below with acetylene:

$$H—C≡C—H \begin{cases} \xrightarrow{H_2,\ Pt} CH_2{=}CH_2 \xrightarrow{H_2,\ Pt} CH_3—CH_3 \\ \\ \xrightarrow{HX} CH_2{=}CH—X \xrightarrow{HX} CH_3—\underset{\displaystyle X}{CH}—X \end{cases}$$

X = Cl, Br, I

These reactions may be terminated at the alkene stage, or they may be allowed to go to completion with the addition of *two* moles of the reagent across the π bonds. The addition of bromine across the π system of an alkyne is used as a diagnostic test for alkynes as with alkenes. The disappearance of the reddish-brown color of the bromine (in carbon tetrachloride) when bromine is added to an alkyne triple bond represents a positive test for the alkynes.

18-7

Aromatic Hydrocarbons

The **aromatic hydrocarbons** are hydrocarbons that include benzene and compounds containing aliphatic or aromatic groups attached to aromatic rings (arenes). Because benzene (C_6H_6) is the simplest aromatic hydrocarbon, we will consider it in some detail.

Benzene

The benzene molecule is planar and has the shape of a regular hexagon. Benzene is a cyclic molecule and consists of six carbon atoms, each using sp^2 hybrid orbitals to form σ bonds with hydrogen and two adjacent carbon atoms as shown in Figure 18-8a. The bond angle between the H—C—C atoms or the C—C—C atoms is 120°. The remaining six nonhybridized p orbitals, one on each carbon, all lie perpendicular to the plane of the carbon ring. These p orbitals overlap with those adjacent to form a delocalized cloud of electrons above and below the plane of the ring. The bonding between adjacent carbon atoms is neither double nor single but something in between. *All* of the *carbons* in benzene are **equivalent** as are *all* of the *hydrogens*. The delocalized π orbitals in benzene are depicted in Figure 18-8b as being located in a pair of doughnut shaped regions lying above and below the plane of the ring. This delocalizing of electrons appears to increase the stability of benzene and make

FIGURE 18-8

The overlap of p-orbital electrons in benzene (C_6H_6). (a) Six p orbitals, one on each of the six carbon atoms; (b) three delocalized π bonds above and below the plane of carbon atoms.

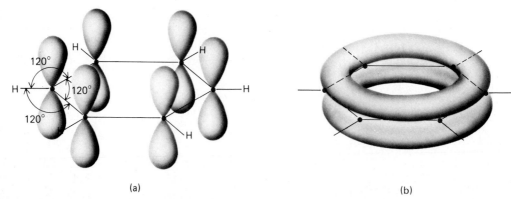

(a)   (b)

it more stable than a hypothetical molecule that consists of fixed bonds, three double bonds and three single bonds alternating around the ring. Molecular models of benzene are shown in Figure 18-9.

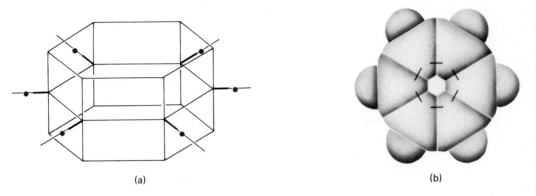

(a)   (b)

FIGURE 18-9

Molecular models of benzene (C_6H_6). (a) Prentice-Hall model; (b) Stuart-Briegleb model.

The structural formula for benzene can be represented in a number of ways. One way is to draw it as containing alternating double and single bonds, but keeping in mind that the π electron cloud is delocalized over all six carbon atoms. Two such formulas can be written, each differing from the other in the positions of the double bonds. These are *Kekulé* structures, named for Friederich August Kekulé[11] who first proposed them in 1865. The Kukulé structural formulas are more conveniently drawn as a simple hexagon containing alternating double bonds, each corner representing a carbon atom bonded to a hydrogen. The actual benzene molecule is thought to be a hybrid of the two and this is indicated by writing a two-headed arrow between them:

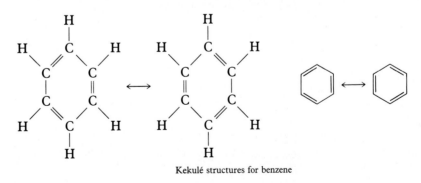

Kekulé structures for benzene

---

[11]Friederich August Kekulé (1829–1896), a German chemist and professor, has been considered the father of the structural theory of organic chemistry.

More recently the Kekulé formulas have been replaced by a formula which depicts the delocalization of the π electrons by using a circle in the center of the hexagon:

This is the structure for benzene we will use in this book.

Nomenclature

The monocyclic aromatic hydrocarbons are named as derivatives of the parent, benzene; however, trivial names are still recognized by the IUPAC for some of them. A few of these trivial names are:

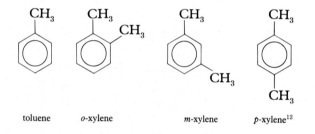

In the IUPAC system, aromatic compounds with **one** *substituent* group are named by combining the name of the substituent group with *benzene*. Since *all* hydrogen atoms are equivalent on benzene, this group may be attached at any position. The following structures are all named *chlorobenzene,* and all represent the same compound.

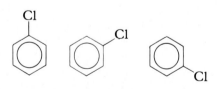

There are three different constitutionally isomeric compounds with **two** *substituents* on the ring. The positions of the substituents on these compounds are indicated as follows:

---

[12]The names for the xylenes are read as follows: ortho-xylene, meta-xylene, and para-xylene.

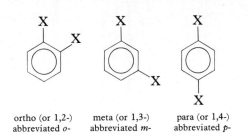

ortho (or 1,2-)   meta (or 1,3-)   para (or 1,4-)
abbreviated *o*-   abbreviated *m*-   abbreviated *p*-

Using either benzene or toluene as the parent structure, the following examples illustrate the use of this system:

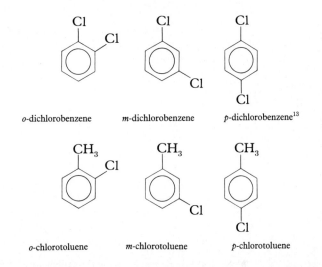

o-dichlorobenzene   *m*-dichlorobenzene   *p*-dichlorobenzene[13]

o-chlorotoluene   *m*-chlorotoluene   *p*-chlorotoluene

For compounds with **three** or **more** *substituents* on the ring, we must use numbers to indicate the position on the ring and apply the following rules:

1. The parent structure is benzene *or* toluene.

2. For benzene, the group name written next to the word "benzene" becomes position number 1 on the ring and the ring is numbered in a direction that will give the lowest possible numbers to the substituents. For toluene, the methyl group becomes position number 1 on the ring and the ring is numbered in a direction that will give the lowest possible numbers to the substituents. The number 1 position is assumed in both benzene and toluene derivatives, but is not indicated in the name.

3. All groups are placed in alphabetical order except the group in the number 1 position.

---

[13]This compound is used as the active ingredient in mothballs.

The following examples illustrate the application of these rules:

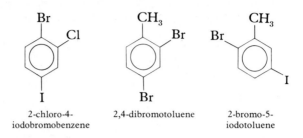

2-chloro-4-
iodobromobenzene

2,4-dibromotoluene

2-bromo-5-
iodotoluene

Problem Example 18-8

Give the IUPAC name for each of the following compounds:

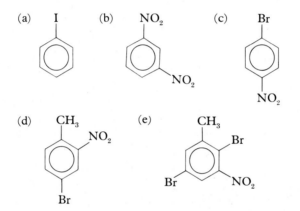

Work Problem 14.

Answers (a) iodobenzene, (b) *m*-dinitrobenzene, (c) *p*-nitrobromobenzene, (d) 4-bromo-2-nitrotoluene, (e) 2,5-dibromo-3-nitrotoluene.

In writing structures of these compounds from the names, we draw the structure of the parent compound and then attach the various groups.

Problem Example 18-9

Write the structural formula for each of the following compounds:

(a) nitrobenzene, (b) *o*-bromochlorobenzene, (c) 2-chloro-4-nitrobromobenzene, (d) 2,6-dinitrotoluene, (e) 2,6-dibromo-3-chloro-5-nitrotoluene.

Answers (a) (b) (c)

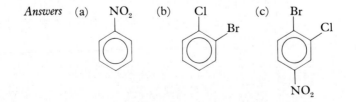

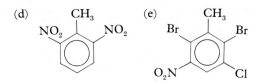

(d)   (e)

Work Problem 15.

Reactions

Aromatic compounds undergo *halogenation* by **substitution,** with a hydrogen atom being replaced by a halogen atom. In our discussion of halogenation of aromatic compounds, we will consider only benzene, which acts as a typical aromatic hydrocarbon.

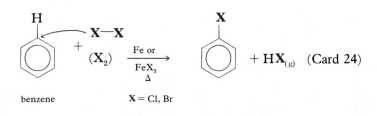

A catalyst in the form of powdered iron or the corresponding iron(III) halide ($FeCl_3$ or $FeBr_3$) is generally required as is heat, especially in the case of the less reactive hydrocarbon derivatives. A gas, HX, is released.

Another **substitution** reaction which takes place readily with aromatic compounds is *nitration*. In this reaction, an aromatic compound such as benzene is treated with a mixture of nitric and sulfuric acids, and a nitro group (NO_2) is substituted for one of the hydrogen atoms on the ring.

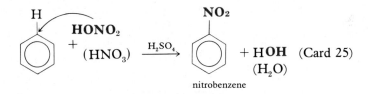

In order to better understand this reaction, we have written nitric acid (HNO_3) as $HONO_2$ to emphasize the nitro group ($-NO_2$) and its substitution for one of the hydrogen atoms on benzene.

Work Problem 16.

An important nitro organic compound is nitroglycerine or glyceryl trinitrate.

$$CH_2-O-NO_2$$
$$CH\ -O-NO_2$$
$$CH_2-O-NO_2$$

nitroglycerine or glyceryl trinitrate

(Note that the —NO_2 is attached to an oxygen atom in nitroglycerine, whereas in nitrobenzene it is attached to a carbon atom.) Nitroglycerine is used as an explosive. It is mixed with sawdust or diatomaceous earth and called "dynamite." Dynamite was invented by the Swedish industrial chemist Alfred Bernhard Nobel (1833–1896). From his great wealth due to his invention of dynamite, he had established from his estate a trust fund that is used to award the highly regarded Nobel Prizes (see footnote 5 in Chapter 1). Nitroglycerine is also used in medicine to save peoples' lives. It acts to dilate the blood vessels in the heart and is used in the treatment of angina pectoris, a disease of the heart in which the coronary blood vessels constrict. By dilating these vessels the heart muscle is able to get nutrients and oxygen and hence repair itself.

Chapter Summary

Organic chemistry is the study of compounds containing the element carbon. In this chapter, only the hydrocarbons—organic compounds that contain only the elements carbon and hydrogen—were considered. The hydrocarbons are divided into *aliphatic hydrocarbons* and *aromatic hydrocarbons;* aliphatic hydrocarbons are further divided into the alkanes, alkenes, and alkynes. The simplest hydrocarbon is the methane molecule. It is composed of four sp^3 hybridized orbitals of the carbon atom bonded to the *s* orbital of four hydrogen atoms to form four single bonds, called *sigma* (σ) *bonds*. The bond angle between all atoms in the methane molecule is 109.5°.

The *alkanes* have the general molecular formula C_nH_{2n+2} (open-chain systems). Table 18-1 gives the first 10 members of the homologous series of alkanes. Compounds that have the same molecular formula but different structural formulas are called *constitutional (structural) isomers.* IUPAC nomenclature for the alkanes, including the various alkyl groups as given in Table 18-2 was considered. Given the name of the compound, the structural formula was written for a number of alkanes. The alkanes are considered relatively inert. The few reactions that the alkanes undergo are *substitution reactions;* halogenation is an illustration of this reaction. Both general and specific equations of monohalogenation of alkanes were given.

The *alkenes* have the general molecular formula C_nH_{2n} (open-chain systems). The simplest alkene is ethylene (C_2H_4). Its structure is described in terms of sigma bonds between carbon and hydrogen atoms and a sigma and a *pi* (π) *bond* between carbons. Nomenclature using the IUPAC system was demonstrated along with writing the structure of a compound, given its name. The alkenes react by *addition reactions;* halogenation is an illustration of this reaction. Both general and specific equations were given. Addition polymers of polyethylene, Teflon, polyvinyl chloride, and Orlon were discussed.

Alkynes have the general molecular formula C_nH_{2n-2} (open-chain systems), with acetylene (C_2H_2) being the simplest alkyne. The structure of acetylene is described in terms of σ bonds between the carbon and hydrogen atoms, and one sigma and *two* π bonds between the carbons. Nomenclature using the

IUPAC system along with writing the structures of alkynes was presented. The alkynes also react by *addition reactions;* halogenation is an illustration of this reaction, except that the alkyne can accept *two* moles of the reagent while the alkene can accept only one. Both general and specific equations were given.

Aromatic hydrocarbons are hydrocarbons that include benzene and compounds that contain aliphatic or aromatic groups attached to aromatic rings. Benzene (C_6H_6) is the simplest example. Each of the six carbons in benzene is composed of three sp^2 hybridized orbitals, with one bonded to an *s* orbital of a hydrogen atom and the other two bonded to adjacent carbon atoms. The remaining *p* orbital of each of the six carbon atoms overlaps with those *p* orbitals from adjacent carbon atoms to form a ring of delocalized clouds of electrons above and below the plane of the ring. The bonding between the carbon atoms is neither single nor double but something in between. All carbons in benzene are equivalent, as are all hydrogens. Structural formulas of benzene, toluene, and *o-*, *m-*, and *p-*xylene were given. Aromatic hydrocarbons were named as derivatives of benzene or toluene using the IUPAC system. Given the name of the compound, the structural formula was written for a number of compounds. Aromatic compounds react by *substitution reactions;* halogenation is an illustration of this reaction. They also react by nitration, which is another type of substitution reaction. Examples of both halogenation and nitration were given.

Table 18-3 summarizes the hydrocarbons (alkanes, alkenes, alkynes, and aromatics).

EXERCISES

1. Define or explain the following terms:

(a)	organic chemistry	(b)	hydrocarbons
(c)	single bond	(d)	sigma (σ) bond
(e)	sp^3 hybrid orbitals	(f)	alkanes
(g)	saturated hydrocarbons	(h)	general molecular formula for alkanes (open-chain systems)
(i)	homologous series	(j)	constitutional isomers
(k)	alkyl groups	(l)	substitution reactions
(m)	alkenes	(n)	unsaturated hydrocarbons
(o)	general molecular formula for alkenes (open-chain systems)	(p)	sp^2 hybrid orbitals
(q)	double bond	(r)	pi (π) bond
(s)	addition reactions	(t)	addition polymers
(u)	alkynes		
(v)	general molecular formula for alkynes (open-chain systems)	(w)	sp hybrid orbitals
(x)	triple bond	(y)	aromatic hydrocarbons

TABLE 18-3 Summary of the Hydrocarbons

Hydrocarbon	General Molecular Formula	Systematic IUPAC Ending	Structural Formula Example	Systematic IUPAC Name	Hybrid-ization of Carbon Atom	Bonding Between Carbon Atoms	Bond Angle	Reactions
Alkane	C_nH_{2n+2}	-ane	CH_3—CH_3	Ethane	sp^3	1σ	109.5°	Substitution
Alkene	C_nH_{2n}	-ene	CH_2=CH_2	Ethene or ethylene	sp^2	1σ, 1π	120°	Addition (1 mole)
Alkyne	C_nH_{2n-2}	-yne	CH≡CH	Ethyne or acetylene	sp	1σ, 2π	180°	Addition (2 moles)
Aromatic	C_6H_6 for benzene	—	⬡	Benzene	sp^2	1σ, 1π delocalized	120°	Substitution

(z)   Kekulé structures   (aa)   *ortho* position
(bb)   *meta* position   (cc)   *para* position

2.   Distinguish between the following:
   (a) sp^3 hybrid orbital and sp^2 hybrid orbital
   (b) sp^3 hybrid orbital and sp hybrid orbital
   (c)   sigma (σ) bond and pi (π) bond
   (d)   general molecular formulas for alkanes and alkenes
   (e)   general molecular formulas for alkenes and alkynes
   (f)   double and triple bond
   (g)   *ortho* and *meta* positions
   (h)   *meta* and *para* positions

PROBLEMS

Constitutional (Structural) Isomers   (See Section 18-3)

3.   Write *condensed* structural formulas for the constitutional (structural) isomers of the following. (The number in parentheses is the number of constitutional isomers for the compound.)
   (a) $C_4H_{10}(2)$ (b) $C_5H_{12}(3)$
   (c) $C_6H_{14}(5)$ (d) $C_7H_{16}(9)$

Alkanes   (See Section 18-3)

4.   Give the IUPAC name for each of the following compounds:

   (a) 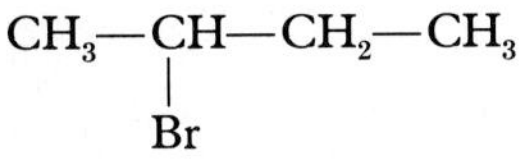 CH_3—CH—CH_2—CH_3
                 |
                 Br

   (b) 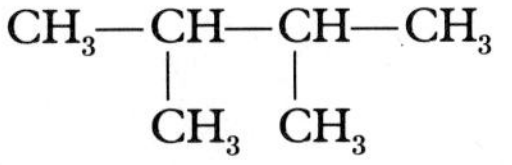 CH_3—CH—CH—CH_3
                 | |
                 CH_3 CH_3

   (c) 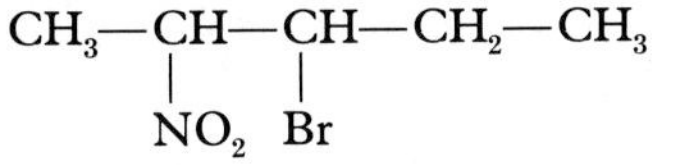 CH_3—CH—CH—CH_2—CH_3
                 | |
                 NO_2 Br

   (d) 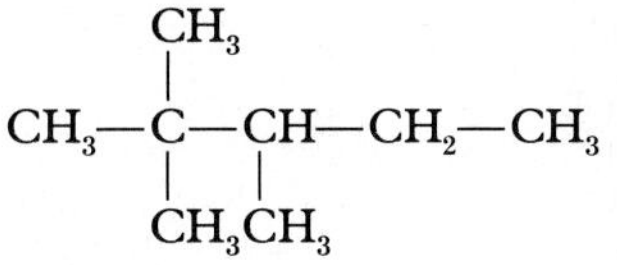 CH_3
                       |
          CH_3—C—CH—CH_2—CH_3
                       | |
                     CH_3 CH_3

   (e) 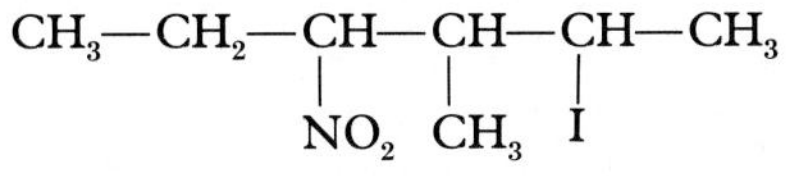 CH_3—CH_2—CH—CH—CH—CH_3
                           | | |
                         NO_2 CH_3 I

   (f) 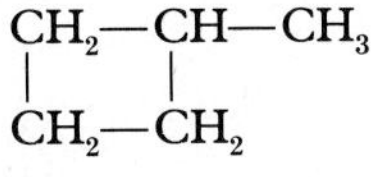 CH_2—CH—CH_3
             | |
          CH_2—CH_2

5.   Write the structural formula for each of the following compounds:
   (a)   2-bromobutane   (b)   3-iodoheptane
   (c)   2-nitropropane   (d)   3-ethylpentane
   (e)   2-bromo-3,3-dimethylhexane   (f)   ethylcyclopentane

6.   Complete the balance the equation for monohalogenation in each of the following reactions:

   (a) CH_4 + Cl_2 $\xrightarrow[\text{light}]{\Delta \text{ or}}$

   (b) CH_3—CH_3 + Br_2 $\xrightarrow[\text{light}]{\Delta \text{ or}}$

Alkenes (See Section 18-4)

7. Give the IUPAC name for each of the following compounds:

(a) $CH_3—CH=CH—CH_3$

(b) $CH_3—C=CH—CH_2—CH_3$
 with CH_3 below the C

(c) $CH_3—CH=C—CH_2—CH_3$
 with CH_3 below the C

(d) $CH_2=CH—CH—CH—CH—CH_3$
 with Cl, CH_3, CH_3 below

(e)
$$CH_2=CH—\underset{\underset{CH_2}{\underset{|}{CH_2}}}{\overset{\overset{Cl}{\overset{|}{C}}}{C}}—CH_3$$
with CH_2, CH_2, CH_3 chain below

(f) $CH_3—CH—CH_2—CH—CH=CH—CH_3$
 with NO_2 and Cl below

8. Write the structural formula for each of the following compounds:
   (a) 1-pentene
   (b) 4-methyl-2-hexene
   (c) 2,3-dimethyl-1-hexene
   (d) cyclobutene
   (e) 1-bromo-3-ethyl-1-heptene
   (f) 2,4,5-trimethyl-5-nitro-2-heptene

9. Complete and balance the equation for each of the following reactions:

(a) $CH_3—CH=CH_2 + Br_2 \rightarrow$
     in CCl_4

(b) $CH_3—C=CH—CH_3 + Br_2 \rightarrow$
     with CH_3 below, in CCl_4

Addition Polymers (see Section 18-5)

10. Give the structural formula of the monomer and the polymer of the following addition polymers:
    (a) polyethylene
    (b) Teflon
    (c) polyvinyl chloride
    (d) Orlon

Alkynes (See Section 18-6)

11. Give the IUPAC name for each of the following compounds:

(a) $CH\equiv C—CH_2—CH_2—CH_3$

(b) $CH_3—C\equiv C—CH—CH_3$
 with CH_3 below

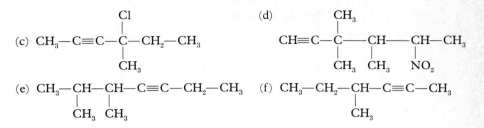

(c) $CH_3—C≡C—\overset{\overset{\displaystyle Cl}{|}}{\underset{\underset{\displaystyle CH_3}{|}}{C}}—CH_2—CH_3$

(d) $CH≡C—\overset{\overset{\displaystyle CH_3}{|}}{\underset{\underset{\displaystyle CH_3}{|}}{C}}—\overset{\overset{\displaystyle }{}}{\underset{\underset{\displaystyle CH_3}{|}}{CH}}—\overset{\overset{\displaystyle }{}}{\underset{\underset{\displaystyle NO_2}{|}}{CH}}—CH_3$

(e) $CH_3—\overset{}{\underset{\underset{\displaystyle CH_3}{|}}{CH}}—\overset{}{\underset{\underset{\displaystyle CH_3}{|}}{CH}}—C≡C—CH_2—CH_3$

(f) $CH_3—CH_2—\overset{}{\underset{\underset{\displaystyle CH_3}{|}}{CH}}—C≡C—CH_3$

12. Write the structural formula for each of the following compounds:
    (a) 1-heptyne
    (b) 2-heptyne
    (c) 4-nitro-1-heptyne
    (d) 1-bromo-4,5-dimethyl-2-hexyne
    (e) 1-chloro-3-methyl-1-hexyne
    (f) 1,2-dichloro-5,5,6-trimethyl-3-octyne

13. Complete and balance the equation for each of the following reactions:

    (a) $CH_3—C≡C—CH_3 + \underset{\substack{\text{excess} \\ \text{in } CCl_4}}{Br_2} →$

    (b) $CH_3—C≡C—\overset{\overset{\displaystyle CH_3}{|}}{\underset{\underset{\displaystyle CH_3}{|}}{C}}—CH_3 + \underset{\substack{\text{excess} \\ \text{in } CCl_4}}{Br_2} →$

Aromatic hydrocarbons (See Section 18-7)

14. Give the IUPAC name for each of the following compounds:

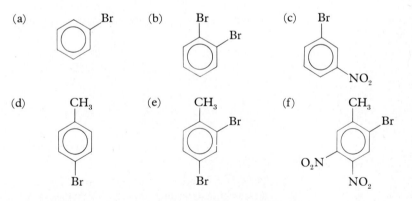

15. Write the structural formula for each of the following compounds:
    (a) toluene
    (b) iodobenzene
    (c) *m*-nitroiodobenzene
    (d) *p*-iodotoluene
    (e) *o*-chlorotoluene
    (f) 2,4,6-trinitrotoluene (the explosive known as TNT)

16. Complete and balance the equation for each of the following reactions:

    (a) ⬡ $+ Br_2 \xrightarrow[\Delta]{Fe}$

    (b) ⬡ $+ HNO_3 \xrightarrow{H_2SO_4}$

General Problems

17. Give the IUPAC name for each of the following compounds:

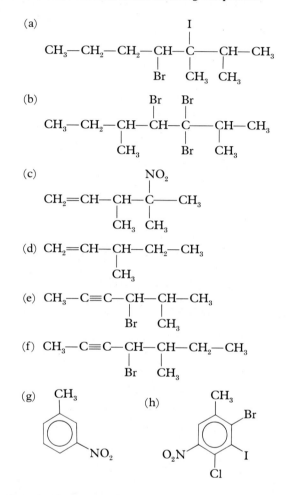

18. The following organic compounds reacted with bromine in various ways; identify the compounds as alkanes, alkenes, alkynes, or aromatic hydrocarbons:
    (a) reacted immediately with 1 mol of bromine in carbon tetrachloride to yield a colorless solution
    (b) reacted with bromine only in the presence of iron(III) bromide and heat; hydrogen bromide was released
    (c) reacted immediately with 2 mol of bromine in carbon tetrachloride to yield a colorless solution
    (d) reacted with bromine only in the presence of light; hydrogen bromide was released

19. Identify the (a) alkene, (b) alkyne, and (c) aromatic groups in each of the following compounds by circling each group and labeling it:

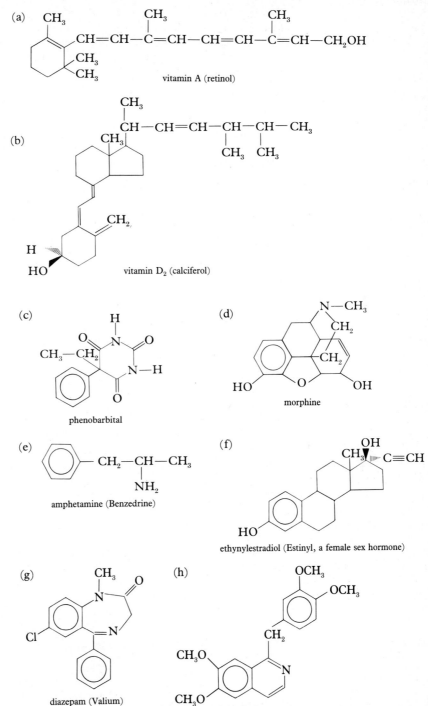

(a) vitamin A (retinol)

(b) vitamin D₂ (calciferol)

(c) phenobarbital

(d) morphine

(e) amphetamine (Benzedrine)

(f) ethynylestradiol (Estinyl, a female sex hormone)

(g) diazepam (Valium)

(h) papaverine

20. A certain aliphatic hydrocarbon gave on analysis 80.0 percent carbon and 20.0 percent hydrogen. At 30°C and 640 torr pressure, 0.285 g of the gas occupied a volume of 281 mL. Calculate the molecular formula, write the structural formula, and give the IUPAC name for the aliphatic hydrocarbon. (*Hint:* See 8-5 and 11-9.)

Readings

Saltzman, Martin D., "Benzene and the Triumph of the Octet Theory." *J. Chem. Educ.*, 1974, v. 51, p. 498. Reviews the historical development of the structure of benzene. It includes theories other than those proposed by Kekulé.

Stille, John K., *Industrial Organic Chemistry*. Englewood Cliffs, N.J.: Prentice-Hall, Inc., 1968. Discusses various useful industrial chemicals and their industrial preparations, and mentions the industrial processes for the preparation of many polymers.

19

Organic Chemistry II: Derivatives of the Hydrocarbons

A nylon parachute. Nylon is formed by the condensation polymerization of hexamethylene diamine with adipic acid. Nylon also has many other useful applications such as nylon hose, jackets, and clothing.

TASKS

1. Memorize the classes of organic compounds, general formulas, and functional groups given in Table 19-1.

2. Memorize the structural formulas for (1) phenol, (2) o-cresol, (3) m-cresol, and (4) p-cresol, as given in Section 19-3.

3. Memorize the trivial names and structural formulas for the aldehydes and carboxylic acids given in Table 19-2.

OBJECTIVES

1. Define each of the following terms and describe the distinguishing characteristics of each:
   (a) functional group (Introduction to Chapter 19)
   (b) condensation polymers (Section 19-11)

2. Given the structural formula for the following types of organic compounds, write an acceptable IUPAC name for each:
   (a) organic halides (Problem Example 19-1, Problem 4)
   (b) alcohols (Problem Example 19-3, Problem 6)
   (c) phenols (Problem Example 19-5, Problem 8)
   (d) ethers (Problem Example 19-7, Problem 10)
   (e) aldehydes (Problem Example 19-9, Problem 12)
   (f) ketones [(1) radicofunctional name, Problem Example 19-11, Problem 14; (2) systematic name, Problem Example 19-12, Problem 15]
   (g) carboxylic acids (Problem Example 19-14, Problem 17)
   (h) esters (Problem Example 19-16, Problem 20)
   (i) amides (Problem Example 19-18, Problem 23)
   (j) amines (Problem Example 19-20, Problem 25)

3. Given the IUPAC names or other acceptable names for the following types of organic compounds, write the structural formula for each:
   (a) organic halides (Problem Example 19-2, Problem 5)
   (b) alcohols (Problem Example 19-4, Problem 7)
   (c) phenols (Problem Example 19-6, Problem 9)
   (d) ethers (Problem Example 19-8, Problem 11)
   (e) aldehydes (Problem Example 19-10, Problem 13)
   (f) ketones (Problem Example 19-13, Problem 16)
   (g) carboxylic acids (Problem Example 19-15, Problem 18)
   (h) esters (Problem Example 19-17, Problem 21)

(i)   amides (Problem Example 19-19, Problem 24)

(j)   amines (Problem Example 19-21, Problem 26)

4.   Given a specific carboxylic acid, complete and balance the equations for the reactions of this acid with appropriate basic compounds, such as sodium or potassium hydroxide, sodium carbonate, or sodium hydrogen carbonate (Section 19-7, Problem 19).

5.   Given a specific ester, complete and balance the equation for its reaction with either aqueous acid or base (Section 19-8, Problem 22).

6.   Given the following condensation polymers:

(a)   Dacron or Mylar

(b)   nylon-66

Give the structural formula for the two monomers and the polymer (Section 19-11, Problem 27).

In Chapter 18, we discussed the four general types of hydrocarbons. In this chapter, we will discuss compounds derived from the hydrocarbons by substituting various functional groups for hydrogen atoms attached to carbon. A **functional group** is an atom or group of atoms (other than hydrogen) that is attached to a hydrocarbon chain and which confers some distinctive chemical and physical properties to the organic compound. In the alkenes the carbon-carbon double bond is the functional group, and in the alkynes the functional group is the carbon-carbon triple bond. Table 19-1 lists some of the more common classes of organic compounds, a general formula for each class, and the functional group.

19-1

Organic Halides

Organic halides consist of *alkyl halides* and *aryl halides*. **Alkyl halides** contain an alkyl group, or substituted alkyl group, that is attached to a halogen atom (F, Cl, Br, or I). The general formula for an alkyl halide is **R—X,** with the **X** representing any halogen atom as the functional group. The **R** may also represent an alkenyl group (carbon-carbon double bond present), or an alkynyl group (carbon-carbon triple bond present).

Aryl halides are compounds having a halogen atom directly attached to an aromatic ring. The general formula for aromatic halides is **Ar—X,** with **Ar** representing the aromatic ring and **X** the halogen atom.

The insecticide DDT (**d**ichloro**d**iphenyl**t**richloroethane) belongs to the class of organic halides. DDT is a very effective insecticide and has been responsible for saving thousands of lives, but is not readily decomposed and accumulates in the environment. In addition, it has been found to produce cancer in laboratory animals. It is also transmitted through the food chain. For example, DDT accumulates in insects and fish, and then fish-eating animals,

TABLE 19-1 Classes of Organic Compounds Derived from Hydrocarbons

Class of Compound	General Formula	Functional Group
Organic halide	R—X, Ar—X[a]	—X[b]
Alcohol	R—OH	—OH
Phenol	Ar—OH	—OH
Ether	R—O—R, Ar—O—R, Ar—O—Ar	—O—
Aldehyde	$R-\overset{O}{\underset{H}{C}}$, $Ar-\overset{O}{\underset{H}{C}}$	$-\overset{O}{\underset{H}{C}}$
Ketone	$R-\overset{O}{\underset{R}{C}}$, $Ar-\overset{O}{\underset{R}{C}}$, $Ar-\overset{O}{\underset{Ar}{C}}$	$C=O$
Carboxylic acid	$R-\overset{O}{\underset{OH}{C}}$, $Ar-\overset{O}{\underset{OH}{C}}$	$-\overset{O}{\underset{OH}{C}}$
Ester	$R-\overset{O}{\underset{OR^c}{C}}$, $Ar-\overset{O}{\underset{OR^c}{C}}$	$-\overset{O}{\underset{OR^c}{C}}$
Amide	$R-\overset{O}{\underset{NH_2}{C}}$, $Ar-\overset{O}{\underset{NH_2}{C}}$	$-\overset{O}{\underset{NH_2}{C}}$
Amine	$R-NH_2$, R_2NH, R_3N^d	$-NH_2$, NH, $-N-$

[a]The symbol Ar designates an aryl group, such as ⬡— or any aromatic group.

[b]The symbol X designates a halogen atom (F, Cl, Br, or I).

[c]R may also be Ar; see 19-8.

[d]R may also be Ar with variations of R and Ar; see Table 19-3.

such as pelicans, are harmed by it in that it is responsible for softening the shell of thin-shelled eggs, thereby permitting them to be easily broken. This endangers the continued reproduction of the animals. For these reasons the Environmental Protection Agency (EPA) has banned its sale except in extreme epidemics. Other insecticides which are more readily decomposed in the atmosphere are being tried, as are other approaches to solving the insect problem. One approach is to infect the insects with a virus which kills the insect, but is not harmful to humans or other animals. The virus preparation Elcar, now on the market, attacks the cotton bollworm, tobacco budworm, and corn earworm.

Another environmental problem posed by organic halides is in the use of chlorofluorocarbons (CFCs, CF_2Cl_2 and $CFCl_3$) in aerosol spray containers. These CFCs appear to remain in the lower atmosphere for at least 30 years. They appear to destroy the earth's ozone layer, which filters out the sun's ultraviolet light. Recently, it has been found that the chlorofluorocarbons are twice as destructive to the ozone layer as the action of supersonic aircraft (see 13-10). Dichlorodifluoromethane (CF_2Cl_2) is also known by its trade name Freon-12 and is used as a refrigerant in refrigerators and freezers.

Nomenclature

In naming both alkyl and aryl halides, the IUPAC system is preferred. The halogen atom is considered to be a substituent on a parent chain or aromatic ring. The IUPAC system also allows limited use of radicofunctional names. **Radicofunctional names** consist of two words. The first word describes the carbon residue attached to the functional group, such as alkyl or aryl in organic halides. The second word is the name of the functional group, such as halide–chloride, bromide, etc. The following problem example illustrates the use of both systems:

Problem Example 19-1

Give the IUPAC name for each of the following compounds:[1]

Answers

(a) 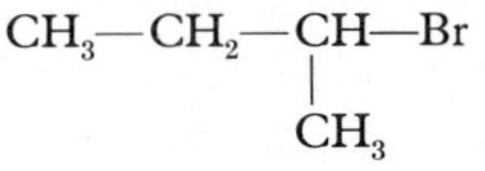 CH_3—CH_2—CH—Br
             |
             CH_3

2-bromobutane
(or *sec*-butyl bromide)

(b) 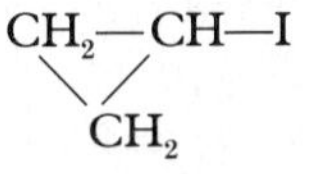 CH_2—CH—I
     \ /
     CH_2

iodocyclopropane
(or cyclopropyl iodide)

(c) 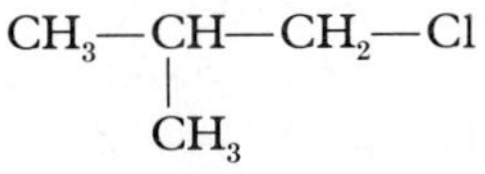 CH_3—CH—CH_2—Cl
         |
         CH_3

1-chloro-2-methylpropane
(or isobutyl chloride)

(d) 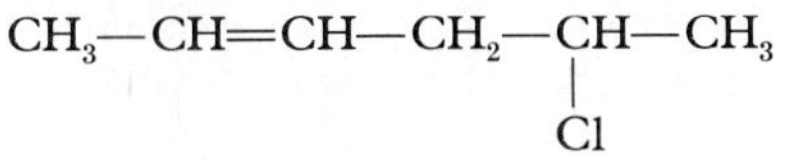 CH_3—CH=CH—CH_2—CH—CH_3
                              |
                              Cl

5-chloro-2-hexene
(the double bond has the lowest possible number)

---

[1]In Problem Examples 19-1 and 19-3, the radicofunctional name is in parentheses for those compounds that can be given radicofunctional names.

(e)

o-difluorobenzene

Work Problem 4.

Problem Example 19-2

Write the structural formula for each of the following compounds:

(a) *tert*-butyl iodide, (b) 3,3-dibromo-1-butene, (c) 4-bromo-2-chloro-toluene.

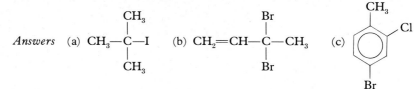

Answers

Work Problem 5.

19-2

Alcohols

The **alcohols** are compounds in which a hydroxyl group (OH) is attached to an *aliphatic* hydrocarbon residue. The general formula for an alcohol is

$$R\text{—}O\text{—}H$$

The hydroxyl group (**OH**) is the functional group and is attached directly to an alkyl group, *not* to an aryl group. The hydroxyl group is not a hydroxide ion (OH^{1-}) as the hydroxyl group is *covalently* bonded to the alkyl group. If the hydroxyl is attached directly to an aromatic ring, a group of compounds called *phenols* are obtained (see 19-3). Alcohols are related to water in that one of the hydrogen atoms of water (**H—O—H**) has been replaced by an **R** group. As you may recall, water ionizes slightly. Alcohols also ionize to yield hydrogen ions (H^{1+}) and are very weak acids, being even *weaker* acids than water.

Gasohol, a mixture of 10 percent ethanol (ethyl alcohol, CH_3CH_2OH) and gasoline, is a fuel for automobiles that has been developed in response to the energy crisis. The ethanol is prepared by fermenting cereal grains, such as corn. It is separated from the fermentation mixture by distillation (see 12-5), all of the water is removed, and the ethanol is then mixed with gasoline. However, extremely wide use of gasohol may possibly create another problem—a shortage of cereal grains. Ethanol may be added to gasoline as a substitute for lead (actually, lead tetraethyl) which is a dangerous air pollutant.

Nomenclature

Radicofunctional names are recognized by IUPAC for some of the simple alcohols. In the radicofunctional name, the alkyl group is named followed by the word "alcohol." However, it is preferred that systematic names be derived according to the following rules:

1. The alcohols all have the ending **-ol.**

2. The longest continuous chain of carbons holding the OH group is determined, and this chain is the parent structure, such as

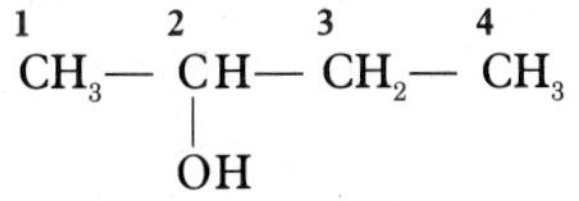

$$\overset{1}{CH_3} - \overset{2}{CH} - \overset{3}{CH_2} - \overset{4}{CH_3}$$
$$| $$
$$OH$$

3. This parent structure is named by dropping the **-e** from -ane, -ene, or -yne of the alkane, alkene, or alkyne, respectively, and adding **-ol.** The above structure would be a butan**ol.**

4. The parent structure is numbered so as to give the **lowest** possible number to the carbon holding the OH group.

5. The number of this carbon is placed before the name of the parent structure. Hence, the alcohol above is 2-butanol. If a carbon-carbon double bond or triple bond is present in the compound, then the number representing the position of this double or triple bond is placed *before* the name of the *parent* structure and the number of the carbon atom holding the OH is placed directly *before* the -ol. Consider the following example:

$$\overset{4}{CH_2} = \overset{3}{CH} - \overset{2}{CH_2} - \overset{1}{CH_2} - OH$$

The correct name is 3-buten-1-ol.

6. All other groups attached to the parent structure are given a number and name and are placed in alphabetical order as prefixes to the parent name.

The following problem example illustrates the use of both of these systems.

Problem Example 19-3

Give the IUPAC name for each of the following compounds:

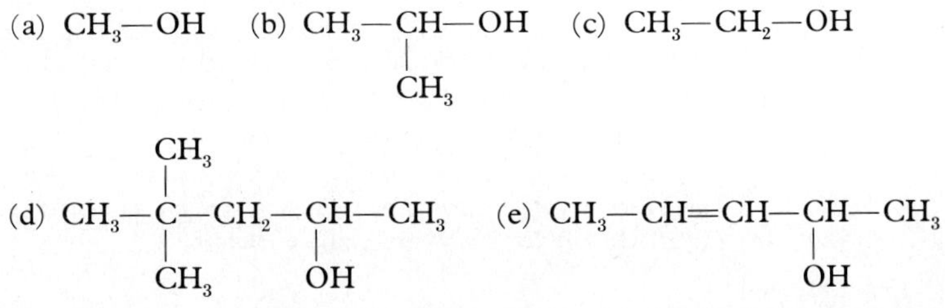

(a) $CH_3 - OH$ (b) $CH_3 - CH - OH$ (c) $CH_3 - CH_2 - OH$
$\qquad\qquad\qquad\qquad\qquad\quad |$
$\qquad\qquad\qquad\qquad\quad CH_3$

(d) $CH_3 - \underset{\underset{CH_3}{|}}{\overset{\overset{CH_3}{|}}{C}} - CH_2 - \underset{\underset{OH}{|}}{CH} - CH_3$ (e) $CH_3 - CH = CH - \underset{\underset{OH}{|}}{CH} - CH_3$

Work Problem 6.

Answers (a) methanol (methyl alcohol), (b) 2-propanol (isopropyl alcohol)[2] (c) ethanol (ethyl alcohol),[3] (d) 4,4,-dimethyl-2-pentanol, (e) 3-penten-2-ol

Problem Example 19-4

Write the structural formula for each of the following compounds:

(a) *sec*-butyl alcohol, (b) isobutyl alcohol, (c) 3-chloro-2-methyl-1-butanol.

Work Problem 7.

$$
\begin{array}{ll}
Answers \quad \text{(a)} \ CH_3{-}CH_2{-}\underset{\underset{\textstyle CH_3}{|}}{CH}{-}OH & \text{(b)} \ CH_3{-}\underset{\underset{\textstyle CH_3}{|}}{CH}{-}CH_2{-}OH
\end{array}
$$

$$
\text{(c)} \ HO{-}CH_2{-}\underset{\underset{\textstyle CH_3}{|}}{CH}{-}\underset{\underset{\textstyle Cl}{|}}{CH}{-}CH_3
$$

19-3

Phenols

The **phenols** (fē′nôls) are compounds that have a hydroxyl group directly attached to an *aromatic* ring. The general formula for a phenol is

$$
\mathbf{Ar{-}O{-}H}
$$

The hydroxyl group (**OH**) is the functional group in phenols, and these compounds, like alcohols and water, ionize to yield hydrogens ions (H^{1+}). The equilibrium constant (K_a, see 17-4) for phenols is about 10^{-10} (mol/L), while the values for water and alcohols are approximately 2×10^{-16} (mol/L) and 10^{-16} to 10^{-18} (mol/L), respectively. Phenols, in general, are stronger acids than water and alcohols. The acidity of these compounds decreases in the order **Ar{-}OH > H$_2$O > R{-}OH**.

Nomenclature

The IUPAC recognizes a few trivial names for some of the common phenols (for example, the cresols); however, it is preferred that phenols be named as derivatives of the parent compound, *phenol*.

---

[2]Both methanol and 2-propanol are highly toxic and may cause blindness. Methanol is also known as wood alcohol. 2-propanol is used as rubbing alcohol.

[3]Ethanol is a drinking alcohol and is commonly referred to as just "alcohol" or "grain alcohol." It is found in various intoxicating beverages and is toxic if taken in excess. Laboratory-grade ethanol may also contain methanol or 2-propanol (added as denaturants) and in this case may be highly toxic. **Do not drink laboratory-grade ethanol.**

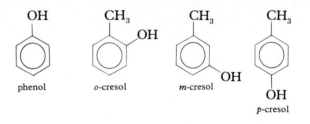

If **two** substituents are attached to the benzene ring (OH group plus one other group), then ortho (*o-*), meta (*m-*), and para (*p-*) are used to indicate the position of the other group. The parent structure is phenol.

When **three** or **more** substituents are on the ring, numbers are used to indicate their positions, according to the following rules:

1. The parent structure is phenol.

2. The hydroxyl group of phenol is at position 1 on the benzene ring, and the ring is numbered in a direction that will give the lowest possible number to the substituents. The number 1 is not included in the name.

3. All groups on the phenol parent structure are placed in alphabetical order as prefixes to the name.

Problem Example 19-5

Give the IUPAC name for each of the following compounds:

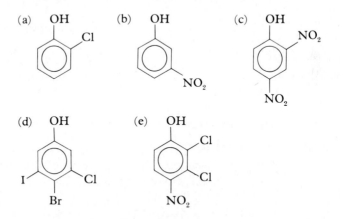

Work Problem 8.

Answers (a) *o*-chlorophenol, (b) *m*-nitrophenol (c) 2,4-dinitrophenol, (d) 4-bromo-3-chloro-5-iodophenol, (e) 2,3-dichloro-4-nitrophenol

Problem Example 19-6

Write the structural formula for each of the following compounds:

(a) *p*-bromophenol, (b) 2,3-dichlorophenol, (c) 2-bromo-3-chloro-4-nitrophenol.

Answers (a) (b) (c)

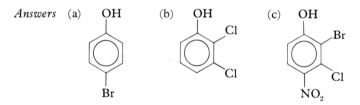

Work Problem 9.

19-4

Ethers

The **ethers** constitute a series of compounds in which *two alkyl* groups (**R,** same or different) or *two aryl* groups (**Ar,** same or different) or *an alkyl* group and *an aryl* group are attached to an oxygen atom. The general formulas for ethers are as follows:

$$\underset{R = R' \;\; or \;\; R \ne R'}{R—O—R'} \qquad R—O—Ar \qquad \underset{Ar = Ar' \;\; or \;\; Ar \ne Ar'}{Ar—O—Ar'}$$

The oxygen atom (—O—) is the functional group. Ethers can be considered as compounds in which both of the H atoms of water (**H—O—H**) have been replaced by an alkyl group (**R**) or an aryl group (**Ar**).

Diethyl ether, or just *ether* ($CH_3—CH_2—O—CH_2—CH_3$), is the most common ether. At one time it was used as an anesthetic, but it has been largely replaced by halothane (Fluothane) because ether is highly flammable. The structural formula of halothane is

$$\begin{array}{c} \quad\; F \quad\; H \\ \quad\; | \qquad | \\ F—C—C—Br \\ \quad\; | \qquad | \\ \quad\; F \quad\; Cl \end{array}$$

Nomenclature

In the IUPAC system it is permissible to use the radicofunctional names for the simpler monoethers of symmetrical structure. For these ethers we use the names of the groups attached to the oxygen atom (—O—) plus the word "ether." The $C_6H_5—$ group, derived from benzene, is named phenyl, not phenol. The ethers $CH_3—O—CH_3$ and ⬡—O—⬡ are named dimethyl ether and diphenyl ether, respectively.

The systematic name is preferred and is named according to the following rules:

1. The alkoxy group (R—O—) is named by naming the group and dropping the *-yl* and adding *-oxy*. For example, CH_3—O— is a meth*oxy* group.

2. For aliphatic hydrocarbons, the position of the alkoxy group is indicated by a number, as is done for other substituents. The smaller group is the alkoxy group.

3. For aromatic hydrocarbons, the prefixes *o-*, *m-*, and *p-* are used for two substituents. If there are three or more substituents attached to the aromatic ring, then numbers are used.

Problem Example 19-7

Give the IUPAC name for each of the following compounds:

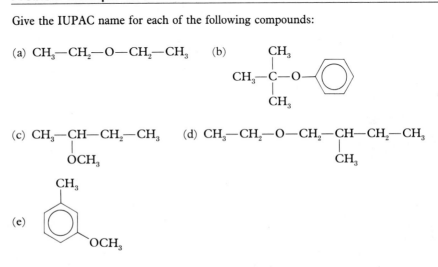

(a) CH_3—CH_2—O—CH_2—CH_3 (b)

(c) CH_3—CH—CH_2—CH_3 (d) CH_3—CH_2—O—CH_2—CH—CH_2—CH_3

Work Problem 10.   (e)

Answers (a) ethoxyethane (diethyl ether), (b) *tert*-butoxybenzene, (c) 2-methoxybutane, (d) 1-ethoxy-2-methylbutane, (e) *m*-methoxytoluene

Problem Example 19-8

Write the structural formula for each of the following compounds:

(a) di-*tert*-butyl ether, (b) 2-chloro-3-methoxytoluene, (c) 4-methoxy-3-methyl-2-pentanol.

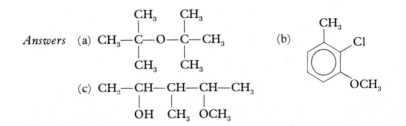

Answers   (a)   (b)

(c)

Work Problem 11.

19-5

Aldehydes

The **aldehydes** (al'de-hīds') are compounds whose structures consist of an alkyl or aryl group attached to a carbonyl group $\left(\diagdown C=O \right)$ to which is also attached a *hydrogen atom*. Formaldehyde, which is the simplest aldehyde, has two hydrogen atoms attached to the carbonyl group. The general formulas for aldehydes are as follows:

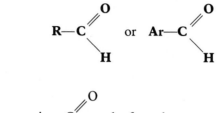

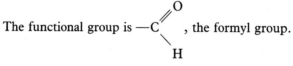

The functional group is , the formyl group.

Nomenclature

Trivial names of a few aldehydes acceptable in the IUPAC system are given in Table 19-2. These names are usually derived from the trivial names of the corresponding carboxylic (kär'bok-sil'ik) acids by replacing the -*ic* ending of the acid with -*aldehyde*; thus, acetic acid becomes acetaldehyde. You must learn the names and structural formulas of the aldehydes and carboxylic acids in Table 19-2.

The carboxylic acids may be formed from the aldehydes by the action of various oxidizing agents (see 16-1), such as O_2, Cu^{2+}, Ag^{1+}, $H_2Cr_2O_7$, or $KMnO_4$.

Systematic names for aldehydes in the IUPAC system are obtained by applying the following rules:

1. The aldehydes all have the ending -**al**.

2. The longest continuous chain of carbons to which the $-\overset{\overset{\displaystyle O}{\|}}{C}-H$ group is attached is considered the parent structure. The carbon of the carbonyl group is *included* in the count of carbons for the longest continuous chain of carbons and is considered as carbon number 1, but this number is *not* indicated in the name. The -*e* of the parent structure, that is, alkan*e*, alken*e*, or alkyn*e*, is dropped and the ending -*al* is added.

3. Aromatic aldehydes are named as derivatives of *benzaldehyde* with the po-

TABLE 19-2   Trivial Names and Structures of a Selected Number of Aldehydes and Carboxylic Acids

Aldehyde	Structure	Acid	Structure
Formaldehyde	$\begin{matrix} & O \\ & \parallel \\ H- & C-H \end{matrix}$	Formic acid[a]	$\begin{matrix} & O \\ & \parallel \\ H- & C-OH \end{matrix}$
Acetaldehyde[b]	$\begin{matrix} & O \\ & \parallel \\ CH_3- & C-H \end{matrix}$	Acetic acid[c]	$\begin{matrix} & O \\ & \parallel \\ CH_3- & C-OH \end{matrix}$
Propionaldehyde	$\begin{matrix} & & O \\ & & \parallel \\ CH_3-CH_2- & C-H \end{matrix}$	Propionic acid	$\begin{matrix} & & O \\ & & \parallel \\ CH_3-CH_2- & C-OH \end{matrix}$
Butyraldehyde	$CH_3-CH_2-CH_2-\overset{\overset{O}{\parallel}}{C}-H$	Butyric acid[d]	$CH_3-CH_2-CH_2-\overset{\overset{O}{\parallel}}{C}-OH$
Valeraldehyde	$CH_3-CH_2-CH_2-CH_2-\overset{\overset{O}{\parallel}}{C}-H$	Valeric acid	$CH_3-CH_2-CH_2-CH_2-\overset{\overset{O}{\parallel}}{C}-OH$
Benzaldehyde[e]		Benzoic acid[f]	
Salicylaldehyde		Salicylic acid	

[a]Formic acid is the active irritant in ant and bee stings.
[b]Acetaldehyde is one of the chief offenders in eye irritation from smog.
[c]Acetic acid is found in vinegar (about 5%). It is also found in sour wine (wine vinegar).
[d]Butyric acid is found in rancid butter.
[e]Benzaldehyde is found in the kernels of bitter almonds.
[f]Benzoic acid is used as a preservative in soft drinks.

sitions of various substituents indicated as in nomenclature of phenols (see 19-3).

4. Numbers are used to indicate the positions of substituents in the parent chain.

5. All substituent groups are placed in alphabetical order before the parent name.

The following problem example illustrates the use of the IUPAC system in naming aldehydes.

Problem Example 19-9

Give the IUPAC name for each of the following compounds:

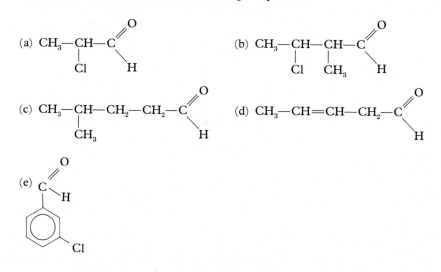

Work Problem 12.

Answers (a) 2-chloropropanal, (b) 3-chloro-2-methyl-butanal, (c) 4-methyl-pentanal, (d) 3-pentenal, (e) *m*-chlorobenzaldehyde

Problem Example 19-10

Write the structural formula for each of the following compounds:

(a) 2-chloro-4-nitrobenzaldehyde, (b) 2,2-dimethylpentanal, (c) 2-chloro-3-methyl-3-hexenal.

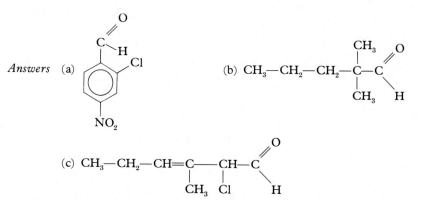

Work Problem 13.

19-6

Ketones

Ketones (kē′tōns) are compounds whose structures consist of *two* alkyl groups, an alkyl and an aryl group, or of *two* aryl groups attached to a carbonyl group ($\diagdown$C=O) . There is *no* hydrogen atom attached to the carbonyl group in ketones, but there is one attached in aldehydes. The general formulas for ketones are as follows:

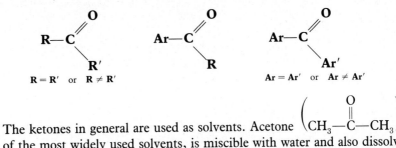

The ketones in general are used as solvents. Acetone $\left(CH_3-\overset{O}{\underset{\|}{C}}-CH_3\right)$, one of the most widely used solvents, is miscible with water and also dissolves many organic substances. Various hormones, such as cortisone, progesterone, and a constituent of oral contraceptives, along with the antibiotic tetracycline, contain the ketone structure.

Nomenclature

Radicofunctional names are recognized in the IUPAC system for a number of simple monoketones. If both groups attached to the carbonyl group are *alkyl* groups, they are named as follows:

1. Name the alkyl groups attached to the carbonyl group. If both groups are the same, the prefix *di-* is used.

2. Add the word "ketone."

The following examples illustrate the use of this system:

$$CH_3-CH_2-\overset{O}{\underset{\|}{C}}-CH_2-CH_3 \quad \text{is diethyl ketone}$$

$$CH_3-CH_2-\overset{O}{\underset{\|}{C}}-CH_3 \qquad \text{is ethyl methyl ketone}$$

$$CH_3-CH_2-\overset{O}{\underset{\|}{C}}-\underset{\underset{CH_3}{|}}{CH}-CH_3 \quad \text{is ethyl isopropyl ketone}$$

$$CH_3-\overset{O}{\underset{\|}{C}}-CH_3 \qquad \text{is dimethyl ketone, but it is usually given the trivial name, } acetone$$

If one of the groups attached to the carbonyl group is a phenyl

(⬡—), these ketones are usually called *phenones*. They have the general structure 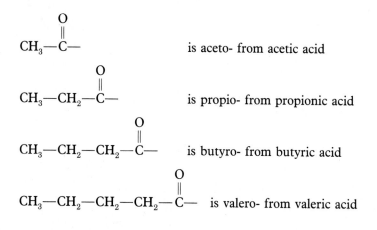 —C—R. The R—C— attached to the benzene ring is

an *acyl group*. The acyl groups are named from the trivial names of the carboxylic acids (see Table 19-2), as follows:

$$CH_3-\overset{\displaystyle O}{\overset{\displaystyle \|}{C}}-$$ is aceto- from acetic acid

$$CH_3-CH_2-\overset{\displaystyle O}{\overset{\displaystyle \|}{C}}-$$ is propio- from propionic acid

$$CH_3-CH_2-CH_2-\overset{\displaystyle O}{\overset{\displaystyle \|}{C}}-$$ is butyro- from butyric acid

$$CH_3-CH_2-CH_2-CH_2-\overset{\displaystyle O}{\overset{\displaystyle \|}{C}}-$$ is valero- from valeric acid

To name compounds as phenones, the following rules should be followed:

1. Name the acyl group as given above.
2. Add "phenone" to the name of the acyl group.

The following examples illustrate the use of this system:

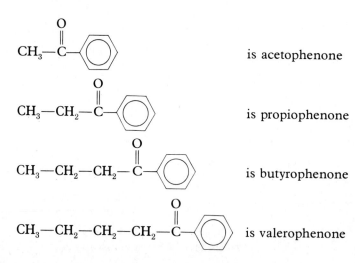

$$CH_3-\overset{\displaystyle O}{\overset{\displaystyle \|}{C}}-$$ is acetophenone

$$CH_3-CH_2-\overset{\displaystyle O}{\overset{\displaystyle \|}{C}}-$$ is propiophenone

$$CH_3-CH_2-CH_2-\overset{\displaystyle O}{\overset{\displaystyle \|}{C}}-$$ is butyrophenone

$$CH_3-CH_2-CH_2-CH_2-\overset{\displaystyle O}{\overset{\displaystyle \|}{C}}-$$ is valerophenone

Problem Example 19-11

Name each of the following compounds by the radicofunctional system:

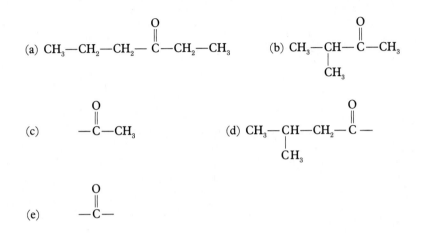

$$\text{(a) } CH_3-CH_2-CH_2-\overset{\overset{\displaystyle O}{\|}}{C}-CH_2-CH_3$$

$$\text{(b) } CH_3-\underset{\underset{\displaystyle CH_3}{|}}{CH}-\overset{\overset{\displaystyle O}{\|}}{C}-CH_3$$

$$\text{(c) } \qquad -\overset{\overset{\displaystyle O}{\|}}{C}-CH_3$$

$$\text{(d) } CH_3-\underset{\underset{\displaystyle CH_3}{|}}{CH}-CH_2-\overset{\overset{\displaystyle O}{\|}}{C}-$$

$$\text{(e) } \qquad -\overset{\overset{\displaystyle O}{\|}}{C}-$$

Work Problem 14.

Answers   (a) ethyl propyl ketone, (b) isopropyl methyl ketone, (c) acetophenone (or methyl phenyl ketone), (d) cyclohexyl isobutyl ketone, (e) benzophenone (or diphenyl ketone)

The systematic names preferred by IUPAC are obtained by applying the following rules.

1.   The ketones all have the ending **-one**.

2.   The longest continuous chain of carbons containing the $\diagdown C{=}O \diagup$ group is considered to be the parent structure. The *-e* of the parent structure, that is, alkan*e*, alken*e*, or alkyn*e*, is dropped and the ending *-one* is added.

3.   The chain is numbered so as to give the lowest possible number to the $\diagdown C{=}O \diagup$ group. This number is placed before the name of the parent structure. If a carbon-carbon double bond or triple bond is present in the compound, the number representing the position of this double or triple bond is placed *before* the name of the *parent structure* and the number of the carbon of the carbonyl group is placed before the *-one*.

4.   Aromatic ketones are frequently named as derivatives of the proper phenone—for example, *acetophenone*—with the positions of the various substituents indicated as previously mentioned in the nomenclature of phenols (see 19-3).

5. Numbers are used to indicate the position of the substituents on the parent chain.

6. All substituent groups are placed in alphabetical order before the parent name.

Problem Example 19-12

Give the systematic IUPAC name for each of the following compounds:

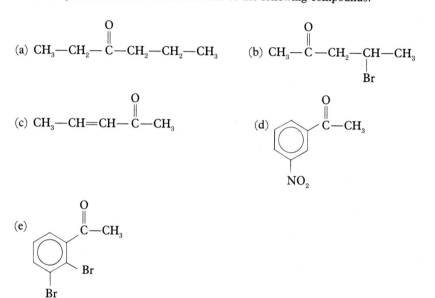

$$
\text{(a) } CH_3-CH_2-\overset{\overset{\textstyle O}{\|}}{C}-CH_2-CH_2-CH_3
$$

$$
\text{(b) } CH_3-\overset{\overset{\textstyle O}{\|}}{C}-CH_2-\underset{\underset{\textstyle Br}{|}}{CH}-CH_3
$$

$$
\text{(c) } CH_3-CH=CH-\overset{\overset{\textstyle O}{\|}}{C}-CH_3
$$

(d)

(e)

Answers (a) 3-hexanone, (b) 4-bromo-2-pentanone, (c) 3-penten-2-one, (d) *m*-nitroacetophenone, (e) 2,3-dibromacetophenone.

Work Problem 15.

Problem Example 19-13

Write structural formulas for each of the following compounds:

(a) diisopropyl ketone, (b) 4,5-dimethyl-2-heptanone, (c) 2-chloro-3-nitroacetophenone, (d) 4-hexyn-2-one.

Answers (a)

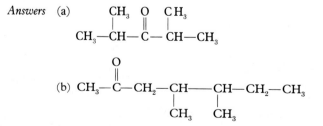

$$
\text{(a) } \underset{CH_3-\underset{\underset{\textstyle CH_3}{|}}{CH}-\overset{\overset{\textstyle O}{\|}}{C}-\underset{\underset{\textstyle CH_3}{|}}{CH}-CH_3}{}
$$

$$
\text{(b) } CH_3-\overset{\overset{\textstyle O}{\|}}{C}-CH_2-\underset{\underset{\textstyle CH_3}{|}}{CH}-\underset{\underset{\textstyle CH_3}{|}}{CH}-CH_2-CH_3
$$

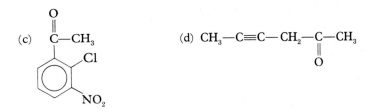

Work Problem 16.

The carbonyl group as in aldehydes or ketones is found in certain sugars, as glucose and fructose. Also, as you will notice from the structure hydroxyl groups as alcohols are also present.

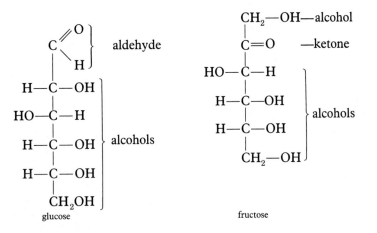

The compound, *sucrose, common table sugar,* is formed by the chemical combination of a glucose unit and a fructose unit. Many glucose units in a certain combination form starch. Another combination of glucose units form cellulose. We are able to metabolize starch, but not cellulose, as we do not possess the enzyme to metabolize it. Animals, such as the cow, sheep, or deer, do possess this enzyme and can metabolize cellulose.

There are three major classes of foods in the diet. These classes are: (1) **carbohydrates,** (2) **fats** and **oils,** and (3) **proteins.** Sugars, such as glucose, fructose, sucrose, cellulose, and starch, belong to the *carbohydrate* group. Fats and oils will be discussed in 19-8, while proteins will be discussed in 19-10.

19-7

Carboxylic Acids

The **carboxylic** (kār'bok·sil'ik) **acids** are compounds that have an alkyl or aryl group attached to a carboxyl group $\left(-C\diagup^{O}_{\diagdown O-H}\right)$. The general formulas for carboxylic acids are as follows:

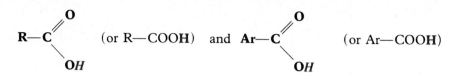

The functional group is the carboxyl group. The acidic hydrogen (H^{1+}) is the *H* attached to the oxygen of the carboxyl group. It is not one of the hydrogens attached to the alkyl or aryl groups. Acetic acid is the most common carboxylic acid. The carboxlic acid, adipic acid, is used in the preparation of nylon-66 (see 19-11). Aspirin contains a carboxylic acid group.

Nomenclature

Trivial names of some of the simpler carboxylic acids are acceptable in the IUPAC system, and a few of these are listed in Table 19-2. The systematic nomenclature of acids by the IUPAC system uses the following rules:

1. The carboxylic acids all have the ending **-oic acid.**

2. The longest continuous chain of carbons including the carboxyl group is considered the parent structure. The carboxyl carbon is numbered 1, but, as in the naming of aldehydes, this number is *not* indicated in the name. The *-e* of the parent structure—that is, the *-e* of alkan*e*, alken*e*, or alkyn*e*—is dropped and the ending *-oic acid* is added.

3. Aromatic acids are named as derivatives of *benzoic acid* with the positions of various substituents indicated as in the discussion of nomenclature of phenols (see 19-3).

4. Numbers are used to indicate the positions of substitutions on the parent chain.

5. All substituent groups on the parent structure are placed in alphabetical order as prefixed to the parent name.

Problem Example 19-14

Give the IUPAC name for each of the following:

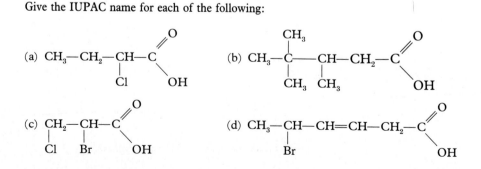

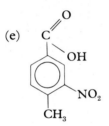

(e)

Answers (a) 2-chlorobutanoic acid, (b) 3,4,4-trimethylpentanoic acid, (c) 2-bromo-3-chloroporpanoic acid, (d) 5-bromo-3-hexenoic acid, (e) 4-methyl-3-nitrobenzoic acid.

Work Problem 17.

Problem Example 19-15

Write the structural formula for each of the following compounds:

(a) 2-methylpropanoic acid, (b) 3-methylbutanoic acid, (c) 5-bromo-2-hexynoic acid.

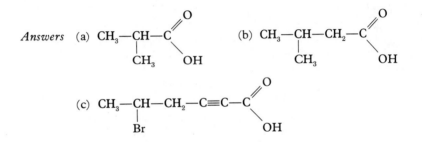

Work Problem 18.

Reactions

These carboxylic acids are weak acids (see 17-4) and readily undergo neutralization (see 9-11) by reaction with appropriate basic compounds. Typical bases used are sodium hydroxide and potassium hydroxide, and the salts, sodium carbonate and sodium hydrogen carbonate (bicarbonate), as illustrated in the following general equations (R = alkyl or aryl) and specific equation.

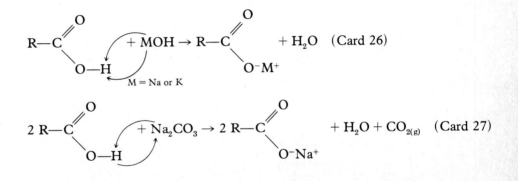

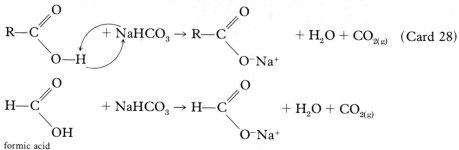

formic acid

Formic acid is the active irritant in ant and bee stings. To neutralize this acid, sodium hydrogen carbonate (bicarbonate, common household baking soda) is often added. The soda neutralizes the acid and prevents the sting from swelling and becoming exceedingly painful.

We mentioned previously that both alcohols and phenols are weakly acidic (see 19-2 and 19-3); however, these compounds are much less acidic than the carboxylic acids. Phenols are less acidic than carbonic acid (H_2CO_3) and do *not*, in general, react with sodium carbonate or sodium hydrogen carbonate to form the sodium salts of the phenols, but they do react with sodium hydroxide or potassium hydroxide. The equilibrium constant (K_a, see 17-4) for carboxylic acids is about 10^{-5} (mol/L), while the values for phenols, water, and alcohols are approximately 10^{-10} (mol/L), 2×10^{-16} (mol/L), and 10^{-16} to 10^{-18} (mol/L), respectively. The strength of these compounds as acids decrease in the following order:

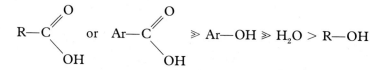

Carboxylic acids are readily converted into *esters* (see 19-8 and Table 19-1) by reaction with alcohols in the presence of an acid catalyst. A general and specific equation for these reactions is as follows:

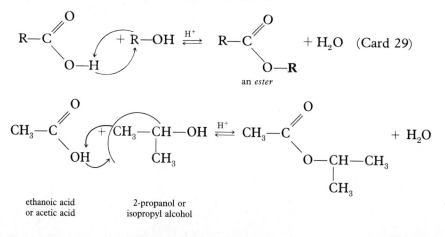

ethanoic acid   2-propanol or
or acetic acid   isopropyl alcohol

Note that the entire isopropyl $\left(\begin{array}{c}CH_3-CH-\\ |\\ CH_3\end{array}\right)$ group replaces the H atom

Work Problem 19.

of the carboxylic acid and in the ester this group is now attached to an oxygen atom. This is an equilibrium reaction which can be driven almost to completion by using excess alcohol or by removing the water formed (Le Châtelier's Principle, 17-3).

19-8

Esters

Esters (es′tērs) are derivatives of carboxylic acids. They are compounds in which an **—OR** or **—OAr** group replaces the **—OH** group of a carboxylic acid. The general formulas for esters are as follows:

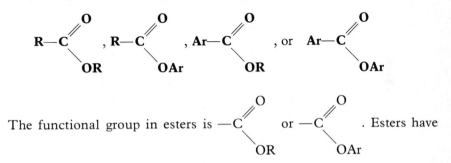

very pleasant odors and some of them are used in perfumery. The aroma and taste of fine wines is produced by the presence of certain esters. The fragrance and flavor of many fruits and flowers are due to the presence of mixtures of esters, and artificial flavoring essences are generally composed of mixtures of selected esters chosen so as to duplicate as closely as possible the flavor and aroma of the natural fruits. The following esters have characteristic aromas (in parentheses): ethyl formate (rum), pentyl acetate (banana), octyl acetate (orange), butyl butyrate (pineapple), methyl salicylate (wintergreen), pentyl butyrate (apricot).

Nomenclature

Names of the simpler esters may be derived from the trivial name of the corresponding carboxylic acid according to the IUPAC system. This is done by naming the alkyl group corresponding to the alcohol-derived fragment, following it with the stem of the trivial name of the carboxylic acid, and replacing its **-ic** ending with **-ate.** The esters listed in the preceding paragraph were named this way. Can you write their structures? Refer to Table 19-2 for the trivial names of the carboxylic acids.

The more systematic IUPAC name is derived in a similar way, except that the systematic IUPAC name of the acid is used; for example, *ethanoate* is derived from *ethanoic, propanoate* from *propanoic, benzoate* from benzoic, etc.

Problem Example 19-16

Give a suitable IUPAC name for each of the following compounds:

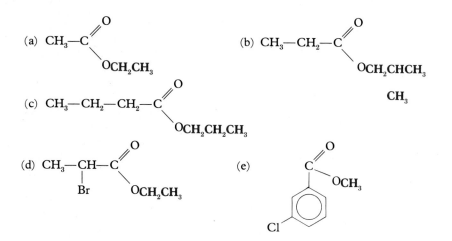

Answers (a) **ethyl** acetate or **ethyl** ethanoate, (b) **isobutyl** propanoate or **isobutyl** propionate, (c) **propyl** butanoate or **propyl** butyrate, (d) **ethyl** 2-bromopropanoate, (e) **methyl** *m*-chlorobenzoate.

Work Problem 20.

Problem Example 19-17

Write the structural formula for each of the following compounds:

(a) **pentyl** 2-butenoate, (b) **isopropyl** 3-methylpentanoate, (c) **butyl** *p*-bromobenzoate.

Answers (a) CH_3—CH=CH—C$\overset{\displaystyle O}{\underset{\displaystyle O-CH_2-CH_2-CH_2-CH_2-CH_3}{\big\langle}}$

(b) CH_3—CH_2—CH—CH_2—C
  | O—CH—CH_3
  CH_3 |
                          CH_3

(c)

Work Problem 21.

Fats and *oils* are the second major food class in the diet. *Fats* and *oils* are *esters* formed from long-chain aliphatic carboxylic acids (C_{12} to C_{22}) and glycerol (1,2,3-propanetriol). These esters are commonly called *glycerides*.

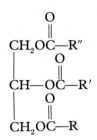

A fat or oil; R, R', and R" may be the same or different alkyl groups.

These naturally occurring fats and oils are not simple esters but generally contain different carboxylic acid residues in the same molecule of the fat or oil. Some of the more common acids involved in the formation of these glycerides are lauric, palmitic, stearic, oleic, and linoleic acids. Some have the general formula $CH_3(CH_2)_nCOOH$: for lauric, $n = 10$; for palmitic, $n = 14$; and for stearic, $n = 16$.

$$CH_3(CH_2)_7CH{=}CH(CH_2)_7COOH$$
oleic acid

$$CH_3(CH_2)_4CH{=}CHCH_2CH{=}CH(CH_2)_7COOH$$
linoleic acid

Fats are glycerides which are generally solid or semisolid at room temperature; *oils* are liquids at room temperature. Oils contain larger amounts of acid residues that are *unsaturated* than do fats. By catalytic hydrogenation a certain number of such unsaturated systems can be converted to saturated acid residues, and vegetable oils are thus converted to higher-melting glycerides or fats. In this manner such products as *Crisco* and *oleomargarine* have been prepared as fat or butter substitutes from vegetable oils such as corn, soybean, peanut, or cottonseed oils. In preparing these fats, the hydrogenation of the double bonds is carried out in such a way as to give a product that has a mixture of glycerides of both saturated and unsaturated acids.

Considerable research has been done on the relationship of fat consumption in humans and diseases of the arteries (primarily atherosclerosis). Unsaturated glycerides (vegetable oils and fats obtained from partial hydrogenation of oils) appear to be beneficial in lowering the incidence of certain heart diseases.

Reactions

Esters may be hydrolyzed to the corresponding acids and alcohols in either aqueous acid or base. The following general equations illustrate this hydrolysis reaction:

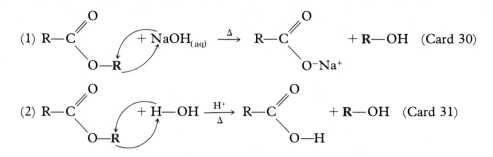

The reaction of sodium hydroxide with animal fat, as shown in (1), is used to prepare soap. The soap is the sodium salt of a long-chain carboxylic acid $(R—COO^-Na^+)$.

Specific equations of this hydrolysis are as follows:

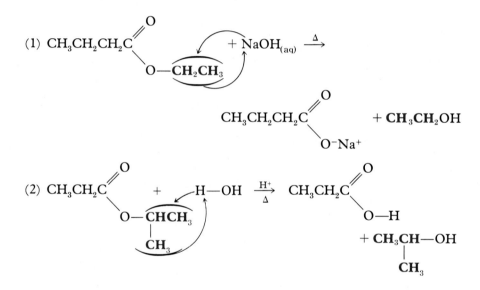

Note that in each case the oxygen-carbon bond in the ester $\left(\begin{array}{c} —O—CH_2—CH_3 \text{ and } —O—CH—CH_3 \\ | \\ CH_3 \end{array}\right)$ is broken and the hydroxyl group in NaOH or H—OH is attached to the alkyl group to form the alcohol.

Work Problem 22.

19-9

Amides

Amides (am'ids) are also derivatives of carboxylic acids. Amides are compounds in which an $—NH_2$ group replaces the $—OH$ group of a carboxylic acid. The general formulas for amides are

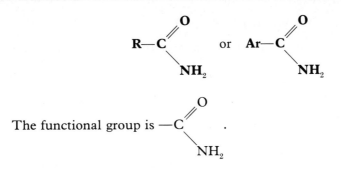

The functional group is —C=O / NH$_2$.

Both proteins and nylon are polyamides. Proteins will be considered in 19-10, while nylon will be considered in 19-11.

Nomenclature

When the IUPAC name of an amide is based on the trivial name of the corresponding carboxylic acid, the *-ic acid* part of the name is replaced by the ending *-amide*. Thus, the amide of acetic acid is *acetamide* and that of valeric acid is *valeramide*. The systematic IUPAC name is obtained by dropping the *-oic acid* ending of the systematic IUPAC name of the acid and replacing it with the ending *-amide*. Thus, *acetamide* is *ethanamide* and *valeramide* is *pentanamide*.

Problem Example 19-18

Give a suitable IUPAC name for each of the following compounds:

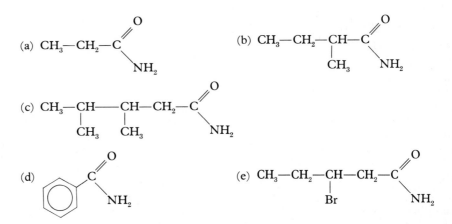

Answers (a) propanamide (propionamide), (b) 2-methylbutanamide, (c) 3,4-dimethylpentanamide, (d) benzamide, (e) 3-bromopentanamide.

Work Problem 23.

Problem Example 19-19

Write the structural formula for each of the following compounds:

(a) 3-methylpentanamide, (b) *p*-chlorobenzamide, (c) 2,2-dimethylpropanamide.

Work Problem 24. *Answers*

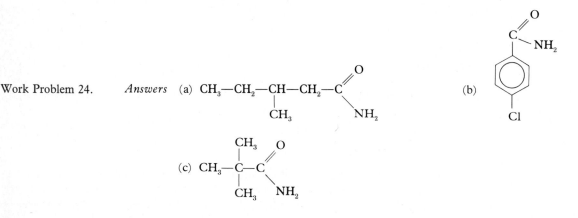

19-10

Amines

Amines (am′ins) are alkyl or aryl derivatives of ammonia (NH_3). *One, two,* or all *three* hydrogen atoms in ammonia may be replaced by alkyl or aryl groups to form amines. The general formulas for amines are shown in Table 19-3. The functional group is $-NH_2$, $-\overset{|}{N}-H$, or $-\overset{|}{\underset{|}{N}}-$. The amine, 1,6-hexanediamine, is used in the preparation of nylon-66 (see 19-11).

TABLE 19-3   General Formulas for Amines

Primary Amines	Secondary Amines	Tertiary Amines
R—NH$_2$	R—$\overset{R}{\underset{R}{N}}$—H	R—$\overset{R}{\underset{R}{N}}$—R, Ar—$\overset{R}{\underset{Ar}{N}}$—R
Ar—NH$_2$	Ar—$\overset{R}{\underset{Ar}{N}}$—H	Ar—$\overset{R}{\underset{Ar}{N}}$—Ar, Ar—$\overset{Ar}{\underset{Ar}{N}}$—Ar

Nomenclature

Amines may be given radicofunctional names according to the IUPAC. The following are the rules:

1. Name the alkyl or aryl groups attached to the nitrogen atom.

2. Add amine to the name of the group or groups.

3. The amine having an amino group ($-NH_2$) attached directly to a benzene ring is called *aniline* (or *benzenamine*).

The following examples illustrate naming by the radicofunctional system:

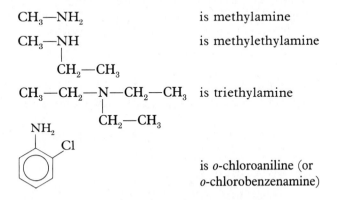

CH_3-NH_2 is methylamine

CH_3-NH
$\quad\;\; |$
$\quad\;\; CH_2-CH_3$ is methylethylamine

$CH_3-CH_2-N-CH_2-CH_3$ is triethylamine
$\qquad\qquad\quad |$
$\qquad\qquad\quad CH_2-CH_3$

is *o*-chloroaniline (or
o-chlorobenzenamine)

Another IUPAC system names amines as follows:

1. Primary amines (see Table 19-3): name the longest parent chain holding the amino group ($-NH_2$), dropping the *-e* and replacing it with *-amine*. A number is placed directly preceding the name of the parent chain, indicating the carbon atom to which the amino group is attached.

2. Secondary and tertiary amines (see Table 19-3): name the longest parent chain holding the amino group, as in primary amines. Use the prefix N- (for nitrogen) plus the name of the alkyl or aryl group (for secondary amines), or N,N-di or N- and N- plus the names of the alkyl or aryl groups (for tertiary amines) before the name of the parent primary amine to indicate the groups attached to the nitrogen atom. Also, a number is placed directly preceeding the name of the parent chain, indicating the carbon atom to which the amino group is attached.

Problem Example 19-20

Give a suitable IUPAC name for each of the following compounds:

(a) CH_3-N-CH_3
$\qquad\quad\;\; |$
$\qquad\quad\;\; H$

(b) $CH_3-CH_2-N-\overset{\displaystyle CH_3}{\underset{\displaystyle CH_3}{CH}}$
$\qquad\qquad\qquad\;\; |$
$\qquad\qquad\qquad\;\; H$

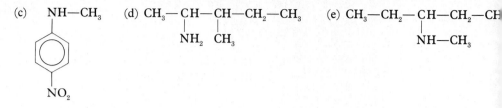

(c) $NH-CH_3$ (d) $CH_3-CH-CH-CH_2-CH_3$ (e) $CH_3-CH_2-CH-CH_2-CH$

Answers (a) N-methylmethanamine (dimethylamine), (b) N-ethyl-2-propanamine (ethyl-isopropylamine), (c) N-methyl-*p*-nitroaniline or N-methyl-*p*-nitrobenzenamine, (d) 3-methyl-2-pentanamine, (e) N-methyl-3-pentanamine.

Work Problem 25.

Problem Example 19-21

Write the structural formula for each of the following compounds:

(a) *sec*-butyldiethylamine (N,N-diethyl-2-butanamine), (b) N,N-dimethyl-1-hexyn-3-amine, (c) 1,2,3-propanetriamine.

Answers (a) 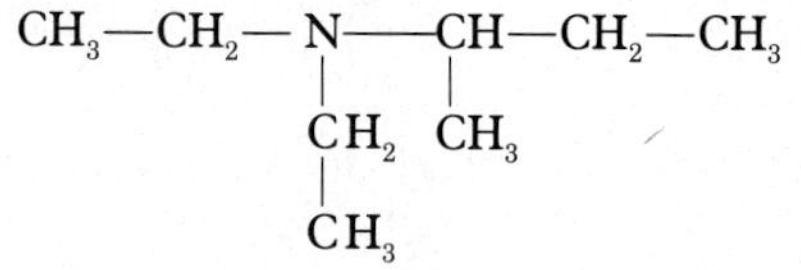 $CH_3-CH_2-N-CH-CH_2-CH_3$
with CH_2 CH_3 and CH_3

(b) 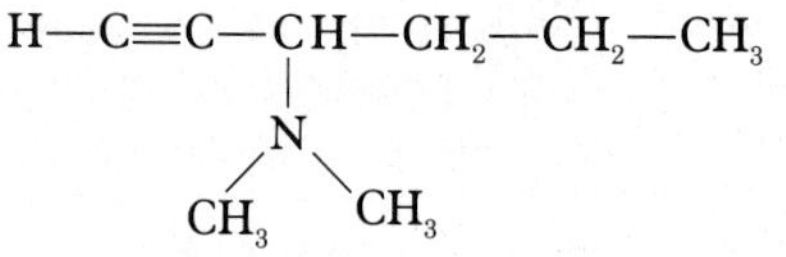 $H-C\equiv C-CH-CH_2-CH_2-CH_3$ with N and CH_3 CH_3

(c) $CH_2-CH-CH_2$ with NH_2 NH_2 NH_2

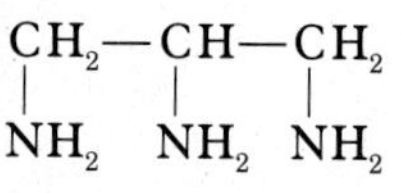

Work Problem 26.

Proteins are the third major food in the diet. Proteins are formed from α-amino acids[4] which have the general formula

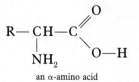

an α-amino acid

Various α-amino acids react with each other to remove molecules of water to form a protein, as follows:

---

[4]In α-amino acids (alpha-amino acids), the α (alpha) means that the amino group (—NH₂) is located on the carbon atom next to the carboxylic acid group.

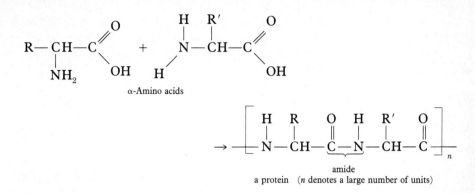

α-Amino acids

amide
a protein   (n denotes a large number of units)

Proteins are polymers consisting of polyamides. Notice the amide structure in the protein. The protein forms a helix or coiled molecular structure that gives it three-dimensional shape. Meat, eggs, and enzymes are examples of proteins.

19-11

Condensation Polymers

In 18-5, we considered addition polymers in which "giant molecules"—polymers—were formed from an alkene monomer by addition across the double bond, breaking the π bond. In this section, we shall consider polymers formed by "condensation" between two molecules, each molecule considered a monomer, through the reaction of certain functional groups. These polymers are called **condensation polymers.** In the condensation reactions normally small molecules are eliminated, such as methanol or water. Two common condensation polymers are: (1) polyester and (2) polyamides. In this section, we shall consider an example of each: (1) polyester—Dacron or Mylar and (2) polyamide—nylon-66.

In the preparation of the polyester, Dacron or Mylar, the two monomers are ethylene glycol and dimethyl terephthalate. The equation for the reaction is as follows:

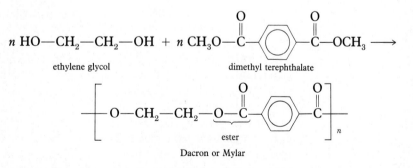

ethylene glycol   dimethyl terephthalate

ester

Dacron or Mylar

A unit of methanol (CH_3—OH) is lost in each condensation. In the formula of the polymer the n denotes a large number of units with the molecular mass being approximately 20,000 amu. Note the ester structure in the polymer and with many of these units the polymer is referred to as a polyester. This polyester polymer can be spun into a fiber as Dacron to produce clothing or made into a film as Mylar with high strength and used as packaging tape.

In the preparation of the polyamide, nylon-66, the two monomers are 1,6-hexanediamine and adipic acid (hexanedioic acid). The equation for the reaction is as follows:

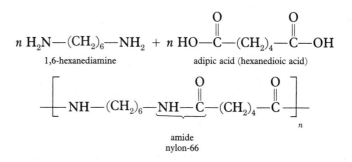

A unit of water is lost in each condensation. The polymer has a molecular mass of approximately 10,000 amu. In the formula of the polymer note the amide structure and with many of these units nylon-66 is referred to as a polyamide. The designation of 66 refers to the carbon atoms between the nitrogen atoms, that is **6** designated in **color** in the amine and 6 in the acid bonded to the next amine unit. Nylon was discovered by Wallace H. Carothers (see **1-3**) while he was working in the Du Pont laboratories in the 1930s. Nylon is used to prepare hose, brushes, parachutes, fishing line, and tire cords.

Work Problem 27.

Chapter Summary

In this chapter, the discussion of organic chemistry was continued, considering derivatives of the hydrocarbons. Derivatives of hydrocarbons are formed by the substitution of a functional group for a hydrogen atom attached to a carbon of a hydrocarbon chain. A *functional group* is an atom or group of atoms (other than hydrogen) attached to a hydrocarbon chain, which confers some distinctive chemical and physical properties to the compound. The functional group, general formula, and IUPAC nomenclature for organic halides, alcohols, phenols, ethers, aldehydes, ketones, carboxylic acid, esters, amides, and amines were considered.

Problem examples of IUPAC nomenclature for the foregoing types of organic compounds were given. Table 19-4 summarizes this IUPAC nomenclature. The radicofunctional names recognized by the IUPAC system were also considered for the appropriate types of organic compounds.

Alcohols, phenols, and carboxylic acids are all slightly acidic. The strength of these compounds in relationship to water in decreasing order of acidity is as follows:

TABLE 19-4 Summary of IUPAC Nomenclature of Functional Groups

Class of Compound	Functional Group	Systematic IUPAC Ending	Structural Formula Example	Systematic IUPAC Name	Radicofunctional Name
Organic Halide	$-X^a$	Halo group	CH_3-CH_2-Cl	Chloroethane	Ethyl chloride
Alcohol	$-OH$	-ol	CH_3-CH_2-OH	Ethanol	Ethyl alcohol
Phenol	$-OH$	—	(phenyl)$-OH$	Phenol	—
Ether	$-O-$	Alkoxy group	$CH_3-CH_2-O-CH_2-CH_3$	Ethoxyethane	Diethyl ether
Aldehyde	$-\overset{O}{\overset{\|}{C}}-H$	-al	$CH_3-\overset{O}{\overset{\|}{C}}-H$	Ethanal	Acetaldehyde
Ketone	$\overset{O}{\overset{\|}{C}}$	-one	$CH_3-\overset{O}{\overset{\|}{C}}-CH_3$	Propanone or acetone	Dimethyl ketone
Carboxylic acid	$-\overset{O}{\overset{\|}{C}}-O-H$	-oic acid	$CH_3-\overset{O}{\overset{\|}{C}}-O-H$	Ethanoic acid or acetic acid	—
Ester	$-\overset{O}{\overset{\|}{C}}-O-R^b$	-ate	$CH_3-\overset{O}{\overset{\|}{C}}-O-CH_3$	Methyl ethanoate or methyl acetate	—
Amides	$-\overset{O}{\overset{\|}{C}}-NH_2$	-amide	$CH_3-\overset{O}{\overset{\|}{C}}-NH_2$	Ethanamide or acetamide	—
Amines	$-NH_2, \; -\overset{}{\underset{}{N}}-H, \; -\overset{}{\underset{}{N}}-$	-amine	$CH_3-CH_2-NH-CH_3$	N-methyl-ethanamine	Ethylmethylamine

[a]The symbol —X designates a halogen atom (F, Cl, Br, or I).
[b]R may be Ar.

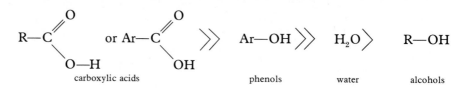

Due to this acidity of carboxylic acids, they react with appropriate basic compounds. Typical bases are sodium hydroxide, potassium hydroxide, and the salts sodium carbonate and sodium hydrogen carbonate. Carboxylic acids also react with alcohols to form esters in the presence of an acid catalyst. Both general and specific equations for these reactions were given.

Esters may be hydrolyzed to the corresponding acids and alcohols in either aqueous acid or base. Both general and specific equations for this hydrolysis were given.

Condensation polymers of Dacron or Mylar (a polyester) and nylon-66 (a polyamide) were discussed.

EXERCISES

1. Define or explain the following terms:
   (a) functional group   (b) alkoxy group
   (c) carbonyl group   (d) carboxyl group

2. Give the functional group for the following:
   (a) organic halides   (b) alcohols
   (c) phenols   (d) ethers
   (e) aldehydes   (f) ketones
   (g) carboxylic acids   (h) esters
   (i) amides   (j) amines

3. Write the general formulas for the following:
   (a) alkyl halide   (b) aryl halide
   (c) alcohol   (d) phenol
   (e) ether   (f) aldehyde
   (g) ketone   (h) carboxylic acid
   (i) ester   (j) amide
   (k) amine

PROBLEMS

Organic halides   (See Section 19-1)

4. Give the IUPAC name for each of the following compounds:

   (a) $CH_3-CH-CH_2-CH_3$
   $\qquad\qquad\;\;|$
   $\qquad\qquad\;\;Br$

   (b) $CH_3-CH-CH_2-CH-CH_2-CH_3$
   $\qquad\qquad\;\;\;|\qquad\qquad|$
   $\qquad\qquad\;\;\;Cl\qquad\quad CH_3$

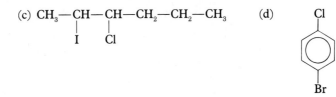

(c) CH_3—CH—CH—CH_2—CH_2—CH_3 (d)

with I and Cl substituents, and a benzene ring with Cl (top) and Br (bottom).

5. Write the structural formula for each of the following compounds:
   (a) *sec*-butyl chloride
   (b) 3,3-dichloro-1-hexyne
   (c) 4-bromo-2,3-dichlorotoluene

Alcohols (See Section 19-2)

6. Give the IUPAC name for each of the following compounds:

(a) CH_3—CH—CH_2—CH_2—CH_3
            |
            OH

(b) CH_3—CH_2—CH—CH_2—CH—CH_3
                   | |
                   CH_3 OH

(c) CH≡C—CH_2—CH—CH_3
                  |
                  OH

(d) CH_3—CH=CH—CH—CH—CH_3
                        | |
                       CH_3 OH

7. Write the structural formula for each of the following compounds:
   (a) isopropyl alcohol
   (b) 4,4-dichloro-1-pentanol
   (c) 2,3-dimethyl-4-hexen-1-ol

Phenols (See Section 19-3)

8. Give the IUPAC name for each of the following compounds:

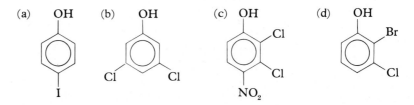

9. Write the structural formula for each of the following compounds:
    (a) *m*-fluorophenol
    (b) 2,4,6-tribromophenol
    (c) 2,6-dibromo-4-nitrophenol

Ethers (See Section 19-4)

10. Give the IUPAC name for each of the following compounds:

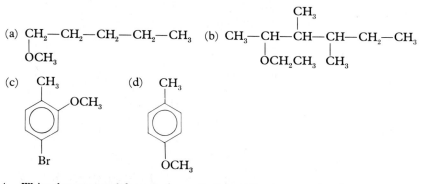

11. Write the structural formula for each of the following compounds:
    (a) dimethyl ether
    (b) 4-chloro-2-methoxy-1-hexanol
    (c) *o*-ethoxytoluene

Aldehydes (See Section 19-5)

12. Give the IUPAC name for each of the following compounds:

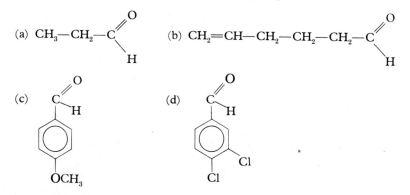

13. Write the structural formula for each of the following compounds:
    (a) 2-bromopropanal
    (b) 3-pentenal
    (c) 2-methoxy-4-nitrobenzaldehyde

Ketones (See Section 19-6)

14. Name each of the following compounds by the radicofunctional system:

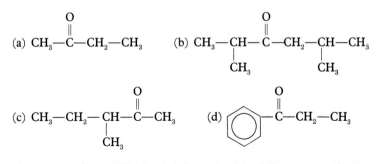

(a) $CH_3-\overset{\overset{\displaystyle O}{\|}}{C}-CH_2-CH_3$ (b) $CH_3-\underset{\underset{\displaystyle CH_3}{|}}{CH}-\overset{\overset{\displaystyle O}{\|}}{C}-CH_2-\underset{\underset{\displaystyle CH_3}{|}}{CH}-CH_3$

(c) $CH_3-CH_2-\underset{\underset{\displaystyle CH_3}{|}}{CH}-\overset{\overset{\displaystyle O}{\|}}{C}-CH_3$ (d) ⬡$-\overset{\overset{\displaystyle O}{\|}}{C}-CH_2-CH_3$

15. Give the systematic IUPAC name for each of the following compounds:

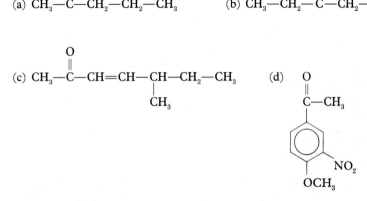

(a) $CH_3-\overset{\overset{\displaystyle O}{\|}}{C}-CH_2-CH_2-CH_3$ (b) $CH_3-CH_2-\overset{\overset{\displaystyle O}{\|}}{C}-CH_2-\underset{\underset{\displaystyle CH_3}{|}}{CH}-CH_3$

(c) $CH_3-\overset{\overset{\displaystyle O}{\|}}{C}-CH=CH-\underset{\underset{\displaystyle CH_3}{|}}{CH}-CH_2-CH_3$ (d)

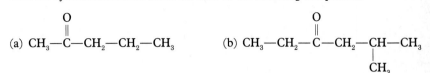

16. Write the structural formula for each of the following compounds:
    (a) acetone
    (b) 3-hexen-2-one
    (c) *p*-chloroacetophenone

Carboxylic acids (See Section 19-7)

17. Give the IUPAC name for each of the following compounds:

(a) $CH_3-\underset{\underset{\displaystyle Cl}{|}}{CH}-C\overset{\displaystyle O}{\underset{\displaystyle OH}{\diagdown}}$ (b) $CH_2=CH-CH_2-C\overset{\displaystyle O}{\underset{\displaystyle OH}{\diagdown}}$

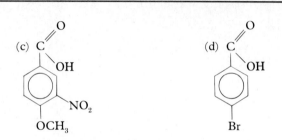

18. Write the structural formula for each of the following compounds:
    (a) 2-chloroethanoic acid
    (b) 2-pentenoic acid
    (c) o-hydroxybenzoic acid

19. Complete and balance the equation for each of the following reactions:

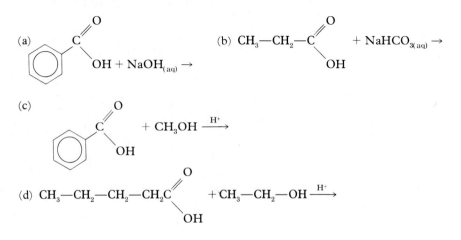

Esters (See Section 19-8)

20. Give a suitable IUPAC name for each of the following compounds:

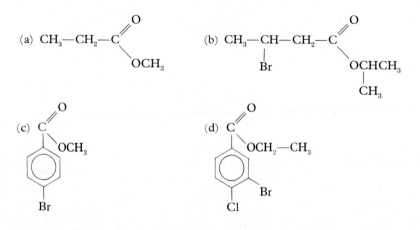

21. Write the structural formula for each of the following compounds:
    (a) methyl valerate
    (b) ethyl 2-methylbutanoate
    (c) methyl *p*-methoxybenzoate

22. Complete and balance the equation for each of the following reactions:

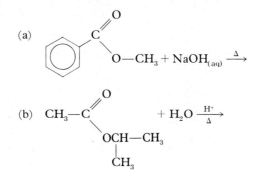

(a)

(b)

Amides (See Section 19-9)

23. Give the IUPAC name for each of the following compounds:

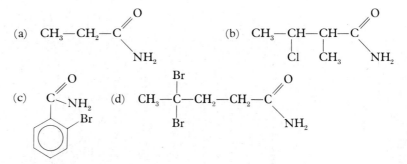

24. Write the structural formula for each of the following compounds:
    (a) propionamide
    (b) 2-butenamide
    (c) 4-chloro-3-methoxybenzamide

Amines (See Section 19-10)

25. Give the IUPAC name for each of the following compounds:

(a) CH₃—CH—CH₂—CH₃
          |
          NH₂

(b) CH₃—CH₂—CH—CH₂—CH₃
                |
                NH—CH₂—CH₃

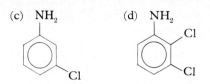

(c) NH_2

(d) NH_2

26. Write the structural formula for each of the following compounds:
    (a) 2-aminopropanoic acid (alanine, an amino acid)
    (b) ethyl 3-aminopentanoate
    (c) ethyl *p*-aminobenzoate (Benzocaine, a local anesthetic)
    (d) N-ethyl-3-hexanamine

Condensation Polymers   (See Section 19-11)

27. Give the structural formula of the two monomers and the polymer for the following condensation polymers:
    (a) Dacron or Mylar   (b) nylon-66

General Problems

28. Give the IUPAC name for each of the following compounds:

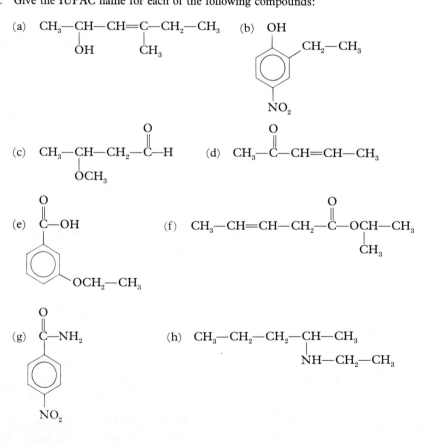

(a) CH_3—CH—CH=C—CH_2—CH_3
         | |
        OH CH_3

(b) OH, CH_2—CH_3, NO_2

(c) CH_3—CH—CH_2—$\overset{O}{\overset{||}{C}}$—$H$
         |
        OCH_3

(d) CH_3—$\overset{O}{\overset{||}{C}}$—$CH$=$CH$—$CH_3$

(e) $\overset{O}{\overset{||}{C}}$—OH, OCH_2—CH_3

(f) CH_3—CH=CH—CH_2—$\overset{O}{\overset{||}{C}}$—$OCH$—$CH_3$
                                        |
                                       CH_3

(g) $\overset{O}{\overset{||}{C}}$—$NH_2$, NO_2

(h) CH_3—CH_2—CH_2—CH—CH_3
                          |
                       NH—CH_2—CH_3

29. Identify the (a) alcohol, (b) phenol, (c) ether, (d) aldehyde, (e) ketone, (f) carboxylic acid, (g) ester, (h) amide, and (i) amine groups in each of the following compounds by circling each group and labeling it:

(a)

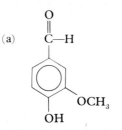

vanillin (a flavoring agent found in vanilla)

(b)

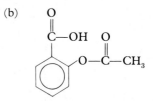

acetylsalicylic acid (aspirin)

(c)

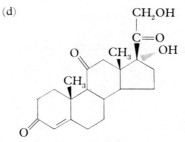

cholesterol (a possible cause of coronary artery disease)

(d)

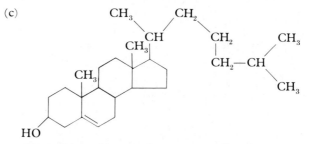

cortisone (an anti-inflammatory drug)

(e)

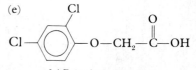

2,4-D (an herbicide which kills broad-leaf plants)

(f)

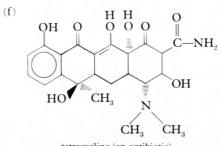

tetracycline (an antibiotic)

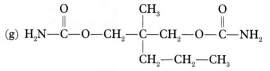

(g) $H_2N-\overset{\displaystyle O}{\overset{\|}{C}}-O-CH_2-\overset{\displaystyle CH_3}{\underset{\displaystyle CH_2-CH_2-CH_3}{\overset{|}{\underset{|}{C}}}}-CH_2-O-\overset{\displaystyle O}{\overset{\|}{C}}-NH_2$

meprobamate (Miltown, Equanil, a minor tranquilizer)

(h)

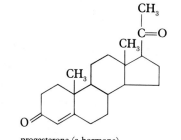

progesterone (a hormone)

(i)

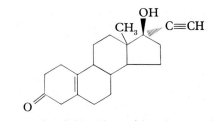

norethynodrel (constituent of the oral contraceptive Enovid)

30. Treatment of acute methanol poisoning uses isotonic sodium hydrogen carbonate given intraveneously to neutralize the action of formic acid formed in the metabolism of methanol. Write a balanced equation for this neutralization reaction.

31. Write condensed structural formulas for the isomers of the following. (The number in parentheses is the number of isomers for the compound.)
    (a) C_3H_7Cl (2)
    (b) $C_6H_4Cl_2$—a substituted benzene (3)
    (c) C_2H_6O—consider different functional groups (2)
    (d) C_3H_8O—consider different functional groups (3)

32. A certain alkene gave on analysis 85.7 percent carbon and 14.3 percent hydrogen. At 27°C and 630 torr, the gas had a density of 1.42 g/L. Calculate the molecular formula, write the structural formula, and give the IUPAC name for the alkene.

Readings

Morrison, Robert T., and Robert A. Boyd, *Organic Chemistry*, 4th ed. Boston: Allyn and Bacon, Inc., 1983.

Solomons, T. W. Graham, *Organic Chemistry*, 3rd ed. New York: John Wiley & Sons, Inc., 1984.

Both of the books above are comprehensive texts on organic chemistry.

20

Nuclear Chemistry

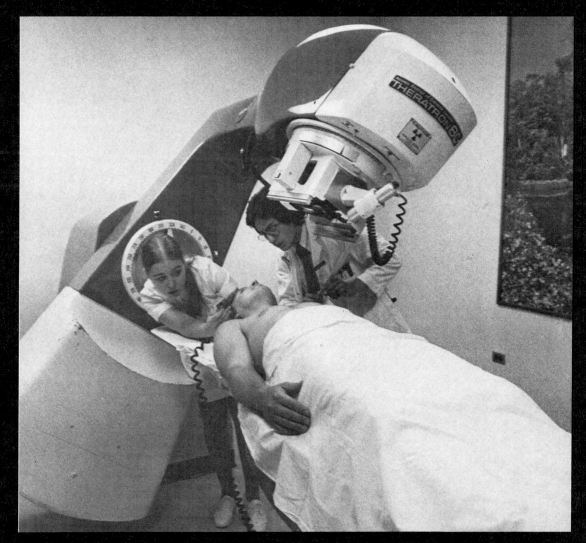

Peaceful use of atomic energy. Patient receiving radiation therapy for treatment of cancer. Humankind makes use of the radiations emitted by radioactive elements in the treatment of cancer and other diseases. (Courtesy Guy Gillette, Photo Researchers, Inc.)

TASK

Memorize the symbols for the particles or rays given in Table 20-2.

OBJECTIVES

1. Given the following terms, define each term and describe the distinguishing characteristics of each:
   (a) radioactivity (Introduction to Chapter 20)
   (b) curie (Section 20-1)
   (c) half-life (Section 20-3)
   (d) nuclear fission (Section 20-5)
   (e) nuclear fusion (Section 20-6)

2. (a) Given a particular radioactive isotope, its method of decay, and the Periodic Table, write a balanced equation for the nuclear reaction.
   (b) Given a particular isotope, a bombarding particle, the particle evolved in the nuclear reaction, and the Periodic Table, write a balanced equation for the nuclear reaction.

   (Problem Examples 20-1, 20-2, 20-3, and 20-4, Problems 9, 10, 11, 12, 13, and 14)

3. Given a particular radioactive isotope, its half-life, the amount of the isotope before decay, and the time of its decay or the amount present at a particular time in its decay, calculate either the amount remaining at a particular time or the time a given amount of the isotope will be present (Problem Examples 20-5, 20-6, and 20-7, Problems 15, 16, 17, 18, 19, and 20).

We previously mentioned that in ordinary chemical reactions the valence electrons of atoms are gained, lost, or shared (6-2) with **no** changes occurring in the nucleus of the atom. Now we are going to consider changes in the nucleus that result in one element being changed into a completely *different* element. Since these reactions involve the nucleus of the atom, they are called **nuclear reactions,** and hence we can write **equations** for them. Accompanying nuclear reactions is radioactivity. **Radioactivity** is a property of certain radioactive isotopes that spontaneously emit from their nuclei certain *radiations*, which can result in the formation of atoms of a *different* element or atoms of an isotope (see 4-5) of the *original* element.

20-1

Nuclear Radiations

Let us consider radiations emitted from the nuclei of radioactive isotopes. The most common of these nuclear radiations are (1) alpha particles, (2) beta particles, (3) gamma rays, (4) positron particles—or, simply, positrons, and (5) neutrons. The first three are named after the first three letters in the Greek alphabet, alpha (α), beta (β), and gamma (γ). Alpha and beta particles and gamma rays are emitted from naturally occurring radioactive isotopes, whereas positrons and neutrons are emitted only by synthetic (human-made) radioactive isotopes. A small amount of a uranium mineral may be placed in a small hole in a lead block, as shown in Figure 20-1. If the emitted radiation is placed under the influence of an electrical field at right angles to the emitted radiation with a photographic plate some distance above the block, the gamma rays will not be deflected, but the alpha and beta particles will be. The alpha particles will be attracted to the negative field and the beta particles to the positive field. We now consider these nuclear radiations in more detail.

FIGURE 20-1

Nuclear radiation from a small amount of uranium mineral under the influence of an electrical field.

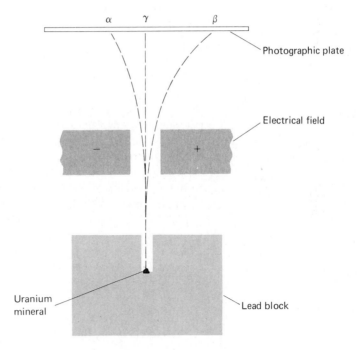

Alpha Particles (α)

Alpha particles are helium nuclei and hence are positively charged; they possess an atomic number of 2 and a mass number of 4. An alpha particle is written ^4_2He, with the 2 representing the atomic number and the 4 representing the mass number (see 4-4 for review). The 2^+ charge is normally omitted. The positively charged helium nucleus ($_2^4\text{He}^{2+}$) will rapidly pick up two electrons from some other atom nearby or from electrons in the atmosphere. This, then, makes the positively charged helium nucleus now a neutral helium atom ($_2^4\text{He}$).

Alpha particles are emitted from the nucleus of an atom with a velocity of about 0.1 times the velocity of light ($c = 3.00 \times 10^{10}$ cm/s). They have very low penetrating power; they can be stopped by such material as a thin sheet of paper, at which point they will be converted to helium gas. If by some accident a radioactive substance gets in the body—for example, by inhalation of radioactive dust particles—the situation is extremely dangerous. Alpha particles injure normal cells, even though they have little penetrating power. In the body, alpha particles are more dangerous than beta particles and gamma rays. The combination of all three radiations produces radiation sickness, which can easily result in death. Alpha particles exhibit a large ionizing effect on gases. This effect is the basis for the detection of *all* nuclear radiation by the Geiger-Müller counter, which we shall discuss at the end of this section.

Beta Particles (β)

Beta particles are identical to electrons and hence have a unit negative charge. Beta particles come from the nucleus, but, as you know, electrons do not exist in the nucleus. Therefore, the beta particles are believed to be produced when a neutron is transformed to a proton, ($n \rightarrow p + \beta$). A beta particle is written $_{-1}^{0}e$, with -1 representing the atomic number of the electron and the 0 indicating a 0 mass number, since an electron has negligible mass (see 4-3). Beta particles are emitted from the nucleus with a velocity of about 0.9 the velocity of light. They have greater penetrating power than alpha particles, but can be stopped by a sheet of aluminum 1 cm thick. As previously mentioned, they are also dangerous if the radioactive substance gets into the body. Beta particles exhibit a small ionizing effect on gases.

Gamma Rays (γ)

Gamma rays are rays of light energy of very short wavelength, similar to, but more energetic than, X rays, and hence without charge or mass. They are written as the Greek letter γ. Gamma rays are emitted from the nucleus and arise from an internal change in the nucleus with a loss in energy. The velocity of release of gamma rays from the nucleus is equal to the velocity of light. They have very high energy and a high penetration power, thereby readily penetrating the body. These rays are stopped by lead and concrete. Low-level gamma rays are believed to have a major effect on the human genetic code by causing mutations (99 percent of which are harmful). This is one of the reasons that many medical doctors use X rays with discretion, especially if the sex organs are exposed to the radiation, since X rays act like low-energy gamma rays. Gamma rays have a very slight ionizing effect, less than beta particles.

Positron Particles or Positrons

Positron particles or positrons are identical to positive electrons and hence have a unit positive charge. Positrons come from the nucleus and are believed to be produced when a proton is transformed to a neutron, ($p \rightarrow$ positron + n). A positron is written $_{+1}^{0}e$, with the 1 representing the atomic number and the 0 indicating a 0 mass number, since an electron has negligible mass. Positrons are similar to beta particles in velocity and ionizing effect, but they have very low penetrating power.

Neutrons

Neutrons ($_0^1 n$) were considered in 4-3. Neutron velocities are variable. They can be adjusted from a very slow velocity (approximately 0.00001 times the velocity of light) to a relatively high velocity (approximately 0.1 times the velocity of light). They have relatively high penetrating power but no ionizing effect, since they have no charge.

Table 20-1 summarizes these radiations.

A radioactive isotope does not emit alpha, beta, and positron particles simultaneously; *only one* of the three is emitted in a given process. Gamma rays, however, are usually emitted along with alpha, beta, or positron particles, due to internal energy adjustments in the nucleus of the atom.

Most nuclear radiation can be detected by the use of the Geiger-Müller counter (Geiger counter), as shown in Figure 20-2. The counter consists of a tube containing two electrodes, a positive anode and a negative cathode, with a high potential of about 1000 volts between them. The tube contains air or argon gas at a pressure of about 5 to 12 torr. Radiation enters the tube through

TABLE 20-1   Summary of Properties of Nuclear Radiations

Nuclear Radiation	Symbol	Mass (amu)	Charge	Approximate Velocity ($\times$ Velocity of Light)	Penetrating Power (Rank)[a]	Ionizing Effect (Rank)[a]
Alpha (α)	$_2^4$He	4	2^+	0.1	4	1
Beta (β)	$_{-1}^{0}$e	Negligible	1^-	0.9	3	3
Gamma (γ)	γ	None	None	1.0	1	4
Positron	$_{+1}^{0}$e	Negligible	1^+	0.9	5	2
Neutron	$_0^1$n	1.0087	0	Slow (0.00001) to fast (~0.1)	2	5

[a]Highest = rank 1.

FIGURE 20-2

A Geiger-Müller counter (Geiger counter).

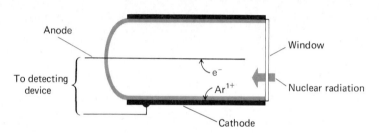

the window, which is usually covered with a thin glass, and ionizes the gas molecules in the tube, forming ions such as Ar^{1+} ions, if the gas is argon. The electrons and the ions produce a short-lived flow of electric current in the circuit. This short flow of current can be amplified to a detecting device to either produce a sound or record counts on an automatic counting device.

One of the units used to measure the quantity of nuclear radiation is the curie, named for Marie Sklodowska Curie (1867–1934) and Pierre Curie (1859–1906),[1] the French husband and wife scientific team who discovered both polonium (Po) and radium (Ra). A **curie** (Ci) is defined as the amount (mass) of a radioactive isotope that will give 3.7×10^{10} disintegrations per second. Since this value is quite large, a microcurie (μCi), 1/1,000,000 of a curie or the amount that would give 3.7×10^4 disintegrations per second ($3.7 \times 10^{10}/10^6$), is often used. A radioactive sample with one or more curies is considered to have high activity. For example, 1.00 g of radium gives 1.00 curie, and is considered to have high activity.

In the eruption of Mount St. Helens, 3 million curies of radon gas were given off. In the Three Mile Island (TMI) incident in Pennsylvania in 1979, 2.5 million curies of radioactive xenon gas was released to the atmosphere. Radon gas is considered 1000 times more hazardous to health than radioactive xenon.[2]

20-2

Nuclear Equations

Since we have introduced the major nuclear radiations, we can use them to write equations showing nuclear reactions. Table 20-2 lists various particles or rays used in nuclear equations, you must know these symbols.

---

[1]In the isolation of radium in 1902, Madame Curie isolated 100 mg of almost pure radium chloride from a ton of uranium ore by repeated crystallizations—truly an arduous task. Professor Curie, a physicist, did most of the physical measurements concerned with their research while Madame Curie did the chemical research. In 1903, they received the Nobel Prize in physics as a team; then, in 1911, Madame Curie received the Nobel Prize in chemistry.

[2]Another unit for measuring radioactivity is the *rem*. The rem takes into consideration the type of radiation, type of tissue exposed, and the total dose. In the United States, the maximum limits of exposure is 500 millirems (0.5 rem) per year per person. For those working in the nuclear industry it is 5000 millirems (5 rems). A chest X ray exposes the lungs to 10 millirems, while living for one year at mile-high elevation exposes the body to 200 millirems.

TABLE 20-2   Symbols for Particles or Rays Used in Nuclear Reactions[a]

Particle or Ray	Symbol
Alpha (α)	^4_2He
Beta (β)	$^0_{-1}\text{e}$
Gamma (γ)	γ
Positron	$^0_{+1}\text{e}$
Neutron	^1_0n
Proton	^1_1H
Deuteron (an isotope of hydrogen)	^2_1H

[a]You may find it helpful to make "flash cards" for these particles or rays and their symbols.

In nuclear equations, the sum of the atomic numbers and the sum of the mass numbers in the reactants must *equal* their corresponding sum of atomic numbers and the sum of the mass numbers in the products. In nuclear reactions, there is some loss in mass, which is converted to energy according to Einstein's equation $E = mc^2$ (see 3-6), but this loss in mass is not shown in the nuclear equation in the mass number because it is a small fraction of an atomic mass unit. One gram of mass converted completely to energy gives off energy equivalent to 2.15×10^{15} cal. This amount of heat energy would raise about 240,000 tons of water (the size of a small mountain lake) from 0°C to 100°C. Since gamma rays are light energy of very short wavelength, their loss does not alter the nuclear equation. Therefore, they also are not shown in the equation, although they *often* accompany a particle loss.

Now let us write the equations for some nuclear reactions:

Problem Example 20-1

Radium-226 (Ra) decays by alpha emission. Write a balanced equation for this nuclear reaction.

SOLUTION: From the Periodic Table (inside front cover), the atomic number for radium is 88. An alpha particle is written ^4_2He (Table 20-2). Decay means "to give off"; therefore, radium is on the reactant side of the equation, and the alpha particle plus some other element must be among the products.

$$^{226}_{88}\text{Ra} \rightarrow {}^4_2\text{He} + \text{?}$$

For the atomic number in the reactant to equal the sum of the atomic numbers in the products, the atomic number of the new element (?) must be 86.

$$88 = 2 + x$$

$$x = 86$$

The same method can be applied to the mass number, and hence the mass number for the new element (?) is 222.

$$226 = 4 + x$$

$$x = 222$$

Refer again to the Periodic Table and look for the symbol of the element with an *atomic number* of 86. The symbol is Rn, and from the list of elements and their symbols (inside front cover), the name of the element is radon, although this last information is not required. Therefore, the equation for this nuclear reaction is

$$^{226}_{88}\text{Ra} \rightarrow {}^{4}_{2}\text{He} + {}^{222}_{86}\text{Rn} \quad \textit{Answer}$$

Problem Example 20-2

Bromine-82 decays by beta emission. Write a balanced equation for this nuclear reaction.

SOLUTION: Using the Periodic Table to determine the atomic number of bromine, and knowing that the symbol for a beta particle is written $_{-1}^{0}\text{e}$ (Table 20-2), you can write an incomplete equation:

$$^{82}_{35}\text{Br} \rightarrow {}^{0}_{-1}\text{e} + ?$$

The atomic number for the new element is 36:

$$35 = -1 + x$$

$$x = 36$$

The mass number for the new element is 82. Refer again to the Periodic Table for the symbol of the element with an *atomic number* of 36. The symbol is Kr and the name of the element is krypton. Therefore, the equation for this nuclear reaction is

$$^{82}_{35}\text{Br} \rightarrow {}^{0}_{-1}\text{e} + {}^{82}_{36}\text{Kr} \quad \textit{Answer}$$

Problem Example 20-3

Oxygen-15 decays by positron emission. Write a balanced equation for this nuclear reaction.

SOLUTION: Using the Periodic Table to determine that atomic number of oxygen, and knowing that the symbol for a positron is written $_{+1}^{0}\text{e}$ (Table 20-2), you can write an incomplete equation:

$$^{15}_{8}\text{O} \rightarrow {}^{0}_{+1}\text{e} + ?$$

The *atomic number* for the new element is 7, and the mass number is 15.

$$8 = 1 + x$$

$$x = 7$$

From the Periodic Table, the new element is nitrogen (N), and the equation for this nuclear reaction is

$$^{15}_{8}O \rightarrow \ ^{0}_{+1}e + \ ^{15}_{7}N \quad \textit{Answer}$$

Problem Example 20-4

Bombardment of uranium-238 with a deuteron results in the formation of another element and the release of two neutrons for each uranium atom. Write a balanced equation for this nuclear reaction.

SOLUTION: Using the Periodic Table to determine the atomic number of uranium, and knowing that the symbol for a deuteron is $^{2}_{1}H$ and the symbol for a neutron is $^{1}_{0}n$ (Table 20-2), you can write an incomplete equation:

$$^{238}_{92}U + \ ^{2}_{1}H \rightarrow 2^{1}_{0}n + \ ?$$

(Note that two neutrons are released for each uranium atom.) The atomic number for the new element is 93.

$$92 + 1 = x + 2(0)$$

$$x = 93$$

The mass number for the new element is 238.

$$238 + 2 = x + 2(1)$$

$$x = 238$$

Work Problems 9, 10, 11, 12, 13, and 14.

Refer again to the Periodic Table for the symbol of the element with an *atomic number* of 93. The symbol is Np and the element is neptunium. Therefore, the equation for this nuclear reaction is

$$^{238}_{92}U + \ ^{2}_{1}H \rightarrow \ ^{238}_{93}Np + 2^{1}_{0}n \quad \textit{Answer}$$

20-3

Natural Radioactivity. Half-Life

Certain isotopes that occur in nature are radioactive; that is, they give off nuclear radiation. The radioactive isotopes found in nature are called *natural radioactive isotopes,* and the process of giving off nuclear radiation spontaneously from these natural radioactive isotopes is called **natural radioactivity.** The exact reason for *certain* isotopes to be radioactive is not completely understood, but one reason is that the nucleus is unstable due to the neutron/proton ratio.

Since the protons are positively charged and hence repel each other, it appears that more neutrons are required to lessen the effect of this proton repulsion. But if there are not enough neutrons to lessen this effect, the isotope is radioactive and emits nuclear radiation. Most isotopes are not radioactive and do not emit nuclear radiation. In this text, we shall not attempt to predict the isotopes that do emit nuclear radiation or the types of radiation they emit.

The rate of disintegration of *any* radioactive isotope, such as a natural radioactive isotope, is in essence *independent* of the external conditions such as temperature and pressure and the oxidation number of the element. The rate is *dependent* on the nature of the element. Not all radioactive isotopes decay at the same rate. This rate of disintegration of various radioactive isotopes is expressed in quantitative values by the term "half-life." The **half-life** of a radioactive isotope is the time required for one-half of any **given amount** of a radioactive element to disintegrate. This value is constant for a given radioactive isotope.

Now let us consider some problems involving half-life calculations:

Problem Example 20-5

An isotope of cesium, cesium-137, has a half-life of 30.0 years. If 1.00 mg of cesium-137 disintegrates over a period of 90.0 years, how many milligrams of cesium-137 will remain?

SOLUTION: At this time (0 year), we have 1.00 mg of the isotope; after 30.0 years we shall have $\frac{1}{2}$ of the original or 0.500 mg, and after 60.0 years we shall have $\frac{1}{2}$ of 0.500 mg, or 0.250 mg. Finally, after 90.0 years, we shall have $\frac{1}{2}$ of 0.250 mg, or 0.125 mg:

Add half- life each time	0 year 30.0 years 60.0 years 90.0 years	1.00 mg 0.500 mg 0.250 mg 0.125 mg *Answer*	Divide the amount by one- half each time

Problem Example 20-6

Approximately 2.50 g of strontium-90, was formed in a 1960 atomic explosion at Johnson Island in the Pacific test site. The half-life of strontium-90 is 28.0 years. In what year will only 0.312 g of this strontium-90 remain?

SOLUTION: In 1960 (0 years), we have 2.50 g; in 1988 (1960 + 28.0 or 28.0 years), we shall have $\frac{1}{2}$ of 2.50 g or 1.25 g; in 2016 (1988 + 28.0 or 56.0 years), we shall have 0.625 g; in 2044 (2016 + 28.0 or 84.0 years), we shall have 0.3125 g or 0.312 g to three significant digits:

Divide the	2.50 g	0 year, 1960		Add half-life each time
amount by	1.25 g	28.0 years, 1988		
one-half	0.625 g	56.0 years, 2016		
each time	0.312 g	84.0 years, 2044	Answer[3,4]	

Problem Example 20-7

A radioactive isotope of xenon, xenon-125, has a half-life of 17 hours. If 0.250 g of xenon-125 disintegrates over a period of 2 days and 3 hours, how many milligrams of xenon-125 will remain?

SOLUTION: The time of 2 days and 3 hours is equivalent to 51 hours:

$$2 \text{ days} \times \frac{24 \text{ hours}}{1 \text{ day}} + 3 \text{ hours} = 51 \text{ hours}$$

At the start (0 hour), we have 0.250 g of xenon-125; after 17 hours we shall have $\frac{1}{2}$ of the original or 0.125 g; after 34 hours we shall have $\frac{1}{2}$ of 0.125 g or 0.0625 g; and finally after 51 hours we shall have $\frac{1}{2}$ of 0.0625 g or 0.03125 g, 0.0312 g to three significant digits or 31.2 mg.

Add half-	0 hours	0.250 g	Divide the amount by one-half each
life each	17 hours	0.125 g	time
time	34 hours	0.0625 g	
	51 hours	0.03125 g	
		31.2 mg Answer	

Work Problems 15, 16, 17, 18, 19, and 20.

In nature there are three radioactive disintegration series, in which naturally occurring radioactive isotopes disintegrate to form other radioactive isotopes. The disintegration continues until a final *nonradioactive* isotope is formed. The three naturally occurring radioactive disintegration series are (1) the uranium-238 series, (2) the uranium-235 series, and (3) the thorium-232 series. All three series end with various nonradioactive isotopes of lead. The uranium-238 series ends with lead-206 ($^{206}_{82}Pb$); the uranium-235 series ends with lead-207 ($^{207}_{82}Pb$); the thorium-232 series ends with lead-208 ($^{208}_{82}Pb$). The uranium-238 disintegration series is given in Table 20-3.

Another radioactive disintegration series exists with the synthetic (human-made) isotope of plutonium, plutonium-241. The plutonium-241 series ends with nonradioactive bismuth-209 ($^{209}_{83}Bi$).

---

[3]A common error is to double the half-life time. Note that you *add* the half-life time to the previous time. The 84.0 years was obtained by adding 28.0 years to 56.0 years, not doubling the 56.0 years.

[4]For our purposes in this text, we shall use half-multiple of the half-life of the amount, since this simplifies the calculations considerably and avoids using more complicated mathematics.

TABLE 20–3 The Uranium-238 Disintegration Series

Isotope	Symbol	Half-Life	Particle Emitted from Nucleus	Equation
Uranium-238	$^{238}_{92}U$	4.15×10^9 yr	α	$^{238}_{92}U \rightarrow \, ^{234}_{90}Th \, + \, ^{4}_{2}He$
Thorium-234	$^{234}_{90}Th$	24.1 days	β	$^{234}_{90}Th \rightarrow \, ^{234}_{91}Pa \, + \, ^{0}_{-1}e$
Protactinium-234	$^{234}_{91}Pa$	1.18 min	β	$^{234}_{91}Pa \rightarrow \, ^{234}_{92}U \, + \, ^{0}_{-1}e$
Uranium-234	$^{234}_{92}U$	2.48×10^5 yr	α	$^{234}_{92}U \rightarrow \, ^{230}_{90}Th \, + \, ^{4}_{2}He$
Thorium-230	$^{230}_{90}Th$	8.00×10^4 yr	α	$^{230}_{90}Th \rightarrow \, ^{226}_{88}Ra \, + \, ^{4}_{2}He$
Radium-226	$^{226}_{88}Ra$	1620 yr	α	$^{226}_{88}Ra \rightarrow \, ^{222}_{86}Rn \, + \, ^{4}_{2}He$
Radon-222	$^{222}_{86}Rn$	3.82 days	α	$^{222}_{86}Rn \rightarrow \, ^{218}_{84}Po \, + \, ^{4}_{2}He$
Polonium-218	$^{218}_{84}Po$	3.05 min	α	$^{218}_{84}Po \rightarrow \, ^{214}_{82}Pb \, + \, ^{4}_{2}He$
Lead-214	$^{214}_{82}Pb$	26.8 min	β	$^{214}_{82}Pb \rightarrow \, ^{214}_{83}Bi \, + \, ^{0}_{-1}e$
Bismuth-214	$^{214}_{83}Bi$	19.7 min	β	$^{214}_{83}Bi \rightarrow \, ^{214}_{84}Po \, + \, ^{0}_{-1}e$
Polonium-214	$^{214}_{84}Po$	1.64×10^{-4} s	α	$^{214}_{84}Po \rightarrow \, ^{210}_{82}Pb \, + \, ^{4}_{2}He$
Lead-210	$^{210}_{82}Pb$	21.0 yr	β	$^{210}_{82}Pb \rightarrow \, ^{210}_{83}Bi \, + \, ^{0}_{-1}e$
Bismuth-210	$^{210}_{83}Bi$	5.00 days	β	$^{210}_{83}Bi \rightarrow \, ^{210}_{84}Po \, + \, ^{0}_{-1}e$
Polonium-210	$^{210}_{84}Po$	138 days	α	$^{210}_{84}Po \rightarrow \, ^{206}_{82}Pb \, + \, ^{4}_{2}He$
Lead-206	$^{206}_{82}Pb$	Nonradioactive	None	None

Since three of the four series end with lead and one with bismuth, it appears that the neutron/proton ratios in lead and bismuth approach the limits for nuclear stability.

20-4

Transmutation of the Elements. Artificial Radioactivity

In 1919, British physicist Ernest Rutherford[5] (1871–1937) found that high-energy alpha particles can collide with a nitrogen nucleus to produce a proton and an atom of oxygen of mass number 17:

$$^{4}_{2}He \, + \, ^{14}_{7}N \rightarrow \, ^{1}_{1}H \, + \, ^{17}_{8}O \tag{20-1}$$

Due to the requirement that the alpha particle must hit the small nucleus of the nitrogen atom for a reaction to occur, you would expect to have very many strikes before a reaction would take place.

---

[5]Rutherford received the Nobel Prize in *chemistry* in 1908, although he was a physicist. Since he had previously worked with various transformations, he remarked during his Nobel Prize speech that the fastest transformation he had ever worked with was his *own*, from a physicist to a chemist! Although we consider Rutherford to be British, he was actually born in New Zealand and received the major portion of his education in that country. The American name for element 104 is rutherfordium (Rf), named in honor of Rutherford.

Rutherford's experiment was the first example of the change of one element to another—*transmutation of an element*. Since then, many transmutations have been made by bombarding various elements with high-energy[6] alpha particles, neutrons, protons, and deuterons.

When very-high-energy particles are needed, an instrument must be used to accelerate the bombarding particles to very high speeds. One such particle accelerator is the *cyclotron*,[7] which consists of two D-shaped electrodes, covered with two oppositely charged electromagnets in a vacuum, as shown in Figure 20-3. The electromagnets act to move the particles in a spiral path.

The particles to be accelerated, such as the alpha particles, are introduced in the center between the two D-shaped electrodes. The electrodes are so adjusted that the charge on them can be altered *rapidly* as the particles cross the gap between the electrodes. For example, suppose that positively charged alpha particles are introduced in the center between the D-shaped electrodes. If one of the electrodes is given a negative charge and the other a positive charge, the alpha particle will travel toward the negative electrode, since negative attracts positive, and will be repelled by the positive electrode, since like charges

FIGURE 20-3

The path (− − −) of a particle in a cyclotron. The entire system is enclosed in a vacuum, and two oppositely charged electromagnets are above and below the D-shaped electrodes.

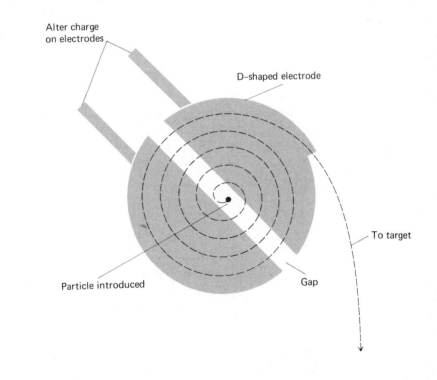

Alter charge on electrodes

D-shaped electrode

To target

Particle introduced

Gap

---

[6]By high-energy particles, we mean particles with high kinetic energy as shown by a high velocity.

[7]Ernest O. Lawrence (1901–1958), American physicist, received the Nobel Prize in 1939 in physics for the development of the cyclotron. Element 103 (lawrencium, Lr) was named in honor of him.

repel each other. The passage of the particle across the gap between the D-shaped electrodes is loosely analogous to a person jumping across a creek, with the assistance of two people. One person pushes him over and the other grabs his hand and pulls him over, as shown in Figure 20-4.

After the alpha particle has completed the half-circle to its oppositely charged electrode, the charge on the electrodes will change and the negative electrode will become positive and the positive, negative. This process will continue until the alpha particle reaches the edge of the electrodes and escapes with a high enough energy to produce a reaction with the target nucleus.

In the process of transmutation of the elements, a radioactive isotope of an element may be formed. This type of radioactivity is called **artificial radioactivity.** The first example of artificial radioactivity was discovered in 1934 by the French scientists Irene Joliot-Curie (1897–1956), daughter of Pierre and Marie Curie, and her husband Frederic Joliot (1900–1958).[8] They discovered that if either boron, magnesium, or aluminum (all non-radioactive) were bombarded with alpha particles, a neutron would be emitted and other elements would form. But these other elements continued to emit nuclear radiation in the form of positrons.

The nuclear equations in the case of boron are as follows:

$$^{10}_{5}\text{B} + {}^{4}_{2}\text{He} \rightarrow {}^{13}_{7}\text{N} + {}^{1}_{0}\text{n} \qquad (20\text{-}2)$$

$$^{13}_{7}\text{N} \rightarrow {}^{13}_{6}\text{C} + {}^{0}_{+1}\text{e}, \qquad \text{half-life} = 9.9 \text{ min} \qquad (20\text{-}3)$$

Nitrogen-13 is radioactive and disintegrates to carbon-13, which is nonradioactive. Therefore, the formation of radioactive nitrogen-13 (artificial) from

FIGURE 20-4

Crossing a creek or the "gap" in a cyclotron. Like charges repel each other, while opposite charges attract.

[8] Both scientists received the Nobel Prize in chemistry in 1935.

nonradioactive boron-10 and the disintegration of radioactive nitrogen-13 to nonradioactive carbon-13 is an example of *artificial radioactivity*.

Another example of artificial radioactivity is the formation of gold from platinum. As you may recall from 1-3, one of the goals of the alchemist was to change various metals into gold. This can now be done, but the process is more expensive than the gold produced. If platinum-196 is bombarded with a deuteron ($^2_1$H), radioactive platinum-197 results and emits a beta particle to form gold-197, the only naturally occurring isotope of gold. The equations for these nuclear reactions are as follows:

$$^{196}_{78}\text{Pt} + {}^{2}_{1}\text{H} \rightarrow {}^{197}_{78}\text{Pt} + {}^{1}_{1}\text{H} \tag{20-4}$$

$$^{197}_{78}\text{Pt} \rightarrow {}^{197}_{79}\text{Au} + {}_{-1}^{0}\text{e}, \quad \text{half-life} = 18.0 \text{ hr} \tag{20-5}$$

The synthesis of new elements is another example of transmutation of the elements followed by artificial radioactivity. Element 105, with the proposed name of hahnium [Ha] in honor of Otto Hahn (see 20-5), is formed by the bombardment of californium-248 with nitrogen-15 nuclei to form element 105 (mass number 260) and the emission of four neutrons. The newly formed element has a half-life of 1.6 seconds and disintegrates to lawrencium-256 and an alpha particle. The equations for these nuclear reactions are as follows:

$$^{249}_{98}\text{Cf} + {}^{15}_{7}\text{N} \rightarrow {}^{260}_{105}[\text{Ha}] + 4{}^{1}_{0}\text{n} \tag{20-6}$$

$$^{260}_{105}[\text{Ha}] \rightarrow {}^{256}_{103}\text{Lr} + {}^{4}_{2}\text{He}, \quad \text{half-life} = 1.6 \text{ s} \tag{20-7}$$

Another more recent example is the formation of the new, but unnamed, element 106. Both Russian and American scientists claim to have prepared it first.[9] The preparation of element 106 was done in the United States at the University of California, Berkeley, under the direction of Albert Ghiorso. Ghiorso also directed the American team in the synthesis of elements 104 (rutherfordium) and 105 (hahnium). The American team bombarded californium-249 with oxygen-18 ions to form element 106 (mass number 263) and the emission of four neutrons. The new element has a half-life of 0.9 second and disintegrates to rutherfordium-259 and an alpha particle. The equations for these nuclear reactions are as follows:

$$^{249}_{98}\text{Cf} + {}^{18}_{8}\text{O} \rightarrow {}^{263}_{106}[\text{element-106}] + 4{}^{1}_{0}\text{n} \tag{20-8}$$

$$^{263}_{106}[\text{element-106}] \rightarrow {}^{259}_{104}\text{Rf} + {}^{4}_{2}\text{He} \quad \text{half-life} = 0.9 \text{ s} \tag{20-9}$$

---

[9]The group that first prepares the element has the honor of naming it. Elements 104 and 105 are also in dispute. The American names for elements 104 and 105 are rutherfordium and hahnium, respectively, which we have adopted in this text. The Soviet names are kurchatovium and niels bohrium, respectively. The IUPAC has recently recommended the following names and symbols for these recently discovered elements: element 104 = unnilquadium—Unq, element 105 = unnilpentium—Unp, element 106 = unnilhexium—Unh, element 107 = unnilseptium—Uns, and element 109 = unnilennium—Une.

The most recent elements 107 and 109 were prepared by a group of West German scientists. Element-109 is prepared by bombarding bismuth-209 with iron-58 and a neutron being released. Element-109 then disintegrates to element-107 and an alpha particle with a half-life of 5 milliseconds. Element-107 disintegrates to hahnium-258 and an alpha particle. The equations for these nuclear reactions are as follows:

$$^{209}_{83}\text{Bi} + {}^{58}_{26}\text{Fe} \rightarrow {}^{266}_{109}[\text{element-109}] + {}^{1}_{0}\text{n} \qquad (20\text{-}10)$$

$$^{266}_{109}[\text{element-109}] \rightarrow {}^{262}_{107}[\text{element-107}] + {}^{4}_{2}\text{He} \qquad (20\text{-}11)$$

$$\text{half-life} = 5 \text{ milliseconds}$$

$$^{262}_{107}[\text{element-107}] \rightarrow {}^{258}_{105}[\text{Ha}] + {}^{4}_{2}\text{He} \qquad (20\text{-}12)$$

20-5

Nuclear Fission

Between 1934 and 1938, two different research groups pioneered the development of nuclear fission, which eventually led to the development of a nuclear fission explosion—the atomic bomb. These two research groups consisted of an Italian group, headed by Enrico Fermi (1901–1954), who later moved to the United States, and a German group, consisting of Otto Hahn (1879–1968), Lise Meitner (1878–1968), and Fritz Strassman (b. 1902).[10]

Nuclear fission involves the *splitting* of an atomic nucleus into two or more smaller nuclei. The reaction involved in a nuclear fission results from bombarding uranium-235 with *neutrons*. A typical equation for this type of reaction is

$$^{235}_{92}\text{U} + {}^{1}_{0}\text{n} \rightarrow {}^{94}_{38}\text{Sr} + {}^{139}_{54}\text{Xe} + 3{}^{1}_{0}\text{n} \qquad (20\text{-}13)$$

During this reaction, released neutrons continue to bombard other uranium-235 atoms. This release of neutrons in a nuclear reaction is called a *chain reaction*, and is like dominos, stacked on end in a solid triangle, falling and knocking over other dominos in progression, as in Figure 20-5.

A nuclear fission explosion is produced by two *subcritical* masses forced together by an ordinary explosive charge to form a *critical mass*. A *critical mass* in this case is a mass of uranium-235 just large enough to sustain a chain reaction. It is a sphere of about an 8-cm radius and 40-kg mass. A neutron then bombards the uranium-235 atom to start the entire chain reaction.

The uranium-235 atom is broken into fragments of smaller atomic numbers. These fragments are radioactive and emit nuclear radiation, accounting for the radiation burns and sickness following a nuclear fission explosion. One fragment from fission eventually formed after nuclear radiation is barium, identified by Hahn and Strassman in 1939.

---

[10]Fermi received the Nobel Prize in physics in 1938. Hahn received it in chemistry in 1944. Element 100 (fermium, Fm) was named in honor of Fermi.

FIGURE 20-5

Dominos stacked on end in a solid triangle, depicting a chain reaction.

Besides uranium-235, which is only 0.7 percent of naturally occurring uranium,[11] plutonium-239 also undergoes fission. Plutonium-239[12] can be formed from uranium-238 (99.3 percent of naturally occurring uranium) according to the following nuclear equations:

$$^{238}_{92}\text{U} + {}^{1}_{0}\text{n} \rightarrow {}^{239}_{92}\text{U} \tag{20-14}$$

$$^{239}_{92}\text{U} \rightarrow {}^{239}_{93}\text{Np} + {}^{0}_{-1}\text{e}, \qquad \text{half-life} = 23.5 \text{ min} \tag{20-15}$$

$$^{239}_{93}\text{Np} \rightarrow {}^{239}_{94}\text{Pu} + {}^{0}_{-1}\text{e}, \qquad \text{half-life} = 2.35 \text{ days} \tag{20-16}$$

Plutonium-239 has an extremely long half-life (24,360 years decaying with alpha emission), and can undergo nuclear fission just as uranium-235 can.

In the process of the nuclear fission explosion, a small but definite loss of mass occurs. This mass loss is a fraction of the mass number and hence is not shown in Equation 20-13. But this loss in mass, converted to energy, is of great consequence in fission explosions. In 1939, Lise Meitner of the German group and her nephew Otto Frisch (1904–1979) calculated the amount of energy released for every uranium atom fissioned as equal to approximately 200 meV (million electron-volts) from Einstein's equation for the conversion of mass to energy, $E = mc^2$. This energy, expressed in more meaningful terms, is 4.6×10^9 kcal per *mol of uranium atoms fissioned*, approximately 2.5 million times that of the energy from a comparable amount by mass of coal. Nuclear fission energy is given off primarily as *heat* energy.

The scientific development of a nuclear fission explosion was carried out primarily at the then isolated areas in the mountains of New Mexico near Santa Fe at Los Alamos; in the mountains of Tennessee near Knoxville at Oak Ridge; at Hanford, Washington; and at the Metallurgical Laboratory of the University of Chicago. The entire project was government supported and was organized by the United States Army. It had the code name "Manhattan Pro-

---

[11]Two large plants were built during World War II to separate the isotopes of uranium by two different processes. Both plants were located at Oak Ridge, Tennessee.

[12]The production of plutonium took place in a plant constructed in 1943 at Hanford, Washington.

ject." Many distinguished scientists were involved in this project, among them Vannevar Bush, James B. Conant, Leslie R. Groves, J. Robert Oppenheimer, and Glenn T. Seaborg, all who have since died except for Seaborg. The first nuclear fission explosion—atomic bomb—was tested in the desert at Trinity Flats near Alamogordo, New Mexico, on July 16, 1945. Then two atomic bombs were dropped on Japan during World War II, one on Hiroshima on August 6, 1945, and the other on Nagasaki on August 9, 1945, bringing an end to World War II on August 14, 1945. From these two bombs, the estimated loss of life to these two Japanese cities was over 200,000.

Nuclear fission appeared to be the first instance in modern times in which the scientific world was awakened to the consequences of its work on the sociological-political level. The story behind the development of the Manhattan Project is an interesting background to this awakening. Lise Meitner, one of the workers in the German group in the pioneering of nuclear fission, was of Jewish ancestry and so was forced to leave Nazi Germany in 1938 or be persecuted. She succeeded in reaching Holland, and, with the assistance of Niels Bohr, eventually got to Stockholm, Sweden. Meitner was aware of the work which the remaining German group of Hahn and Strassman were carrying out, and conveyed her knowledge of the group's work to her nephew Otto Frisch, who was working in Bohr's laboratory in Copenhagen, Denmark, at the time. They realized the tremendous military potential nuclear fission might have. During a trip to the United States, Bohr discussed the possibility with many European scientists then working in the United States. One of these scientists, Leo Szilard (1898–1964), a Hungarian, spoke to Albert Einstein about the military potential of nuclear fission and about the possibility that the Germans might at that time be developing it. Einstein wrote a letter to President Franklin Delano Roosevelt in August 1939, expressing his concern, and Szilard transmitted the letter to the president. President Roosevelt subsequently established the Manhattan Project.

Although Hahn and Strassman continued their work, they did not develop the atomic bomb, for Hitler was believed to have considered this research "Jewish-tainted" and did not give it financial support. Neither Hahn nor Strassman supported Hitler, but they were permitted to continue their work during the war years.

The interweaving of a person's ancestry with scientific discovery provides an interesting footnote to history. If this interweaving had not occurred and Nazi Germany had been the first to develop nuclear fission explosives, the consequences to the rest of the world may have been disastrous.

20-6

Nuclear Fusion

In **nuclear fusion,** energy is produced by combining two or more nuclei to form a heavier nucleus. Two examples of nuclear fusion explosions are (1) the sun and (2) the hydrogen or thermonuclear bomb.

Sun

The nuclear reaction on the sun probably involves the fusion of two isotopes of hydrogen—deuterium, $_1^2$H, and tritium, $_1^3$H, to form helium, $_2^4$He, according to the following overall nuclear equation:

$$_1^2\text{H} + _1^3\text{H} \rightarrow _2^4\text{He} + _0^1\text{n} \tag{20-17}$$

The deuterium and tritium isotopes are formed indirectly by bombardment of hydrogen nuclei ($_1^1$H) with each other. During the fusion process, there is considerable loss in mass, which is converted to energy, primarily in the form of heat. But, a considerable amount of heat energy must be supplied for the occurrence of one of these nuclear fusion reactions. This type of heat energy is available in the sun, since the sun's interior temperature is estimated to be 2×10^7°C. Therefore, we owe our continued existence to the nuclear fusion reaction occurring on the sun, which heats our planet.

Hydrogen or Thermonuclear Bomb

The hydrogen or thermonuclear bomb consists of a reaction similar to that given in Equation 20-17. The heat energy required to obtain such a reaction is supplied by a nuclear fission explosion—atomic bomb—which acts as the fuse for the nuclear fusion explosion. The energy released from a nuclear fusion explosion (hydrogen bomb) is estimated at about 15 times the energy released from a large nuclear fission explosion (atomic bomb).

20-7

Peaceful Uses of Nuclear Reactions

The two types of nuclear reactions (fission and fusion) may be used to produce power in nuclear reactors. Nuclear fission technology is well developed, and nuclear reactors producing electrical power have been in continuous operation in the United States since 1957, although recently no new reactors have been ordered.[13]

The first sustained nuclear fission reaction was carried out by the director of the Italian research group, Enrico Fermi, on December 2, 1942, in the squash courts below the Alonzo Stagg Field at the University of Chicago. The energy produced in this reaction was in the form of heat energy and was quite small, being discharged to the surrounding air. Since then, considerable development has occurred in nuclear fission power plants. The plant consists of three essential parts: (1) nuclear fission reactor; (2) turbine and generator; and (3) condenser, as shown in Figure 20-6.

---

[13]A *natural* fusion reaction occurred in Gabon, western Africa, about 2 billion years ago. It lasted 100,000 years to 1 million years and produced 10 tons of fission products with 4 tons of plutonium-239.

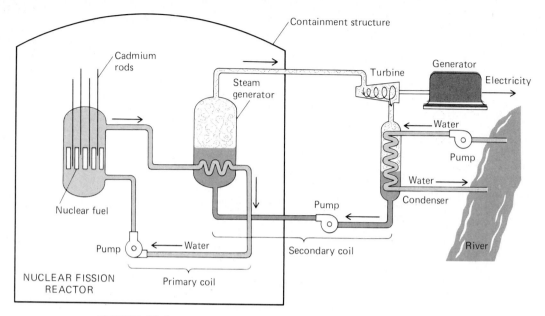

FIGURE 20-6

A nuclear fission power plant (water cooled).

Nuclear Fission Reactor

The uranium used in the nuclear fission reactor is uranium dioxide (UO_2), with the uranium slightly enriched to 2 or 3 percent in uranium-235. Due to this low concentration of uranium-235 in the reactor and the fact that the concentration must be around 90 percent for a nuclear fission bomb, a nuclear explosion cannot occur in a nuclear fission reactor. (Natural uranium consists of only 0.7 percent uranium-235, with nearly all of the remainder uranium-238.) Most of the energy is produced by the uranium-235, but uranium-238 can be converted to plutonium-239, which is also fissionable (see Equations 20-14 through 20-16). Both uranium-235 and plutonium-239 can be used in nuclear reactors. Water is usually used as a moderator to slow down the neutrons, and cadmium rods are inserted into the reactor to absorb the neutrons and to control the production of neutrons for a slow evolution of heat and prevent overheating. Large amounts of heat are liberated from the reactor, which heats the water and converts it to steam (see Figure 20-6).

Turbine and Generator

The steam from the secondary coil (see Figure 20-6) then passes to the turbine, as in a nonnuclear power plant. The turbine drives the generator, which produces electrical power. The heat energy from a nuclear reaction is thus converted to mechanical energy in the turbine and to electrical energy in the generator.

The difference between a nuclear power plant and a nonnuclear power plant is the source of the steam. In a nuclear power plant, the steam is produced by heat from a nuclear reaction, whereas in a nonnuclear power plant, it is produced by heat from burning fossil fuels, such as coal, oil, or gas.

Condenser

The steam from the secondary coil, after turning the turbine, is liquefied in the condenser and recycled passing over the primary coil. The water used to cool the condenser is obtained from nearby rivers. After some cooling in cooling towers the excess heat in the water is dissipated into the river, producing some thermal pollution. A possible use for this heated water may be to irrigate farm crops and lengthen the growing season.

Problems with Nuclear Fission

Nuclear fission power plants have been used to supply electricity to many cities throughout the world. In the United States about 13 percent of the electrical power is produced by nuclear fission. This may increase in the future if the problems associated with nuclear fission power plants can be solved.

These problems are (1) disposal of the radioactive waste, (2) the possibility of a nuclear accident, and (3) the possibility of theft of fissionable material during transportation. At present there is no permanent site for storage of the radioactive waste that accumulates through the nuclear fission process. The Department of Energy (DOE) has under construction a site in southern New Mexico near Carlsbad to store certain defense nuclear waste by solidifying the material and storing it as fused glass in salt domes (waste isolation pilot plant, WIPP). Their plans call for the site to be operational by 1998. Congress has also approved the building of a permanent underground storage facility. This *first* facility may be located in one of a number of places: a salt formation in Utah, Louisiana, or Mississippi; a volcanic rock structure in Nevada; or a basalt rock formation in Washington (Hanford). The *second* facility may be built in granite rock formations in Wisconsin or New England. Plans call for the first permanent site to be operational by 1997 to 2006.

There is always the possibility of a nuclear accident in a nuclear fission power plant. This was vividly brought to our attention by the TMI (Three Mile Island) incident in Pennsylvania in 1979. If the reactor vessel (see Figure 20-6) becomes too hot and the method used to cool it does not work or is not used, a "meltdown" could occur. The reactor bottom could melt, diffusing radioactive waste into the atmosphere and spilling it into the ground, the latter sometimes referred to as the "China Syndrome." The probability of this happening is remote (1 chance in 1000 years). Adequate safeguards against a serious accident like this should be developed if nuclear fission power plants are to continue to be used to meet the world's energy needs.

Both uranium-235 and plutonium-239 can be used to make atomic bombs. If this material gets into the hands of terrorists, there is a possibility that they could develop an atomic bomb and bring about the downfall of governments through blackmail or widespread physical destruction through its use. The fuel in a reactor would have to be processed to enrich its concentration for use in an atomic bomb. Strict security measures are needed for the transportation of all nuclear material.

Nuclear Fusion

Nuclear fusion might also be used to produce electrical power in the future. Scientists believe that nuclear fusion power plants would produce less thermal pollution and nuclear radiation and have less risk of nuclear accident than nuclear fission power plants.

There are two approaches being used to develop nuclear fusion to produce electricity: (1) magnetic confinement of the hydrogen and (2) use of lasers or electron beams to unite the hydrogen atoms. The two isotopes of hydrogen that are used are deuterium ($_1^2$H) and tritium ($_1^3$H), as shown in Equation 20-17. The deuterium is readily available from sea water and the tritium can be produced by bombarding lithium with neutrons. Lithium salts can be obtained from rocks or sea water. To obtain a fusion of the hydrogen isotopes three conditions must be met: (1) high temperatures—about 100 million degrees Celsius, (2) high density—about 10^{14} to 10^{16} particles per cubic centimeter, and (3) confinement of the hydrogen isotopes long enough, about 1 second, at high temperatures and density to enable the fusion reaction to occur and sustain itself. So far, all three of these conditions have not been met at the same time by either the magnetic confinement or laser or electron beam approaches. The magnetic confinement approach is being developed at Princeton, New Jersey. A new facility, called Tokamak Fusion Test Reactor (TFTR), has been completed. Recent tests have reached a temperature of 100 million degrees Celsius and a confinement of 50 milliseconds. The laser approach is being developed at Los Alamos National Laboratory in New Mexico and Lawrence Livermore Laboratory in California. The California laboratory is now working with the world's largest and most powerful laser, called Shiva. Nova, a new laser, at Livermore is now under construction. Scientists hope that by the late 1980s they can achieve the "break-even" point or produce as much power as the fusion reactor uses and by 2000 actually build a pilot plant that converts the heat from fusion to electrical power, as is now done in nuclear fission power plants (see Figure 20-6).

20-8

Uses of Radioactive Isotopes

We shall now point out a few of the varied uses that have been made of radioactive isotopes.

By using the half-life of the radioactive isotope carbon-14, we can determine the age of various objects.[14] In living material, the ratio of carbon-14 to carbon-12 (nonradioactive) remains relatively constant. The carbon-14 in our atmosphere arises from the bombardment of a nitrogen-14 atom with a neutron from the upper atmosphere, according to the following nuclear equation:

$$^{14}_{7}N + ^{1}_{0}n \rightarrow ^{14}_{6}C + ^{1}_{1}H \qquad (20\text{-}18)$$

Carbon-14 disintegrates to form a beta particle, according to the following nuclear equation:

$$^{14}_{6}C \rightarrow ^{0}_{-1}e + ^{14}_{7}N, \qquad \text{half-life} = 5770 \text{ yr} \qquad (20\text{-}19)$$

The ratio of carbon-14 to carbon-12 in living tissue is constant. When the tissue in an animal or plant dies, the carbon-14 decreases because the intake and utilization of carbon-14 does not occur. Therefore, in dead tissue the ratio of carbon-14 to carbon-12 would decrease, depending on the age of the tissue. The age of the dead tissue is determined in the following manner: A sample of the dead tissue is burned to carbon dioxide, and the carbon dioxide is analyzed for the ratio of carbon-14 to carbon-12; the ratio of carbon-14 to carbon-12 in the dead tissue is compared to that in normal living tissue by considering a decrease of the ratio in dead tissue by one-half for every 5770 years, the half-life for carbon-14; the age of the dead tissue can then be determined.

The entire photosynthesis scheme in plants has been worked out by Melvin Calvin[15] (see 1-3) using carbon-14 and tracing the absorption of $^{14}CO_2$ in the plant to its synthesis of various sugars, starch, and cellulose.

Various radioactive isotopes have been used as nuclear radiation sources in cancer therapy by directing the radiation on the cancer growth. Some of these radioactive isotopes are radium-226, cobalt-60, and sodium-24. Chromium-51, as a wire, has been implanted *into* the cancerous tissue. Chromium-51 has a half-life of just 27.8 days, which is just sufficient for an effective dose but not long enough to produce an overdose. Iridium-192 (half-life = 74

---

[14]Willard F. Libby (1908–1980), American chemist, received the Nobel Prize in 1960 in chemistry for his work in this field.

[15]Calvin received the Nobel Prize in 1961 in chemistry for this work. Calvin is currently doing research to help solve our energy problem: he is growing a plant that produces a hydrocarbon which could be used as a fuel. On a 1-acre plot he was able to grow the equivalent of about 10 barrels of oil in a 7-month period.

days), in the form of tiny seeds, has been used in the treatment of breast cancer. The iridium is placed in thin plastic tubes which are inserted with steel needles into the breast tumor area. The needles are then removed. The iridium-192 releases gamma rays which destroy the tumor. After about three to five days the seeds of iridium are removed, along with the tubes. This technique has produced about the same survival rate (75 to 80 percent after 10 years) as radical mastectomies, with none of the mutilation that results from this surgery. The problem with cancer therapy using nuclear radiation is that the radiation must be of sufficient intensity to destroy the cancerous tissue but not so great as to destroy completely normal tissue. The chromium-51 wire meets this criterion. Iridium-192 seeds must be placed precisely into the tumor so as to avoid damage to the healthy tissue. Therefore, radioactive material besides possibly causing cancer can be used to *cure* cancer.

A very specific application of nuclear energy is in the treatment of brain tumors, which have the unusual property of absorbing boron compounds much faster than normal tissue does. Therefore, a boron compound is injected into a person with a brain tumor, and after 10 minutes, a stream of neutrons from a nuclear reaction is directed at the exact region of the tumor. In the brain tumor, the boron atoms hit with neutrons give off alpha particles, which destroy the tumor cells but which are stopped by their low penetrating power from further entry into the normal tissue.

We have mentioned only a few of the many uses and applications of radioactive isotopes. Many more could be mentioned and more remain to be discovered.

Chapter Summary

In this chapter, reactions involving the nucleus of an atom—*nuclear reactions*—were considered. Accompanying nuclear reactions is radioactivity. *Radioactivity* is a property of certain radioactive isotopes that spontaneously emit from their nucleus certain *radiations* that can result in the formation of atoms of a different element or atoms of an isotope of the original element. These radiations are (1) alpha particles, (2) beta particles, (3) gamma rays, (4) positron particles, and (5) neutrons. Using these radiations (see Table 20-2), equations for the nuclear reactions were written.

Certain isotopes that occur in nature are radioactive and give off nuclear radiations. These radioactive isotopes are called natural radioactive isotopes, and the process of giving off nuclear radiation spontaneously from natural radioactive isotopes is called *natural radioactivity*. The rate of disintegration of any radioactive isotope is, in essence, independent of external conditions but is dependent on the nature of the element. This rate is expressed quantitatively by the term "half-life." The *half-life* of a radioactive isotope is the time required for one-half of any given amount of a radioactive element to disintegrate. Various examples of half-life calculations were given.

Elements can be changed from one element to another using a particle accelerator such as a cyclotron. This is called transmutation of an element. In the process, a radioactive isotope of an element may be formed; this type of radioactivity is called *artificial radioactivity*. Various examples of artificial radioactivity were given.

Nuclear fission is the splitting of an atomic nucleus into two or more nuclei, while *nuclear fusion* is the combination of two or more nuclei to form a heavier nucleus. The atomic bomb is an example of nuclear fission; the reaction on the sun and the hydrogen bomb are examples of nuclear fusion. A peaceful use of nuclear fission is to produce electricity in nuclear power plants. The parts of nuclear fission power plants along with some problems involved in their use were briefly discussed. Nuclear fusion might also be used to produce electricity, but the technology has not yet been developed.

Radioactive isotopes have been used in a variety of ways. A few of these are the dating of various objects, research into the process of photosynthesis, and the treatment of various types of cancer.

In this chapter and throughout the book, our goal has been to introduce you to what we consider the fascinating science of chemistry and to point out how many people from diverse backgrounds, nationalities, and races have been able to work together to solve *some* of the chemical problems related to our existence on the planet Earth.

EXERCISES

1. Define or explain
   (a) nuclear reactions
   (b) radioactivity
   (c) alpha particles
   (d) beta particles
   (e) gamma rays
   (f) positron particles
   (g) curie
   (h) neutron
   (i) proton
   (j) deuteron
   (k) natural radioactivity
   (l) half-life
   (m) transmutation of an element
   (n) cyclotron
   (o) artificial radioactivity
   (p) nuclear fission
   (q) chain reaction
   (r) subcritical mass
   (s) critical mass
   (t) nuclear fusion

2. Distinguish between
   (a) an alpha particle and a beta particle
   (b) an alpha particle and a gamma ray
   (c) an alpha particle and a positron
   (d) a beta particle and a positron
   (e) a proton and a deuteron
   (f) natural and artificial radioactivity

Nuclear Radiation (See Section 20-1)

3. In your own words, explain the operation of a Geiger-Müller counter.

4. Give the symbols for the following particles or rays:
   (a) alpha
   (b) beta
   (c) gamma
   (d) proton
   (e) neutron
   (f) positron
   (g) deuteron

Natural Radioactivity (See Section 20-3)

5. List the three naturally occurring radioactive disintegration series and give their final nonradioactive isotopes.

6. List one synthetic radioactive disintegration series and give its final nonradioactive isotope.

Transmutation of the Elements (See Section 20-4)

7. In your own words, explain the operation of a cyclotron.

Peaceful Uses of Nuclear Reactions (See Section 20-7)

8. In your own words, list and explain the operation of the three essential parts of a nuclear fission power plant.

PROBLEMS

Nuclear Equations (See Section 20-2)

(Refer to the Periodic Table on the inside front cover of this text to answer these questions.)

9. Write a balanced equation for each of the following nuclear reactions:
   (a) Krypton-87 (Kr) decays by beta emission.
   (b) Curium-240 (Cm) decays by alpha emission.
   (c) Uranium-232 (U) decays by alpha emission.
   (d) Zinc-71 decays by beta emission.
   (e) Praseodymium-140 (Pr) decays by positron emission.
   (f) Oxygen-16 plus a neutron results in the formation of another element and the release of an alpha particle.
   (g) Boron-10 plus a neutron results in the formation of another element and the release of an alpha particle.
   (h) Rhenium-187 (Re) plus a deuteron results in the formation of an isotope of rhenium and the release of a proton.
   (i) Beryllium-9 plus a proton results in the formation of another element and the release of an alpha particle.
   (j) Lutetium-176 (Lu) plus a bombarding particle results in the formation of lutetium-177 and gamma rays.

10. Write a balanced equation for each of the following nuclear reactions:
    (a) Neon-18 decays by positron emission.

(b) Silicon-32 decays by beta emission.

(c) Copper-59 decays by positron emission.

(d) Americium-243 (Am) decays by alpha emission.

(e) Cobalt-60 decays by beta emission.

(f) Einsteinium-253 (Es) plus an alpha particle results in the formation of another element and the release of a neutron.

(g) Cadmium-113 in nuclear reactors absorbs a neutron to form an isotope of cadmium and gamma rays.

(h) Lithium-7 plus a proton results in the formation of another element and the release of a neutron.

(i) Gold-197 plus a deuteron results in the formation of another element and a neutron.

(j) Chlorine-36 is formed in sodium chloride after a nuclear fission explosion. Chlorine-36 decays by beta emission.

11. Nickel-58 is bombarded with a proton and an alpha particle is emitted in the transmutation process. Write the equation for this nuclear reaction.

12. Palladium-108 (Pd) is bombarded with a high-speed alpha particle and a proton is emitted in the transmutation process. Write the equation for this nuclear reaction.

13. Tritium ($^3_1$H), a radioactive isotope of hydrogen, is prepared by bombarding lithium-6 with a neutron. Write the equation for this nuclear reaction.

14. Phosphorus-32 is formed in milk, butter, seafood, baby food, pork, and beans, if the food is exposed to a nuclear fission explosion one-quarter of a mile from ground zero. Phosphorus-32 decays by beta emission. Write the equation for this nuclear reaction.

Half-life (See Section 20-3)

15. Actinium-226 has a half-life of 29.0 hours. If $\overline{100}$ mg of actinium-226 disintegrates over a period of 58.0 hours, how many milligrams of actinium-226 will remain?

16. Thallium-201 has a half-life of 73.0 hours. If 4.00 mg of thallium-201 disintegrates over a single period of 6.00 days and 2.00 hours, how many milligrams of thallium-201 will remain?

17. Sodium-25 was to be used in an experiment, but it took 3.00 minutes to get the sodium from the reactor to the laboratory vessel. If 5.00 mg of sodium-25 was removed from the reactor, how many milligrams of sodium-25 was placed in the reaction vessel if the half-life of sodium-25 is 60.0 seconds?

18. The half-life of a radioactive isotope, X, is 2.00 years. How many years would it take a 4.00-g sample of X to decay and have only 0.500 g of X remain?

19. Selenium-83 has a half-life of 25.0 minutes. How many hours would it take a 10.0-mg sample of selenium-83 to decay and have only 1.25 mg of it remain?

20. Tin-113 is believed to be formed on "tin cans" (used to store food) in a nuclear fission explosion. If 8.00×10^{-6} g of tin-113 is formed on a "tin can" following a nuclear fission explosion, how many days will it take to have just 1.00×10^{-6} g of tin-113, if the half-life of tin-113 is 118 days?

General Problems

21. Following a nuclear fission explosion, the sodium in glass containers formed sodium-24. Sodium-24 decays with emission of a beta particle and has a half-life of 15.0 hours. (a) Write the equation for this nuclear reaction. (b) If 5.00×10^{-6} g of sodium-24 is formed in a glass container, how many grams will remain in 2.50 days? (c) What percent activity remained at the end of this time?

22. Zinc-65 accounts for 50 percent or more of the total radioactivity in fish exposed to a nuclear fission explosion. Zinc-65 decays by positron emission and has a half-life of 245 days. (a) Write the equation for this nuclear reaction. (b) If 1.00 μg of zinc-65 is formed in a fish, how many days will it take to have just 0.250 μg of zinc-65? Note that the 0.250 μg of zinc-65 remains if the fish lives or dies, but if it is caught and eaten by people, the zinc-65 resides in the people who ate the fish.

23. A 4.00×10^{-6} percent solution of sodium-24 chloride is prepared at 9:00 A.M. on Tuesday. What is the percent of sodium-24 chloride at 3:00 P.M. on Wednesday, if sodium-24 has a half-life of 15.0 hours?

24. A 1.50×10^{-6} M solution of cesium-137 chloride was prepared by a research chemist in 1955. What is the molar concentration of cesium-137 chloride in 1985? (The half-life of cesium-137 is 30.0 years.)

Readings

Bethe, H. A., "The Necessity of Fission Power." *Sci. Am.*, Jan. 1976, v. 234, p. 21. Discusses some of the problems encountered in nuclear fission power plants, such as the possibilities of accidents, storage of waste, and plant security.

Cowan, George A., "A Natural Fission Reactor." *Sci. Am.*, July 1976, v. 235, p. 36. Describes a natural reactor in West Africa that became critical and then shut down; this occurred about 2 billion years ago.

Emmett, John L., John Nuckolls, and Lowell Wood, "Fusion Power by Laser Implosion," *Sci. Am.*, June 1974, v. 230, p. 24. Discuss the history of laser development and the use of lasers in nuclear fusion power plants.

Furth, Harold P., "Progress Toward a Tokamak Fusion Reactor." *Sci. Am.*, Aug. 1979, v. 241, p. 50. Recent reports show that the possibility of developing a fusion reactor should be known in a few years.

Hahn, Otto, *Otto Hahn: A Scientific Autobiography*. New York: Charles Scribner's Sons, 1966. An autobiography of Otto Hahn, considered by some to be the father of nuclear chemistry. Discusses some of the problems encountered by the scientific community in Germany during the Nazi regime.

Kaku, Michio, and Jennifer Trainer, *Nuclear Power: Both Sides*. New York: W. W. Norton & Company, Inc., 1982. This book is a collection of essays by prominent scientists and concerned citizens, presenting *both* sides of the nuclear power issue. It is written for the lay person. You will find it very interesting and informative.

Appendices

SI Units and Some Conversion Factors

The International System of Units (SI) has been introduced to you in this text; this system will probably be used throughout the scientific community within a few years. The SI is derived from seven *base units* and these are shown in the following table.

Base Units

Quantity Measured	Unit Name	SI Symbol for Unit
Length	meter	m
Mass	kilogram	kg
Time	second	s
Electric current	ampere	A
Thermodynamic temperature	kelvin	K
Amount of substance	mole	mol
Luminous intensity	candela	cd

In addition to these base units, there are two *supplementary units*, the radian (rad) and steradian (sr), used to define angular measurement.

Multiple and submultiple *prefixes* are used to indicate orders of magnitude. These prefixes define either a fractional or multiple value of the base unit; thus, a kilometer is 1000 meters, and a millimeter is 0.001 (or 10^{-3}) meter. These prefixes and their orders of magnitude are shown in the following table.

SI Prefixes

Factor	Prefix	SI Symbol
10^{12}	tera	T
10^{9}	giga	G
10^{6}	mega	M
10^{3}	kilo	k
10^{2}	hecto[a]	h
10^{1}	deka[a]	da
10^{-1}	deci[a]	d
10^{-2}	centi[a]	c
10^{-3}	milli	m
10^{-6}	micro	μ
10^{-9}	nano	n
10^{-12}	pico	p
10^{-15}	femto	f
10^{-18}	atto	a

[a]These prefixes are to be avoided where possible.

A series of *derived units* which define various physical quantities used in scientific measurements are derived from the seven base and two supplementary units. Some of these are shown in the following table.

Derived Units

Physical Quantity	Unit	SI Symbol	Definition
Acceleration	meter/second^2	m/s^2	m/s^2
Area	meter^2	m^2	m^2
Density	kilogram/meter^3	kg/m^3	kg/m^3
Electric capacitance	farad	F	A·s/V
Electric potential difference	volt[a]	V	J/A·s = W/A
Electric resistance	ohm	Ω	V/A = kg·m^2/s^3·A^2
Energy	joule	J	N·m = kg·m^2/s^2
Force	newton	N	kg·m/s^2
Power	watt	W	J/s = kg·m^2/s^3
Quantity of electricity	coulomb	C	A·s
Pressure	pascal	Pa	N/m^2 = kg/m·s^2
Quantity of heat	joule	J	N·m = kg·m^2/s^2
Specific heat	joule/kilogram·kelvin	J/kg·K	J/kg·K
Velocity	meter/second	m/s	m/s
Volume[b]	cubic meter	m^3	m^3

[a]Also the unit for expressing electromotive force.

[b]In 1964, the liter was adopted as a special name for the cubic decimeter (dm^3), but its use for measurement of extremely precise volumes is discouraged.

Units currently being used in chemical measurements (and which have been used in this text) that are not *exactly* defined in terms of SI units are the *atmosphere, torr, mm Hg,* and *calorie*. It is recommended that these units be abandoned; however, it is likely that their use by many chemists will continue for some time.

The following table lists conversion factors which may be used to convert from the non-SI unit to the recognized SI unit.

Selected Conversion Factors

To Convert From	To	Multiply By
calorie (cal.)	joule (J)	4.184
atmosphere (atm)	pascal (Pa)	1.013×10^5
torr (torr)	pascal (Pa)	1.333×10^2
inch (in.)	meter (m)	2.54×10^{-2}
pound-mass (lbm)	kilogram (kg)	4.536×10^{-1}

The English-Based Units. Conversion from the Metric System to the English-Based Units, and Vice Versa

The English-based units are also used in the measurement of matter. These units are used primarily in the United States with the metric system used throughout almost all of the rest of the world, even in Great Britain. Gradually, these English-based units are being replaced in the United States with metric measurements.

II-1

The English-Based Units

The English-based units consist of the ounce, pound, and ton (mass); the fluid ounce, pint, quart, and gallon (volume); the inch, foot, yard, and mile (length); and the second (time). Table II-1 summarizes the English-based units of measurements.

TABLE II-1 Summary of the English-Based Units of Measurements

Mass		Length		Volume	
16 ounces (oz)	= 1 pound (lb)	12 inches (in.)	= 1 foot (ft)	16 fluid ounces (fl oz)	= 1 pint (pt)
2000 pounds	= 1 ton	3 feet	= 1 yard (yd)	2 pints	= 1 quart (qt)
		5280 feet	= 1 mile (mi)	4 quarts	= 1 gallon (gal)

II-2

Conversion from the Metric to the English-Based Units, and Vice Versa

In order to convert from the metric system to the English-based units and vice versa, we need to know certain conversion factors. You must either memorize these factors or have them given to you to solve problems involving these units.

TABLE II-2 Metric-English Unit Equivalents

Dimension	Metric Unit	English Equivalent
Mass[a]	454 grams (g)	= 1 pound (lb)
Volume	1 liter (L)	= 1.06 quarts (qt)
Length	2.54 centimeters (cm)	= 1 inch (in.)
Time	1 second (s)	= 1 second (sec)

[a]Note: Although gram is a unit of mass and pound is a unit of weight, the two units are used interchangeably in most calculations in chemistry.

Problem Example II-1

Convert 385 g to pounds.

SOLUTION: Using the conversion factor 454 g = 1 lb, the problem is solved as follows:

$$385 \text{ g} \times \frac{1 \text{ lb}}{454 \text{ g}} = 0.848 \text{ lb} \quad \textit{Answer}$$

Problem Example II-2

Convert 1.00 gal to liters.

SOLUTION: To use the conversion factor 1 L = 1.06 qt, the gallon must be converted to quarts, using 4 qt = 1 gal. The solution is as follows:

$$1.00 \text{ gal} \times \frac{4 \text{ qt}}{1 \text{ gal}} \times \frac{1 \text{ L}}{1.06 \text{ qt}} = 3.77 \text{ L} \quad \textit{Answer}$$

Problem Example II-3

Convert 3.00 ft to meters.

SOLUTION: To use the factor 2.54 cm = 1 in., the feet must be converted to inches using 12 in. = 1 ft. The centimeters must then be converted to meters, using the metric conversion factor 100 cm = 1 m (see 2-8). The solution is as follows:

$$3.00 \; \cancel{ft} \times \frac{12 \; \cancel{in.}}{1 \; \cancel{ft}} \times \frac{2.54 \; \cancel{cm}}{1 \; \cancel{in.}} \times \frac{1 \; m}{100 \; \cancel{cm}} = 0.914 \; m \quad Answer$$

Problem Example II-4

A box has the following dimensions: 15.0 in., 20.0 cm, and 2.00 ft. Calculate its volume in (a) cubic centimeters and (b) cubic inches.

SOLUTION

(a) Convert the 15.0 in. and 2.00 ft to centimeters and then calculate the volume in cubic centimeters as side × side × side:

$$15.0 \; \cancel{in.} \times \frac{2.54 \; cm}{1 \; \cancel{in.}} \times 20.0 \; cm \times 2.00 \; \cancel{ft} \times \frac{12 \; \cancel{in.}}{1 \; \cancel{ft}} \times \frac{2.54 \; cm}{1 \; \cancel{in.}}$$

$$= 46,500 \; cm^3, \text{ to three significant digits} \quad Answer$$

Note that $cm^1 \times cm^1 \times cm^1 = cm^3$, with the exponents being added algebraically as in multiplication of exponential numbers, (see 2-3).

(b) Convert the 20.0 cm and 2.00 ft to inches and then calculate the volume in cubic inches:

$$15.0 \; in. \times 20.0 \; \cancel{cm} \times \frac{1 \; in.}{2.54 \; \cancel{cm}} \times 2.00 \; \cancel{ft} \times \frac{12 \; in.}{1 \; \cancel{ft}} = 2,830 \; in.^3,$$

to three significant digits   *Answer*

Problem Example II-5

Carry out the following conversions:

(a) a density of 2.60 g/mL to lb/gal
(b) a density of 5.00 g/cm³ to lb/ft³

SOLUTION

(a) The mass in grams must be converted to mass in pounds using the factor 1 lb = 454 g and the volume in milliliters must be converted to volume in gallons using the factors: 1000 mL = 1 L, 1 L = 1.06 qt, and 4 qt = 1 gal. The solution is as follows:

$$\frac{2.60 \; \cancel{g}}{\cancel{mL}} \times \frac{1 \; lb}{454 \; \cancel{g}} \times \frac{1000 \; \cancel{mL}}{1 \; \cancel{L}} \times \frac{1 \; \cancel{L}}{1.06 \; \cancel{qt}} \times \frac{1 \; \cancel{qt}}{1 \; gal} = 21.6 \; lb/gal \quad Answer$$

(b) The mass in grams must be converted to mass in pounds using the factor 1 lb = 454 g and the volume in cm³ must be converted to volume in ft³, but this involves $(1 \; in.)^3 = (2.54 \; cm)^3$ and then $(12 \; in.)^3 = (1 \; ft)^3$. To perform this operation the

2.54 is multiplied by itself three times, that is, $2.54 \times 2.54 \times 2.54$, and the same is done for the 12, that is, $12 \times 12 \times 12$ (see positive powers of exponential numbers, 2-3). The solution is as follows:

$$\frac{5.00\,g}{cm^3} \times \frac{1\ lb}{454\,g} \times \frac{(2.54)^3\ cm^3}{1\ in.^3} \times \frac{(12)^3\ in.^3}{1\ ft^3} = 312\ lb/ft^3 \quad Answer$$

Problems

1. Carry out each of the following conversions:
   (a)   2.00 kg to pounds
   (b)   3.25 lb to kilograms
   (c)   250 mL to pints
   (d)   6.00 qt to liters
   (e)   16.8 in. to meters
   (f)   6.00 m to feet

2. Mt. Everest on the Nepal-Tibet border is 29,028 ft above sea level. What is this height in meters to three significant digits?

3. A box has the following dimensions: 25.0 cm, 11.0 in., and 2.00 ft. Calculate its volume in (a) cubic centimeters and (b) cubic inches.

4. Carry out the following conversions:
   (a) a density of 3.65 g/mL to lb/gal
   (b) a density of 295 lb/ft^3 to g/mL

5. Calculate the density in lb/ft^3 of a piece of metal having a mass of 246 kg and the following dimensions: 15.0 cm, 30.0 in., and 0.500 ft.

Some Naturally Occurring Isotopes

Atomic Number	Isotope	Percent Abundance	Atomic Number	Isotope	Percent Abundance	Atomic Number	Isotope	Percent Abundance
1	$^{1}_{1}\text{H}$	99.98		$^{30}_{14}\text{Si}$	3.09		$^{51}_{23}\text{V}$	99.76
	$^{2}_{1}\text{H}$	0.02	15	$^{31}_{15}\text{P}$	100.00	24	$^{50}_{24}\text{Cr}$	4.31
2	$^{3}_{2}\text{He}$	Trace	16	$^{32}_{16}\text{S}$	95.00		$^{52}_{24}\text{Cr}$	83.76
	$^{4}_{2}\text{He}$	100.00		$^{33}_{16}\text{S}$	0.76		$^{53}_{24}\text{Cr}$	9.55
3	$^{6}_{3}\text{Li}$	7.42		$^{34}_{16}\text{S}$	4.22		$^{54}_{24}\text{Cr}$	2.38
	$^{7}_{3}\text{Li}$	92.58		$^{36}_{16}\text{S}$	0.01	25	$^{55}_{25}\text{Mn}$	100.00
4	$^{9}_{4}\text{Be}$	100.00	17	$^{35}_{17}\text{Cl}$	75.53	26	$^{54}_{26}\text{Fe}$	5.82
5	$^{10}_{5}\text{B}$	19.6		$^{37}_{17}\text{Cl}$	24.47		$^{56}_{26}\text{Fe}$	91.66
	$^{11}_{5}\text{B}$	80.4	18	$^{36}_{18}\text{Ar}$	0.34		$^{57}_{26}\text{Fe}$	2.19
6	$^{12}_{6}\text{C}$	98.89		$^{38}_{18}\text{Ar}$	0.06		$^{58}_{26}\text{Fe}$	0.33
	$^{13}_{6}\text{C}$	1.11		$^{40}_{18}\text{Ar}$	99.60	27	$^{59}_{27}\text{Co}$	100.00
7	$^{14}_{7}\text{N}$	99.63	19	$^{39}_{19}\text{K}$	93.10	28	$^{58}_{28}\text{Ni}$	67.88
	$^{15}_{7}\text{N}$	0.37		$^{40}_{19}\text{K}$	0.01		$^{60}_{28}\text{Ni}$	26.23
8	$^{16}_{8}\text{O}$	99.76		$^{41}_{19}\text{K}$	6.88		$^{61}_{28}\text{Ni}$	1.19
	$^{17}_{8}\text{O}$	0.04	20	$^{40}_{20}\text{Ca}$	96.97		$^{62}_{28}\text{Ni}$	3.66
	$^{18}_{8}\text{O}$	0.20		$^{42}_{20}\text{Ca}$	0.64		$^{64}_{28}\text{Ni}$	1.08
9	$^{19}_{9}\text{F}$	100.00		$^{43}_{20}\text{Ca}$	0.14	29	$^{63}_{29}\text{Cu}$	69.09
10	$^{20}_{10}\text{Ne}$	90.92		$^{44}_{20}\text{Ca}$	2.06		$^{65}_{29}\text{Cu}$	30.91
	$^{21}_{10}\text{Ne}$	0.26		$^{46}_{20}\text{Ca}$	Trace	30	$^{64}_{30}\text{Zn}$	48.89
	$^{22}_{10}\text{Ne}$	8.82		$^{48}_{20}\text{Ca}$	0.18		$^{66}_{30}\text{Zn}$	27.81
11	$^{23}_{11}\text{Na}$	100.00	21	$^{45}_{21}\text{Sc}$	100.00		$^{67}_{30}\text{Zn}$	4.11
12	$^{24}_{12}\text{Mg}$	78.70	22	$^{46}_{22}\text{Ti}$	7.93		$^{68}_{30}\text{Zn}$	18.57
	$^{25}_{12}\text{Mg}$	10.13		$^{47}_{22}\text{Ti}$	7.28		$^{70}_{30}\text{Zn}$	0.62
	$^{26}_{12}\text{Mg}$	11.17		$^{48}_{22}\text{Ti}$	73.94	31	$^{69}_{31}\text{Ga}$	60.4
13	$^{27}_{13}\text{Al}$	100.00		$^{49}_{22}\text{Ti}$	5.51		$^{71}_{31}\text{Ga}$	39.6
14	$^{28}_{14}\text{Si}$	92.21		$^{50}_{22}\text{Ti}$	5.34	32	$^{70}_{32}\text{Ge}$	20.52
	$^{29}_{14}\text{Si}$	4.70	23	$^{50}_{23}\text{V}$	0.24		$^{72}_{32}\text{Ge}$	27.43

Continued

Atomic Number	Isotope	Percent Abundance	Atomic Number	Isotope	Percent Abundance	Atomic Number	Isotope	Percent Abundance
	$^{73}_{32}\text{Ge}$	7.76	46	$^{102}_{46}\text{Pd}$	0.96		$^{134}_{54}\text{Xe}$	10.44
	$^{74}_{32}\text{Ge}$	36.54		$^{104}_{46}\text{Pd}$	10.97		$^{136}_{54}\text{Xe}$	8.87
	$^{76}_{32}\text{Ge}$	7.76		$^{105}_{46}\text{Pd}$	22.23	55	$^{133}_{55}\text{Cs}$	100.00
33	$^{75}_{33}\text{As}$	100.00		$^{106}_{46}\text{Pd}$	27.33	56	$^{130}_{56}\text{Ba}$	0.10
34	$^{74}_{34}\text{Se}$	0.87		$^{108}_{46}\text{Pd}$	26.71		$^{132}_{56}\text{Ba}$	0.10
	$^{76}_{34}\text{Se}$	9.02		$^{110}_{46}\text{Pd}$	11.81		$^{134}_{56}\text{Ba}$	2.42
	$^{77}_{34}\text{Se}$	7.58	47	$^{107}_{47}\text{Ag}$	51.82		$^{135}_{56}\text{Ba}$	6.59
	$^{78}_{34}\text{Se}$	23.52		$^{109}_{47}\text{Ag}$	48.18		$^{136}_{56}\text{Ba}$	7.81
	$^{80}_{34}\text{Se}$	49.82	48	$^{108}_{48}\text{Cd}$	1.22		$^{137}_{56}\text{Ba}$	11.32
	$^{82}_{34}\text{Se}$	9.19		$^{108}_{48}\text{Cd}$	0.88		$^{138}_{56}\text{Ba}$	71.66
35	$^{79}_{35}\text{Br}$	50.54		$^{110}_{48}\text{Cd}$	12.39	57	$^{138}_{57}\text{La}$	0.09
	$^{81}_{35}\text{Br}$	49.46		$^{111}_{48}\text{Cd}$	12.75		$^{139}_{57}\text{La}$	99.91
36	$^{78}_{36}\text{Kr}$	0.35		$^{112}_{48}\text{Cd}$	24.07	58	$^{136}_{58}\text{Ce}$	0.19
	$^{80}_{36}\text{Kr}$	2.27		$^{113}_{48}\text{Cd}$	12.26		$^{138}_{58}\text{Ce}$	0.25
	$^{82}_{36}\text{Kr}$	11.56		$^{114}_{48}\text{Cd}$	28.86		$^{140}_{58}\text{Ce}$	88.48
	$^{83}_{36}\text{Kr}$	11.55		$^{116}_{48}\text{Cd}$	7.58		$^{142}_{58}\text{Ce}$	11.07
	$^{84}_{36}\text{Kr}$	56.90	49	$^{113}_{49}\text{In}$	4.28	59	$^{141}_{59}\text{Pr}$	100.00
	$^{86}_{36}\text{Kr}$	17.37		$^{115}_{49}\text{In}$	95.72	60	$^{142}_{60}\text{Nd}$	27.11
37	$^{85}_{37}\text{Rb}$	72.15	50	$^{112}_{50}\text{Sn}$	0.96		$^{143}_{60}\text{Nd}$	12.17
	$^{87}_{37}\text{Rb}$	27.85		$^{114}_{50}\text{Sn}$	0.66		$^{144}_{60}\text{Nd}$	23.85
38	$^{84}_{38}\text{Sr}$	0.56		$^{115}_{50}\text{Sn}$	0.35		$^{145}_{60}\text{Nd}$	8.30
	$^{86}_{38}\text{Sr}$	9.86		$^{116}_{50}\text{Sn}$	14.30		$^{146}_{60}\text{Nd}$	17.22
	$^{87}_{38}\text{Sr}$	7.02		$^{117}_{50}\text{Sn}$	7.61		$^{148}_{60}\text{Nd}$	5.73
	$^{88}_{38}\text{Sr}$	82.56		$^{118}_{50}\text{Sn}$	24.03		$^{150}_{60}\text{Nd}$	5.62
39	$^{89}_{39}\text{Y}$	100.00		$^{119}_{50}\text{Sn}$	8.58	62	$^{144}_{62}\text{Sm}$	3.09
40	$^{90}_{40}\text{Zr}$	51.46		$^{120}_{50}\text{Sn}$	32.85		$^{147}_{62}\text{Sm}$	14.97
	$^{91}_{40}\text{Zr}$	11.23		$^{122}_{50}\text{Sn}$	4.92		$^{148}_{62}\text{Sm}$	11.24
	$^{92}_{40}\text{Zr}$	17.11		$^{124}_{50}\text{Sn}$	5.94		$^{149}_{62}\text{Sm}$	13.83
	$^{94}_{40}\text{Zr}$	17.40	51	$^{121}_{51}\text{Sb}$	57.25		$^{150}_{62}\text{Sm}$	7.44
	$^{96}_{40}\text{Zr}$	2.80		$^{123}_{51}\text{Sb}$	42.75		$^{152}_{62}\text{Sm}$	26.72
41	$^{93}_{41}\text{Nb}$	100.00	52	$^{120}_{52}\text{Te}$	0.09		$^{154}_{62}\text{Sm}$	22.71
42	$^{92}_{42}\text{Mo}$	15.84		$^{122}_{52}\text{Te}$	2.46	63	$^{151}_{63}\text{Eu}$	47.82
	$^{94}_{42}\text{Mo}$	9.04		$^{123}_{52}\text{Te}$	0.87		$^{153}_{63}\text{Eu}$	52.18
	$^{95}_{42}\text{Mo}$	15.72		$^{124}_{52}\text{Te}$	4.61	64	$^{152}_{64}\text{Gd}$	0.20
	$^{96}_{42}\text{Mo}$	16.53		$^{125}_{52}\text{Te}$	6.99		$^{154}_{64}\text{Gd}$	2.15
	$^{97}_{42}\text{Mo}$	9.46		$^{126}_{52}\text{Te}$	18.71		$^{155}_{64}\text{Gd}$	14.73
	$^{98}_{42}\text{Mo}$	23.78		$^{128}_{52}\text{Te}$	31.79		$^{156}_{64}\text{Gd}$	20.47
	$^{100}_{42}\text{Mo}$	9.13		$^{130}_{52}\text{Te}$	34.48		$^{157}_{64}\text{Gd}$	15.68
44	$^{96}_{44}\text{Ru}$	5.51	53	$^{127}_{53}\text{I}$	100.00		$^{158}_{64}\text{Gd}$	24.87
	$^{98}_{44}\text{Ru}$	1.87	54	$^{124}_{54}\text{Xe}$	0.10		$^{160}_{64}\text{Gd}$	21.90
	$^{99}_{44}\text{Ru}$	12.72		$^{126}_{54}\text{Xe}$	0.09	65	$^{159}_{65}\text{Tb}$	100.00
	$^{100}_{44}\text{Ru}$	12.62		$^{128}_{54}\text{Xe}$	1.92	66	$^{156}_{66}\text{Dy}$	0.05
	$^{101}_{44}\text{Ru}$	17.07		$^{129}_{54}\text{Xe}$	26.44		$^{158}_{66}\text{Dy}$	0.09
	$^{102}_{44}\text{Ru}$	31.61		$^{130}_{54}\text{Xe}$	4.08		$^{160}_{66}\text{Dy}$	2.29
	$^{104}_{44}\text{Ru}$	18.58		$^{131}_{54}\text{Xe}$	21.18		$^{161}_{66}\text{Dy}$	18.88
45	$^{103}_{45}\text{Rh}$	100.00		$^{132}_{54}\text{Xe}$	26.89		$^{162}_{66}\text{Dy}$	25.53

Atomic Number	Isotope	Percent Abundance	Atomic Number	Isotope	Percent Abundance	Atomic Number	Isotope	Percent Abundance
	$^{163}_{66}Dy$	24.97		$^{179}_{72}Hf$	13.75		$^{194}_{78}Pt$	32.90
	$^{164}_{66}Dy$	28.18		$^{180}_{72}Hf$	35.24		$^{195}_{78}Pt$	33.80
67	$^{165}_{67}Ho$	100.00	73	$^{180}_{73}Ta$	0.01		$^{196}_{78}Pt$	25.30
68	$^{162}_{68}Er$	0.14		$^{181}_{73}Ta$	99.99		$^{198}_{78}Pt$	7.21
	$^{164}_{68}Er$	1.56	74	$^{180}_{74}W$	0.14	79	$^{197}_{79}Au$	100.00
	$^{166}_{68}Er$	33.41		$^{182}_{74}W$	26.41	80	$^{196}_{80}Hg$	0.15
	$^{167}_{68}Er$	22.94		$^{183}_{74}W$	14.40		$^{198}_{80}Hg$	10.02
	$^{168}_{68}Er$	27.07		$^{184}_{74}W$	30.64		$^{199}_{80}Hg$	16.84
	$^{170}_{68}Er$	14.88		$^{186}_{74}W$	28.41		$^{200}_{80}Hg$	23.13
69	$^{169}_{69}Tm$	100.00	75	$^{185}_{75}Re$	37.07		$^{201}_{80}Hg$	13.22
70	$^{168}_{70}Yb$	0.14		$^{187}_{75}Re$	62.93		$^{202}_{80}Hg$	29.80
	$^{170}_{70}Yb$	3.03	76	$^{184}_{76}Os$	0.02		$^{204}_{80}Hg$	6.85
	$^{171}_{70}Yb$	14.31		$^{186}_{76}Os$	1.59	81	$^{203}_{81}Tl$	29.50
	$^{172}_{70}Yb$	21.82		$^{187}_{76}Os$	1.64		$^{205}_{81}Tl$	70.50
	$^{173}_{70}Yb$	16.13		$^{188}_{76}Os$	13.30	82	$^{204}_{82}Pb$	1.48
	$^{174}_{70}Yb$	31.84		$^{189}_{76}Os$	16.10		$^{206}_{82}Pb$	23.60
	$^{176}_{70}Yb$	12.73		$^{190}_{76}Os$	26.40		$^{207}_{82}Pb$	22.60
71	$^{175}_{71}Lu$	97.41		$^{192}_{76}Os$	41.00		$^{208}_{82}Pb$	52.30
	$^{176}_{71}Lu$	2.59	77	$^{191}_{77}Ir$	37.3	83	$^{209}_{83}Bi$	100.00
72	$^{174}_{72}Hf$	0.18		$^{193}_{77}Ir$	62.7	92	$^{234}_{92}U$	0.01
	$^{176}_{72}Hf$	5.20	78	$^{190}_{78}Pt$	0.01		$^{235}_{92}U$	0.72
	$^{177}_{72}Hf$	18.50		$^{192}_{78}Pt$	0.78		$^{238}_{92}U$	99.27
	$^{178}_{72}Hf$	27.14						

Electronic Configuration of the Elements Showing Sublevels

Atomic Number	Element Symbol	1 s	2 s p	3 s p d	4 s p d f	5 s p d f	6 s p d f	7 s
1	H	1						
2	He	2						
3	Li	2	1					
4	Be	2	2					
5	B	2	2 1					
6	C	2	2 2					
7	N	2	2 3					
8	O	2	2 4					
9	F	2	2 5					
10	Ne	2	2 6					
11	Na	2	2 6	1				
12	Mg	2	2 6	2				
13	Al	2	2 6	2 1				
14	Si	2	2 6	2 2				
15	P	2	2 6	2 3				
16	S	2	2 6	2 4				
17	Cl	2	2 6	2 5				
18	Ar	2	2 6	2 6				
19	K	2	2 6	2 6	1			
20	Ca	2	2 6	2 6	2			
21	Sc	2	2 6	2 6 1	2			
22	Ti	2	2 6	2 6 2	2			
23	V	2	2 6	2 6 3	2			
24	Cr	2	2 6	2 6 5	1			
25	Mn	2	2 6	2 6 5	2			
26	Fe	2	2 6	2 6 6	2			
27	Co	2	2 6	2 6 7	2			
28	Ni	2	2 6	2 6 8	2			
29	Cu	2	2 6	2 6 10	1			

Continued

Atomic Number	Element Symbol	1 s	2 s p	3 s p d	4 s p d f	5 s p d f	6 s p d f	7 s
30	Zn	2	2 6	2 6 10	2			
31	Ga	2	2 6	2 6 10	2 1			
32	Ge	2	2 6	2 6 10	2 2			
33	As	2	2 6	2 6 10	2 3			
34	Se	2	2 6	2 6 10	2 4			
35	Br	2	2 6	2 6 10	2 5			
36	Kr	2	2 6	2 6 10	2 6			
37	Rb	2	2 6	2 6 10	2 6	1		
38	Sr	2	2 6	2 6 10	2 6	2		
39	Y	2	2 6	2 6 10	2 6 1	2		
40	Zr	2	2 6	2 6 10	2 6 2	2		
41	Nb	2	2 6	2 6 10	2 6 4	1		
42	Mo	2	2 6	2 6 10	2 6 5	1		
43	Tc	2	2 6	2 6 10	2 6 6	1		
44	Ru	2	2 6	2 6 10	2 6 7	1		
45	Rh	2	2 6	2 6 10	2 6 8	1		
46	Pd	2	2 6	2 6 10	2 6 10			
47	Ag	2	2 6	2 6 10	2 6 10	1		
48	Cd	2	2 6	2 6 10	2 6 10	2		
49	In	2	2 6	2 6 10	2 6 10	2 1		
50	Sn	2	2 6	2 6 10	2 6 10	2 2		
51	Sb	2	2 6	2 6 10	2 6 10	2 3		
52	Te	2	2 6	2 6 10	2 6 10	2 4		
53	I	2	2 6	2 6 10	2 6 10	2 5		
54	Xe	2	2 6	2 6 10	2 6 10	2 6		
55	Cs	2	2 6	2 6 10	2 6 10	2 6	1	
56	Ba	2	2 6	2 6 10	2 6 10	2 6	2	
57	La	2	2 6	2 6 10	2 6 10	2 6 1	2	
58	Ce	2	2 6	2 6 10	2 6 10 1	2 6 1	2	
59	Pr	2	2 6	2 6 10	2 6 10 3	2 6	2	
60	Nd	2	2 6	2 6 10	2 6 10 4	2 6	2	
61	Pm	2	2 6	2 6 10	2 6 10 5	2 6	2	
62	Sm	2	2 6	2 6 10	2 6 10 6	2 6	2	
63	Eu	2	2 6	2 6 10	2 6 10 7	2 6	2	
64	Gd	2	2 6	2 6 10	2 6 10 7	2 6 1	2	
65	Tb	2	2 6	2 6 10	2 6 10 9	2 6	2	
66	Dy	2	2 6	2 6 10	2 6 10 10	2 6	2	
67	Ho	2	2 6	2 6 10	2 6 10 11	2 6	2	
68	Er	2	2 6	2 6 10	2 5 10 12	2 6	2	
69	Tm	2	2 6	2 6 10	2 6 10 13	2 6	2	
70	Yb	2	2 6	2 6 10	2 6 10 14	2 6	2	
71	Lu	2	2 6	2 6 10	2 6 10 14	2 6 1	2	
72	Hf	2	2 6	2 6 10	2 6 10 14	2 6 2	2	
73	Ta	2	2 6	2 6 10	2 6 10 14	2 6 3	2	
74	W	2	2 6	2 6 10	2 6 10 14	2 6 4	2	
75	Re	2	2 6	2 6 10	2 6 10 14	2 6 5	2	
76	Os	2	2 6	2 6 10	2 6 10 14	2 6 6	2	
77	Ir	2	2 6	2 6 10	2 6 10 14	2 6 7	2	
78	Pt	2	2 6	2 6 10	2 6 10 14	2 6 9	1	
79	Au	2	2 6	2 6 10	2 6 10 14	2 6 10	1	
80	Hg	2	2 6	2 6 10	2 6 10 14	2 6 10	2	

Continued

Atomic Number	Element Symbol	1 s	2 s	2 p	3 s	3 p	3 d	4 s	4 p	4 d	4 f	5 s	5 p	5 d	5 f	6 s	6 p	6 d	6 f	7 s
81	Tl	2	2	6	2	6	10	2	6	10	14	2	6	10		2	1			
82	Pb	2	2	6	2	6	10	2	6	10	14	2	6	10		2	2			
83	Bi	2	2	6	2	6	10	2	6	10	14	2	6	10		2	3			
84	Po	2	2	6	2	6	10	2	6	10	14	2	6	10		2	4			
85	At	2	2	6	2	6	10	2	6	10	14	2	6	10		2	5			
86	Rn	2	2	6	2	6	10	2	6	10	14	2	6	10		2	6			
87	Fr	2	2	6	2	6	10	2	6	10	14	2	6	10		2	6			1
88	Ra	2	2	6	2	6	10	2	6	10	14	2	6	10		2	6			2
89	Ac	2	2	6	2	6	10	2	6	10	14	2	6	10		2	6	1		2
90	Th	2	2	6	2	6	10	2	6	10	14	2	6	10		2	6	2		2
91	Pa	2	2	6	2	6	10	2	6	10	14	2	6	10	2	2	6	1		2
92	U	2	2	6	2	6	10	2	6	10	14	2	6	10	3	2	6	1		2
93	Np	2	2	6	2	6	10	2	6	10	14	2	6	10	4	2	6	1		2
94	Pu	2	2	6	2	6	10	2	6	10	14	2	6	10	6	2	6			2
95	Am	2	2	6	2	6	10	2	6	10	14	2	6	10	7	2	6			2
96	Cm	2	2	6	2	6	10	2	6	10	14	2	6	10	7	2	6	1		2
97	Bk	2	2	6	2	6	10	2	6	10	14	2	6	10	9	2	6			2
98	Cf	2	2	6	2	6	10	2	6	10	14	2	6	10	10	2	6			2
99	Es	2	2	6	2	6	10	2	6	10	14	2	6	10	11	2	6			2
100	Fm	2	2	6	2	6	10	2	6	10	14	2	6	10	12	2	6			2
101	Md	2	2	6	2	6	10	2	6	10	14	2	6	10	13	2	6			2
102	No	2	2	6	2	6	10	2	6	10	14	2	6	10	14	2	6			2
103	Lr	2	2	6	2	6	10	2	6	10	14	2	6	10	14	2	6	1		2
104	[Rf]a	2	2	6	2	6	10	2	6	10	14	2	6	10	14	2	6	2		2
105	[Ha]a	2	2	6	2	6	10	2	6	10	14	2	6	10	14	2	6	3		2
106	[ ]a	2	2	6	2	6	10	2	6	10	14	2	6	10	14	2	6	4		2
107	[ ]a	2	2	6	2	6	10	2	6	10	14	2	6	10	14	2	6	5		2
109	[ ]a	2	2	6	2	6	10	2	6	10	14	2	6	10	14	2	6	7		2

aName and symbol not officially approved.

V

Vapor Pressure of Water at Various Temperatures

Temperature	Pressure[a]		
(°C)	torr	atm	Pa
0	4.6	0.0061	610
5	6.5	0.0086	872
10	9.2	0.0121	1227
15	12.8	0.0168	1705
16	13.6	0.0179	1818
17	14.5	0.0191	1937
18	15.5	0.0204	2063
19	16.5	0.0217	2197
20	17.5	0.0231	2338
21	18.6	0.0245	2486
22	19.8	0.0261	2643
23	21.1	0.0277	2809
24	22.4	0.0294	2983
25	23.8	0.0313	3167
26	25.2	0.0332	3360
27	26.7	0.0352	3564
28	28.3	0.0373	3779
29	30.0	0.0395	4005
30	31.8	0.0419	4242
31	33.7	0.0443	4492
32	35.7	0.0469	4755
33	37.7	0.0496	5030
34	39.9	0.0525	5319

[a]Units in torr to nearest tenth; atm to nearest ten-thousandth; Pa to nearest unit.

Temperature	Pressure[a]		
(°C)	torr	atm	Pa
35	42.2	0.0555	5622
36	44.6	0.0586	5941
37	47.1	0.0619	6275
38	49.7	0.0654	6625
39	52.4	0.0690	6992
40	55.3	0.0728	7376
45	71.9	0.0946	9583
50	92.5	0.1217	12,333
55	118.0	0.1553	15,737
60	149.4	0.1965	19,915
65	187.5	0.2468	25,002
70	233.7	0.3075	31,157
75	289.1	0.3804	38,543
80	355.1	0.4672	47,342
85	433.6	0.5705	57,808
90	525.8	0.6918	70,094
95	633.9	0.8341	84,512
100	760.0	1.0000	101,325

Logarithms of Numbers

N	0	1	2	3	4	5	6	7	8	9
1.0	.0000	.0043	.0086	.0128	.0170	.0212	.0253	.0294	.0334	.0374
1.1	.0414	.0453	.0492	.0531	.0569	.0607	.0645	.0682	.0719	.0755
1.2	.0792	.0828	.0864	.0899	.0934	.0969	.1004	.1038	.1072	.1106
1.3	.1139	.1173	.1206	.1239	.1271	.1303	.1335	.1367	.1399	.1430
1.4	.1461	.1492	.1523	.1553	.1584	.1614	.1644	.1673	.1703	.1732
1.5	.1761	.1790	.1818	.1847	.1875	.1903	.1931	.1959	.1987	.2014
1.6	.2041	.2068	.2095	.2122	.2148	.2175	.2201	.2227	.2253	.2279
1.7	.2304	.2330	.2355	.2380	.2405	.2430	.2455	.2480	.2504	.2529
1.8	.2553	.2577	.2601	.2625	.2648	.2672	.2695	.2718	.2742	.2765
1.9	.2788	.2810	.2833	.2856	.2878	.2900	.2923	.2945	.2967	.2989
2.0	.3010	.3032	.3054	.3075	.3096	.3118	.3139	.3160	.3181	.3201
2.1	.3222	.3243	.3263	.3284	.3304	.3324	.3345	.3365	.3385	.3404
2.2	.3424	.3444	.3464	.3483	.3502	.3522	.3541	.3560	.3579	.3598
2.3	.3617	.3636	.3655	.3674	.3692	.3711	.3729	.3747	.3766	.3784
2.4	.3802	.3820	.3838	.3856	.3874	.3892	.3909	.3927	.3945	.3962
2.5	.3979	.3997	.4014	.4031	.4048	.4065	.4082	.4099	.4116	.4133
2.6	.4150	.4166	.4183	.4200	.4216	.4232	.4249	.4265	.4281	.4298
2.7	.4314	.4330	.4346	.4362	.4378	.4393	.4409	.4425	.4440	.4456
2.8	.4472	.4487	.4502	.4518	.4533	.4548	.4564	.4579	.4594	.4609
2.9	.4624	.4639	.4654	.4669	.4683	.4698	.4713	.4728	.4742	.4757
3.0	.4771	.4786	.4800	.4814	.4829	.4843	.4857	.4871	.4886	.4900
3.1	.4914	.4928	.4942	.4955	.4969	.4983	.4997	.5011	.5024	.5038
3.2	.5051	.5065	.5079	.5092	.5105	.5119	.5132	.5145	.5159	.5172
3.3	.5185	.5198	.5211	.5224	.5237	.5250	.5263	.5276	.5289	.5302

Continued

N	0	1	2	3	4	5	6	7	8	9
3.4	.5315	.5328	.5340	.5353	.5366	.5378	.5391	.5403	.5416	.5428
3.5	.5441	.5453	.5465	.5478	.5490	.5502	.5514	.5527	.5539	.5551
3.6	.5563	.5575	.5587	.5599	.5611	.5623	.5635	.5647	.5658	.5670
3.7	.5682	.5694	.5705	.5717	.5729	.5740	.5752	.5763	.5775	.5786
3.8	.5798	.5809	.5821	.5832	.5843	.5855	.5866	.5877	.5888	.5899
3.9	.5911	.5922	.5933	.5944	.5955	.5966	.5977	.5988	.5999	.6010
4.0	.6021	.6031	.6042	.6053	.6064	.6075	.6085	.6093	.6107	.6117
4.1	.6128	.6138	.6149	.6160	.6170	.6180	.6191	.6201	.6212	.6222
4.2	.6232	.6243	.6253	.6263	.6274	.6284	.6294	.6304	.6314	.6325
4.3	.6335	.6345	.6355	.6365	.6375	.6385	.6395	.6405	.6415	.6425
4.4	.6435	.6444	.6454	.6464	.6474	.6484	.6493	.6503	.6513	.6522
4.5	.6532	.6542	.6551	.6561	.6571	.6580	.6590	.6599	.6609	.6618
4.6	.6628	.6637	.6646	.6656	.6665	.6675	.6684	.6693	.6702	.6712
4.7	.6721	.6730	.6739	.6749	.6758	.6767	.6776	.6785	.6794	.6803
4.8	.6812	.6821	.6830	.6839	.6848	.6857	.6866	.6875	.6884	.6893
4.9	.6902	.6911	.6920	.6928	.6937	.6946	.6955	.6964	.6972	.6981
5.0	.6990	.6998	.7007	.7016	.7024	.7033	.7042	.7050	.7059	.7067
5.1	.7076	.7084	.7093	.7101	.7110	.7118	.7126	.7135	.7143	.7152
5.2	.7160	.7168	.7177	.7185	.7193	.7202	.7210	.7218	.7226	.7235
5.3	.7243	.7251	.7259	.7267	.7275	.7284	.7292	.7300	.7308	.7316
5.4	.7324	.7332	.7340	.7348	.7356	.7364	.7372	.7380	.7388	.7396
5.5	.7404	.7412	.7419	.7427	.7435	.7443	.7451	.7459	.7466	.7474
5.6	.7482	.7490	.7497	.7505	.7513	.7520	.7528	.7536	.7543	.7551
5.7	.7559	.7566	.7574	.7582	.7589	.7597	.7604	.7612	.7619	.7627
5.8	.7634	.7642	.7649	.7657	.7664	.7672	.7679	.7686	.7694	.7701
5.9	.7709	.7716	.7723	.7731	.7738	.7745	.7752	.7760	.7767	.7774
6.0	.7782	.7789	.7796	.7803	.7810	.7818	.7825	.7832	.7839	.7846
6.1	.7853	.7860	.7868	.7875	.7882	.7889	.7896	.7903	.7910	.7917
6.2	.7924	.7931	.7938	.7945	.7952	.7959	.7966	.7973	.7980	.7987
6.3	.7993	.8000	.8007	.8014	.8021	.8028	.8035	.8041	.8048	.8055
6.4	.8062	.8069	.8075	.8082	.8089	.8096	.8102	.8109	.8116	.8122
6.5	.8129	.8136	.8142	.8149	.8156	.8162	.8169	.8176	.8182	.8189
6.6	.8195	.8202	.8209	.8215	.8222	.8228	.8235	.8241	.8248	.8254
6.7	.8261	.8267	.8274	.8280	.8287	.8293	.8299	.8306	.8312	.8319
6.8	.8325	.8331	.8338	.8344	.8351	.8357	.8363	.8370	.8376	.8382
6.9	.8388	.8395	.8401	.8407	.8414	.8420	.8426	.8432	.8439	.8445
7.0	.8451	.8457	.8463	.8470	.8476	.8483	.8488	.8494	.8500	.8506
7.1	.8513	.8519	.8525	.8531	.8537	.8543	.8549	.8555	.8561	.8567
7.2	.8573	.8579	.8585	.8591	.8597	.8603	.8609	.8615	.8621	.8627
7.3	.8633	.8639	.8645	.8651	.8657	.8663	.8669	.8675	.8681	.8686
7.4	.8692	.8698	.8704	.8710	.8716	.8722	.8727	.8733	.8739	.8745
7.5	.8751	.8756	.8762	.8768	.8774	.8779	.8785	.8791	.8797	.8802

n	0	1	2	3	4	5	6	7	8	9
7.6	.8808	.8814	.8820	.8825	.8831	.8837	.8842	.8848	.8854	.8859
7.7	.8865	.8871	.8876	.8882	.8887	.8893	.8899	.8904	.8910	.8915
7.8	.8921	.8927	.8932	.8938	.8943	.8949	.8954	.8960	.8965	.8971
7.9	.8976	.8982	.8987	.8993	.8998	.9004	.9009	.9015	.9020	.9025
8.0	.9031	.9036	.9042	.9047	.9053	.9058	.9063	.9069	.9074	.9079
8.1	.9085	.9090	.9096	.9101	.9106	.9112	.9117	.9122	.9128	.9133
8.2	.9138	.9143	.9149	.9154	.9159	.9165	.9170	.9175	.9180	.9186
8.3	.9191	.9196	.9201	.9206	.9212	.9217	.9222	.9227	.9232	.9238
8.4	.9243	.9248	.9253	.9258	.9263	.9269	.9274	.9279	.9284	.9289
8.5	.9294	.9299	.9304	.9309	.9315	.9320	.9325	.9330	.9335	.9340
8.6	.9345	.9350	.9355	.9360	.9365	.9370	.9375	.9380	.9385	.9390
8.7	.9395	.9400	.9405	.9410	.9415	.9420	.9425	.9430	.9435	.9440
8.8	.9445	.9450	.9455	.9460	.9465	.9469	.9474	.9479	.9484	.9489
8.9	.9494	.9499	.9504	.9509	.9513	.9518	.9523	.9528	.9533	.9538
9.0	.9542	.9547	.9552	.9557	.9562	.9566	.9571	.9576	.9581	.9586
9.1	.9590	.9595	.9600	.9605	.9609	.9614	.9619	.9624	.9628	.9633
9.2	.9638	.9643	.9647	.9652	.9657	.9661	.9666	.9671	.9675	.9680
9.3	.9685	.9689	.9694	.9699	.9703	.9708	.9713	.9717	.9722	.9727
9.4	.9731	.9736	.9741	.9745	.9750	.9754	.9759	.9763	.9768	.9773
9.5	.9777	.9782	.9786	.9791	.9795	.9800	.9805	.9809	.9814	.9818
9.6	.9823	.9827	.9832	.9836	.9841	.9845	.9850	.9854	.9859	.9863
9.7	.9868	.9872	.9877	.9881	.9886	.9890	.9894	.9899	.9903	.9908
9.8	.9912	.9917	.9921	.9926	.9930	.9934	.9939	.9943	.9948	.9952
9.9	.9956	.9961	.9965	.9969	.9974	.9978	.9983	.9987	.9991	.9996

APPENDIX **VII**

Linear Equations

A linear equation is an equation that usually contains one unknown (variable) whose highest power is equal to 1. A general example is $ax^{"1"} = b$, where the "1" is the first power of the unknown (x), a is a coefficient $(a \neq 0)$, and b is a number. The solution of linear equations can be used in chemistry for calculating oxidation numbers of elements in compounds and ions (see 6-1) and for substituting into the ideal gas equation to solve for the variables P (pressure), V (volume), n (moles), and T (temperature) (see 11-8).

VII-1

Solving a Linear Equation

To solve for the unknown quantity in a linear equation, we need to carry out algebraic transformations or changes.

1. Clear any parentheses.

2. Collect similar terms. Place all unknowns on one side of the equation (usually the left) and all numbers on the other side (usually the right).

3. Rearrange. To place unknowns on one side of the equation and numbers on the other side, we must rearrange the equation as follows.
   (a) Adding or subtracting the same number on *both* sides of the equation does not change the equation.
   (b) Multiplying or dividing *both* sides of the equation by the same number does not change the equation.

4. Solve for *one* unit of the unknown.

Problem Example VII-1

Solve the following linear equations for the unknown $(x.)$.

(a) $2x = 4$
(b) $2x = 5 - 1$
(c) $2x + 1 = 5$
(d) $2(x + 1) = 8$
(e) $3x = -18$

SOLUTION

(a) $2x = 4$ Divide both sides of the equation by 2.

$$\frac{\cancel{2}x}{\cancel{2}} = \frac{4}{2}$$

$x = 2$ *Answer*

(b) $2x = 5 - 1$ Collect similar terms.

$2x = 5 - 1 = 4$

$$\frac{\cancel{2}x}{\cancel{2}} = \frac{4}{2}$$

$x = 2$ *Answer*

(c) $2x + 1 = 5$ Subtract 1 from both sides of the equation.

$2x + \cancel{1} - \cancel{1} = 5 - 1$

$2x = 4$

$$\frac{\cancel{2}x}{\cancel{2}} = \frac{4}{2}$$

$x = 2$ *Answer*

(d) $2(x + 1) = 8$ Clear the parentheses.

$2x + 2 = 8$ Subtract 2 from both sides of the equation.

$2x + \cancel{2} - \cancel{2} = 8 - 2$

$2x = 6$ Divide both sides of the equation by 2.

$$\frac{\cancel{2}x}{\cancel{2}} = \frac{6}{2}$$

$x = 3$ *Answer*

(e) $3x = -18$ Divide both sides of the equation by 3.

$$\frac{\cancel{3}x}{\cancel{3}} = \frac{-18}{3}$$

$x = -6$ *Answer*

VII-2

Substituting into a Linear Equation

A linear equation can be checked by substituting the value obtained for the unknown into the original equation. Checking the linear equations of Problem Example VII-1 gives the following solutions:

(a) $2x = 4$; $x = 2$
$2(2) = 4$
$4 = 4$

(b) $2x = 5 - 1$; $x = 2$
$2(2) = 5 - 1$
$4 = 4$

(c) $2x + 1 = 5$; $x = 2$
$2(2) + 1 = 5$
$4 + 1 = 5$
$5 = 5$

(d) $2(x + 1) = 8$; $x = 3$
$2(3 + 1) = 8$
$2(4) = 8$
$8 = 8$

(e) $3x = -18$; $x = -6$
$3(-6) = -18$
$-18 = -18$

Problem Example VII-2

Give the following ideal-gas equation:

$$PV = nRT$$

(a) Solve for pressure in atmospheres (P) if the volume (V) is 6.00 L, the temperature (T) is 300 K, the moles of gas (n) is 0.900 mol, and the universal gas constant (R) is 0.0821 atm · L/mol · K.

(b) Solve for the moles of a gas (n) if the pressure (P) is 1.25 atm, the volume (V) is 2.00 L, the temperature (T) is 273 K, and the universal gas constant (R) is 0.0821 atm · L/mol · K.

(c) Solve for the temperature in °C if the pressure (P) is 1.10 atm, the volume (V) is 30.0 L, the moles of gas (n) is 1.25 mol, and the universal gas constant (R) is 0.0821 atm · L/mol · K.

SOLUTION

(a) Use the ideal-gas equation and solve for P as follows:

$$PV = nRT \quad \text{Divide both sides of the equation by } V.$$
$$\frac{P\cancel{V}}{\cancel{V}} = \frac{nRT}{V}$$
$$P = \frac{nRT}{V}$$

Substitution into the linear equation above for $V = 6.00$ L, $T = 300$ K, $n = 0.900$ mol, and $R = 0.0821$ atm · L/mol · K gives the following equation:

$$P = \frac{0.900 \cancel{\text{mol}} \times 0.0821 \dfrac{\text{atm} \cdot \cancel{L} \times 300 \cancel{K}}{\cancel{\text{mol}} \cdot \cancel{K}}}{6.00 \cancel{L}}$$
$$= 3.69 \text{ atm} \quad \textit{Answer}$$

Note that the units of R cancel out, leaving only atmospheres, the units of pressure.

(b) Use the ideal-gas equation and solve for n as follows:

$$PV = nRT \qquad \text{Place the unknown } n \text{ on the left.}$$
$$nRT = PV \qquad \text{Divide both sides of the equation by } RT.$$
$$\frac{n\cancel{RT}}{\cancel{RT}} = \frac{PV}{RT}$$
$$n = \frac{PV}{RT}$$

Substitution into the linear equation above for P = 1.25 atm, V = 2.00 L, T = 273 K, and R = 0.0821 atm · L/mol · K gives the following equation:

$$n = \frac{1.25 \,\cancel{atm} \times 2.00 \,\cancel{L}}{0.0821 \dfrac{\cancel{atm} \cdot \cancel{L}}{mol \cdot \cancel{K}} \times 273 \,\cancel{K}}$$
$$= 0.112 \text{ mol} \quad \textit{Answer}$$

Note that the units of R cancel out again, leaving only mole. (In division of fractions, you invert and multiply; therefore, $\dfrac{\dfrac{1}{1}}{\dfrac{1}{mol}} = 1 \times \dfrac{mol}{1} = mol.$)

(c) Use the ideal-gas equation and solve for T as follows:

$$PV = nRT \qquad \text{Place the unknown } T \text{ on the left.}$$
$$nRT = PV \qquad \text{Divide both sides of the equation by } nR.$$
$$\frac{\cancel{n}R T}{\cancel{n}\cancel{R}} = \frac{PV}{nR}$$
$$T = \frac{PV}{nR}$$

Substitution into the linear equation above for P = 1.10 atm, V = 30.0 L, n = 1.25 mol, and R = 0.0821 atm · L/mol · K gives the following equation:

$$T = \frac{1.10 \,\cancel{atm} \times 30.0 \,\cancel{L}}{1.25 \,\cancel{mol} \times 0.0821 \dfrac{\cancel{atm} \cdot \cancel{L}}{\cancel{mol} \cdot K}}$$
$$= 322 \text{ K}$$

Note that the units of R cancel out again, leaving only K. The temperature requested is to be expressed in °C. Therefore, °C = K − 273 from Equation 2-4 in 2-11; the temperature in °C is as follows:

$$322 \text{ K} = (322 - 273)°C = 49°C \quad \textit{Answer}$$

PROBLEMS

1. Solve the following linear equations for the unknown (x):

   (a) $2x = 5 + 3$ (b) $5x + 6 = 4x + 2$

   (c) $2(x + 2) = 10$ (d) $3(x + 1) = 2(x + 4)$

2. Given the following ideal-gas equation:

$$PV = nRT$$

   (a) Solve for volume in liters (V) if the pressure (P) is 2.00 atm, the temperature is 300 K, the moles of gas (n) is 0.750 mol, and the universal gas constant is 0.0821 atm $\cdot$ L/ mol $\cdot$ K.

   (b) Solve for the temperature in °C if the pressure (P) is 1.50 atm, the volume (V) is 38.0 L, the moles of gas (n) is 1.10 mol, and the universal gas constant is 0.0821 atm $\cdot$ L/mol $\cdot$ K.

   (c) Solve for the moles of gas (n) if the pressure (P) is 0.970 atm, the volume (V) is 3.00 L, the temperature is 303 K, and the universal gas constant is 0.0821 atm $\cdot$ L/mol $\cdot$ K.

Answers to Selected Exercises and Problems

Chapter 1

3. (a) analytical;
   (b) organic;
   (c) physical
4. (a) physical;
   (b) biochemical;
   (c) inorganic;
   (d) inorganic;
   (e) physical;
   (f) analytical;
   (g) organic

Chapter 2

5. (a) 3; (b) 2; (c) 6; (d) 3;
   (e) 5; (f) 1
6. (a) 1; (b) 3; (c) 4; (d) 4;
   (e) 3; (f) 4
7. (a) 1.37; (b) 3.37; (c) 16.4;
   (d) 2.66; (e) 16.6;
   (f) 0.0366
8. (a) 7.27; (b) 0.00332;
   (c) 0.0544; (d) 0.646;
   (e) 0.533; (f) 9.74
9. 0.454
10. (a) 11.5; (b) 81.612;
    (c) 13.14; (d) 0.304; (e) 12;
    (f) 280; (g) 0.010; (h) 7.2;
    (i) 24

11. 2.75×10^4
12. 3.25×10^{-2}
13. (a) 4.74×10^3;
    (b) 5.39×10^2;
    (c) 1.01×10^3;
    (d) 6.52×10^5;
    (e) 9.16×10^5;
    (f) 4.07×10^4;
    (g) 2.74×10^4;
    (h) 2.90×10^{-3}
14. (a) 3.00×10^4;
    (b) 6.00×10^2;
    (c) 7.00×10^4;
    (d) 4.47×10^{-4}
15. (a) 4.45×10^6;
    (b) 1.54×10^{16};
    (c) 3.05×10^6;
    (d) 8.12×10^9
16. (a) 8.72×10^6;
    (b) 7.45×10^{-2};
    (c) 7.27×10^3;
    (d) 3.28×10^{-2}
17. (a) 7.64×10^{-3};
    (b) 7.25×10^5;
    (c) 9.74×10^3;
    (d) 6.28×10^{-3}
18. (a) 6.5×10^6 mg;
    (b) 12 km;
    (c) 3.5×10^{-5} kg;

   (d) 0.3 L; (e) 764 L;
   (f) $3.5 \times 10^{-3}\ \mu$
19. (a) 3200 m; (b) 15 kg;
    (c) 8.5×10^{-5} kg;
    (d) 0.675 L; (e) $35\bar{0}$ Å;
    (f) 6.75 cc
20. 4.975 g
21. 2010.7004 m
22. 91.4 cm/yd
23. (a) 86.0°F, 303.0 K;
    (b) $23\bar{0}$°F, 383 K;
    (c) −94.0°F, 203.0 K;
    (d) −202°F, 143 K
24. (a) 18.9°C, 291.9 K;
    (b) −23.3°C, 249.7 K;
    (c) −37°C, 236 K;
    (d) 232°C, 505 K
25. −196°C, −321°F
26. −40.0°C or °F
27. −62.1°C
28. 58.0°C, 331.0 K
29. 0.200 cal/g · °C
30. 14.0 kcal
31. 3040 cal
32. 7.35 g
33. 72.6 g
34. 2.28×10^3 J
35. (a) 5.8 g/mL; (b) 3.1 g/mL;
    (c) 12 g/mL; (d) 7.0 g/mL

36. (a) 46.9 mL; (b) 214 mL;
    (c) 25.8 mL;
    (d) 1.9×10^3 mL
37. (a) 17.7 g; (b) 403 g;
    (c) 7.60 g; (d) 686 g
38. (a) 0.145 L; (b) 3.35 L;
    (c) 0.299 L; (d) 1.25 L
39. (a) 22.0 g; (b) 160; g;
    (c) 2.61×10^3 g;
    (d) 5.20×10^3 g
40. (a) 423 cal; (b) 1.77×10^3 J
41. (a) 0.400 cal/g · °C;
    (b) 1.67×10^3 J/kg · K

Chapter 3 _____

3. (a) solid; (c) liquid; (e) gas
4. O, Si, Al, Fe, Ca, Na, K,
   Mg, H, Ti
5. mp = −38.87°C;
   bp = 356.58°C
6. (a) chemical; (c) physical;
   (e) chemical; (g) chemical
7. (a) physical; (c) chemical;
   (e) physical
8. (a) chemical; (c) chemical;
   (e) chemical
9. (a) 1 atom carbon,
       4 atoms hydrogen; 5
   (c) 1 atom carbon,
       2 atoms chlorine,
       2 atoms fluorine; 5
   (e) 34 atoms carbon,
       32 atoms hydrogen,
       1 atom iron, 4 atoms
       nitrogen, 4 atoms
       oxygen; 75
10. (a) SO_2; (c) Ag_2S;
    (e) $C_{10}H_{16}N_5O_{13}P_3$
11. (a) element; (c) element;
    (e) compound; (g) element;
    (h) mixture
20. 0.164 L
21. 13.2 lb
22. (a) 5.42×10^4 J;
    (b) 13.0 kcal
23. 0.673 kg

Chapter 4 _____

5. (a)
5p 6n | 5e⁻

   (c)
22p 24n | 22e⁻

   (e)
27p 32n | 27e⁻

   (f)
44p 52n | 44e⁻

6. (a)
11p 12n | 11e⁻

   (c)
46p 56n | 46e⁻

   (e)
58p 84n | 58e⁻

   (f)
92p 143n | 92e⁻

7. (b)
8. 10.811 is nearer to 11.009
   than 10.013, hence boron-
   11 must predominate in
   nature.
9. 69.72 amu
10. 121.8 amu
11. (a) 2; (b) 8; (c) 32; (d) 18;
    (e) 72; (f) 98

12. (a)
3p 4n | 2e⁻ 1e⁻ (1)

   (c)
8p 10n | 2e⁻ 6e⁻ (6)

   (e)
15p 16n | 2e⁻ 8e⁻ 5e⁻ (5)

13. (a)
4p 7n | 2e⁻ 2e⁻ (2)

   (c)
9p 10n | 2e⁻ 7e⁻ (7)

   (e)
12p 14n | 2e⁻ 8e⁻ 2e⁻ (2)

14. (a) He:; (c) Be·; (e) :F·;
    (f) :Ar:
15. (a) ·C·; (c) Na·; (e) ·P·;
    (f) ·S:
16. (a) $1s^2, 2s^1$ (1);
    (c) $1s^2, 2s^2 2p^2$ (4);
    (e) $1s^2, 2s^2 2p^6, 3s^2 3p^3$ (5);
    (f) $1s^2, 2s^2 2p^6, 3s^2 3p^6 3d^{10}$,
        $4s^2 4p^4$ (6), or $1s^2, 2s^2 2p^6$,
        $3s^2 3p^6, 4s^2, 3d^{10}, 4p^4$
17. (a) $1s^2, 2s^2 2p^6, 3s^2 3p^1$ (3);
    (c) $1s^2, 2s^2 2p^6, 3s^2 3p^6 3d^{10}$,
        $4s^2 4p^5$ (7), or $1s^2, 2s^2 2p^6$,
        $3s^2 3p^6, 4s^2, 3d^{10}, 4p^5$;
    (e) $1s^2, 2s^2 2p^6, 3s^2 3p^6 3d^{10}$,
        $4s^2 4p^6 4d^{10}, 5s^2 5p^1$ (3), or
        $1s^2, 2s^2 2p^6, 3s^2 3p^6, 4s^2$,
        $3d^{10}, 4p^6, 5s^2, 4d^{10}, 5p^1$;

(f) $1s^2$, $2s^22p^6$, $3s^23p^63d^{10}$,$4s^24p^64d^{10}4f^7$, $5s^25p^65d^1$, $6s^2$ (usually 2, actually 3 as the one $5d$ electron also acts as a valence electron), or $1s^2$, $2s^22p^6$, $3s^23p^6$, $4s^2$, $3d^{10}$, $4p^6$, $5s^2$, $4d^{10}$, $5p^6$, $6s^2$, $5d^1$, $4f^7$

18. (a) 5513°F; (b) 2.257 × 10^4 kg/m³; (c) 1.408 × 10^3 lb/ft³

19. $1s^2$, $2s^22p^6$, $3s^23p^63d^{10}$, $4s^24p^64d^{10}4f^{14}$, $5s^25p^65d^6$, $6s^2$, or $1s^2$, $2s^22p^6$, $3s^23p^6$, $4s^2$, $3d^{10}$, $4p^6$, $5s^2$, $4d^{10}$, $5p^6$, $6s^2$, $4f^{14}$, $5d^6$

Chapter 5 ───────

3. (a) metal; (c) metalloid; (d) nonmetal
4. (a) nonmetal; (c) metalloid; (d) metal
5. (a) 1; (b) 4; (c) 6; (d) 8
6. (a) 8; (b) 7; (c) 3; (d) 5
7. (a) and (c); (b) and (d)
8. (a) Te; (c) S;
9. (a) and (d); (b) and (c)
10. (a) P; (c) K
11. (a) arsenic; (c) aluminum
12. (a) barium; (c) lead
13. (a) chlorine; (c) barium
14. (a) phosphorus; (c) barium
15. (a) metalloid; (b) 4; (c) Ge: $1s^2$, $2s^22p^6$, $3s^23p^63d^{10}$, $4s^24p^2$, or $1s^2$, $2s^22p^6$, $3s^23p^6$, $4s^2$, $3d^{10}$, $4p^2$; Si: $1s^2$, $2s^22p^6$, $3s^23p^2$; (d) more metallic; (e) 5.5 × 10^3 kg/m³, 5.3 × 10^3 kg/m³

Chapter 6 ───────

4. (a) +1 or 1^+; (b) +4 or 4^+; (c) +3 or 3^+; (d) +7 or 7^+; (e) +4 or 4^+; (f) +5 or 5^+; (g) +6 or 6^+; (h) +5 or 5^+; (i) +3 or 3^+; (j) +5 or 5^+

5. (a) +7 or 7^+; (b) +1 or 1^+; (c) +5 or 5^+; (d) +3 or 3^+; (e) +7 or 7^+; (f) +3 or 3^+; (g) +6 or 6^+; (h) +6 or 6^+; (i) +1 or 1^+; (j) +6 or 6^+

6. (1) Loss of third principal energy level and (2) greater nuclear attraction.

7. Smaller nuclear attraction.

8. (a)

1p
0n   1+

(c)

12p
12n $2e^-$ $8e^-$ 2+

(e)

13p
14n $2e^-$ $8e^-$ 3+

(g)

8p
8n $2e^-$ $8e^-$ 2−

(i)

7p
7n $2e^-$ $8e^-$ 3−

(j)

15p
16n $2e^-$ $8e^-$ $8e^-$ 3−

9. (a) $1s^2$; (c) $1s^2$, $2s^22p^6$; (e) $1s^2$, $2s^22p^6$, $3s^23p^6$; (g) $1s^2$, $2s^22p^6$, $3s^23p^6$; (i) $1s^2$, $2s^22p^6$, $3s^23p^6$; (j) $1s^2$, $2s^22p^6$, $3s^23p^63d^{10}$, $4s^24p^6$, or $1s^2$, $2s^22p^6$, $3s^23p^6$, $4s^2$, $3d^{10}$, $4p^6$

10. 3.42 × 10^3 J

11. (a) $\overset{\delta(+)}{H}\ \overset{\delta(-)}{F}$;
(c) $\overset{\delta(+)}{H_2}\ \overset{\delta(-)}{O}$;

(e) $\overset{\delta(+)}{B}\ \overset{\delta(-)}{Cl_3}$;
(g) $\overset{\delta(+)}{P}\ \overset{\delta(-)}{Cl_5}$;
(h) $\overset{\delta(-)}{N}\ \overset{\delta(+)}{H_3}$;
(i) $\overset{\delta(+)}{O}\ \overset{\delta(-)}{F_2}$

12. (Structural formulas only)

(a) H—F; (c) Cl—C—Cl with Cl above and Cl below;

(e) N≡N;

(g) H—C≡C—H;

(i) [C≡N]$^{1-}$;

(j) $\left[\text{O—S—O}\right]^{2-}$ with O below;

13. (Structural formulas only)

(a) Cl—Cl;

(c) Cl—C—Cl with H above and Cl below; (e) F—F;

(g) H—O—C—O—H with O double-bonded above C;

(i) $\left[\text{O—P—O}\right]^{3-}$ with O above and O below;

(j) $\left[\text{O—P—O—P—O}\right]^{4-}$ with O above each P and O below each P;

14. (a) KBr; (c) Mg_3N_2; (e) CdO; (g) LiH; (i) $Al(ClO_4)_3$; (j) $Ba(MnO_4)_2$

15. (a) AgI; (c) CuBr; (e) $Zn(HCO_3)_2$; (g) $Fe_3(PO_4)_2$; (i) $Hg(CN)_2$; (j) $(NH_4)_2Cr_2O_7$

16. (a) 2^+; (b) 1^+; (c) 6^+, 2^-;
(d) 7^+, 1^-; (e) 3^+; (f) 6^+,
2^-; (g) 7^+, 1^-; (h) 5^+, 3^-;
(i) 3^+; (j) 8^+

17. (a) CaS; (c) Na_3N;
(e) In_2O_3; (g) Al_2S_3;
(i) Tl_2S_3

18. (a) 2.30 Å; (b) 3.66 g/mL;
(c) 778°C; (d) 4.63 g/mL

19. (a) Cs_2SO_4; (c) $Mg_3(AsO_4)_2$;
(e) Na_2SeO_4; (g) Tl_2S_3;
(i) Rb_2SeO_4

20. (a) ionic; (c) covalent;
(e) covalent; (g) ionic;
(i) covalent

21. (a) IVA; (b) 4; (c) more
metallic; (d) Pb;
(e)

(Structural formula only)

22. (a) IA; (b) 1; (c) 1^+;
(d) Fr; (e) YBr, ionic

Chapter 7 _____

6. Question 7 in the order j-a.
7. Question 6 in the order j-a.
8. Question 9 in the order j-a.
9. Question 8 in the order j-a.
10. Question 11 in the order
j-a.
11. Question 10 in the order
j-a.
12. Question 13 in the order
j-a.
13. Question 12 in the order
j-a.
14. Question 15 in the order
j-a.
15. Question 14 in the order
j-a.
16. (a) (1); (b) (4); (c) (3);
(d) (3); (e) (1); (f) (5);
(g) (1); (h) (2); (i) (5);
(j) (5)
17. (a) (5); (b) (3); (c) (4);
(d) (2); (e) (3); (f) (5);

(g) (5); (h) (1); (i) (5);
(j) (2)

18. (a) $HC_2H_3O_2$; (b) $CaCO_3$;
(c) NaCl; (d) $NaHCO_3$;
(e) $Mg(OH)_2$; (f) NH_3

19. (a) $Sn_3(PO_4)_2$; (b) $AgMnO_4$;
(c) $Ca(IO)_2$; (d) $MgSO_3$;
(e) $Na_2C_2O_4$; (f) $HClO_4$ in
water; (g) PF_3;
(h) $Cd(NO_3)_2$;
(i) $Pb_3(PO_4)_2$;
(j) $Sr(HCO_3)_2$; (k) Ca_3N_2;
(l) $HgCl_2$; (m) Au_2O_3;
(n) PbO_2; (o) PCl_5;
(p) Na_2HPO_4

20. (a) calcium hydrogen
phosphate; (b) magnesium
chloride; (c) lead(II) sulfate
or plumbous sulfate;
(d) calcium dichromate;
(e) barium hydroxide;
(f) tin(II) or stannous
hydrogen carbonate or
bicarbonate; (g) potassium
oxalate; (h) lithium
dichromate; (i) acetic acid
or vinegar; (j) zinc chloride;
(k) bismuth fluoride;
(l) gold(III) bromide or
auric bromide;
(m) cadmium phosphide;
(n) lithium hydroxide;
(o) calcium oxalate;
(p) tin(IV) sulfide or
stannic sulfide

21. (NaCl) (Na_2CO_3) (Na_2SO_4)
(Na_3PO_4); ($CaCl_2$) ($CaCO_3$)
($CaSO_4$) [$Ca_3(PO_4)_2$]; ($FeCl_3$)
[$Fe_2(CO_3)_3$] [$Fe_2(SO_4)_3$]
($FePO_4$); ($CuCl_2$) ($CuCO_3$)
($CuSO_4$) [$Cu_3(PO_4)_2$]

Chapter 8 _____

3. (a) 137 amu; (b) $18\bar{0}$ amu;
(c) 17.0 amu; (d) 16.0
amu; (e) 80.1 amu;
(f) 76.0 amu
4. (a) 56.1 amu; (b) 103.4

amu; (c) 74.1 amu;
(d) 324.6 amu; (e) $31\bar{0}$
amu; (f) 342 amu

5. (a) 1 mol C atoms,
2 mol O atoms;
(b) 2 mol N atoms,
4 mol O atoms;
(c) 2 mol Al atoms,
3 mol S atoms,
12 mol O atoms;
(d) 1 mol Ca atoms,
2 mol O atoms,
2 mol H atoms;
(e) 2 mol K atoms,
1 mol C atoms,
3 mol O atoms;
(f) 1 mol Ba atoms,
4 mol C atoms,
6 mol H atoms,
4 mol O atoms

6. (a) 0.237 mol;
(b) 3.38 mol;
(c) 1.69 mol;
(d) 0.356 mol;
(e) 0.0255 mol;
(f) 1.53 mol;
(g) 1.25 mol Fe atoms,
3.75 mol Cl atoms;
(h) 10.5 mol Mg^{2+} ions,
7.00 mol P atoms,
28.0 mol O atoms;
(i) 0.274 mol;
(j) 14.4 mol

7. (a) 1.04 mol;
(b) 2.71 mol;
(c) 6.56 mol;
(d) 0.0744 mol;
(e) 45.0 mol;
(f) 0.467 mol;
(g) 1.05 mol;
(h) 9.39 mol;
(i) 2.49 mol;
(j) 114 mol

8. (a) 54.1 g; (b) 221 g;
(c) 25.3 g; (d) 7.90 g;
(e) 383 mg; (f) 445 g;
(g) $34\bar{0}$ mg; (h) 43.4 g;
(i) 80.0 g; (j) 13.9 g

9. (a) 98.0 g; (b) 152 g;

(c) 69.1 g; (d) 76.8 mg;
(e) 9.46 g; (f) 173 mg;
(g) 196 g; (h) 0.500 kg;
(i) 0.0338 g

10. (a) 2.41×10^{23} atoms;
(b) 1.81×10^{22} atoms;
(c) 4.70×10^{24} molecules;
(d) 2.05×10^{23} molecules

11. (a) 1.38×10^{24} molecules;
(b) 4.82×10^{24} molecules;
(c) 3.01×10^{24} atoms;
(d) 2.13×10^{23} atoms

12. (a) 6.64×10^{-24} g/atom;
(b) 1.03×10^{-22} g/atom;
(c) 1.41×10^{-22} g/atom;
(d) 3.39×10^{-22} g/atom

13. (a) 0.580 mol;
(b) 0.0388 mol;
(c) 2.01 mol;
(d) 27.5 g; (e) 4.64 g;
(f) 7.06 g

14. (a) 3.88 amu;
(b) 28.6 amu;
(c) 69.5 amu;
(d) 16.2 amu;
(e) 37.0 amu;
(f) 40.3 amu

15. (a) 0.759 g/L; (b) 1.34 g/L;
(c) 1.16 g/L; (d) 1.96 g/L;
(e) 5.71 g/L; (f) 0.500 g/L

16. (a) 6.40 L; (b) 5.08 L;
(c) 1.84 L; (d) 1.45 L;
(e) 5.16 L; (f) 3.21 L

17. (a) 13.6% Ca, 86.4% I;
(b) 5.9% H, 94.1%S;
(c) 69.6% Ba, 6.08% C,
24.3% O;
(d) 38.7% Ca. 20.0% P,
41.3% O;
(e) 52.2% C, 13% H,
34.8% O;
(f) 24.0% Fe, 30.9% C,
3.9% H, 41.2% O

18. (a) 57.0%; (b) 47.9%;
(c) 58.7%; (d) 67.1%

19. 31.0% Na

20. (a) 19.8 g; (b) 47.0 g;
(c) 12.1 g; (d) 6.91 g

21. (a) $ZnCl_2$; (b) SnI_4;

(c) $FeBr_2$; (d) $Pb(NO_3)_2$;
(e) $MgCO_3$; (f) $Ca_3(PO_4)_2$;
(g) Na_2SO_3; (h) K_2SO_4;
(i) PBr_5; (j) Ga_2O_3

22. (a) Al_2O_3; (b) CO;
(c) Na_2S; (d) SO_2

23. (a) C_2H_6; (b) C_6H_{14};
(c) C_2H_2; (d) $C_4H_4O_4$;
(e) $C_4H_8Cl_2$

24. $C_{10}H_{10}N_4SO_2$

25. $C_{18}H_{22}O_2$

26. $C_{10}H_{14}N_2$

27. 3.28 mol

28. 259 mL

29. 1.31×10^{19} molecules/mL

30. Ammonia

31. (a) 3.68 mmol; (b) 3.83 g;
(c) 4.97×10^{21} molecules

32. (a) 0.00500 mmol/mL;
0.00778 mmol/mL;
3.01×10^{18} molecules/
mL, 4.68×10^{18}
molecules/mL
(b) 27.5 mmol, 42.8 mmol;
4.95 g, 7.70 g

33. (a) 0.00750 mmol/mL;
0.0128 mmol/mL;
4.52×10^{18} molecules/
mL, 7.71×10^{18}
molecules/mL
(b) 41.3 mmol, 70.4 mmol;
7.43 g, 12.7 g

34. C_2H_4

35. C_3H_6

36. C_2N_2

Chapter 9 _____

(The numbers represent the coefficients in front of the formulas in the balanced equation.)

4. (a) $1 + 1 \rightarrow 1 + 2$;
(b) $2 \rightarrow 2 + 3$;
(c) $1 + 1 \rightarrow 1 + 2$;
(d) $2 + 3 \rightarrow 1 + 6$;
(e) $2 + 2 \rightarrow 2 + 1$;
(f) $3 + 1 \rightarrow 1$;
(g) $2 + 1 \rightarrow 2$;

(h) $2 + 1 \rightarrow 2 + 1$;
(i) $1 + 3 \rightarrow 2 + 3$;
(j) $3 + 2 \rightarrow 1 + 6$

5. (a) $1 + 2 \rightarrow 1 + 1$;
(b) $3 + 4 \rightarrow 2 + 3$;
(c) $3 + 2 \rightarrow 1 + 3 + 3$;
(d) $2 + 3 \rightarrow 1 + 3$;
(e) $1 + 6 \rightarrow 4$;
(f) $1 + 5 \rightarrow 3 + 4$;
(g) $6 + 1 \rightarrow 4$;
(h) $1 + 4 \rightarrow 1 + 5$;
(i) $1 + 2 \rightarrow 2 + 1$;
(j) $1 + 2 \rightarrow 1 + 4$

6. (a) $2NaCl + Pb(NO_3)_2 \rightarrow$
$PbCl_2 + 2 NaNO_3$;
(c) $3 NaHCO_3 + H_3PO_4 \rightarrow$
$Na_3PO_4 + 3 CO_2 +$
$3 H_2O$;
(e) $CaI_2 + H_2SO_4 \rightarrow 2 HI$
$+ CaSO_4$;
(g) $Mg(CN)_2 + 2 HCl \rightarrow$
$2 HCN + MgCl_2$;
(i) $2 NaHSO_3 + H_2SO_4 \rightarrow$
$Na_2SO_4 + 2 SO_2 +$
$2 H_2O$;
(j) $Al_2(SO_4)_3 + 6 NaOH$
$\rightarrow 2 Al(OH)_3 +$
$3 Na_2SO_4$;

7. (a) $2 Fe + 3 Cl_2 \rightarrow$
$2 FeCl_3$;
(c) $Ba + 2 H_2O \rightarrow$
$Ba(OH)_2 + H_2$;
(e) $(NH_4)_2S + HgBr_2 \rightarrow$
$2 NH_4Br + HgS$;
(g) $SnO + 2 HCl \rightarrow$
$SnCl_2 + H_2O$;
(i) $2 HBr + Ca(OH)_2 \rightarrow$
$CaBr_2 + 2 H_2O$

8. (a) $2 + 1 \rightarrow 2 CaO$;
(c) $2 + 3 \rightarrow 2 SO_2$;
(e) $1 + 1 \rightarrow Ca(OH)_2$;
(g) $1 + 1 \rightarrow H_2SO_3$;
(i) $1 + 1 \rightarrow NH_4Br$

9. (a) $4 + 3 \rightarrow 2 Al_2O_3$;
(c) $1 + 1 \rightarrow SiO_2$;
(e) $2 + 1 \rightarrow 2 AlN$;
(g) $1 + 3 \rightarrow 2 Al(OH)_3$;
(i) $1 + 1 \rightarrow CaSO_4$

10. (a) $2 \rightarrow 2 Hg + O_2$;

(c) $2 \rightarrow 2\ H_2 + O_2$;
(e) $1 \rightarrow CdO + CO_2$;
(g) $1 \rightarrow CaSO_4 + 2\ H_2O$;
(i) $1 \rightarrow 6\ C + 6\ H_2O$

11. (a) $2 \rightarrow 2\ H_2O + O_2$;
(c) $1 \rightarrow MgO + CO_2$;
(e) $1 \rightarrow Na_2CO_3 + H_2O$;
(g) $1 \rightarrow 12\ C + 11\ H_2O$;
(i) $1 \rightarrow CaCO_3 + H_2O + CO_2$

12. (a) $1 + 1 \rightarrow CdSO_4 + H_2$;
(c) $1 + 2 \rightarrow PbCl_2 + H_2$;
(e) $1 + 2 \rightarrow Ca(OH)_2 + H_2$;
(g) $1 + 1 \rightarrow FeCl_2 + Cu$;
(i) $1 + 2 \rightarrow 2\ NaCl + Br_2$

13. (a) $2 + 6 \rightarrow 2\ AlCl_3 + 3\ H_2$;
(c) $2 + 6 \rightarrow 2\ Al(C_2H_3O_2)_3 + 3\ H_2$;
(e) $2 + 3 \rightarrow 2\ AlCl_3 + 3\ Sn$;
(f) NR;
(g) $1 + 1 \rightarrow PbBr_2 + Hg$;
(i) NR; (j) $1 + 2 \rightarrow 2\ NaBr + I_2$

14. (a) $1 + 2 \xrightarrow{\text{cold}} PbCl_{2(s)} + 2\ HNO_3$;
(c) $1 + 3 \rightarrow Bi(OH)_{3(s)} + 3\ NaNO_3$;
(e) $1 + 1 \rightarrow PbSO_{4(s)} + 2\ KC_2H_3O_2$;
(g) $1 + 2 \rightarrow CaCl_2 + H_2O + CO_2$;
(h) $3 + 2 \rightarrow Zn_3(PO_4)_{2(s)} + 3\ H_2O + 3\ CO_{2(g)}$;
(i) $1 + 2 \rightarrow Ba(NO_3)_2 + H_2O + CO_{2(g)}$

15. (a) $2 + 1 \rightarrow Ag_2S_{(s)} + 2\ HNO_3$;
(c) $1 + 1 \rightarrow PbCrO_{4(s)} + 2\ KNO_3$;
(e) $2 + 3 \rightarrow Bi_2S_{3(s)} + 6\ HNO_3$;
(g) $1 + 1 \rightarrow FeSO_4 + H_2O + CO_{2(g)}$;
(h) $1 + 2 \rightarrow 2\ NaC_2H_3O_2 + H_2O + CO_{2(g)}$;
(i) $1 + 1 \rightarrow PbS_{(s)} + 2\ HNO_3$

16. (a) $1 + 1 \rightarrow ZnSO_4 + 2\ H_2O$;
(c) $1 + 1 \rightarrow BaSO_{4(s)} + H_2O$;
(e) $1 + 1 \rightarrow CaCO_{3(s)} + H_2O$ [excess CO_2 forms $Ca(HCO_3)_2$];
(g) $1 + 3 \rightarrow AlCl_3 + 3\ H_2O$;
(i) $2 + 1 \rightarrow K_2CO_3 + H_2O$

17. (a) $1 + 2 \rightarrow Zn(NO_3)_2 + 2\ H_2O$;
(c) $1 + 2 \rightarrow Ca(C_2H_3O_2)_2 + 2\ H_2O$;
(e) $1 + 2 \rightarrow K_2CO_3 + H_2O$;
(g) $1 + 2 \rightarrow BaCl_2 + H_2O$;
(i) $1 + 2 \rightarrow K_2SO_3 + H_2O$

18. (1) (a) $1 + 1 \rightarrow CdCl_{2(s)}$;
(c) $1 + 2 \rightarrow Ca(NO_3)_2 + H_2O + CO_{2(g)}$;
(d) NR
(2) (a) combination;
(c) double replacement

19. (1) (a) $2\ Al + 3\ PbCl_2 \rightarrow 2\ AlCl_3 + 3\ Pb_{(s)}$;
(c) $BaCl_2 + Na_2CO_3 \rightarrow 2\ NaCl + BaCO_{3(s)}$;
(2) (a) single replacement;
(c) double replacement

20. $H_3C_6H_5O_7 + 3\ NaHCO_3 \rightarrow 3\ CO_{2(g)} + Na_3C_6H_5O_7 + 3\ H_2O$

21. (1) $CaCO_3 + H_2SO_4 \rightarrow CaSO_4 + H_2O + CO_2$;
(2) $SO_3 + H_2O \rightarrow H_2SO_4$

22. $PCl_3 + 3\ H_2O \rightarrow H_3PO_3 + 3\ HCl$

23. $TiCl_4 + 2\ BCl_3 + 5\ H_2 \rightarrow TiB_2 + 10\ HCl$

Chapter 10

3. 70.9 g
4. 0.756 g

5. 65.8 g
6. 0.885 kg
7. 44.4 g
8. 0.155 kg
9. 43.6 g
10. 2.98 g
11. 0.288 mol
12. 0.823 mol
13. 0.400 mol
14. 0.127 mol
15. 218 g
16. 1.70 mol
17. 4.40 mol
18. 91.8 g
19. (a) 48.7 g; (b) 92.8%
20. (a) 0.525 mol; (b) 95.2%; (c) 0.085 mol
21. (a) 0.12 g; (b) 50%; (c) 0.027 mol
22. (a) 38.2 g; (b) 89.0%; (c) 0.057 mol
23. 10.2 L
24. 4.42 L
25. 0.368 mol
26. 13.0 g, 0.385 mol
27. 12.6 L
28. 19.2 L
29. (a) 15.1 L; (b) 0.205 mol
30. (a) 44.3 g; (b) 0.121 mol
31. 2.10 L
32. 33.0 L
33. 76.6 L
34. 2.75 L
35. (a) 3.00 L; (b) 2.75 L
36. (a) 8.25 L; (b) 0.18 L
37. (a) exothermic; (b) 1.25×10^2 kcal
38. (a) endothermic; (b) 14.4 g
39. (a) $CH_{4(g)} + 2\ O_{2(g)} \rightarrow CO_{2(g)} + 2\ H_2O_{(g)}$;
(b) 2.00 mol;
(c) 12.5 mol; (d) 32.0 g;
(e) 16.8 L; (f) 11.2 L;
(g) 48.4 g;
(h) 84.1%
40. 95.0 mL
41. 194 mL
42. (a) 0.030 g; (b) 83%
43. (a) 0.26 g; (b) 85%; (c) 0.03 mol

Chapter 11

8. 1520 mL
9. 102 mL
10. 238 torr
11. 21.0 atm
12. 228 mL
13. 83.4 mL
14. $-73°C$
15. $25\overline{0}°F$
16. 147 torr
17. $69\overline{0}$ torr
18. 94°C
19. $-5°C$
20. 443 mL
21. 281 mL
22. 177 mL
23. 1930 mm Hg
24. 0.815 atm
25. 0.793 atm
26. 455°C
27. 518°C
28. $-64°C$
29. (a) 746 torr; (b) 1.23 L
30. (a) 385 torr; (b) 163 mL
31. $12\overline{2}$ mL
32. $20\overline{0}$ mL
33. 565 mL
34. 1.26 atm
35. $9\overline{0}°C$
36. 7.17 g
37. 24.9 amu
38. 15.5 amu
39. 5.21 g/L
40. 0.337 g/L
41. 621 mL
42. 0.320 mol
43. 90.4 amu
44. 1.16×10^{23} molecules
45. 86.0 amu; C_6H_{14}
46. 30.0 amu; C_2H_6
47. 197 amu; $C_2HBrClF_3$
48. 8.31 Pa $\cdot$ m^3/mol $\cdot$ K

Chapter 12

4. Higher kinetic energy molecules escape from surface of the liquid.
5. At higher elevations (mountains), atmospheric pressure is less.
6. Lower surface tension.
7. Temperature increases; average KE increases; breaks attractive forces between molecules; viscosity decreases.
8. Oil and water immiscible with oil on top of water.
9. Glycerine.
10. Strongest-solid; weakest-gas.
11. (1) Increased pressure and (2) friction.
12. Alcohol has lower melting point than mercury.
13. Sublimation occurs at slower rate in a covered box.
14. Solid sublimed.
15. (a) 90°; (b) 85°; (c) 70°; (d) 65°
16. 6.48 kcal
17. 2.19×10^5 J
18. $52\overline{0}0$ cal for CCl_4; 26,500 cal for NaCl
19. $-0.22°C$
20. 2.5×10^4 J
21. 1010 cal
22. 9.28 cal
23. 1.84 kcal
24. 14.4 kcal
25. $12\overline{0}$ kJ
26. 10,896 cal
27. $3\overline{0}00$ g
28. (a) 1.07×10^{12} m^3; (b) 1.7×10^{11} kcal

Chapter 13

3. (a) 90°; (b) $2s$ and $2p$; (c) No
4. (a) 109.5°; (b) 105°; (c) Unshared pairs of electrons repel each other.
5. (a) and (b): No; (c) Yes
10. (a) H—O $\quad$; $\quad$ (b) 0;
$\quad\quad$ ↗ H

(c) H$\overset{\leftrightarrow}{—}$Cl; $\quad$ (d) Br$\overset{\leftrightarrow}{—}$Cl

11. HF is highly hydrogen bonded while HCl and HBr are not.
12. (a) $1 + 6 \rightarrow 6\ CO_{2(g)} + 6\ H_2O_{(g)}$;
(c) $1 + 1 \rightarrow 2\ H_2O_{(\ell)} + CaSO_{4(s)}$;
(e) $1 + 2 \rightarrow K_2CO_3 + H_2O$;
(g) $2 + 2 \rightarrow 2\ KOH + H_{2(g)}$;
(i) $1 + 1 \xrightarrow{\Delta} MgO_{(s)} + H_{2(g)}$;
(j) $2 + 3 \xrightarrow{\Delta} Al_2O_{3(s)} + 3\ H_{2(g)}$
13. (a) $2 + 7 \xrightarrow{\Delta} 4\ CO_{2(g)} + 6\ H_2O_{(g)}$;
(c) $1 + 1 \rightarrow CaCO_{3(s)} + H_2O$;
(e) $2 + 1 \rightarrow K_2SO_4 + H_2O$;
(g) $1 + 2 \rightarrow ZnCl_2 + H_2O$;
(h) $1 + 2 \rightarrow Ca(OH)_2 + H_{2(g)}$;
(i) $2 + 2 \rightarrow 2\ LiOH + H_{2(g)}$;
(j) $1 + 1 \xrightarrow{\Delta} ZnO_{(s)} + H_{2(g)}$
14. (a) 51.1%; (b) 20.9%; (c) 43.8%; (d) 44.8%; (e) 24.5%
15. (a) $CuSO_4 \cdot H_2O$;
(b) $CuSO_4 \cdot 3\ H_2O$;
(c) $CuSO_4 \cdot 5\ H_2O$;
(d) $Na_2CO_3 \cdot H_2O$;
(e) $Na_2CO_3 \cdot 10\ H_2O$
16. $CaCl_2 \cdot 6\ H_2O$
17. (Structural formulas only)
(a) H—O
$\quad\quad\quad$ \
$\quad\quad\quad\quad$ H ;

(b) H
$\quad$ \
$\quad\quad$ O—O
$\quad\quad\quad\quad$ \
$\quad\quad\quad\quad\quad$ H
18. (a) $2 \rightarrow 2\ H_2O + O_{2(g)}$;
(c) $3 \rightarrow 2\ O_{3(g)}$
19. 7.03 g

20. 3.86 L
21. 5.54 L
22. 39.5 mL

Chapter 14

3. (a) solid in liquid;
   (c) solid in liquid;
   (e) solid in liquid;
   (g) gas in liquid;
   (j) solid in liquid
4. (a) water; (b) water;
   (c) carbon tetrachloride;
   (d) water
6. Add crystal of solute.
7. (a) solid in liquid—sol;
   (c) liquid in gas—liquid aerosol
8. A = solution
9. The colloid comes in contact with ocean water containing ionic compounds at the river delta.
10. (a) 12 g/100 g;
    (b) 5 g/100 g;
    (c) 36 g/100 g;
    (d) 34 g/100 g;
    (e) 50 g/100 g;
    (f) 45 g/100 g
11. 0.0400 g/L
12. 0.294 g/L
13. (a) 8.95%; (b) 20.1%;
    (c) 4.71%
14. (a) 12.5%; (b) 13.3%;
    (c) 1.02%
15. (a) 71.7 g; (b) 2.05 g
16. (a) 53.6 g; (b) 207 g
17. (a) 569 g; (b) 20.0 g
18. (a) 171 ppm; (b) 202 ppm;
    (c) 1.7 ppm
19. (a) 650 ppm; (b) 130 ppm;
    (c) 6.0×10^{-6} ppm
20. (a) 48 mg; (b) 640 mg;
    (c) 4.5×10^{-3} mg
21. (a) 3.65 M; (b) 1.13 M;
    1.13 M Cl^{1-}; (c) 0.221 M
22. (a) 0.252 M; 0.252 M Br^{1-};

(b) 0.169 M; 0.338 M Cl^{1-};
(c) 0.0439 M; 0.0878 M
Br^{1-}
23. (a) 2.20 g; the NaOH (2.20 g) is dissolved in sufficient water to make the total volume of the solution equal to $50\overline{0}$ mL; (b) 6.11 g;
   (c) 2.20g
24. (a) 496 mL; (b) 49.2 mL;
   (c) $20\overline{0}$ mL
25. (a) 0.492 N; (b) 0.0491 N;
   (c) 8.01 N
26. (a) 0.329 N; (b) 0.318 N;
   (c) 0.288 N
27. (a) 0.127 g; (b) 3.79 g;
   (c) 0.709 g
28. (a) 316 mL; (b) 706 mL;
   (c) $54\overline{0}$ mL
29. (a) 5.69 m; (b) 2.97 m;
   (c) 0.131 m
30. (a) 0.419 m; (b) 0.406 m;
   (c) 2.79 m
31. (a) 7.23 g; (b) 32.7 g;
   (c) 70.2 g
32. (a) 181 g; (b) 34.7 g;
   (c) 342 g
33. (a) 10.8%; (b) 43.9%;
   (c) 14.7%
34. (a) 1.84 M; (b) 0.100 M;
   (c) 0.827 M
35. (a) 6.10 N; (b) 1.50 N;
   (c) 6.90 N
36. (a) 6.85 m; (b) 14.0 m;
   (c) 6.31 m
37. (a) 100.86°C, -3.07°C;
   (b) 100.21°C, -0.74°C;
   (c) 79.31°C, -118.62°C
38. (a) 100.39°C, -1.40°C;
   (b) 100.34°C, -1.23°C;
   (c) 79.36°C, -118.71°C
39. (a) 97.6 amu; (b) 43.3 amu; (c) 111 amu
40. (a) 75.0 amu; (b) 41.1 amu; (c) 132 amu
41. $50\overline{0}$ m, 18.3 M, 36.6 N, 36.6 M H^{1+}
42. 16.1 m, 12.0 M, 12.0 N
43. $C_6H_{12}O_6$

Chapter 15

6. (a), (d), (e): acid
7. (a), (d), (e): acid; (b), (c), (f): amphiprotic
8. 0.0455 M
9. 0.0194 M
10. 0.123 M
11. (a) 0.752 M; (b) 2.71%
12. (a) 0.792 M; 4.72%
13. 0.383 M
14. (a) 0.877 M; (b) 3.15%
15. 0.160 N
16. 35.3 mL
17. (a) 9.0, 5.0; (b) 11.7, 2.3;
   (c) 6.7, 7.3; (d) 6.2, 7.8;
   (e) 3.6, 10.4
18. (a) 4.6, 9.4; (b) 3.1, 10.9;
   (c) 2.9, 11.1; (d) 8.3, 5.7;
   (e) 5.1, 8.9
19. (a) 7.5 to 8.0 (b) Bitter soapy taste due to properties of bases.
20. (a) 4.5, 9.5; (b) 10.0, 4.0
21. (a) 6.3×10^{-6} mol/L;
   (b) 2.5×10^{-10} mol/L;
   (c) 3.2×10^{-12} mol/L;
   (d) 1.0×10^{-2} mol/L;
   (e) 5.9×10^{-10} mol/L;
   (f) 5.0×10^{-9} mol/L
22. (a) 1.0×10^{-5} mol/L;
   (b) 2.5×10^{-10} mol/L;
   (c) 5.0×10^{-9} mol/L;
   (d) 1.6×10^{-6} mol/L;
   (e) 2.5×10^{-9} mol/L
23. 1.0×10^{-4} mol/L. Nonmetal oxides dissolved in water form acids.
24. (Net ionic equations only)
   (a) $Ba^{2+} + CO_3^{2-} \rightarrow BaCO_{3(s)}$;
   (b) $Fe^{3+} + 3\ NH_3 + 3\ H_2O \rightarrow Fe(OH)_{3(s)} + 3\ NH_4^{1+}$;
   (c) $Sr^{2+} + CO_3^{2-} \rightarrow SrCO_{3(s)}$;
   (d) $CO_3^{2-} + 2\ H^{1+} \rightarrow H_2O + CO_{2(g)}$;
   (e) $Cl^{1-} + Ag^{1+} \rightarrow AgCl_{(s)}$;

(f) $2 Al_{(s)} + 6 H^{1+} \rightarrow$
$2 Al^{3+} + 3 H_{2(g)}$;

(g) $CO_{2(g)} + Ca^{2+} +$
$2 OH^{1-} \rightarrow CaCO_{3(s)} +$
H_2O;

(h) $SrCO_{3(s)} + 2 HC_2H_3O_2$
$\rightarrow Sr^{2+} + 2 C_2H_3O_2^{1-}$
$+ H_2O + CO_{2(g)}$;

(i) $Fe + Cu^{2+} \rightarrow Fe^{2+} +$
$Cu_{(s)}$;

(j) $Cd^{2+} + H_2S \rightarrow CdS_{(s)}$
$+ 2 H^{1+}$

25. (Net ionic equations only)
(a) $HgCl_2 + H_2S \rightarrow HgS_{(s)}$
$+ 2 H^{1+} + 2 Cl^{1-}$;

(b) $Mg^{2+} + 2 OH^{1-} \rightarrow$
$Mg(OH)_{2(s)}$;

(c) $CaO_{(s)} + 2 H^{1+} \rightarrow$
$Ca^{2+} + H_2O$;

(d) $Fe^{3+} + 3 NH_3 +$
$3 H_2O \rightarrow Fe(OH)_{3(s)} +$
$3 NH_4^{1+}$;

(e) $Fe^{2+} + S^{2-} \rightarrow FeS_{(s)}$;

(f) $Al(OH)_{3(s)} + 3 H^{1+} \rightarrow$
$Al^{3+} + 3 H_2O$;

(g) $H_3PO_4 + 3 OH^{1-} \rightarrow$
$PO_4^{3-} + 3 H_2O$;

(h) $Mg^{2+} + CO_3^{2-} \rightarrow$
$MgCO_{3(s)}$;

(i) $Cl_{2(g)} + 2 Br^{1-} \rightarrow$
$2 Cl^{1-} + Br_2$;

(j) no ionic equation

26. (a) $2.14 M$; (b) 7.53%

27. Net: $AlZ_{3(s)} + 3 H^{1+} \rightarrow$
$Al^{3+} + 3HZ$

Chapter 16 _____

(The numbers represent the coefficients in front of the formulas in the balanced equation.)

8. (a) $2 + 6 \rightarrow 2 + 3 + 4$;
(b) $8 + 5 \rightarrow 1 + 4 + 4 + 4$;
(c) $3 + 8 \rightarrow 3 + 2 + 4$;
(d) $1 + 7 + 8 \rightarrow 8 + 4 + 4$;
(e) $10 + 1 \rightarrow 10 + 4 + 2$;
(f) $2 + 2 \rightarrow 1 + 1 + 2$;

(g) $1 + 6 + 14 \rightarrow 2 + 6$
$+ 7$;

(h) $1 + 4 + 1 \rightarrow 1 + 1 +$
2;

(i) $1 + 1 + 2 \rightarrow 2 + 1 +$
1;

(j) $3 + 1 + 14 \rightarrow 3 + 2$
$+ 7$

9. Oxidized, reducing agent:
(a) HI; (b) KI; (c) Cu;
(d) KI; (e) I_2; (f) Ag;
(g) Fe^{2+}; (h) Cu: (i) CdS;
(j) Zn

10. (a) $2 + 3 \rightarrow 2 + 3 + 3$;
(b) $6 + 10 \rightarrow 3 + 10 + 5$;
(c) $1 + 4 + 6 \rightarrow 6 + 1 +$
$3 + 2$;
(d) $2 + 5 + 16 \rightarrow 2 + 5$
$+ 5 + 7$;
(e) $3 + 5 + 1 + 3 \rightarrow 3 +$
$3 + 3$;
(f) $1 + 5 \rightarrow 1 + 5$;
(g) $1 + 14 + 6 \rightarrow 2 + 3$
$+ 7$;
(h) $10 + 2 + 16 \rightarrow 2 + 5$
$+ 8$;
(i) $2 + 5 + 2 \rightarrow 2 + 5 +$
4;
(j) $3 + 1 + 8 \rightarrow 3 + 2 +$
7

11. Oxidized, reducing agent:
(a) K_2SnO_2; (b) Sb;
(c) NaI; (d) $Mn(NO_3)_2$;
(e) KI; (f) CO; (g) Cl^{1-};
(h) I^{1-}; (i) SO_2; (j) H_2S

12. (a) $3 + 1 + 6H^{1+} \rightarrow 3 +$
$1 + 3$;
(b) $5 + 2 + 6H^{1+} \rightarrow 5 +$
$2 + 3$;
(c) $5 + 2 + 16H^{1+} \rightarrow 10$
$+ 2 + 8$;
(d) $2 + 5 + 14H^{1+} \rightarrow 2$
$+ 5 + 7$;
(e) $2 + 5 + 6H^{1+} \rightarrow 2 +$
$8 + 5$;
(f) $1 + 1 + 4H^{1+} \rightarrow 1 +$
$1 + 2$;
(g) $3 + 6OH^{1-} \rightarrow 1 + 5$
$+ 3$;

(h) $1 + 2OH^{1-} \rightarrow 1 + 1$
$+ 1$;
(i) $2 + 1 + 4OH^{1-} \rightarrow 2$
$+ 2$;
(j) $1 + 4 \rightarrow 1 + 4$

13. Oxidized, reducing agent;
(a) Sn^{2+}; (b) AsO_2^{1-};
(c) $C_2O_4^{2-}$; (d) Mn^{2+};
(e) H_2O_2; (f) Fe; (g) Cl_2;
(h) Cl_2; (i) MnO_2; (j) PbS

14. (a) $1 + 3 + 14 H^{1+} \rightarrow 2$
$+ 6 + 7$;
(b) $3 + 2 + 8 H^{1+} \rightarrow 3 +$
$2 + 4$;
(c) $5 + 2 + 6 H^{1+} \rightarrow 5 +$
$2 + 3$;
(d) $1 + 1 + 1 \rightarrow 1 + 2 +$
$2 H^{1+}$;
(e) $1 + 2 + 2 H^{1+} \rightarrow 1 +$
$1 + H_2O$;
(f) $2 + 3 \rightarrow 3 + 2 + 3$;
(g) $1 + 1 + 2 OH^{1-} \rightarrow 1$
$+ 2$;
(h) $4 + 3 + 2 \rightarrow 4 + 3 +$
$4 OH^{1-}$;
(i) $3 + 8 H^{1+} + 2 NO_3^{1-}$
$\rightarrow 3 Ni^{2+} + 2 + 3 +$
4;
(j) $SbCl_6^{1-} + 1 \rightarrow SbCl_6^{3-}$
$+ 1 + 2 H^{1+}$

15. Oxidized, reducing agent:
(a) $C_2O_4^{2-}$; (b) S^{2-};
(c) SO_3^{2-}; (d) AsO_3^{3-};
(e) I^{1-}; (f) SnO_2^{2-};
(g) Mn^{2+}; (h) ClO_2^{1-};
(i) NiS; (j) H_2S

16. (a), (b), and (d): Yes

17. (a) $+2.36$ V; (b) -0.14 V;
(c) -0.44 V; (d) -1.36 V;
(e) $+1.06$ V;

18. (a) F_2; (b) Ca

19. Small positive oxidation E^0 value and hence not very active.

20. (a) 35.7 g; (b) 26.7 L;
(c) 31.9 L; (d) 21.9 g

21. 7.78 g

22. (a) $2 + 10 + 8 \rightarrow 2 + 5$
$+ 1 + 8$;

(b) 0.00421 mol;
(c) 42.1 mL
23. (a) $3 + 2 + 8 \rightarrow 3 + 2 + 2 + 11$;
(b) $K_2Cr_2O_7$

Chapter 17 _____

4. A catalyst does not alter the point of equilibrium.
7. (a) second order NO, first order Br_2, overall third order; (b) second order A, zero order B, overall second order
8. (a) 4 times; (b) 9 times; (c) 3 times; (d) 6.25 times
9. Sulfurous acid >
   chloroacetic acid >
   formic acid >
   lactic acid >
   barbituric acid >
   acetic acid
10. Silver hydroxide >
    ammonia > novocain >
    codeine > nicotine > urea
11. (a) $K = \dfrac{[CH_3Cl]\,[HCl]}{[CH_4]\,[Cl_2]}$;

    (c) $K = \dfrac{[NO_2]^2}{[NO]^2\,[O_2]}$;

    (e) $K = \dfrac{[H_2]^4}{[H_2O]^4}$;

    (g) $K = \dfrac{[SO_2]\,[Cl_2]}{[SO_2Cl_2]}$;

    (i) $K = [NO_2]^4\,[O_2]$;

    (j) $K = \dfrac{[H_2O]^2}{[H_2]^2\,[O_2]}$

12. (a) $K = \dfrac{[NO]^2}{[N_2]\,[O_2]}$;

    (c) $K = [SO_2]$;

    (e) $K = \dfrac{[NO]^2\,[Cl_2]}{[NOCl]^2}$;

    (g) $K = [Hg]^2\,[O_2]$;

    (i) $K = \dfrac{[CO_2]\,[H_2O]^2}{[CH_4]\,[O_2]^2}$;

    (j) $K = \dfrac{1}{[Cl_2]}$

13. (a), (c), (f), (h), (i): right
14. (a), (c), (f), (h): right; (i) and (j): no effect
15. (a) (1) left, (2) right, (3) right, (4) left;
    (b) (1) left, (2) right, (3) no effect, (4) no effect;
    (c) (1) right, (2) left, (3) no effect, (4) no effect;
    (d) (1) right, (2) left, (3) right, (4) left;
    (e) (1) left, (2) right, (3) right, (4) left
16. (a) (1) left, (2) right, (3) right, (4) left;
    (b) (1) left, (2) right, (3) no effect, (4) no effect;
    (c) (1) left, (2) right, (3) right, (4) left;
    (d) (1) right, (2) left, (3) right, (4) left;
    (e) (1) left, (2) right, (3) no effect, (4) no effect
17. (a) 1.81×10^{-5} (mol/L);
    (b) 7.38×10^{-4} (mol/L);
    (c) 7.20×10^{-10} (mol/L);
    (d) 9.14×10^{-5} (mol/L)
18. (a) 1.81×10^{-5} (mol/L);
    (b) 1.97×10^{-4} (mol/L);
    (c) 5.00×10^{-5} (mol/L);
    (d) 1.78×10^{-4} (mol/L)
19. (a) 1.84×10^{-3} mol/L, 0.920%;
    (b) 2.92×10^{-3} mol/L, 0.584%;
    (c) 1.82×10^{-2} mol/L, 0.910%;
    (d) 4.00×10^{-4} mol/L, 0.160%
20. (a) 1.87×10^{-5} mol/L, 0.0187%;
    (b) 8.00×10^{-7} mol/L, 0.00288%;
    (c) 2.35×10^{-4} mol/L, 0.470%;

(d) 9.49×10^{-5} mol/L, 0.0316%
21. Silver acetate > silver bromate > silver chloride > silver bromide > silver iodide
22. (a) $K_{sp} = [Ag^{1+}][Cl^{1-}]$;
    (c) $K_{sp} = [Sn^{4+}][S^{2-}]^2$;
    (e) $K_{sp} = [As^{5+}]^2[S^{2-}]^5$
23. (a) $K_{sp} = [Hg^{2+}][S^{2-}]$;
    (c) $K_{sp} = [Fe^{3+}][OH^{1-}]^3$;
    (e) $K_{sp} = [Cu^{1+}]^2[S^{2-}]$
24. (a) 7.73×10^{-13} (mol²/L²);
    (b) 8.10×10^{-9} (mol²/L²);
    (c) 6.97×10^{-30} (mol²/L²);
    (d) 2.07×10^{-10} (mol²/L²)
25. (a) 1.60×10^{-9} (mol²/L²);
    (b) 2.99×10^{-26} (mol²/L²);
    (c) 3.70×10^{-19} (mol²/L²);
    (d) 5.70×10^{-7} (mol²/L²)
26. (a) 6.00×10^{-15} M, 8.67×10^{-13} g/L;
    (b) 8.37×10^{-5} M, 1.65×10^{-2} g/L;
    (c) 2.00×10^{-3} M, 0.569 g/L;
    (d) 1.18×10^{-12} M, 1.07×10^{-10} g/L, Yes
27. (a) 1.22×10^{-8} M, 2.86×10^{-6} g/L;
    (b) 1.82×10^{-7} M, 4.86×10^{-5} g/L;
    (c) 0.152 M, 23.7 g/L;
    (d) 1.55×10^{-5} M, 3.93×10^{-3} g/L, No
28. (a) 2.5, 11.5; (b) 11.1, 2.9
29. (a) 3.4, 10.6; (b) 11.5, 2.5
30. Lower pH shifts equilibrium to left increasing concentration of Cl_2.

Chapter 18 _____

3. (a) $CH_3(CH_2)_2CH_3$,
   CH_3CHCH_3;
       |
       CH_3

(c) $CH_3(CH_2)_4CH_3$,

$CH_3CH(CH_2)_2CH_3$,
    $|$
    CH_3

$CH_3CH_2CHCH_2CH_3$,
       $|$
       CH_3

$CH_3CH—CHCH_3$,
   $|$ $|$
   CH_3 CH_3

   CH_3
   $|$
$CH_3C—CH_2CH_3$;
   $|$
   CH_2

(d) $CH_3(CH_2)_5CH_3$,

$CH_3CH(CH_2)_3CH_3$,
    $|$
    CH_3

$CH_3CH_2CH(CH_2)_2CH_3$,
          $|$
          CH_3

$CH_3CH—CH—CH_2CH_3$,
   $|$ $|$
   CH_3 CH_3

$CH_3CHCH_2CHCH_3$,
   $|$ $|$
   CH_3 CH_3

   CH_3
   $|$
$CH_3C—(CH_2)_2CH_3$,
   $|$
   CH_3

        CH_3
        $|$
$CH_3CH_2C—CH_2CH_3$,
        $|$
        CH_3

        CH_3
        $|$
$CH_3CH—C—CH_3$,
   $|$ $|$
   CH_3 CH_3

$CH_3CH_2CHCH_2CH_3$
       $|$
       CH_2CH_3

4. (a) 2-bromobutane;
   (c) 3-bromo-2-nitropentane;
   (e) 2-iodo-3-methyl-4-nitrohexane;
   (f) methylcyclobutane

5. (a) CH_3CH_2CHBr;
        $|$
        CH_3
   (c) CH_3CHCH_3;
        $|$
        NO_2
            CH_3
            $|$
   (e) $CH_3CH—C—(CH_2)_2CH_3$;
        $|$ $|$
        Br CH_3
   (f) [cyclopentane ring with]
        H
        $|$
        CH_2CH_3

6. (a) $1 + 1 \rightarrow CH_3—Cl + HCl_{(g)}$;
   (b) $1 + 1 \rightarrow CH_3CH_2Br + HBr_{(g)}$

7. (a) 2-butene;
   (c) 3-methyl-2-pentene;
   (e) 3-chloro-3-methyl-1-hexene;
   (f) 4-chloro-6-nitro-2-heptene

8. (a) $CH_2=CH(CH_2)_2CH_3$;
   (c) $CH_2=C—CH(CH_2)_2CH_3$;
         $|$ $|$
         CH_3 CH_3
   (e) $CH=CH—CH—(CH_2)_3CH_3$;
       $|$ $|$
       Br CH_2CH_3
                      NO_2
                      $|$
   (f) $CH_3C=CH—CH—C—CH_2CH_3$
         $|$ $|$ $|$
         CH_3 CH_3 CH_3

9. (a) $1 + 1 \rightarrow CH_3CH—CH_2$;
                         $|$ $|$
                         Br Br
                    Br Br
                    $|$ $|$
   (b) $1 + 1 \rightarrow CH_3C—CHCH_3$
                    $|$
                    CH_3

10. (a) $CH_2=CH_2$

    $-[CH_2—CH_2]_n-$;

(b) $CF_2=CF_2$

   $\begin{array}{ccc} & F & F \\ & | & | \\ -[& C—C &]- \\ & | & | \\ & F & F \end{array}_n$;

(c) $CH_2=CH$
          $|$
          Cl

   $-[CH_2—CH]-$;
           $|$
           Cl $_n$

(d) $CH_2=CH$
          $|$
          CN

   $-[CH_2—CH]-$
           $|$
           CN $_n$

11. (a) 1-pentyne;
    (c) 4-chloro-4-methyl-2-hexyne;
    (e) 5, 6-dimethyl-3-heptyne;
    (f) 4-methyl-2-hexyne

12. (a) $H—C\equiv C—(CH_2)_4—CH_3$;

    $H—C\equiv C—CH_2—CH—(CH_2)_2—CH_3$;
                     $|$
                     NO_2
    (e) $Cl—C\equiv C—CH—(CH_2)_2—CH_3$;
                    $|$
                    CH_3
                                  CH_3
                                  $|$
    (f) $Cl—CH_2CH—C\equiv C—C—CH—CH_2—CH_3$
                $|$ $|$ $|$
                Cl CH_3 CH_3

13. (a) $1 + 2 \rightarrow CH_3—\underset{\underset{Br}{|}}{\overset{\overset{Br}{|}}{C}}—\underset{\underset{Br}{|}}{\overset{\overset{Br}{|}}{C}}—CH_3$;

    (b) $1 + 2 \rightarrow CH_3—\underset{\underset{Br}{|}}{\overset{\overset{Br}{|}}{C}}—\underset{\underset{Br}{|}}{\overset{\overset{Br}{|}}{C}}—\underset{\underset{CH_3}{|}}{\overset{\overset{CH_3}{|}}{C}}—CH_3$

14. (a) bromobenzene;
    (c) *m*-bromonitrobenzene;
    (e) 2, 4-dibromotoluene;
    (f) 2-bromo-4,5-dinitrotoluene

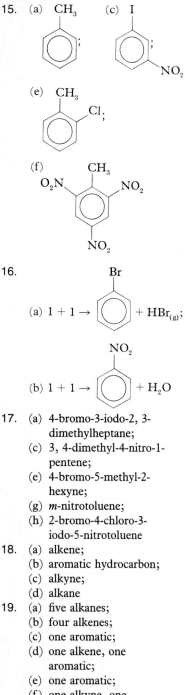

15. (a) CH₃ ; (c) I ;

(e) CH₃ Cl ;

(f) O₂N CH₃ NO₂ NO₂

16. Br

(a) 1 + 1 → + HBr₍g₎ ;

NO₂

(b) 1 + 1 → + H₂O

17. (a) 4-bromo-3-iodo-2, 3-dimethylheptane;
(c) 3, 4-dimethyl-4-nitro-1-pentene;
(e) 4-bromo-5-methyl-2-hexyne;
(g) *m*-nitrotoluene;
(h) 2-bromo-4-chloro-3-iodo-5-nitrotoluene

18. (a) alkene;
(b) aromatic hydrocarbon;
(c) alkyne;
(d) alkane

19. (a) five alkanes;
(b) four alkenes;
(c) one aromatic;
(d) one alkene, one aromatic;
(e) one aromatic;
(f) one alkyne, one aromatic;

(g) two aromatics;
(h) two aromatics

20. C₂H₆, CH₃—CH₃, ethane

Chapter 19 _____

4. (a) 2-bromobutane;
(c) 3-chloro-2-iodohexane;
(d) *p*-bromochlorobenzene

5. (a) CH₃—CH₂—CH—Cl;
|
CH₃

(c) CH₃ Cl Cl Br

6. (a) 2-pentanol;
(c) 4-pentyn-2-ol;
(d) 3-methyl-4-hexen-2-ol

7. (a) CH₃CH—OH;
|
CH₃

(c) CH₂CH—CH—CH=CH—CH₃
| | |
OH CH₃ CH₃

8. (a) *p*-iodophenol;
(c) 2,3-dichloro-4-nitrophenol;
(d) 2-bromo-3-chlorophenol

9. (a) OH ; F

(c) OH Br Br NO₂

10. (a) 1-methoxypentane;
(c) 4-bromo-2-methoxytoluene;
(d) *p*-methoxytoluene

11. (a) CH₃—O—CH₃;

(c) CH₃ OCH₂—CH₃

12. (a) propanal;
(c) *p*-methoxybenzalde-hyde;
(d) 3,4-dichlorobenzalde-hyde

13. (a) CH₃—CH—C
| \\O
Br H;

(c)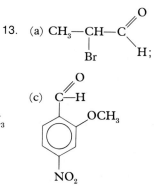

14. (a) methyl ethyl ketone;
(c) methyl *sec*-butyl ketone;
(d) propiophenone (ethyl phenyl ketone)

15. (a) 2-pentanone;
(c) 5-methyl-3-hepten-2-one;
(d) 4-methoxy-3-nitroacetophenone

16. O
||
(a) CH₃—C—CH₃ ;

(c)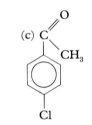

17. (a) 2-chloropropanoic acid;
    (c) 4-methoxy-3-nitrobenzoic acid;
    (d) *p*-bromobenzoic acid

18. (a)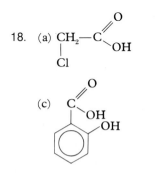

    (c)

19. (a) 1 + 1 →

    (b) 1 + 1 →

    (d) 1 + 1 →

20. (a) methyl propanoate;
    (c) methyl *p*-bromobenzoate;
    (d) ethyl 3-bromo-4-chlorobenzoate

21. (a) 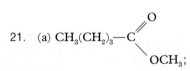 $CH_3(CH_2)_3$—C $\underset{OCH_3;}{\overset{O}{\|}}$

(c)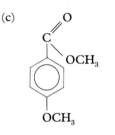

22. (a) 1 + 1 →
    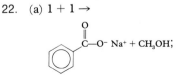 C—O⁻ Na⁺ + CH_3OH;

    (b) 1 + 1 →
    CH_3—C—OH + CH_3CH—OH
    $\overset{O}{\|}$ $\underset{CH_3}{}$

23. (a) propanamide;
    (c) *o*-bromobenzamide;
    (d) 4,4-dibromopentanamide

24. (a) CH_3CH_2C—NH_2;
    (c)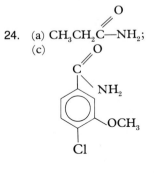

25. (a) 2-butanamine;
    (c) *m*-chlorobenzenamine or *m*-chloroaniline;
    (d) 2,3-dichlorobenzenamine or 2-3-dichloroaniline

26. (a) CH_3—CH—C—OH;
    $\underset{NH_2}{}$ $\overset{O}{\|}$

(c)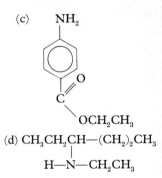

(d) CH_3CH_2CH—$(CH_2)_2CH_3$
    $\underset{H—N—CH_2CH_3}{}$

27. (a)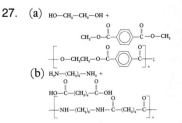

    (b)

28. (a) 4-methyl-3-hexen-2-ol;
    (c) 3-methoxybutanal;
    (e) *m*-ethoxybenzoic acid;
    (f) isopropyl 3-pentenoate;
    (g) *p*-nitrobenzamide;
    (h) N-ethyl-2-pentanamine

29. (a) one phenol, one ether, one aldehyde;
    (b) one carboxylic acid, one ester;
    (c) one alcohol;
    (d) two alcohols, three ketones;
    (e) one ether, one carboxylic acid;
    (f) four alcohols, one phenol, two ketones, one amide, one amine;
    (g) two amides, two esters;
    (h) two ketones;
    (i) one alcohol, one ketone

30. 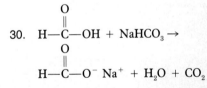 H—C—OH + $NaHCO_3$ →
    $\overset{O}{\|}$
    H—C—O⁻ Na⁺ + H_2O + CO_2

31. (a) $CH_3-CH_2-CH_2-Cl$,
$CH_3-CH-CH_3$;
with Cl substituent

(b)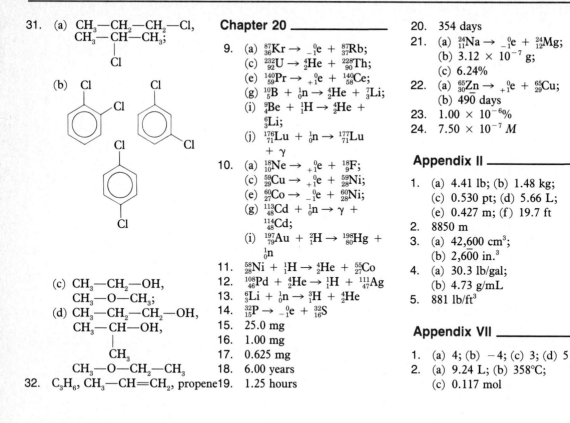

(c) CH_3-CH_2-OH,
CH_3-O-CH_3;

(d) $CH_3-CH_2-CH_2-OH$,
$CH_3-CH-OH$,
$\quad\quad |$
$\quad\quad CH_3$
$CH_3-O-CH_2-CH_3$

32. C_3H_6, $CH_3-CH=CH_2$, propene

Chapter 20

9. (a) $^{87}_{36}Kr \rightarrow {}^{0}_{-1}e + {}^{87}_{37}Rb$;
(c) $^{232}_{92}U \rightarrow {}^{4}_{2}He + {}^{228}_{90}Th$;
(e) $^{140}_{59}Pr \rightarrow {}^{0}_{+1}e + {}^{140}_{58}Ce$;
(g) $^{10}_{5}B + {}^{1}_{0}n \rightarrow {}^{4}_{2}He + {}^{7}_{3}Li$;
(i) $^{9}_{4}Be + {}^{1}_{1}H \rightarrow {}^{4}_{2}He + {}^{6}_{3}Li$;
(j) $^{176}_{71}Lu + {}^{1}_{0}n \rightarrow {}^{177}_{71}Lu + \gamma$

10. (a) $^{18}_{10}Ne \rightarrow {}^{0}_{+1}e + {}^{18}_{9}F$;
(c) $^{59}_{29}Cu \rightarrow {}^{0}_{+1}e + {}^{59}_{28}Ni$;
(e) $^{60}_{27}Co \rightarrow {}^{0}_{-1}e + {}^{60}_{28}Ni$;
(g) $^{113}_{48}Cd + {}^{1}_{0}n \rightarrow \gamma + {}^{114}_{48}Cd$;
(i) $^{197}_{79}Au + {}^{2}_{1}H \rightarrow {}^{198}_{80}Hg + {}^{1}_{0}n$

11. $^{58}_{28}Ni + {}^{1}_{1}H \rightarrow {}^{4}_{2}He + {}^{55}_{27}Co$
12. $^{108}_{46}Pd + {}^{4}_{2}He \rightarrow {}^{1}_{1}H + {}^{111}_{47}Ag$
13. $^{6}_{3}Li + {}^{1}_{0}n \rightarrow {}^{3}_{1}H + {}^{4}_{2}He$
14. $^{32}_{15}P \rightarrow {}^{0}_{-1}e + {}^{32}_{16}S$
15. 25.0 mg
16. 1.00 mg
17. 0.625 mg
18. 6.00 years
19. 1.25 hours
20. 354 days
21. (a) $^{24}_{11}Na \rightarrow {}^{0}_{-1}e + {}^{24}_{12}Mg$;
(b) 3.12×10^{-7} g;
(c) 6.24%
22. (a) $^{65}_{30}Zn \rightarrow {}^{0}_{+1}e + {}^{65}_{29}Cu$;
(b) 490 days
23. 1.00×10^{-6}%
24. 7.50×10^{-7} M

Appendix II

1. (a) 4.41 lb; (b) 1.48 kg;
(c) 0.530 pt; (d) 5.66 L;
(e) 0.427 m; (f) 19.7 ft
2. 8850 m
3. (a) 42,600 cm³;
(b) 2,600 in.³
4. (a) 30.3 lb/gal;
(b) 4.73 g/mL
5. 881 lb/ft³

Appendix VII

1. (a) 4; (b) −4; (c) 3; (d) 5
2. (a) 9.24 L; (b) 358°C;
(c) 0.117 mol

Glossary

acid (Arrhenius definition) A substance that yields hydrogen ions (H^{1+}) when dissolved in water (see 15-1).

acid (Brønsted-Lowry definition) A substance that can give or donate a proton (H^{1+}) to some other substance (see 15-1).

acid oxide A nonmetal oxide (see 9-7).

actual yield Amount of product that is actually obtained in a given reaction (see 10-4).

addition polymers Polymers formed from alkene monomers which react with each other by "addition" across the double bond breaking the pi (π) bond (see 18-5).

alkanes (saturated hydrocarbons) Aliphatic hydrocarbons that have the general molecular formula C_nH_{2n+2} for open-chain systems (see 18-3).

alkenes Aliphatic hydrocarbons that have the general molecular formula C_nH_{2n} for open-chain systems (see 18-4).

alkynes Aliphatic hydrocarbons that have the general molecular formula C_nH_{2n-2} for open-chain systems (see 18-5).

amorphous solid A solid whose particles are arranged in an irregular manner and hence lacks shape or form (see 12-8).

amphiprotic A substance capable of behaving either as a Brønsted-Lowry acid or base (see 15-1).

analytical chemistry A study of qualitative and quantitative analysis (examination) of elements and compounds (see 1-2).

anions Ions with a negative charge (see 6-1).

aromatic hydrocarbons Hydrocarbons that include benzene and compounds containing aliphatic or aromatic groups attached to aromatic rings (see 18-7).

atom   The smallest particle of an element that can undergo chemical changes in a reaction (see 3-7).

atomic mass scale   Relative scale of atomic masses, based on an arbitrarily assigned value of exactly 12 atomic mass units (amu) for the mass of carbon-12 (see 4-1).

atomic number   Number of protons found in the nucleus of an atom of an element (see 4-4).

Avogadro's   (ä′vo·gä′dro) number (N) 6.02×10^{23}. The number of elementary units such as atoms, formula units, molecules, or ions that constitutes one mole of the said particle (see 8-2).

base   (Arrhenius definition) A substance that yields hydroxide ions (OH^{1-}) when dissolved in water (see 15-1).

base   (Brønsted-Lowry) A substance capable of receiving or accepting a proton (H^{1+}) from some other substance (see 15-1).

basic oxide   A metal oxide (see 9-7).

biochemistry   A study of the chemistry of biological processes, such as the utilization of foods that produce energy and the synthesis of biologically active compounds in living organisms (see 1-2).

boiling point   The temperature at which the vapor pressure of the liquid is equal to the external pressure acting upon the surface of the liquid. The *normal* boiling point of a liquid is the temperature at which the vapor pressure of the liquid is $76\overline{0}$ torr (see 12-4).

bond angle   The angle formed between three atoms in a molecule (see 6-6).

bond length   The distance between the nuclei of covalently bonded atoms (see 6-4).

Boyle's Law   At constant temperature, the volume of a fixed mass of a given gas is inversely proportional to the pressure it exerts (see 11-3).

calorie   The amount of heat required to raise the temperature of 1 g of water from 14.5 to 15.5°C (see 2-12).

catalyst   A substance that speeds up a chemical reaction but is recovered relatively unchanged at the end of the reaction (see 9-2).

cations   Ions with a positive charge (see 6-1).

Charles' Law   At constant pressure, the volume of a fixed mass of a given gas is directly proportional to the kelvin temperature (see 11-4).

chemical changes   Changes in substances that result in a change in the composition of the substance. New substances are formed (see 3-5).

chemical kinetics   The study of reaction rates and the mechanism or path of a reaction from its reactants to its products. This is a special subdivision of most chemistry divisions (physical, inorganic, organic, biochemistry, etc.) (see**introduction to Chapter 17**).

chemical properties   Properties of substances that can be observed only when a substance undergoes a change in composition (see 3-4).

chemistry   A study of the composition of matter and the changes it undergoes (see 1-1).

colloid A dispersed mixture in which the particles are dispersed without appreciable bonding to solvent molecules and do not settle out on standing (see 14-13).

compound A pure substance that can be broken down by various chemical means into two or more different substances (see 3-3).

condensation The reverse of evaporation, that is, the return of molecules from the vapor state to the liquid state (see 12-2).

condensation polymers Polymers formed by "condensation" between two molecules, each molecule considered a monomer, through the reaction of certain functional groups (see 19-11).

Conservation of Energy, Law of Energy can be neither created nor destroyed, but may be transformed from one form to another (see 3-6).

Conservation of Mass, Law of Mass can be neither created nor destroyed (see 3-6).

constitutional (structural) isomers Compounds having the same molecular formula but different structural formulas (see 18-3).

coordinate covalent bond A bond formed when *both* of the electrons of the electron-pair bond are supplied by *one* atom (see 6-5).

covalent bond A bond formed by the sharing of electrons between atoms (see 6-4).

crystalline solid A solid whose particles are arranged in a definite geometric shape or form that is distinctive for a given solid (see 12-8).

curie (Ci) The amount (mass) of a radioactive isotope that will give 3.7×10^{10} disintegrations per second (see 20-1).

Dalton's Law of Partial Pressures Each gas in a mixture of gases exerts a partial pressure equal to the pressure it would exert if it were the only gas present in the same volume; the total pressure of the mixture is then the sum of the partial pressures of all the gases present (see 11-7).

Definite Proportions or Constant Composition, Law of A given pure compound always contains the same elements in exactly the same proportions by mass (see 3-8).

deliquescent substance A substance that absorbs enough moisture from the air to form a solution, such as calcium chloride (see 13-8).

density Mass of a substance occupying a unit volume (see 2-13).

$$\text{Density} = \frac{\text{mass}}{\text{volume}}$$

dispersed particles (dispersed phase) The colloidal particles in a colloid, comparable to the solute in a solution, with a range from 10 to 2000 Å in diameter (see 14-13).

dispersing medium (dispersing phase) The substance in a colloid in which the colloidal particles are distributed, comparable to the solvent in a solution (see 14-13).

dissociation A process referring to the separation of ionic substances into ions by the action of the solvent (see 15-1).

distillation A process used in the purification of liquids (see 12-5).

efflorescent substance   A hydrate that loses its water of hydration when simply exposed to the atmosphere, such as washing soda—$Na_2CO_3 \cdot 10\ H_2O$ (see 13-8).

electrolytes   Substances whose aqueous solutions or melted salts conduct an electric current, as observed by the glowing of a standard light bulb, because they release ions in the solution (see 15-5).

electrolytes, strong   Substances whose aqueous solutions conduct an electric current to produce a *bright* glow in a standard light bulb. Most salts; some acids—such as sulfuric acid (H_2SO_4), hydrochloric acid (HCl), nitric acid (HNO_3), and perchloric acid ($HClO_4$); some bases—such as group IA hydroxides like sodium and potassium hydroxide (NaOH and KOH); and other bases, such as barium, strontium, and calcium hydroxide [$Ba(OH)_2$, $Sr(OH)_2$, and $Ca(OH)_2$] are classed as strong electrolytes (see 15-5).

electrolytes, weak   Substances whose aqueous solutions conduct an electric current to produce a *dull* glow in a standard light bulb. Most acids and bases and the salts lead(II) acetate [$Pb(C_2H_3O_2)_2$] and mercury(II) or mercuric chloride ($HgCl_2$) are classed as weak electrolytes (see 15-5).

electron   A particle having a relative unit negative charge (actual charge $= -1.602 \times 10^{-19}$ coulomb) with a mass of 9.109×10^{-28} g or 5.486×10^{-4} amu (relatively negligible) (see 4-3).

electrovalent or ionic bond   A bond formed by the transfer of one or more electrons from one atom to another (see 6-3).

element   A pure substance that cannot be decomposed into simpler substances by ordinary chemical means. All of its atoms have the same atomic number (see 3-3).

empirical formula   The formula of a compound that contains the smallest whole number ratio of atoms present in a molecule or formula unit of a compound (see 8-5).

endothermic reaction   A reaction in which heat is absorbed (see 10-7).

energy   The capacity for doing work or to transfer heat (see 2-10).

equation, chemical   A shorthand way of expressing a chemical change (reaction) in terms of symbols and formulas (see 9-1).

equation, ionic   Expresses a chemical change (reaction) in terms of ions for those compounds existing mostly in ionic form in aqueous solution (see 15-6).

equation, net ionic   Shows only those ions that have actually undergone a chemical change (see 15-6).

equation, word   Expresses the chemical equation in words instead of symbols and formulas (see 9-5).

evaporation   The actual escape of molecules (the most energetic) from the surface of the liquid (below the boiling point) to form a vapor in the surrounding space above the liquid (see 12-2).

exothermic reaction   A reaction in which heat is evolved (see 10-7).

exponent   A whole number or symbol written as a superscript above another number or symbol, the base, denoting the number of times the base must be repeated as a factor (see 2-3).

exponential notation   A form for expressing a number using a product of two numbers; one of the numbers is a decimal and the other is a power of 10 (see 2-3).

fission, nuclear   The splitting of an atomic nucleus into two or more smaller nuclei (see 20-5).

formula unit   Generally, the smallest combination of charged particles (ions) in which the opposite charges present balance each other so that the overall compound has a net charge of zero, such as NaCl (see 3-8 and 6-3).

freezing point (melting point)   The temperature at which the liquid and solid forms are in dynamic equilibrium with each other. At dynamic equilibrium, the rate of melting is equal to the rate of freezing (see 12-9).

functional group   An atom or group of atoms that is attached to a hydrocarbon chain and in general determines the chemical and physical properties of the organic compound (see **introduction to Chapter 19**).

fusion, nuclear   A process in which energy is produced by combining two or more nuclei to form a heavier nucleus (see 20-6).

Gay-Lussac's (gā′lü·sak′) **Law**   At constant volume, the pressure of a fixed mass of a given gas is directly proportional to the kelvin temperature (see 11-5).

Gay-Lussac's Law of Combining Volumes   At the same temperature and pressure, whenever gases react or gases are formed they do so in the ratio of small numbers by volume (see 10-6).

groups   The 18 vertical columns in the Periodic Table (see 5-2).

half-life   The time required for one-half of any given amount of a radioactive element to disintegrate (see 20-3).

heat of condensation   The quantity of heat liberated when a unit mass of vapor (usually at the normal boiling point) is condensed to the liquid; the heat of vaporization has the same numerical value (see 12-4).

heat of fusion   The quantity of heat necessary to convert a unit mass of a solid substance to the liquid state at the melting point of the substance. *Specific* heat of fusion is expressed in calories per gram or joules per kilogram; *molar* heat of fusion is expressed in kilocalories per mole or kilojoules per mole (see 12-9).

heat of reaction   The number of calories or joules of heat energy evolved or absorbed in a given chemical reaction per given amount of reactants and/or products (see 10-7).

heat of solidification   (crystallization) The quantity of heat evolved in going from the liquid state to the solid state of a unit mass of liquid at the melting point; the heat of fusion has the same numerical value (see 12-9).

heat of vaporization   The quantity of heat required to evaporate a unit mass of a given liquid at constant temperature and pressure (usually the normal boiling point). *Specific* heat of vaporization is expressed in calories per gram or joules per kilogram; *molar* heat of vaporization is expressed in kilocalories per mole or kilojoules per mole (see 12-4).

Henry's Law   The solubility of gas in a liquid is directly proportional to the partial pressure of the gas above the liquid (see 14-3).

heterogeneous matter Matter not uniform in composition and properties and consisting of two or more physically distinct portions or phases unevenly distributed (see 3-2).

homogeneous matter Matter uniform in composition and properties throughout (see 3-2).

homogeneous mixture Matter homogeneous throughout and composed of two or more pure substances whose proportions may be *varied,* in some cases without limit (see 3-2).

homologous series A series of compounds in which each compound in the series differs from the next compound by a methylene ($-CH_2-$) group (see 18-3).

Hund's Rule One electron is placed in each orbital, such as p_x, p_y, and p_z, before pairing up the electrons to a maximum of two electrons in each of these orbitals. This rule also applies to the filling of d and f orbitals (see 13-4).

hybrid orbitals Mixed orbitals, such as sp^3, consisting in this case of one part s and three parts p character (see 13-1).

hydrates Crystalline substances that contain chemically bound water in definite proportions. An example is Epsom salts, magnesium sulfate **hepta**hydrate ($MgSO_4 \cdot 7H_2O$) (see 13-8).

hydrocarbons Organic compounds consisting of only the elements carbon and hydrogen (see 18-2).

hydrogen bond A type of bond resulting when a hydrogen atom bonded to a highly electronegative atom (F, O, and N) becomes bonded additionally to another electronegative atom. An example is water (see 13-3).

$$\text{H}-\overset{..}{\underset{|}{\text{O}}}:\text{----H}-\overset{..}{\underset{|}{\text{O}}}:\text{----H}-\overset{..}{\underset{|}{\text{O}}}:\text{----}$$
$$\quad\ \text{H}\qquad\qquad\text{H}\qquad\qquad\text{H}$$

hygroscopic substance A substance that readily absorbs moisture from the air, such as sugar (see 13-8).

ideal-gas equation $PV = nRT$ (see 11-8). (P = pressure, V = volume, n = quantity of gas in moles, T = temperature, R = universal gas constant)

indicators Compounds whose color is affected by acid and base (see 15-1).

inorganic chemistry A study of all elements and compounds other than organic compounds (see 1-2).

ionization A process referring to the formation of ions from atoms or molecules by the transfer of electrons (see 15-1).

ions Charged species (atoms or groups of atoms) with positive or negative ionic charges (see 6-1).

isotopes Atoms having different atomic masses or mass numbers but the same atomic number (see 4-5).

joule   A unit of energy equal to a force of one newton(N) acting through a distance of one meter in the direction of the force $(kg \cdot m^2/s^2)$ (see 2-12).

kinetic energy   Energy possessed by a substance by virtue of its motion (see 3-6).

Le Châtelier's (le·shä′te·lyäs′) **Principle**   If an equilibrium system is subjected to a change in conditions of concentration, temperature, or pressure, the system will change in a direction that will tend to restore the original conditions (see 17-3).

mass   The quantity of matter in a particular body (see 2-3).

Mass Action, Law of   The rate of a chemical reaction is proportional to the "active masses" of the reactants. The "active masses" have been found related to the relative molar concentration of the reactants in moles per liter for solutions or gases (see 17-1).

mass number   Sum of the number of protons and neutrons in the nucleus of an atom of an element (see 4-4).

matter   Anything that has mass and occupies space (see 2-3).

melting point (freezing point)   The temperature at which the liquid and solid forms are in dynamic equilibrium with each other. At dynamic equilibrium, the rate of melting is equal to the rate of freezing (see 12-9).

mixture   Matter composed of two or more substances, each of which retains its identity and specific properties (see 3-2).

molality (m)   The concentration of solute in a solution expressed as the number of moles of solute per *kilogram* of **solvent** (see 14-10).

$$m = \text{molality} = \frac{\text{moles of solute}}{\text{kilogram of \textbf{solvent}}}$$

molarity (M)   The concentration of solute in a solution expressed as the number of moles of solute per *liter* of *solution* (see 14-8).

$$M = \text{molarity} = \frac{\text{moles of solute}}{\text{liter of \textit{solution}}}$$

molar volume of a gas   The volume occupied by 1 mol of any gas; 22.4 L of gas molecules at 0°C and 760 torr (see 8-3).

mole (mol)   The amount of a substance containing the same number of elementary units (atoms, formula units, molecules, or ions) as there are atoms in exactly 12 g of carbon-12. One mole of elementary units consists of 6.02×10^{23} units, such as atoms, formula units, molecules, or ions, and this number of units has a mass equal to the atomic, molecular, or formula mass of the units expressed in grams (see 8-2).

molecular formula   A formula composed of an appropriate number of symbols of elements representing one molecule of the given compound. Also defined as the true formula and containing the actual number of atoms of each element in one molecule of the compound (see 3-8 and 8-5).

molecule   Generally, the smallest particle of a pure substance (element or compound) that can exist and still retain the physical and chemical properties of the substance, such as O_2 and H_2O (see 3-8).

neutron   A particle having no charge and with a mass of 1.6748×10^{-24} g or 1.0087 amu (approximately 1 amu) (see 4-3).

newton (N)   A unit of force which gives a mass of one kilogram an acceleration of one meter per second, per second ($kg \cdot m/s^2$) (see 2-12).

nonelectrolytes   Substances whose aqueous solutions do not conduct an electric current. Examples of nonelectrolytes are sugar (sucrose, $C_{12}H_{22}O_{11}$), ethyl alcohol (C_2H_6O), glycerine ($C_3H_8O_3$), and water (H_2O) (see 15-5).

normality (N)   The concentration of a solute in a solution expressed as the number of equivalents of solute per *liter* of *solution* (see 14-9).

$$N = \text{normality} = \frac{\text{equivalents of solute}}{\text{liter of } solution}$$

orbital   A region of space within an atom in which there can be no more than two electrons (see 4-9).

organic chemistry   A study of compounds containing the element carbon (see 1-2 and **introduction to Chapter 18**).

oxidation   A chemical change in which a substance loses electrons or one or more elements in it increase in oxidation number (see 16-1).

oxidation number   A positive or negative whole number used to describe the combining capacity of an element in a compound (see 6-1 and 16-1).

oxidizing agent   The substance reduced (see 16-1).

parts per million (ppm)   The concentration of a solute in a solution expressed as parts by mass of solute per 1,000,000 parts by mass of *solution* (see 14-7).

$$\text{parts per million (ppm)} = \frac{\text{mass of solute}}{\text{mass of } solution} \times 1,000,000$$

percent by mass   The concentration of a solute in a solution expressed as parts by mass of solute per 100 parts by mass of *solution* (see 14-6).

$$\text{\% by mass} = \frac{\text{mass of solute}}{\text{mass of } solution} \times 100$$

percent yield   Percent of the theoretical yield that is actually obtained in a chemical reaction (see 10-4).

$$\text{\% yield} = \frac{\text{actual yield}}{\text{theoretical yield}} \times 100$$

Periodic Law   The chemical properties of the elements are periodic functions of their atomic numbers (see 5-1).

periods   The seven horizontal rows in the Periodic Table (see 5-2).

physical changes   Changes in the properties of a substance that occur with no change in the composition of the substance. (see 3-5).

physical chemistry   A study of reaction rates, mechanisms, bonding and structure of compounds, and thermodynamics (see 1-2).

physical properties   Properties of substances that can be observed without changing the composition of the substance (see 3-4).

pi (π) bond   A covalent bond formed by a side-to-side overlap of p orbitals on adjacent carbon atoms. It has its maximum strength when the axes of these p orbitals are parallel (see 18-4).

polar bond   The unequal sharing of electrons in a covalent bond due to the differences in electronegativities of the atoms (see 6-4 and 13-2).

polyatomic ions   Ions consisting of two or more atoms with a net negative or positive charge on the ion (see 6-6).

potential energy   Energy possessed by a substance by virtue of its position in space or composition (see 3-6).

pressure   Force per unit area (see 11-2).

proton   A particle having a relative unit positive charge (actual charge $= 1.602 \times 10^{-19}$ coulomb) and with a mass of 1.6725×10^{-24} g or 1.0073 amu (approximately 1 amu) (see 4-3).

radioactivity   A property of certain radioactive isotopes, which spontaneously emit from their nuclei certain radiations that can result in the formation of atoms of a different element or atoms of an isotope of the original element (see **introduction to Chapter 20**).

reaction rate   The rate or speed at which the products are produced or the reactants consumed in a given reaction (see **introduction to Chapter 17**).

reactions, combination   Two or more substances (either elements or compounds) react to produce one substance (see 9-7).

$$A + Z \rightarrow A Z \text{ where } A \text{ and } Z \text{ are elements or compounds}$$

reactions, decomposition   One substance undergoes a reaction to form two or more substances (see 9-8).

$$AZ \rightarrow A + Z \text{ where } A \text{ and } Z \text{ are elements or compounds}$$

reactions, double replacement   Two compounds involved in a reaction with the positive ion (cation) of one compound being exchanged with the positive ion (cation) of another compound (see 9-10).

$$AX + BZ \rightarrow AZ + BX$$

reactions, neutralization   An acid (HX) or an acid oxide reacts with a base (MOH) or a basic oxide. In most of these reactions, water is one of the products (see 9-11).

$$\overset{\frown}{HX + MOH} \rightarrow MX + HOH$$

reactions, single replacement   One element reacts by replacing another element in a compound (see 9-9).

1. A metal replacing a metal ion in its salt or hydrogen ion in an acid.

$$\underset{\smile}{A + BZ} \rightarrow AZ + B$$

2. A nonmetal replacing a nonmetal ion in its salt or acid.

$$\underset{\smile}{X + BZ} \rightarrow BX + Z$$

reducing agent   The substance oxidized (see 16-1).

reduction   A chemical change in which a substance gains electrons or one or more elements in it decreases in oxidation number (see 16-1).

representative elements   The elements in the group A elements and group 0 elements (see 5-2).

salt   A compound formed when one or more of the hydrogen ions of an acid is replaced by a cation (metal or positive polyatomic ion) or when one or more of the hydroxide ions of a base is replaced by an anion (nonmetal or negative polyatomic ion) (see 7-6).

salt, acid   A salt that contains one or more hydrogen atoms bonded to the anion (see 7-6).

salt, hydroxy   A salt that contains one or more hydroxide ions (see 7-6).

salt, normal   A simple salt that does not contain hydrogen atoms bonded to the anion (acid salts) or hydroxide ions (hydroxy salts) (see 7-6).

saturated solution   A solution that is in dynamic equilibrium with undissolved solute ($\rightleftarrows$). The rate of dissolution of undissolved solute is equal to the rate of crystallization of dissolved solute, as shown (see 14-4):

$$\text{undissolved solute} \underset{\text{rate of crystallization}}{\overset{\text{rate of dissolution}}{\rightleftarrows}} \text{dissolved solute}$$

science   A study of organized or systematized knowledge (see 1-1).

scientific notation   A more systematic form of exponential notation in which the decimal part is from one to less than 10 (see 2-4).

sigma (σ) bond   A covalent bond that is cylindrically symmetrical about the line joining the nuclei (see 18-1).

significant digits   Number of digits that give reasonably reliable information (see 2-1).

solute   The component of a solution that is in lesser quantity (see 14-1).

solution   Homogeneous mixture which is composed of two or more pure substances; its *composition* can be *varied* usually **within certain limits** (see 3-2 and 14-1).

solvent   The component of a solution that is in greater quantity (see 14-1).

specific gravity   Density of a substance divided by the density of some substance taken as a standard, usually water at 4°C for solids and liquids.

$$\text{Specific gravity} = \frac{\text{density of substance}}{\text{density of water at 4°C}},$$

that is, the ratio of the density of the substance to that of the standard (see 2-14).

specific heat   The number of calories required to raise the temperature of 1.00 g of a substance 1.00°C or the number of joules required to raise the temperature of 1.00 kg of the substance 1.00 K (see 2-12).

stoichiometry   Measurement based on the quantitative laws of chemical combination (see **introduction to Chapter 10**).

structural formula   Formula showing the arrangement of atoms within a molecule, using a dash for each pair of electrons shared between atoms (see 6-6).

sublimation   The direct conversion of a solid to the vapor without passing through the liquid state (see 12-10).

substance, pure   Homogeneous matter characterized by definite and constant composition and definite and constant properties under a given set of conditions (see 3-2).

supersaturated solution   A solution in which the concentration of solute is *greater* than that possible in a saturated (equilibrium) solution under the *same* conditions. This solution is unstable and will revert to a saturated solution if a "seed" crystal of solute is added; the excess solute crystallizes out of solution (see 14-4).

surface tension   The property of a liquid that tends to draw the surface molecules into the body of the liquid and hence to reduce the surface to a minimum (see 12-6).

theoretical yield   Amount of product obtained when we assume that all of the limiting reagent forms products with none of it left over and that none of the product is lost in its isolation and purification (see 10-4).

titration   (with reference to neutralization) A process for determining the concentration of an acid or base in a solution through the addition of a base or an acid of known concentration, respectively, until the neutralization point or end point is reached, as shown by an indicator or by an instrument such as the pH meter (see 15-2).

transition elements   The elements in the group B elements and the group VIII elements (see 5-2).

unsaturated solution   A solution in which the concentration of solute is *less* than that of the saturated (equilibrium) solution under the *same* conditions (see 14-4).

valence   A whole number used to describe the combining capacity of an element in a compound (see 6-1).

vapor pressure   The pressure exerted by the molecules in the vapor (at constant temperature) in dynamic equilibrium with the liquid in a closed system. Dynamic equilibrium is established when the rate of molecules leaving the surface of the liquid (evaporation) is equal to the rate of the molecules reentering the liquid (condensation) (see 12-3).

viscosity   The resistance of liquids to flow (see 12-6).

weight   The gravitational force of attraction between the body's mass and the mass of the planet or satellite on which it is weighed (see 2-3).

Index

Rules for the Solubility of Inorganic Substances in Water

1. Nearly all *nitrates* (NO_3^{1-}) and *acetates* ($C_2H_3O_2^{1-}$) are *soluble.*

2. All *chlorides* (Cl^{1-}) are *soluble*, except $AgCl$, Hg_2Cl_2, and $PbCl_2$. ($PbCl_2$ is soluble in hot water.)

3. All *sulfates* (SO_4^{2-}) are *soluble*, except $BaSO_4$, $SrSO_4$, and $PbSO_4$. ($CaSO_4$ and Ag_2SO_4 are only slightly soluble.)

4. Most of the *alkali metals* (Group IA, Li, Na, K, etc.) salts and *ammonium* (NH_4^{1+}) salts are *soluble.*

5. All the common *acids* are *soluble.*

6. All *oxides* (O^{2-}) and *hydroxides* (OH^{1-}) are **insoluble,** except those of the alkali metals and certain alkaline earth metals (Group IIA, Ca, Sr, Ba, Ra). [$Ca(OH)_2$ is only moderately soluble.]

7. All *sulfides* (S^{2-}) are **insoluble,** except those of the alkali metals, alkaline earth metals, and ammonium sulfide.

8. All *phosphates* (PO_4^{3-}) and *carbonates* (CO_3^{2-}) are **insoluble,** except those of the alkali metals and ammonium salts.

Electromotive Series or Activity Series

Li
K
Ba
Ca
Na
Mg
Al
Zn
Fe
Cd
Ni
Sn
Pb
(H)
Cu
Hg
Ag
Au